Lentils

Lentils

Production, Processing Technologies, Products, and Nutritional Profile

Edited by

Jasim Ahmed
Food and Nutrition Program
Environment and Life Sciences Research Center
Kuwait Institute for Scientific Research
Safat, Kuwait

Muhammad Siddiq
Food Science Consultant
Windsor, ON, Canada

Mark A. Uebersax
Department of Food Science and Human Nutrition
Michigan State University
East Lansing, MI, USA

WILEY

This edition first published 2023

Registered Office(s)
John Wiley & Sons, Inc., 111 River Street, Hoboken, NJ 07030, USA
John Wiley & Sons Ltd, The Atrium, Southern Gate, Chichester, West Sussex, PO19 8SQ, UK

For details of our global editorial offices, customer services, and more information about Wiley products visit us at www.wiley.com.

Wiley also publishes its books in a variety of electronic formats and by print-on-demand. Some content that appears in standard print versions of this book may not be available in other formats.

Library of Congress Cataloging-in-Publication Data applied for:
Hardback ISBN: 9781119866893

Cover Design: Tricia Principe
Cover Images: Courtesy of Jasim Ahmed; Courtesy of Muhammad Siddiq; © New Africa/Adobe Stock Photos; Stepan Popov/Adobe Stock Photos; julia_gr/Adobe Stock Photos

Set in 9.5/12.5pt STIXTwoText by Straive, Pondicherry, India

Contents

Preface

Lentils are one of the oldest crop species in the world, produced in more than 45 countries, and consumed widely in various forms. Legumes, like lentils, are an excellent source of protein, low digestible carbohydrates, dietary fiber, and selected minerals and vitamins. Recent consumer interest in plant proteins as a substitute for meat proteins has been dramatic. Thus, food processors, nutrition and health professionals, and policymakers have directed an increased attention to lentils as a potential candidate for meat-alternate protein source. Plant breeding experts endeavor to exploit the latest genetic tools and biotechnological approaches to improve the yield and quality of lentils as a sustainable crop with demonstrated environmental benefits. Innovative lentil breeding methods can potentially improve the economic return for the farmers and bring additional proteins to the food supply chain. Lentil continues to be one of the most important food legumes in the world due to its nutrient-dense properties. The superior functionality of lentils provide an excellent source of natural products from which a wide range of value-added products and by-products can be prepared. The extraction of a huge amounts of proteins from lentils results in an equal or greater amount of starch in the production facility. Lentil starch can also be exploited for nonfood uses for expanding lentils use beyond food applications. Technological advancements in lentil-based product formulations can improve functionality, remove antinutritional factors, and produce healthy food products.

The book is unique in that it provides topical coverage across lentil value-chain, from breeding methods and postharvest handling to global consumption trends, value-added processing, nonfood uses, nutritional significance, functional properties, and sensory attributes. This book should serve as a useful reference for food scientists, food technologists, food industry professionals, and graduate students. In addition, nutritionists, dietitians, and policy makers and nongovernmental organization working in the food security and international development areas can equally benefit from information presented in this book. The contents of the book have been arranged in three separate sections so that readers can find their choice of subjects easily. The first section, *Overview, Breeding Practices, Postharvest Handling, and Storage* contains three chapters describing global production and trade, breeding and genetics, and postharvest handling and storage of lentils. The second section, *Processing, Physical and Functional Properties, and Food and Nonfood Applications*, has nine chapters. Particular attention is given to the nutritional and antinutritional factors of lentils, value-added processing, functional properties including rheological and pasting properties, the development of lentil-based snack products, and the role of legume proteins in developing emulsions for food and drug delivery. The milling of lentils is included in the book to apprise the reader of the processes associated with isolating major components and the importance of size reduction. The most relevant trend in today's world is the meat analog using legume/lentil proteins, which is covered in a dedicated chapter. Applications of innovative processing technologies are included in a complete chapter focusing on details of high-pressure processing and dielectric heating. The third section, *Nutrition, Antinutrients, Sensory Properties, and Global Consumption Trends*, has four chapters, which cover nutritional profiles, antinutritional factors in legumes/lentils, sensory evaluation, global consumption, and culinary trends in lentils. Over 40 researchers with diverse subject–matter background have contributed to this book.

The editors acknowledge many individuals for their support from conception through final development of this book. Foremost is our sincere thanks and gratitude to all authors for their contributions and for bearing with us during the review and finalization process of their chapters. Thanks are due to Amaan Thasin for providing library and literature search support. We are grateful to our family members for their understanding and support enabling us to complete this work. The editors thank Wiley for their diligent support in publishing this book.

Jasim Ahmed
Muhammad Siddiq
Mark A. Uebersax

List of Contributors

George O. Abong
Department of Food Science, Nutrition & Technology, University of Nairobi
Kangemi, Kenya

Jasim Ahmed
Food and Nutrition Program, Environment and Life Sciences Research Center
Kuwait Institute for Scientific Research
Safat, Kuwait

Tahira M. Ali
Department of Food Science and Technology, University of Karachi
Karachi, Pakistan

Nandika Bandara
Department of Food and Human Nutritional Sciences, Faculty of Agricultural and Food Sciences
University of Manitoba, Winnipeg
MB, Canada

and

Richardson Centre for Food Technology and Research (RCFTR), University of Manitoba
Winnipeg, MB, Canada

Santanu Basu
Department of Molecular Sciences
Swedish University of Agricultural Sciences, Uppsala, Sweden

Fatema H. Brishti
Department of Food Technology
Faculty of Food Science & Technology
Universiti Putra Malaysia, Serdang
Selangor, Malaysia

Natasha A. Butt
Department of Biomedical Engineering
Faculty of Engineering, Science, Technology and Management, Ziauddin University, Karachi, Pakistan

Anup Chandra
Division of Crop Protection, ICAR-Indian Institute of Pulses Research, Kanpur
UP, India

Sanju Bala Dhull
Department of Food Science and Technology, Chaudhary Devi Lal University, Sirsa, Haryana, India

Thilini Dissanayake
Department of Food and Human Nutritional Sciences, Faculty of Agricultural and Food Sciences
University of Manitoba, Winnipeg
MB, Canada

and

Richardson Centre for Food Technology and Research (RCFTR), University of Manitoba
Winnipeg, MB, Canada

Madhuresh Dwivedi
Department of Food Process Engineering
National Institute of Technology
Rourkela, Orissa, India

Ayca A. Emir
Department of Food Technology
Faculty of Canakkale Applied Sciences
Çanakkale Onsekiz Mart University
Canakkale, Turkey

Tadesse S. Gela
Department of Plant Sciences
University of Saskatchewan
Saskatoon, SK, Canada

Debjyoti S. Gupta
Division of Crop Improvement, ICAR-Indian Institute of Pulses Research
Kanpur, UP, India

Nazimah Hamid
Department of Food Science and Microbiology, Auckland University of Technology, Auckland, New Zealand

Abid Hasnain
Department of Food Science and Technology, University of Karachi
Karachi, Pakistan

Lilian D. Kaale
Department of Food Science and Technology, University of Dar es Salaam
Dar es Salaam, Tanzania

Kevin Kantono
Department of Food Science and Microbiology, Auckland University of Technology, Auckland, New Zealand

Kulwinder Kaur
Department of Processing and Food Engineering, Punjab Agricultural University, Ludhiana, India

Preetinder Kaur
Department of Processing and Food Engineering, Punjab Agricultural University, Ludhiana, India

Semin O. Keskin
Department of Food Processing, Kocaeli University, Kocaeli, Turkey

Hamid Khazaei
Natural Resources Institute Finland (Luke), Helsinki, Finland

and

Department of Agricultural Sciences
University of Helsinki, Helsinki
Finland

Joyce Kinabo
Department of Human Nutrition and Consumer Sciences, Sokoine University of Agriculture, Morogoro, Tanzania

Gaurav Kumar
Department of Food Engineering
National Institute of Food Technology Entrepreneurship and Management
Kundli, India

Jitendra Kumar
Division of Crop Improvement,
ICAR-Indian Institute of Pulses Research
Kanpur, UP, India

Ye Liu
Department of Food Science and Microbiology, Auckland University of Technology, Auckland
New Zealand

Tareq Mzek
School of Business and Economics
Universiti Putra Malaysia, Serdang
Selangor, Malaysia

Harshani Nadeeshani
Department of Food and Human Nutritional Sciences, Faculty of Agricultural and Food Sciences
University of Manitoba, Winnipeg
MB, Canada

and

Richardson Centre for Food Technology and Research (RCFTR), University of Manitoba
Winnipeg, MB, Canada

Najmun Nahar
Department of Food Science, Maulana Abul Kalam Azad University of Technology, Haringhata, West Bengal, India

Charlotte Oduro-Yeboah
Food Technology Research Division
CSIR – Food Research Institute
Accra, Ghana

Priya Pal
Division of Food Science and Postharvest Technology, ICAR-Indian Agricultural Research Institute, New Delhi, India

Natalie Pettitt
Department of Food Science and Microbiology, Auckland University of Technology, Auckland, New Zealand

Pramod K. Prabhakar
Department of Food Science and Technology, National Institute of Food Technology Entrepreneurship and Management, Kundli, India

Elenjikkal J. Rifna
Department Food Process Engineering
National Institute of Technology, Rourkela
Odisha, India

Shalini G. Rudra
Division of Food Science and Postharvest Technology, ICAR-Indian Agricultural Research Institute, New Delhi, India

Nazamid Saari
Department of Food Science, Faculty of Food Science and Technology, Universiti Putra Malaysia, Serdang, Selangor
Malaysia

Muhammad Siddiq
Food Science Consultant, Windsor
ON, Canada

Akanksha Singh
Amity Institute of Organic Agriculture
Amity University, Noida, India

Rabiha B. Sulaiman
Department of Food Technology
University Putra Malaysia (UPM)
Serdang, Selangor, Malaysia

Gulum Sumnu
Department of Food Engineering, Middle East Technical University, Ankara, Turkey

Rahul K. Thakur
Division of Food Science and Postharvest Technology, ICAR-Indian Agricultural Research Institute, New Delhi, India

Mark A. Uebersax
Department of Food Science and Human Nutrition, Michigan State University
East Lansing, MI, USA

Eda Yildiz
Department of Food Engineering
Middle East Technical University
Ankara, Turkey

Part I

Overview, Breeding Practices, Postharvest Handling, and Storage

1

An Overview of Lentil Production, Trade, Processing, and Nutrient Profile

Mark A. Uebersax[1], *Muhammad Siddiq*[2], *and Lilian D. Kaale*[3]

[1] *Department of Food Science and Human Nutrition, Michigan State University, East Lansing, MI, USA*
[2] *Food Science Consultant, Windsor, ON, Canada*
[3] *Department of Food Science and Technology, University of Dar es Salaam, Dar es Salaam, Tanzania*

1.1 Introduction

Lentil (*Lens culinaris* Medik.) is an important legume crop with respect to global production, trade, and consumption patterns. Lentil belongs to the *Leguminosae* family and is a self-pollinated crop. It is one of the highly valued pulse crops in farming systems. Lentil is an annual legume plant with its lens-shaped seed and is grown in more than 45 countries (Khazaei et al. 2019). An artificial intelligence (AI)-based research data search engine (Dimensions 2022) indicates there are more than 147,000 publications, 75,000 patents, and over 150 clinical trials on lentils. These data clearly demonstrate the importance of lentils in today's world.

The total world production of lentils was 5.61 million metric tons (MT) in 2021, which has more than doubled from 2.66 million MT since 1991. Lentil production has shown a faster growth from 2011 to 2021 (an increase of 1.16 million MT) as compared to the preceding two decades combined total increase of 1.80 million MT. However, global lentil production experienced a significant drop in 2021 compared to 2020, when it had reached an all-time high of 6.54 million metric tons. Traditionally, India had been the leading lentil producer in the world but Canada has emerged as a leader in recent years. During the last four decades, lentil has become a major pulse crop in the United States, mainly in the Pacific Northwest and Midwestern states (Siva et al. 2017). Canada has also become the leading lentil exporting country, while India ranks first in lentil imports. Although there have been wide variations in cultivated area of lentils in recent years, overall, it has also increased significantly from 3.27 to 5.59 million hectares for the 1991–2021 period. The yield of lentils was 1004 kg/ha in 2021, which also showed a significant increase from 813 kg/ha in 1991. The average yield has shown mixed trends over the last three decades, from a low of 780 kg/ha in 1992 to the highest figure of 1305 kg/ha in 2020 (FAO 2022).

Lentils have a long ancient and modern history of food use reported in the literature. Lentils originated in Turkey, with their consumption dating back to early civilization, as

Lentils: Production, Processing Technologies, Products, and Nutritional Profile, First Edition.
Edited by Jasim Ahmed, Muhammad Siddiq, and Mark A. Uebersax.

they have been an important food since prehistoric times and among one of the first food crops to have been cultivated (Sarker and Erskine 2006; Chelladurai and Erkinbaev 2020). It was produced in the Near East more than 8500 years ago (Asakereh et al. 2010). Lentils dating back to about 6800 BCE were found at an archeological site Yiftahel in lower Galilee area of Israel (Kislev 1985; Liber et al. 2021). Lentils were highly valued by the Egyptians. Later, during the Roman Empire, legumes were so highly valued that many affluent and prestigious Roman families adopted the name of the legumes as part of their names (Waines and Price 1975; Czapp 2007; Amin and Borchgrevink 2022; Guerra-Garcia et al. 2022).

Due to its good protein content, carbohydrates, calories, essential minerals and vitamins, dietary fiber, soluble fiber, antioxidants, phytoestrogens, and folate compared to other legumes, lentils supplement cereal diets. It provides dietary amino acids and bioactive peptides that play major role in health. Its protein contents range from 21% to 36% (Oplinger et al. 1990; Jarpa-Parra et al. 2014; Joshi et al. 2017; Jarpa-Parra 2018; Dhull et al. 2022), with a balanced amino acid profile, significant content of low-digestible carbohydrates, dietary fiber, and selected minerals and vitamins. Lentils have significant contents of resistant starch and a number of bioactive phytochemicals (Fouad and Rehab 2015; Morales et al. 2015a; Ma et al. 2018; Rezvankhah et al. 2021). Lentil crop promotes sustainable cereal-based production systems with a potential of fixing free nitrogen (Matny 2015), which replaces inorganic fertilizers and thereby provides an attractive agricultural system and controls environmental pollution by reducing greenhouse gas emissions.

Traditionally, lentils have been considered as major source of protein and consumed regularly in many South Asian and Middle Eastern countries. Though the popularity of lentils has been growing rapidly in North America (United States and Canada), Australia, and many European countries. In South Asia (India, Pakistan, and Bangladesh), lentils are eaten with staple foods like *chapati* or flat bread and rice. Further, dehulled and split lentils or lentil flour are often used to make soups, stews, and fried products (Dagher 1991; Xu et al. 2019; Sidhu et al. 2022), and mixed with cereals to make bakery products and infant foods. Fried and roasted whole lentils are also processed and consumed, though on a smaller scale, in some countries. Lentils are also used as a source of starch for the textile industry.

Due to the inexpensive source of proteins, vitamins, and minerals, legumes including lentils are healthy choice for all consumers, and in particular for those who are vegetarians or vegans as well as those opting for environment friendly and sustainable meat alternatives of protein (Hill 2022; Oduro-Yeboah et al. 2022). This chapter presents an overview of lentil production/trade, lentil types, value-added processing, nutrients and antinutrients, and health benefits. The role of lentils in sustainable agriculture systems and world food security is also discussed.

1.2 Lentil Plant and Seed Characteristics

1.2.1 Plant Characteristics and Growth

The lentil plant is indeterminate, as it has considerable variation in its growth habit, which mainly depends on genotype (Saxena 2009). Plants are generally short (20–76 cm), with height depending on the climatic conditions and plant strains. It grows well in cool season, on leveled or partially rolling land having a pH of 6.0–8.0, under irrigated conditions (MFAL 2011). Saxena (2009) reported that under optimum environmental conditions, typically coinciding with late-winter and early-spring planting, lentil plants exhibit rapid vegetative/reproductive growth. Maturity is generally reached within 75–100 days after sowing date. Overall, height of stem and number and angle of branches, which determine the width of plant canopy, are the key traits contributing to plant structure.

Monteith (1981) noted that lentil plant growth is mainly affected by climatic conditions, especially variations in air and soil temperature, sunlight, moisture availability, and wind intensity. Intercropping is an option, as lentil is frequently grown alone. However, lentils can be intercropped with a variety of other crops, e.g., wheat, barley, rice, sugarcane, mustard, and linseed (McPhee et al. 2004). Lentil plant growth and crop yields are affected by abiotic stresses (drought, high temperature, rainfall, flooding) and biotic stresses (primarily diseases). Developing lentil genotypes resistant to these abiotic and biotic stresses is an important area of research in plant breeding and genetics. Kumar et al. (2021) reported that, in recent years, genomics-assisted plant breeding practices have become a powerful tool to develop high-yielding crops that adapt to abiotic stress, e.g., heat and drought stresses.

1.2.2 Seed Characteristics

Lentil seeds appear as oblong lens in shape and come in a variety of sizes. The seeds are produced within rhomboidal pods and have a diverse range of seed coat colors: red-orange, purple, brown, green, or blackish-brown, and spotted/dotted color patterns (Khazaei et al. 2019). Figure 1.1 shows images of major lentil types, whereas Table 1.1 lists selected physical characteristics of different lentil types.

1.3 Global Production and Trade

The total world production of lentils was 5.61 million metric tons (MT) in 2021, which represented an increase of about 111.1% as compared to 2.66 million MT in 1991 (Figure 1.2). During the period from 2011 to 2021, lentil production increased by 1.16 million MT or almost 26%, whereas in the preceding two decades (1991–2011), global production had

Figure 1.1 Whole lentil seeds and split/dehulled *dhal*. *Source:* Sidhu et al. (2022), with permission of John Wiley & Sons.

Table 1.1 Attributes of different lentil types.

Lentil type	Color	Shape/surface	100-seed weight (g)	Seed size, D × T[a] (mm)
Large lentil, whole	Greenish-brown	Ovoid, convex, smooth	5.7	6.3 × 2.9
Large lentil, *dhal*[b]	Light-red	Flat, smooth	2.5	6.2 × 1.4
Small lentil, whole	Brownish-red	Ovoid, convex, smooth	3.6	4.9 × 2.8
Small lentil, *dhal*[b]	Red	Flat, smooth	1.5	4.7 × 1.3

[a] Diameter × thickness.
[b] Split, dehulled.
Source: Adapted from Sidhu et al. (2022).

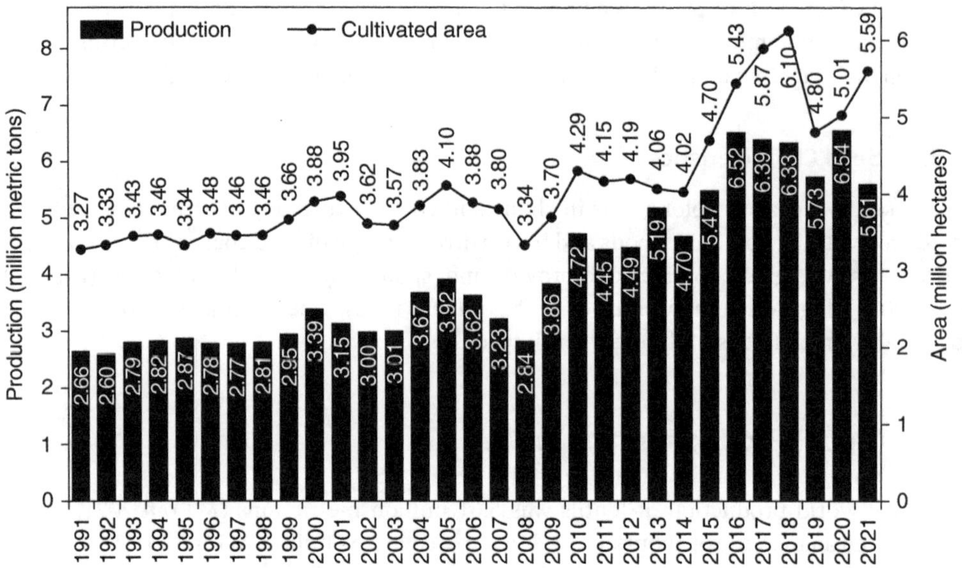

Figure 1.2 Word production and cultivated area of lentils (1991–2021). *Source:* Adapted from FAO (2022).

increased by only 1.80 million MT. It is noteworthy that global lentil production peaked to 6.54 million MT in 2020 but declined significantly in 2021. The total area under lentil cultivation was 5.59 million hectares in 2021, which represented an increase of 71% compared to 3.27 million hectares in 1991. Further, during 2011–2021, area under lentils cultivation has ranged between 4.02 million hectares in 2014 and 6.10 million hectares in 2018, which represents wide variations in cultivated area of lentils in recent years.

The average yield of lentil was 1004 kg/ha in 2021, which exhibited a 23.5% increase compared to that in 1991, which was 813 kg/ha. During the 2011–2021 period, average yield ranged between 1038 kg/ha in 2018 and 1305 kg/ha in 2020. The year 2021 represented a sharp decrease of over 20% in average yield compared to the preceding year. It is noteworthy that year 2020 was the first time that average yield has exceeded 1300 kg/ha. Lentil

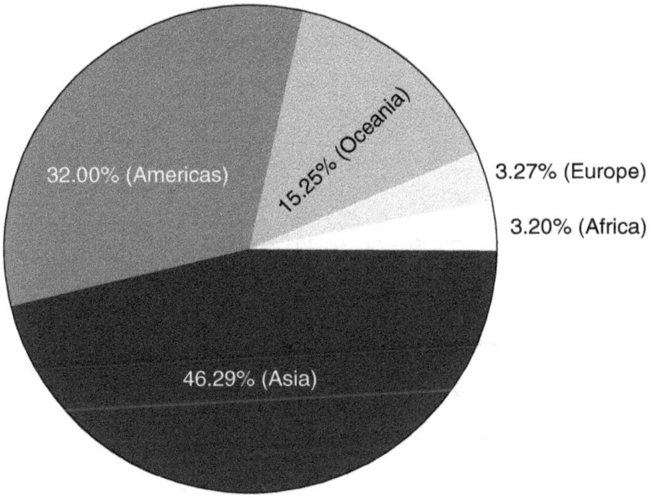

Figure 1.3 Percent share of lentil production by region in 2021. *Source:* Adapted from FAO (2022).

yields, like any other field crop, are affected by climatic conditions or abiotic stresses (i.e., drought, high temperature, rainfall, and flooding) and biotic stresses (mainly diseases).

Lentils are produced in over 45 countries; however, top 10 countries produce about 95% of total world production (FAO 2022). Significance of lentils, as a major legume crop, is demonstrated by the wide distribution of production in different agro-ecological regions throughout the world (Figure 1.3). With respect to regional distribution, lentil production is led by Asia with 46.29% share of world production, followed by Americas (32.00%), Oceania (15.25%), Europe (3.27%), and Africa (3.20%). Lentil producing countries, further divided into subregions, are summarized below. Countries are listed in decreasing order of production by subregion and country-wise within each subregion:

- *North America:* Canada, USA, Mexico
- *South Asia:* India, Bangladesh, Nepal, Pakistan
- *Oceania:* Australia, New Zealand
- *West Asia/Middle East:* Turkey, Syria, Iran, Yemen, Lebanon, Palestine, Jordan, Israel, Iraq
- *East Asia:* China, Myanmar
- *South America:* Argentina, Ecuador, Peru, Colombia, Chile
- *Central Asia:* Kazakhstan, Uzbekistan, Azerbaijan, Armenia
- *East Africa:* Eritrea, Ethiopia, Kenya
- *West Africa:* Algeria, Egypt, Libya, Madagascar, Malawi, Morocco, Tunisia
- *Europe:* Russian Federation, Ukraine, Belarus, North Macedonia, Bosnia, and Herzegovina

The top 10 lentil producing, exporting, and importing countries in 2021 and their percent global share in each category are listed in Table 1.2. Canada ranks first in global lentil production with 1,606,441 metric tons (MT), which accounted for 28.63% of world's total production. India (1,490,000 MT), Australia (853,642 MT), Turkey (263,000 MT), and Nepal (246,092 MT) were among the top 5 lentil producers, which represented 26.56%, 15.22%,

Table 1.2 Top 10 lentil producing, exporting, and importing countries in 2021, by quantity (metric tons, MT) and respective global share (%) in each category.

Lentil producers			Lentil exporters			Lentil importers		
Country	Quantity (MT)	Global share (%)	Country	Quantity (MT)	Global share (%)	Country	Quantity (MT)	Global share (%)
Canada	1,606,441	28.63	Canada	1,928,933	50.90	India	724,537	17.79
India	1,490,000	26.56	Australia	840,230	22.17	Turkey	536,702	13.18
Australia	853,642	15.22	Turkey	289,418	7.64	Bangladesh	455,298	11.18
Turkey	263,000	4.69	UAE	268,969	7.10	U.A.E.	244,420	6.00
Nepal	246,092	4.39	United States	201,674	5.32	Sri Lanka	205,281	5.04
Bangladesh	185,500	3.31	Russian Fed.	80,979	2.14	Pakistan	164,911	4.05
Russian Fed.	176,132	3.14	Kazakhstan	48,083	1.27	Egypt	145,852	3.58
China	165,158	2.94	Egypt	15,272	0.40	Ethiopia	103,000	2.53
United States	150,910	2.69	Belgium	12,219	0.32	Venezuela	100,000	2.46
Ethiopia	122,766	2.19	Syria	9936	0.26	Colombia	85,232	2.09

Source: Adapted from FAO (2022).

4.69%, and 4.39% share globally, respectively. It is noted that Canada had a significant decrease in lentil production from 2,867,800 MT in 2020, whereas India's production increased by about 26% from 1,180,00 MT in 2020. Overall, the top 10 lentil producing countries accounted for over 95% of the total global production of lentils in 2021 (FAO 2022).

Traditionally, India had been the leading lentil producer in the world, but its lentil production has increased by only 1.4-fold since the 2000. By contrast, some other countries, e.g., Australia and Canada have emerged as major producers of lentils since 2000, with 5.2-fold and 1.8-fold increases, respectively. Turkey has shown decreasing production trends, especially, since 2005, from 570,000 MT to 263,000 MT in 2021. The consumer demand for meat alternatives, i.e., non-animal protein sources, as well as trends in greater diversity in exotic cuisines could be the most likely reasons for significant increases in lentil production in developed countries, e.g., Australia, Canada, and United States, which have also emerged as major exporters of lentils.

In 2021, the total global exports of lentil were 3,779,792 MT, equivalent to US $2.83 billion. Canada led global exports with 1,928,933 MT (Table 1.2) or 50.90% of global exports, followed by Australia (840,230 MT), Turkey (289,414 MT), UAE (268,969 MT), and United States (201,674 MT), which accounted for 22.17%, 7.64%, 7.10%, and 5.32% of global exports, respectively. Canada and the United States exported more lentils than it produced in 2021, due possibly to the export of previous year's stocks and including some imports. Nearly all of the global lentil exports (about 98%) were by the top 10 lentil exporting countries. It is noted that UAE does not produce lentils locally; therefore, the exports from UAE primarily constitute re-packaged imported lentils. With respect to global imports of lentils, India led all countries with 724,537 MT or 17.79% of global imports, followed by Turkey (536,702 MT or 13.18%), Bangladesh (455,298 MT or 11.18%), UAE (244,420 MT or 6.00%), and Sri Lanka (205,281 MT or 5.04%). The top 10 lentil importing countries had about 68% of the global lentil imports.

1.4 Preharvest and Preharvest Quality Management

As is the case with other grain crops, the preharvest crop management has a substantial impact on the postharvest quality of lentils. Therefore, to ensure a good quality of harvested lentil crop, appropriate preharvest practices are followed in developed countries (e.g., Canada, Australia, and United States). Among these practices, the use of crop desiccants is commonly practiced, with an objective to ensure uniform maturity (or plant dry-down) and maximize yields through improved harvest efficiency. Generally, preharvest herbicides are also applied when the crop is mature (i.e., when the seed color changes). The herbicide use is highly regulated since each herbicide has specific restrictions with respect to the end-use of the crop after the harvest (Bruce 2008; Bertholet 2019). Lentil crop production practices in the least developed countries rarely employ the use of preharvest desiccants and herbicides, which negatively impacts the lentil crop yields and its postharvest quality.

Typically, harvesting of lentils at a moisture content of 18–20% is recommended, which is higher than the optimum postharvest storage moisture content of 13–14% but is necessary for avoiding shattered seeds of <18% moisture. Upon delivery to the elevator (storage facility), lentils are cleaned and sorted to remove field debris and damaged seeds.

Prior to postharvest storage, lentils are dried to 13–14% moisture content to optimize yield of high quality, i.e., seed color uniformity and for safe long-term storage without subsequent seed breakage or damage. The removal of field stones, split/damaged lentils, and light and heavy debris can be achieved by de-stoners, air aspirators, and gravity separators. Moreover, mesh screens of different sizes are used to remove extraneous organic/inorganic materials, such as immature seeds, stalks, stones, and other field debris. At 14% moisture and 15 °C (recommended conditions), lentils could be stored safely for up to 40 weeks, without any damage to the seed coat and overall quality (Tang et al. 1991; Joshi et al. 2017; Oduro-Yeboah et al. 2022). Grading of lentils for quality standards is commonly done before bulk and retail sales (lentil grades are well established in Canada, Australia, and United States).

A careful monitoring of storage temperature and relative humidity (RH) is important during postharvest storage of lentils for optimum quality preservation since any fluctuations in the storage conditions impact the quality negatively. In the humid climates, high RH is common for most of the year, which can increase the moisture content of grains (like lentils) stored in porous woven bags (i.e., gunny/jute bags). Such conditions allow fungal and insect infestations. Therefore, the use of polypropylene bags with adequate moisture barrier is recommended for lentils packaging and shipping. Oduro-Yeboah et al. (2022) reported that in the developing countries, a lack of appropriate storage facilities is the major cause of postharvest losses in lentils in terms of both quantity and quality. Furthermore, poor packaging, transportation, and handling are other significant factors contributing to postharvest losses in developing countries.

1.5 Nutritional Profile and Health Benefits

Lentils have a nutrient-dense profile and offer a good balance between protein and carbohydrates. For example, when compared to three major staple foods (Table 1.3), wheat, corn, and rice, lentils have 1.8-fold, 2.6-fold, and 3.5-fold more protein content, respectively (USDA 2022). On the other hand, carbohydrate content in these three cereal crops is 12–26% higher than that in lentils. Further, ash content is also significantly higher than that in wheat, corn, and rice, which is comprised of the reported minerals: calcium, potassium, phosphorus, iron, manganese, copper, molybdenum, boron, iodine, cobalt, zinc (Mamakhai and Zagoruiko 2022). Lentils also contain fatty acids from the omega-3 and

Table 1.3 Composition of lentil compared to major cereals.

Composition	Lentil	Wheat	Corn	Rice
Protein	24.6	13.7	9.4	7.1
Carbohydrate	63.4	71.1	74.3	80
Total lipids	1.1	2.5	4.7	0.7
Ash	2.7	1.8	1.2	0.6

Source: Adapted from USDA (2022).

omega-6 groups. This important balance between relatively high-quality protein and mineral content and lower carbohydrate content is an important feature of lentils from a human health perspective (Dhull et al. 2022).

1.5.1 Nutritional Profile

The composition of raw and cooked lentils is shown in Table 1.4. Based on per cup basis, lentil's protein content is 45.90, 17.90, 9.65, and 8.81 g in raw, cooked/boiled, soup, and curry, respectively (USDA 2022). During hydration and cooking, lentils absorb >100% water, which lowers nutrient content on prepared or ready-to-eat weight basis. Nonetheless, lentils still offer a balanced and healthy nutritional profile on per serving basis. It is noted that a considerable variation in nutrient composition of lentil seeds is reported in the literature due mainly to varietal differences (Thavarajah et al. 2008; Joshi et al. 2017).

The amino acid profile of lentil seeds is among the richest in cool season pulses, as all lentil types provide sufficient amounts of the most essential amino acids to meet the

Table 1.4 Composition of raw and cooked lentils (per cup).

Composition	Unit	Raw, red or pink (192 g)	Cooked/boiled, no salt (198 g)	Soup (248 g)	Curry (240 g)
Proximate					
Water	g	15	138	210	184
Energy	kcal/kJ	687/2880	230/964	159/667	264/1107
Protein	g	45.90	17.90	9.65	8.81
Ash	g	5.76	1.64	—	—
Total lipid (fat)	g	4.17	0.75	4.24	13.5
Carbohydrate, by difference	g	121	39.8	21.9	29.5
Dietary fiber	g	20.7	15.6	7.94	8.16
Minerals					
Calcium	mg	92.2	37.6	29.8	60
Iron	mg	14.2	6.59	2.9	3.46
Magnesium	mg	113	71.3	32.2	55.2
Phosphorus	mg	564	356	166	187
Potassium	mg	1280	731	342	701
Sodium	mg	13.4	3.96	464	919
Zinc	mg	6.91	2.52	1.24	1.42
Selenium	μg	0	5.54	3.22	3.6
Vitamins[a]					
Vitamin C	mg	3.26	2.97	3.22	17.8
Niacin	mg	2.88	2.1	1.06	1.9

[a] Other vitamins <1.0 mg content.
Source: Adapted from USDA (2022).

nutrient requirements. However, similar to most other pulses, lentils are deficient in sulfur-containing amino acid, methionine (Iqbal et al. 2006; Yadav et al. 2007). When lentils are consumed with cereal-based foods, which have methionine (e.g., flat wheat bread, rice), lentils provide complete proteins with adequate amounts of all the essential amino acids (Sidhu et al. 2022). It is important to fully understand the quantity/quality of proteins in the legume ingredients that are used in the formulations of various foods. Complementary blending of proteins will reduce the limiting amino acids; however, it is essential for specific complimentary protein formulations to be fully assessed prior to adoption for use in commercially prepared foods directed for use in supplemental feeding programs.

Lentils are a rich source of resistant starch, which varies between 11.4% and 14.9% for different cultivars (Perera et al. 2010). From a rate of digestion perspective, starches are classified into different groups, namely rapidly digestible starch (RDS), slowly digestible starch (SDS), and resistant starch (RS). The RDS, commonly referred to as "bad" starch, is readily digested in the small intestine, therefore eliciting a high glycemic response. On the other hand, the SDS is digested at a slower rate, which results in a relatively lower glycemic response in blood. By contrast, the RS is not digested in the small intestine but it undergoes fermentation by the microorganisms in the large intestine (Englyst et al. 1992). Ahmed and Auras (2011) suggested that, besides traditional uses, lentil starch has potential in the formulation of manufactured foods for diabetics.

Lentils are rich in phenolic contents, which demonstrates its relatively higher antioxidant activity. Lentils also possess numerous biologically active phytochemicals, i.e., flavonoids, phytic acid, saponins, and tannins (Oomah et al. 2011; Zou et al. 2011; Bautista-Expósito et al. 2018; Yeo and Shahidi 2020; Paranavitana et al. 2021). Some of these compounds (e.g., tannins) are considered as antinutrients due to their interference with minerals' absorption, especially iron. Aguilera et al. (2010) studied the effects of soaking, cooking, and dehydration on the phenolic profile and the antioxidant properties in Pardina lentil. Higher antioxidant activity was observed in raw or control lentils than that in processed samples. The industrial processing (hydration) reduced catechins and procyanidins, flavonoids, and flavanone, while hydroxybenzoic compounds were shown to increase significantly.

1.5.2 Antinutritional Factors

Lentils, like other legumes, contain a variety of bioactive compounds that have been traditionally referred to as *antinutrients* or *antinutritional factors* due to a negative impact on nutritional quality, specifically digestibility, bioavailability, and bioaccessibility if pulses are not fully cooked (Nosworthy et al. 2018). These compounds include trypsin inhibitors (TIs), chymotrypsin inhibitors, α-amylase inhibitor, phenolics/tannins, saponins, phytohemagglutinins (lectins), phytic acid/phytate, flavonoids, goitrogens, and oxalic acid (Ruiz et al. 1996; Shnyrov et al. 1996; Ragg et al. 2006; Thavarajah et al. 2010; Mirali et al. 2016; Ganesan and Xu 2017). In the past, some of these compounds were considered to impart only negative health effects but they are now being re-evaluated and recognized as having important health benefits, e.g., antioxidant and anti-inflammatory activities (Campos-Vega et al. 2010; Ganesan and Xu 2017; Wiesinger et al. 2022). It has also been reported that improved processing through application of heat and incorporating organic acids, e.g., ascorbic acid, significantly reduces antinutritional factors. Even pre-soaking legumes in hot water prior to processing can lower the antinutrient concentration, e.g., lectin, and flatulence causing oligosaccharides, i.e., raffinose and stachyose (Khattab and Arntfield 2009; Alphonce et al. 2020).

1.5.3 Health Benefits

Regular consumption of lentils, similar to other pulses, offers a variety of health benefits. Benefits of consuming lentils have been reported in many other biological processes, i.e., lowering blood cholesterol levels and acting as an antioxidant (Khandelwal et al. 2010; Patterson et al. 2017; Kamboj and Nanda 2018; Johnson et al. 2020; Didinger and Thompson 2022). Lentil consumption is also effective against various chronic disease conditions such as cardiovascular diseases, hypertension, hypercholesterolemia, and cancer (Alshikh et al. 2015; Moreno-Valdespino et al. 2020: Verni et al. 2020). Higher protein content and significant dietary fiber content in lentils lead to a slower digestion of available carbohydrates and hence a lower glycemic response. One cup of cooked lentils contains about 15.6g of fiber, which fulfills about 60% daily fiber requirement. Dietary fiber has been shown to impart beneficial effects, e.g., regulating bowel function and probiotics' growth stimulation (Johnson et al. 2020; Sidhu et al. 2022).

Besides modern research, health benefits of consuming lentils are mentioned in ancient treatment remedies as well (Lardos 2006; Faris and Attlee 2017). Lentil soup was consumed as a staple meal in the ancient world, as it was also prepared for ailing individuals as a health remedy (Totelin 2015). Lentil seeds have been reported to be used in the folk medicine to treat different illnesses, e.g., against diabetes, for treating skin infections using lentil-water paste topically (Houshmand et al. 2016), and for treating burns by direct application of roasted and milled lentil flour (Sezik et al. 2001; Giday et al. 2007).

1.6 Lentil Processing and Emerging Trends

1.6.1 Value-Added Processing

Value-added processing of lentils involves diverse processing methods, which produce a variety of end products. The processing of lentils can be divided into three levels: (i) *Primary processing*, e.g., cleaning, sorting, grading, and packaging to market whole lentils for domestic cooking and commercial processing; (ii) *secondary processing*, e.g., dehulling, splitting, and polishing of the whole/split lentils; and (iii) *tertiary processing* that involves grinding/milling of whole or dehulled seeds and separating protein- and starch-rich fractions for use in a variety of food products. Further, a very limited supply of thermally processed (canned) lentils are available in commercial retail food distribution (Joshi et al. 2017; Oduro-Yeboah et al. 2022). Boye et al. (2010b) reported that besides balanced nutrition, lentil and other legume proteins also possess a variety of functional properties, which are known to play an important role in food formulations, processing, and new product development. Some examples of these functional properties include solubility, water and fat binding capacity, gelation, and foaming, all of which play an important role in the development of diverse food products (e.g., bakery products, soups, pastas, and extruded products).

Dehulling of lentils is a commonly used method to process split lentils or dhal. It is common to soak and temper pulses with water to facilitate hulling process. Milling, either dry or wet, is used to produce legume flour and protein- or starch-rich fractions and isolates (Ahmed et al. 2016; Ma et al. 2016; Aryee and Boye 2017). It is to be noted that the type of milling method used has significant effect on the functional and sensory properties of flours and fractions. The functional properties of flours and fractions must be assessed fully before their application in diverse foods. Additionally, extrusion processing can be used to prepare lentil flour by optimizing feed-rate, moisture content, and process temperature parameters (Dhull et al. 2022).

Various cooking or processing methods, besides improving palatability and nutrient availability, also affect the protein digestibility and level of antinutritional factors in lentils. Soaking in water is shown to effect significant reduction in the phytic acid content, while the addition of sodium bicarbonate ($NaHCO_3$) in soak water results in the reduction of phenolics and tannins (Aguilera et al. 2009; Ma et al. 2011; Martín-Cabrejas et al. 2009; Morales et al. 2015b; Joehnke et al. 2021; Dhull et al. 2020, 2022). Different processing methods have variable effect on nutritional profile of lentils and reduce antinutrients as well. Autoclaving improves digestibility, starch gelatinization, and physical disintegration of lentil seed (Hernández-Nava et al. 2011; Świeca 2015; Dhull et al. 2022). Extrusion cooking is also reported to be efficient in improving nutritional quality and increases shelf life of lentils with enhanced flavor and textural attributes (Rathod and Annapure 2017). Germination/sprouting of lentils has significant positive effect on compositional/nutritional quality, e.g., increased crude protein and reduced lipids, carbohydrates, phytic acid, and tannins (Ghumman et al. 2016).

Sozer et al. (2017) reported that the use of pulses is projected to expand as plant-based protein alternatives for meat and as environmentally sustainable food options. Moreover, pulses or pulse ingredients, combined with cereal grain ingredients, would find potential new applications to offer nutritious and healthy food choices for worldwide consumers. Siddiq et al. (2022) indicated that the popularity of convenience foods, e.g., dehydrated, extruded, frozen, and microwavable food products, has provided multiple venues for the development of new lentil-based, gluten-free products.

1.6.2 Emerging Research and Development Trends

Lentil is a least researched legume crop; therefore, the emphasis should be on lentil value chain (LVC), especially in breeding strategies to develop varieties with high yield (Kaale et al. 2022). A number of innovative technologies have been introduced for processing and preservation of food products, which could be applied for grain legume processing. As previously discussed, legumes, including lentils, possess some antinutrient factors, e.g., TIs; new processing techniques have been researched to eliminate or significantly reduce TIs in protein concentrates or protein isolates. Boye et al. (2010a) investigated ultrafiltration and isoelectric-precipitation application and evaluated the effect on the functional properties of protein concentrates from lentil and two other pulses. The ultrafiltration treatment yielded significantly higher protein content in the concentrates (from 69.1% to 88.6%), than with the isoelectric-precipitation method that yielded 63.9–81.7% protein. Furthermore, concentrate of red and green lentils showed higher protein solubility of 70–77% using ultrafiltration isolation.

Ahmed et al. (2019) investigated the effect of high pressure (HP, 300, 450, and 600 MPa) for 15 minutes on the alcalase-induced enzymatic hydrolysis (AEH) of lentil protein (LP). The HP application prior to AEH resulted in significant changes to the secondary structure of proteins. The foaming properties and antioxidant activities were improved by 50% and 100%, respectively. In another study, Ahmed et al. (2016) noted that the HP treatment of lentil starch can be potentially utilized to formulate new products with optimally desired functional and sensory properties. Primozic et al. (2018) used HP-homogenization at 5000–15,000 psi (34–103 MPa) to produce nano-emulsions from lentil protein isolate (LPI) for improved digestibility in the gastrointestinal tract. HP homogenization significantly reduced the particle size of LPI, interfacial storage moduli, and surface hydrophobicity. It was concluded that physical modification of LPI by HP homogenization prior to emulsification could be a potential novel method to utilize pulse proteins with an improved digestibility.

The ultrasound treatment or sonication, as a low-energy means to improve pulse protein functionality, offers a method to cleave disulfide bonds without the chemical reductants. However, the process must be optimized to improve the gel strength and thermal stability. Gharibzahedi and Smith (2020) reported that the use of ultrasound resulted in reduced particle size, turbidity, surface tension, and viscosity of proteins solutions and increased the sulfhydryl and hydrophobic functional groups.

1.7 Role of Lentil in Sustainable Agriculture Systems

Legumes provide a number of various ecosystem benefits and also contribute to the sustainability of a cropping system primarily through (i) their ability to contribute nitrogen (N) to the system by biological N_2 fixation; (ii) diversification of cropping systems (including crop rotation) that results in decreased disease, pest, and weed incidences and potential improvement in biodiversity; (iii) a significantly reduced fossil energy usage in plant production/protection practices; and (iv) providing protein-rich foods for human and animal feed (Jensen et al. 2010). These benefits contribute to sustainable agriculture systems with reduced negative impacts to the environment.

Lentil, being an important legume crop, can offer a valuable ecosystem service to cropping systems with its ability to symbiotically fix atmospheric N_2 through root nodules. However, the legumes' plant depends on intrinsic seed N and soil mineral N initially, i.e., from seedling emergence through nodule development. Jensen et al. (2010) also reported that a legume crop's consequent contribution to the N-economy in a cropping system is achieved mainly from (i) unused or spared soil N and rhizodeposits of N remaining after plant growth, and (ii) mineralized N from above-ground organic residues and the nodulated roots after crop is harvested. As shown in Figure 1.4, Lindström and Mousavi (2020) illustrated a general mechanism for symbiotic nitrogen fixation (SNF) in legume root nodules, wherein legumes can convert atmospheric N_2 (nitrogen) to NH_3 (ammonia) by attracting specialized bacteria to the root hairs (Akter et al. 2013; Kamfwa et al. 2019). The SNF ability distinguishes legumes from other field crops and, therefore, plays a major role in sustainable soil fertility (Heilig et al. 2017).

Knight (2012) reported that when incorporated into crop rotations, legumes significantly boost the yields of subsequent grain crops (up to ~20%). In the life cycle assessment (LCA)

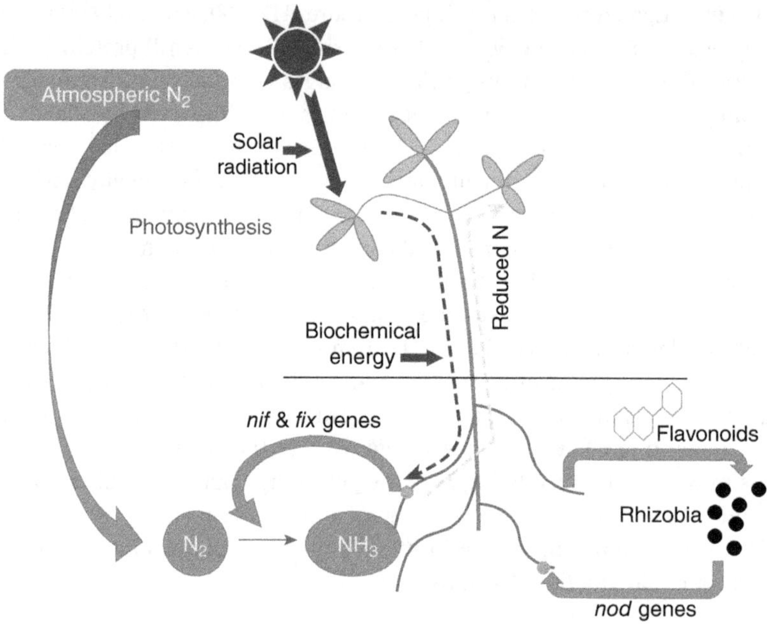

Figure 1.4 General mechanism for symbiotic nitrogen fixation in legume root nodules by rhizobia bacteria. *Source:* Reproduced from Lindström and Mousavi (2020)/John Wiley & Sons/CC BY 4.0.

studies, legumes have been shown to reduce energy consumption and emission of greenhouse gases compared to agricultural systems solely based on mineral-N fertilization (Jensen et al. 2010). Given the widespread use of monocultures in most crop production systems, legumes can potentially help drive crop diversification and protect biodiversity, which promotes more sustainable agricultural systems (Redden et al. 2007; IPCC 2019; Warne et al. 2019; DiverIMPACTS 2020).

A number of studies have emphasized the positive effects of plant-based diets, which not only have benefits for human health but are also environment friendly (Auestad and Fulgoni 2015; Tilman and Clark 2014; Marette and Roosen 2022). Among healthy food choices, legumes are prominent as useful substitutes for healthy and sustainable diet, due to being a rich source of proteins, dietary fibers, minerals, vitamins, and other bioactive compounds, e.g., antioxidants (González et al. 2011).

1.8 Research on Lentil Crop

Research and promotion of lentils are carried out by a number of regional, national, and international organizations. Among these, the most prominent ones are International Crops Research Institute for the Semi-Arid Tropics (ICRISAT), International Center for Agriculture Research in the Dry Areas (ICARDA), Food and Agriculture Organization (FAO), Indian Institute of Pulses Research (IIPR), Pakistan Agricultural Research Council (PARC), Pulse Australia, Australian Centre for International Agricultural Research

(ACIAR), USA Dry Pea and Lentil Council, and Pulse Canada (Siddiq et al. 2022; Sidhu et al. 2022). Some of these organizations are primarily involved in research (e.g., ICRISAT, ICARDA, IIPR, PARC, ACIAR), while others are involved in promotion of legumes, including lentils as healthy food choice. According to the Global Crop Diversity Trust (GCDT 2008), a global composite collection of 1000 lentil accessions has been established representing the overall genetic diversity and the agro-climatological range of lentil.

1.9 Conclusion

Lentils have been an important food since prehistoric times and among one of the first food crops to have been cultivated. Lentils are good source of proteins, carbohydrates (including dietary fiber), B-complex vitamins, minerals, and a number of health-promoting phytochemicals. A large majority of the population in South Asia, especially India and Pakistan, get almost 70% of their protein and energy (calories) from pulses and cereals, of which lentils are a significant part. A variety of cuisines are prepared using lentils in major lentil consuming countries, such as India and Pakistan. Lentil consumption has been growing in recent years beyond its traditional regions of consumption since lentils are a healthy food choice for diverse and environmentally conscious consumers, including vegetarians and vegans.

Management of lentil quality, both at the pre- and postharvest stages, is imperative to maximize this legume crop's economic value. Poor postharvest storage and handling practices can result in quantitative and qualitative losses. For improving marketability and consumer acceptance, clearly defined quality standards and grades have been established. Besides traditional domestic soaking and cooking, a number of processing techniques are used to prepare a variety of products with improved nutritional and sensory quality. Value-added processing of lentils and development of new products are helpful in expanding economic significance and consumer acceptance of this important food legume. Lentil flours and starch- and protein-rich fractions offer a great potential for gluten-free product development. Lentils-based food products fit well with the changing consumer preferences for healthy and sustainable foods.

References

Aguilera, Y., Esteban, R.M., Benitez, V. et al. (2009). Starch, functional properties, and microstructural characteristics in chickpea and lentil as affected by thermal processing. *Journal of Agricultural and Food Chemistry* 57: 10682–10688.

Aguilera, Y., Dueñas, M., Estrella, I. et al. (2010). Evaluation of phenolic profile and antioxidant properties of Pardina lentil as affected by industrial dehydration. *Journal of Agricultural and Food Chemistry* 58: 10101–10108.

Ahmed, J. and Auras, R. (2011). Effect of acid hydrolysis on rheological and thermal characteristics of lentil starch slurry. *LWT Food Science and Technology* 44: 976–983.

Ahmed, J., Taher, A., Mulla, M.Z. et al. (2016). Effect of sieve particle size on functional, thermal, rheological and pasting properties of Indian and Turkish lentil flour. *Journal of Food Engineering* 186: 34–41.

Ahmed, J., Mulla, M., Al-Ruwaih, N., and Arfat, Y.A. (2019). Effect of high-pressure treatment prior to enzymatic hydrolysis on rheological, thermal, and antioxidant properties of lentil protein isolate. *Legume Science* 1: e10.

Akter, Z., Pageni, B.B., Lupwayi, N.Z., and Balasubramanian, P.M. (2013). Biological nitrogen fixation and *nif*H gene expression in dry beans (*Phaseolus vulgaris* L.). *Canadian Journal of Plant Science* 94: 203–212.

Alphonce, S., Kaale, L.D., Millinga, F., and Rweyemamu, L.M.P. (2020). Enrichment of fermented cassava meal 'mchuchume' with micronutrient ingredients from soya bean flour and *Moringa oleifera* leaves powder. *Journal of the Science of the Food and Agriculture* 101: 3575–3581.

Alshikh, N., Costa de Camargo, A., and Shahidi, F. (2015). Phenolics of selected lentil cultivars: antioxidant activities and inhibition of low-density lipoprotein and DNA damage. *Journal of Functional Foods* 18: 1022–1038.

Amin, S. and Borchgrevink, C.P. (2022). A Culinology® perspective of dry beans and other pulses. In: *Dry Beans and Pulses: Production, Processing, and Nutrition*, 2e (ed. M. Siddiq and M.A. Uebersax), 453–480. Hoboken, NJ: Wiley.

Aryee, A.N. and Boye, J.I. (2017). Comparative study of the effects of processing on the nutritional, physicochemical and functional properties of lentil. *Journal of Food Processing and Preservation* 41: e12824.

Asakereh, A., Shiekhdavoodi, M.J., and Safaieenejad, M. (2010). Energy consumption pattern of organic and conventional lentil in Iran. A case study: Kuhdasht County. *Asian Journal of Agricultural Sciences* 2: 111–116.

Auestad, N. and Fulgoni, V.L. (2015). What current literature tells us about sustainable diets: emerging research linking dietary patterns, environmental sustainability, and economics. *Advances in Nutrition* 6: 19–36.

Bautista-Expósito, S., Peñas, E., Silván, J.M. et al. (2018). pH-controlled fermentation in mild alkaline conditions enhances bioactive compounds and functional features of lentil to ameliorate metabolic disturbances. *Food Chemistry* 248: 262–271.

Bertholet, E. (2019). Evaluation of efficacy of new and existing desiccants in lentil (*Lens culinaris* Medik). Doctoral dissertation. University of Saskatchewan, Saskatoon, Canada.

Boye, J.I., Aksay, S., Roufik, S. et al. (2010a). Comparison of the functional properties of pea, chickpea and lentil protein concentrates processed using ultrafiltration and isoelectric precipitation techniques. *Food Research International* 43: 537–546.

Boye, J.I., Zare, F., and Pletch, A. (2010b). Pulse proteins: processing, characterization, functional properties and applications in food and feed. *Food Research International* 43: 414–431.

Bruce, J.L. (2008). The effects of preharvest treatments on the milling efficiency of red lentil. MS thesis. University of Saskatchewan, Saskatoon, Canada.

Campos-Vega, R., Loarca-Pina, G., and Oomah, B.D. (2010). Minor components of pulses and their potential impact on human health. *Food Research International* 43: 461–482.

Chelladurai, V. and Erkinbaev, C. (2020). Lentils. In: *Pulses* (ed. A. Manickavasag and P. Thirunathan), 129–143. Cham, Switzerland: Springer.

Czapp, K. (2007). Putting the polish on those humble beans. http://www.westonaprice.org/food-features/putting-the-polish-on-those-humble-beans (accessed 18 September 2021).

Dagher, S.M. (1991). *Traditional Foods in the Near East*. Rome, Italy: The Food and Agriculture Organization (FAO).

Dhull, S.B., Punia, S., Kidwai, M.K. et al. (2020). Solid-state fermentation of lentil (*Lens culinaris* L.) with *Aspergillus awamori*: effect on phenolic compounds, mineral content, and their bioavailability. *Legume Science* 2020: e37.

Dhull, S.B., Kinabo, J., and Uebersax, M.A. (2022). Nutrient profile and effect of processing methods on the composition and functional properties of lentils (*Lens culinaris* Medik) – a review. *Legume Science* 4: e156.

Didinger, C. and Thompson, H.J. (2022). The role of pulses in improving human health: a review. *Legume Science* 2022: e147.

Dimensions (2022). Lentils. https://app.dimensions.ai/discover/publication (accessed 21 December 2022).

DiverIMPACTS (2020). European Project H2020. https://www.diverimpacts.net (accessed 23 March 2022).

Englyst, H.N., Kingman, S.M., and Cummings, J.H. (1992). Classification and measurement of nutritionally important starch fractions. *European Journal of Clinical Nutrition* 46: S33–S50.

FAO (Food and Agriculture Organization) (2022). Crop production and trade data. http://www.fao.org/faostat/en/#home (accessed 30 December 2022).

Faris, M.A.E. and Attlee, A. (2017). Lentils (*Lens culinaris* L.): a novel functional food. In: *Exploring the Nutrition and Health Benefits of Functional Foods* (ed. H.U. Shekhar, Z.H. Howlader, and Y. Kabir), 42–72. Hershey, PA: ICI Global.

Fouad, A.A. and Rehab, F.M. (2015). Effect of germination time on proximate analysis, bioactive compounds and antioxidant activity of lentil (*Lens culinaris* Medik.) sprouts. *Acta Scientiarum Polonorum Technologia Alimentaria* 14: 233–246.

Ganesan, K. and Xu, B. (2017). Polyphenol-rich dry common beans (*Phaseolus vulgaris* L.) and their health benefits. *International Journal of Molecular Sciences* 18: 2331.

GCDT (Global Crop Diversity Trust) (2008). Global strategy for the *ex situ* conservation of lentil (*Lens* Miller). https://www.genebanks.org/resources/publications/lens-strategy-2008 (accessed 16 December 2022).

Gharibzahedi, S.M.T. and Smith, B. (2020). The functional modification of legume proteins by ultrafiltration: a review. *Trends in Food Science & Technology* 98: 107–116.

Ghumman, A., Kaur, A., and Singh, N. (2016). Impact of germination on flour, protein and starch characteristics of lentil (*Lens culinari*) and horsegram (*Macrotyloma uniflorum* L.) lines. *LWT Food Science & Technology* 65: 137–144.

Giday, M., Teklehaymanot, T., Animut, A., and Mekonnen, Y. (2007). Medicinal plants of the Shinasha, Agew-awi and Amhara peoples in Northwest Ethiopia. *Journal of Ethnopharmacology* 110: 516–525.

Guerra-Garcia, A., Haile, T., Ogutcen, E. et al. (2022). An evolutionary look into the history of lentil reveals unexpected diversity. *Evolutionary Applications* 15: 1313–1325.

Heilig, J.A., Beaver, J.S., Wright, E.M. et al. (2017). QTL analysis of symbiotic nitrogen fixation in a black bean population. *Crop Science* 57: 118–129.

Hernández-Nava, R.G., Bello-Pérez, L.A., Martín-Martínez, S. et al. (2011). Effect of extrusion cooking on the functional properties and starch components of lentil/banana blends: response surface analysis. *Revista Mexicana de Ingeniería Química* 10: 409–419.

Hill, H. (2022). Utilization of dry beans and other pulses as ingredients in diverse food products. In: *Dry Beans and Pulses: Production, Processing, and Nutrition*, 2e (ed. M. Siddiq and M.A. Uebersax), 307–329. Hoboken, NJ: Wiley.

Houshmand, G., Tarahomi, S., Arzi, A. et al. (2016). Red lentil extract: neuroprotective effects on perphenazine induced catatonia in rats. *Journal of Clinical and Diagnostic Research: JCDR* 10: FF05–FF08.

IPCC (Intergovernmental Panel on Climate Change) (2019). *Climate Change and Land: An IPCC Special Report on Climate Change, Desertification, Land Degradation, Sustainable Land Management, Food Security, and Greenhouse Gas Fluxes in Terrestrial Ecosystems*. Geneva, Switzerland: IPCC. https://www.ipcc.ch/srccl (accessed 15 April 2022).

Iqbal, A., Khalil, I.A., Ateeq, N., and Khan, M.S. (2006). Nutritional quality of important food legumes. *Food Chemistry* 97: 331–335.

Jarpa-Parra, M. (2018). Lentil protein: a review of functional properties and food application. An overview of lentil protein functionality. *International Journal of Food Science & Technology* 53: 892–903.

Jarpa-Parra, M., Bamdad, F., Wang, Y. et al. (2014). Optimization of lentil protein extraction and the influence of process pH on protein structure and functionality. *LWT Food Science and Technology* 57: 461–469.

Jensen, E.S., Peoples, M.B., and Hauggaard-Nielsen, H. (2010). Faba bean in cropping systems. *Field Crops Research* 115: 203–216.

Joehnke, M.S., Jeske, S., Ispiryan, L. et al. (2021). Nutritional and anti-nutritional properties of lentil (*Lens culinaris*) protein isolates prepared by pilot-scale processing. *Food Chemistry: X* 9: 100112.

Johnson, N., Johnson, C.R., Thavarajah, P. et al. (2020). The roles and potential of lentil prebiotic carbohydrates in human and plant health. *Plants, People, Planet* 2: 310–319.

Joshi, M., Timilsena, Y., and Adhikari, B. (2017). Global production, processing and utilization of lentil: a review. *Journal of Integrative Agriculture* 16: 2898–2913.

Kaale, L.D., Siddiq, M., and Hooper, S. (2022). Lentil (*Lens culinaris* Medik) as nutrient-rich and versatile food legume: a review. *Legume Science* 4: e169.

Kamboj, R. and Nanda, V. (2018). Proximate composition, nutritional profile and health benefits of legumes: a review. *Legume Research-An International Journal* 41: 325–332.

Kamfwa, K., Cichy, K.A., and Kelly, J.D. (2019). Identification of quantitative trait loci for symbiotic nitrogen fixation in common bean. *Theoretical and Applied Genetics* 132: 1375–1387.

Khandelwal, S., Udipi, S.A., and Ghugre, P. (2010). Polyphenols and tannins in Indian pulses: effect of soaking, germination and pressure cooking. *Food Research International* 43: 526–530.

Khattab, R.Y. and Arntfield, S.D. (2009). Nutritional quality of legume seeds as affected by some physical treatments 2. Antinutritional factors. *LWT Food Science and Technology* 42: 1113–1118.

Khazaei, H., Subedi, M., Nickerson, M. et al. (2019). Seed protein of lentils: current status, progress, and food applications. *Foods* 8: 1–23.

Kislev, M.L. (1985). Early neolithic horsebean from Yiftah'el, Israel. *Science* 228: 319–320.

Knight, J.D. (2012). Frequency of field pea in rotations impacts biological nitrogen fixation. *Canadian Journal of Plant Science* 92: 1005–1011.

Kumar, J., Sen Gupta, D., Baum, M. et al. (2021). Genomics-assisted lentil breeding: current status and future strategies. *Legume Science* 3: e71.

Lardos, A. (2006). The botanical materiamedica of the Iatrosophikon – a collection of prescriptions from a monastery in Cyprus. *Journal of Ethnopharmacology* 104: 387–406.

Liber, M., Duarte, I., Maia, A.T., and Oliveira, H.R. (2021). The history of lentil (*Lens culinaris* subsp. culinaris) domestication and spread as revealed by genotyping-by-sequencing of wild and landrace accessions. *Frontiers in Plant Science* 12: 628439.

Lindström, K. and Mousavi, S.A. (2020). Effectiveness of nitrogen fixation in rhizobia. *Microbial Biotechnology* 13: 1314–1335.

Ma, Z., Boye, J.I., Simpson, B.K. et al. (2011). Thermal processing effects on the functional properties and microstructure of lentil, chickpea, and pea flours. *Food Research International* 44: 2534–2544.

Ma, Z., Boye, J.I., Swallow, K. et al. (2016). Techno-functional characterization of salad dressing emulsions supplemented with pea, lentil and chickpea flours. *Journal of the Science of Food and Agriculture* 96: 837–847.

Ma, Z., Yin, X., Hu, X. et al. (2018). Structural characterization of resistant starch isolated from *Laird lentils* (*Lens culinaris*) seeds subjected to different processing treatments. *Food Chemistry* 263: 163–170.

Mamakhai, A.K. and Zagoruiko, M.G. (2022). Territorial prospects for growing lentils. *IOP Conference Series: Earth and Environmental Science* 988: 1–6.

Marette, S. and Roosen, J. (2022). Just a little bit more legumes! Results of an online survey in Europe. *International Food and Agribusiness Management Review* 25: 329–345.

Martín-Cabrejas, M.A., Aguilera, Y., Pedrosa, M.M. et al. (2009). The impact of dehydration process on antinutrients and protein digestibility of some legume flours. *Food Chemistry* 114: 1063–1068.

Matny, O.N. (2015). Lentil (*Lens culinaris* Medikus) current status and future prospect of production in Ethiopia. *Advances in Plants and Agriculture Research* 2: 45–53.

McPhee, K., Miller, P., Chen, C., and Muehlbauer, F. (2004). Adaptation of winter legumes to direct seeding in northern climates. In: *Proceeding of the 5th European Conference on Grain Legumes/2nd International Conference on Legume Genomics and Genetics, AEP – l'Association Européenne de Recherche sur les Protéagineuse*, 55–56. Paris, France.

MFAL (Ministry of Food, Agric and Livestock) (2011). *Pulses Varieties of Pakistan*. Islamabad, Pakistan: Federal Seed Certification and registration Department.

Mirali, M., Purves, R.W., Stonehouse, R. et al. (2016). Genetics and biochemistry of zero-tannin lentils. *PLoS One* 11: e0164624.

Monteith, J.L. (1981). Coupling of plants to the atmosphere. In: *Plants and their Atmospheric Environment* (ed. J. Grace, E.D. Ford, and P.G. Jarvis), 1–29. London, UK: Blackwell Publishing.

Morales, P., Berrios, J.D.J., Varela, A. et al. (2015a). Novel fiber-rich lentil flours as snack-type functional foods: an extrusion cooking effect on bioactive compounds. *Food & Function* 6: 3135–3143.

Morales, P., Cebadera-Miranda, L., Cámara, R.M. et al. (2015b). Lentil flour formulations to develop new snack-type products by extrusion processing: phytochemicals and antioxidant capacity. *Journal of Functional Foods* 19: 537–544.

Moreno-Valdespino, C.A., Luna-Vital, D., Camacho-Ruiz, R.M., and Mojica, L. (2020). Bioactive proteins and phytochemicals from legumes: mechanisms of action preventing obesity and type-2 diabetes. *Food Research International* 130: 108905.

Nosworthy, M.G., Medina, G., Franczyk, A.J. et al. (2018). Effect of processing on the *in vitro* and *in vivo* protein quality of red and green lentils (*Lens culinaris*). *Food Chemistry* 240: 588–593.

Oduro-Yeboah, C., Sulaiman, R., Uebersax, M.A., and Dolan, K.D. (2022). A review of lentil (*Lens culinaris* Medik) value-chain: postharvest handling, processing and processed products. *Legume Science* 4: e171.

Oomah, B.D., Caspar, F., Malcolmson, L.J., and Bellido, A.S. (2011). Phenolics and antioxidant activity of lentil and pea hulls. *Food Research International* 44: 436–441.

Oplinger, E.S., Hardman, L.L., Kaminski, A.R. et al. (1990). Lentil. https://hort.purdue.edu/newcrop/afcm/lentil.html (accessed 15 May 2022).

Paranavitana, L., Oh, W.Y., Yeo, J., and Shahidi, F. (2021). Determination of soluble and insoluble-bound phenolic compounds in dehulled, whole, and hulls of green and black lentils using electrospray ionization (ESI)-MS/MS and their inhibition in DNA strand scission. *Food Chemistry* 361: 130083.

Patterson, C.A., Curran, J., and Der, T. (2017). Effect of processing on antinutrient compounds in pulses. *Cereal Chemistry* 94: 2–10.

Perera, A., Meda, V., and Tyler, R.T. (2010). Resistant starch: a review of analytical protocols for determining resistant starch and of factors affecting the resistant starch content of foods. *Food Research International* 43: 1959–1974.

Primozic, M., Duchek, A., Nickerson, M., and Ghosh, S. (2018). Formation, stability and *in vitro* digestibility of nanoemulsions stabilized by high-pressure homogenized lentil proteins isolate. *Food Hydrocolloids* 77: 126–141.

Ragg, E.M., Galbusera, V., Scarafoni, A. et al. (2006). Inhibitory properties and solution structure of a potent Bowman-Birk protease inhibitor from lentil (*Lens culinaris*, L) seeds. *The FEBS Journal* 273: 4024–4039.

Rathod, R.P. and Annapure, U.S. (2017). Physicochemical properties, protein and starch digestibility of lentil based noodle prepared by using extrusion processing. *LWT Food Science and Technology* 80: 121–130.

Redden, B., Maxted, N., Furman, B., and Coyne, C. (2007). Lens biodiversity. In: *Lentil, an Ancient Crop for Modern Times* (ed. S. Yadav, D.L. McNeil, and P.C. Stevenson) 462 pp. New York, USA: Springer.

Rezvankhah, A., Yarmand, M.S., Ghanbarzadeh, B., and Mirzaee, H. (2021). Characterization of bioactive peptides produced from green lentil (*Lens culinaris*) seed protein concentrate using Alcalase and Flavourzyme in single and sequential hydrolysis. *Journal of Food Processing and Preservation* 45: e15932.

Ruiz, R.G., Price, K.R., Arthur, A.E. et al. (1996). Effect of soaking and cooking on the saponin content and composition of chickpeas (*Cicer arietinum*) and lentils (*Lens culinaris*). *Journal of Agricultural and Food Chemistry* 44: 1526–1530.

Sarker, A. and Erskine, W. (2006). Recent progress in the ancient lentil. *Journal of Agricultural Science* 144: 19–29.

Saxena, M.C. (2009). Plant morphology, anatomy and growth habit. In: *The Lentil: Botany, Production and Uses* (ed. W. Erskine, F.J. Muehlbauer, A. Sarker, and B. Sharma), 34–66. Wallingford, Oxfordshire, UK: CABI.

Sezik, E., Yeşilada, E., Honda, G. et al. (2001). Traditional medicine in Turkey X. Folk medicine in central Anatolia. *Journal of Ethnopharmacology* 75: 95–115.

Shnyrov, V.L., Marcos, M.J., and Villar, E. (1996). Kinetic study on the irreversible thermal denaturation of lentil lectin. *IUBMB Life* 39: 647–656.

Siddiq, M., Uebersax, M.A., and Siddiq, F. (2022). Global production, trade, processing and nutritional profile of dry beans and other pulses. In: *Dry Beans and Pulses: Production, Processing, and Nutrition* (ed. M. Siddiq and M.A. Uebersax), 1–28. Hoboken, NJ: Wiley.

Sidhu, J.S., Zafar, T., Benyathiar, P., and Nasir, M. (2022). Production, processing, and nutritional profile of chickpeas and lentils. In: *Dry Beans and Pulses: Production, Processing, and Nutrition*, 2e (ed. M. Siddiq and M.A. Uebersax), 383–407. Hoboken, NJ: Wiley.

Siva, N., Johnson, C.R., Duckett, S. et al. (2017). Can lentil (*Lens culinaris* Medikus) reduce the risk of obesity? *Journal of Functional Foods* 38: 706–715.

Sozer, N., Holopainen-Mantila, U., and Poutanen, K. (2017). Traditional and new food uses of pulses. *Cereal Chemistry* 94: 66–73.

Świeca, M. (2015). Production of ready-to-eat lentil sprouts with improved antioxidant capacity: optimization of elicitation conditions with hydrogen peroxide. *Food chemistry* 180: 219–226.

Tang, J., Sokhansanj, S., and Sosulski, F. (1991). Determination of the breakage susceptibility of lentil seed. *Cereal Chemistry* 68: 647–650.

Thavarajah, D., Ruszkowski, J., and Vandenberg, A. (2008). High potential for selenium biofortification of lentils (*Lens culinaris* L.). *Journal of Agricultural and Food Chemistry* 56: 10747–10753.

Thavarajah, P., See, C.T., and Vandenberg, A. (2010). Phytic acid and Fe and Zn concentration in lentil (*Lens culinaris* L.) seeds is influenced by temperature during seed filling period. *Food Chemistry* 122: 254–259.

Tilman, D. and Clark, M. (2014). Global diets link environmental sustainability and human health. *Nature* 515: 518–522.

Totelin, L. (2015). When foods become remedies in ancient Greece: the curious case of garlic and other substances. *Journal of Ethnopharmacology* 167: 30–37.

USDA (US Dept of Agriculture) (2022). Food Data Central. https://fdc.nal.usda.gov (accessed 23 March 2022).

Verni, M., Demarinis, C., Rizzello, C.G., and Baruzzi, F. (2020). Design and characterization of a novel fermented beverage from lentil grains. *Foods* 9: 893.

Waines, J.G. and Price, N.P.S. (1975). Plant remains from Khirokitia in Cyprus. *Paléorient* 3: 281–284.

Warne, T., Ahmed, S., Byker Shanks, C., and Miller, P. (2019). Sustainability dimensions of a North American lentil system in a changing world. *Frontiers in Sustainable Food Systems* 3: 88.

Wiesinger, J.A., Marsolais, F., and Glahn, R.P. (2022). Health implications and nutrient bioavailability of bioactive compounds in dry beans and other pulses. In: *Dry Beans and Pulses: Production, Processing, and Nutrition*, 2e (ed. M. Siddiq and M.A. Uebersax), 505–529. Hoboken, NJ: Wiley.

Xu, M., Jin, Z., Simsek, S. et al. (2019). Effect of germination on the chemical composition, thermal, pasting, and moisture sorption properties of flours from chickpea, lentil, and yellow pea. *Food Chemistry* 295: 579–587.

Yadav, S.S., Stevenson, P., Rizvi, A.H. et al. (2007). Uses and consumption. In: *Lentils – an Ancient Crop for Modern Time* (ed. S.S. Yadav, D. McNeil, and P. Stevenson), 33–46. Dordrecht, the Netherlands: Springer.

Yeo, J. and Shahidi, F. (2020). Identification and quantification of soluble and insoluble-bound phenolics in lentil hulls using HPLC-ESI-MS/MS and their antioxidant potential. *Food Chemistry* 315: 126202.

Zou, Y., Chang, S.K., Gu, Y., and Qian, S.Y. (2011). Antioxidant activity and phenolic compositions of lentil (*Lens culinaris* var. Morton) extract and its fractions. *Journal of Agricultural and Food Chemistry* 59: 2268–2276.

2

Recent Advances in Lentil Genetics, Genomics, and Molecular Breeding

Jitendra Kumar[1], *Tadesse S. Gela*[2], *Debjyoti S. Gupta*[1], *Anup Chandra*[3], *and Hamid Khazaei*[4,5]

[1] *Division of Crop Improvement, ICAR-Indian Institute of Pulses Research, Kanpur, UP, India*
[2] *Department of Plant Sciences, University of Saskatchewan, Saskatoon, SK, Canada*
[3] *Division of Crop Protection, ICAR-Indian Institute of Pulses Research, Kanpur, UP, India*
[4] *Natural Resources Institute Finland (Luke), Helsinki, Finland*
[5] *Department of Agricultural Sciences, University of Helsinki, Helsinki, Finland*

Jitendra Kumar and Tadesse S. Gela were contributed equally to this work.

2.1 Introduction

Cultivated lentil (*Lens culinaris* Medik.) is a self-pollinated cool-season grain legume crop (pulse). It is a diploid species with seven chromosomes ($2n = 2x = 14$) and with a genome size of ~4.3 Gbp (Bett et al. 2016; Ogutcen et al. 2018). It is grown in more than 45 countries, covering an area of 5.0 million hectares, with a global production of 6.7 million metric tons (FAOSTAT 2022). Lentil seeds are a rich source of protein, dietary fiber, prebiotic carbohydrates, minerals, and vitamins (Kumar et al. 2016; Khazaei et al. 2019; Subedi et al. 2021). It has been proposed to help combat malnutrition and micronutrient deficiencies in the world due to the presence of high concentrations of iron (Fe), zinc (Zn), and folate in its seed (Kumar et al. 2016). Globally many people are preferring nutritious plant-based protein in their diets, leading to increased per capita consumption of lentils. Compared to other grain legumes, lentil seeds cook faster. Moreover, lentil-based cropping systems help improve soil health similar to other legumes through biological nitrogen fixation (Stagnari et al. 2017) and are useful commodities for improving the livelihood of smallholder farmers in dryland areas of developing countries (Kumar et al. 2013).

Traditional crop breeding methods have been used to develop improved cultivars of lentil with moderate genetic gain (Sadras et al. 2022). Genomic tools could be used to enhance the productivity of lentils, particularly under biotic and abiotic constraints

Lentils: Production, Processing Technologies, Products, and Nutritional Profile, First Edition.
Edited by Jasim Ahmed, Muhammad Siddiq, and Mark A. Uebersax.

(Dissanayake et al. 2021; Tiwari et al. 2022). They can also precisely and rapidly accumulate superior alleles in breeding programs (Kumar et al. 2015).

In lentil, genomic resources such as genetic maps, sequencing-based trait mapping, genome and germplasm sequencing, and transcriptomes have been developed and are being integrated into breeding programs globally (Kumar et al. 2015, 2018a,b,c; Singh et al. 2017, 2019a). This has been further accelerated by next-generation sequencing (NGS) technologies (Singh et al. 2017, 2019a). These genomic resources are widely used for genetic diversity and relationship studies (Lombardi et al. 2014; Kumar et al. 2015; Wong et al. 2015; Khazaei et al. 2016; Pavan et al. 2019; Dissanayake et al. 2020; Liber et al. 2021). Recently, the genome assemblies of a cultivated lentil and a wild relative (*L. ervoides*) have been reported, which will further advance lentil genomics and breeding (Ramsay et al. 2021). The present chapter discusses the classical genetics and its application in breeding, and current status and future opportunities for lentil genomics-based breeding (Figure 2.1).

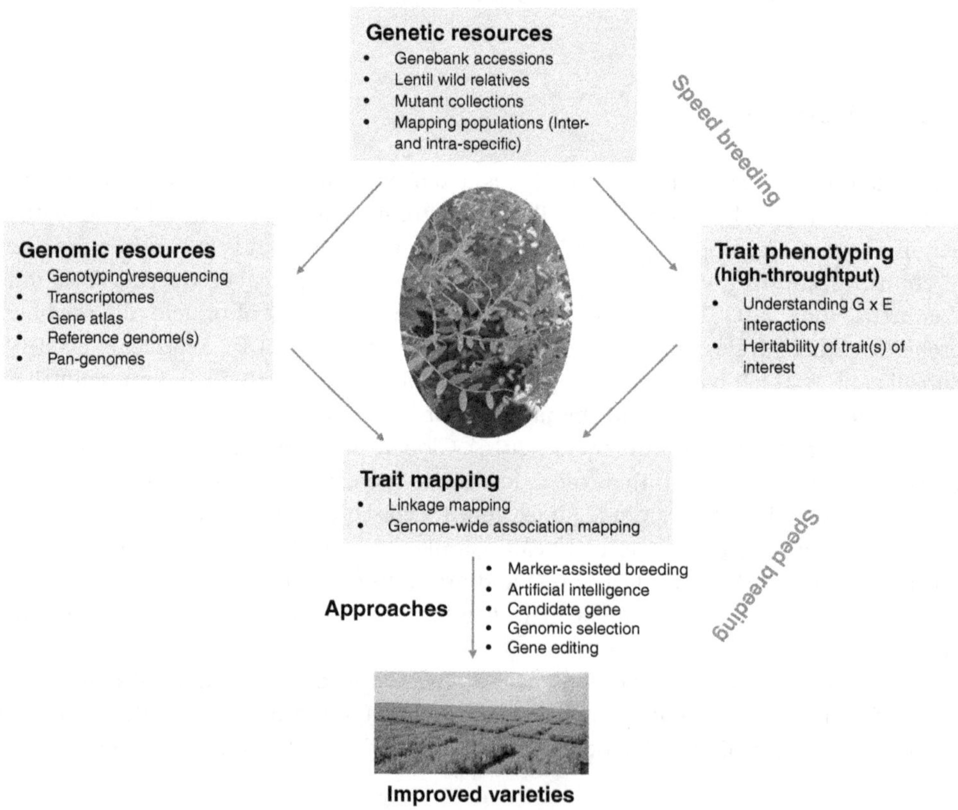

Figure 2.1 A schematic overview of genomics-based breeding in lentils. *Source:* Hamid Khazaei (Co-author).

2.2 Lentil Genetic Resources

The availability of diverse plant genetic resources is essential for any breeding program to benefit from diverse sources of alleles or genes and enrich the diversity. In lentil, over 58,000 germplasm accessions are stored in different genebanks globally (Khazaei et al. 2016). The International Center for Agricultural Research in the Dry Areas (ICARDA) holds a collection of 13,907 and 603 accessions of cultivated and wild species of lentil, respectively (Singh and Chung 2016). The second largest collection of lentil germplasm, comprising 5254 accessions, is stored in the Australian Temperate Field Crop Collection (FAO 2010). The Seed and Plant Improvement Institute (Iran) and the United States Department of Agriculture (USDA) are maintaining 3000 and 2875 accessions of lentil, respectively (Malhotra et al. 2019). In India, the ICAR-National Bureau of Plant Genetic Resources (NBPGR) hosts 2655 accessions (2083 indigenous and 572 exotic) in its gene bank (Singh and Chung 2016). These collections consist of a considerable number of wild accessions of lentil, which can be useful resources for unknown traits or even for regular breeding traits.

Various researchers have employed morphological, phenological, agronomic, and nutritional traits to characterize lentil germplasm collections (Khazaei et al. 2017; Malhotra et al. 2019; Singh et al. 2020a; Hussain et al. 2022). The core collections and FIGS (Focused Identification of Germplasm Strategy) sets were created to make this germplasm more accessible and useful in breeding programs (Khazaei et al. 2013). The USDA developed a core set of 287 accessions on the basis of country of origin. This set was screened for agro-morphological traits (Tullu et al. 2001). A core collection was recently developed for the lentil accessions collection of the Indian national genebank, and it was characterized for agro-morphological traits (Tripathi et al. 2022). The FIGS strategy implemented at ICARDA employs robust geographical data sets and has proven successful for various adaptive traits in lentils, including heat and drought tolerance, and disease resistance. A FIGS set of 162 lentil accessions identified few heat and drought tolerant lentils (El Haddad et al. 2020). In another study, 320 global lentil accessions were evaluated across multiple locations and years worldwide to assess their morphological and phenological responses in different environments (Wright et al. 2021).

Wild relatives of lentil provide opportunities to broaden the genetic base of cultivated lentil species and mine useful alleles that could not be used during domestication. The wild *Lens* species consist of *L. culinaris, L. orientalis, L. odemensis Ladiz., L. tomentosus, L. nigricans* (M. Bieb.) Grand., *L. ervoides* (Brign.) Grand., and *L. lamottei* Ladiz. With the aid of DNA markers, they were divided into four gene pools, with the primary gene pool containing *L. culinaris, L. orientalis, and L. tomentosus*; these species can be easily crossed and produce fertile offspring. The remaining species are classified as secondary (*L. lamottei* and *L. odemensis*), tertiary (*L. ervoides*), and quaternary (*L. nigricans*) gene pools (Wong et al. 2015; Dissanayake et al. 2020). Embryo/ovule rescue techniques are required to overcome hybridization barriers, particularly for interspecies crosses of the primary gene pool species with the tertiary and quaternary gene pools. Singh et al. (2014) evaluated 405 accessions of the wild lentil species for agronomic traits and disease reactions and identified useful sources for these traits. Further, a core set of 96 accessions was developed from these wild accessions and used to validate the nutritional and key biotic/abiotic traits

(Kumar et al. 2018d; Singh et al. 2020a). Several other researchers also assessed the genetic diversity available within wild and cultivated lentil germplasm for nutrition uptake and concentration (Se, Fe, and Zn), biotic and abiotic stresses, and agronomic traits (Thavarajah et al. 2017; Khazaei et al. 2017; Kumar et al. 2018a,b,c; Bhadauria et al. 2017; Guerra-García et al. 2021).

DNA markers have been widely deployed to assess the degree of genetic diversity in cultivated and wild relatives of lentil (Aboelwafa et al. 1995; Ahmad and McNeil 1996; Ford et al. 1997; Havey and Muehlbauer 1989; Sharma et al. 1995, 1996; Kumar et al. 2014; Hamwieh et al. 2009; Alo et al. 2011; Yadav et al. 2016). Lentil genetic resources have also been used in genome-wide association mapping studies (GWAS) (Khazaei et al. 2017; Kumar et al. 2018a,b; Polanco et al. 2019; Dissanayake et al. 2020; Johnson et al. 2021; Gela et al. 2021a; Neupane et al. 2022).

2.3 Lentil Classical Genetics and Its Application in Breeding

Classical genetics was applied to lentil breeding in the early 1900s with an easy-to-manage qualitative trait approach. Since then, several morphological traits have been explored following Mendelian genetics and have been utilized in lentil breeding programs for trait improvement until the advent of molecular markers. The definitive works on the pre-molecular lentil breeding era identified both independent and linked inheritance modes in various characters of interest (Muehlbauer and Slinkard 1981). In addition to the qualitative traits, attempts were made to investigate the inheritance of quantitative traits that are under polygenic control, such as plant height (Tahir et al. 1994), flowering time (Sarker et al. 1999), seed size (Abbo et al. 1991), winter hardiness (Kahraman et al. 2004), and proteins (Chauhan and Singh 1995). However, polygenic traits always require a QTL (quantitative trait loci) mapping course to place medium-sized or even small-sized genetic effects within a dense phenotype-independent DNA marker framework. The economically important traits have been discussed in this section, for which segregation ratios can be determined without the use of complicated progeny testing or molecular markers.

The first report of a genetic study on lentil was on the inheritance of cotyledon color (Tschermak-Seysenegg 1928) by crossing lentil genotypes with red and yellow cotyledon colors. Following that, many geneticists and lentil breeders became interested in cotyledon color genetics due to its market value and use as a morphological marker for the hybridity of F_1 seeds in the crossing blocks (Wilson et al. 1970; Slinkard 1978; Singh 1978; Sinha et al. 1987; Emami 1996; Kumar 2002; Thomas 2016). It was observed that the red cotyledon was dominant over the yellow cotyledon in F_1 lentil seeds when the yellow × red cotyledon parents were crossed. The segregation of the F_2 seeds fits a three red to one yellow ratio, indicating monogenic dominance. Slinkard (1978) designated the red cotyledon gene as *Yc*, although Singh (1978) described it as an orange (*O*) cotyledon. Subsequently, Muehlbauer and Slinkard (1981) proposed the gene symbol *Yc* for the cotyledon color class red or orange. Singh (1978) also found that the intensity of the orange cotyledon color in homozygous plants was indistinguishable from reciprocal crosses between orange and yellow (F_1 seeds) and concluded that the trait had no cytoplasmic maternal effect. Furthermore,

Slinkard (1978) investigated the inheritance of the green cotyledon in lentils and reported that the red cotyledon was completely dominant over both yellow and green types, whereas yellow was completely dominant over green. *Yc* allele was genetically mapped in lentil (Subedi et al. 2018).

Seed coat (testa) color is another important trait of lentil for which genetic control has been investigated (Erdoğan 2015). Seed coat color inheritance is less clear than cotyledon color inheritance, which could be owing to the influence of cotyledon pigments. Although there are various names for seed coat color in the literature, four basic color classes are currently accepted: black, brown, gray, and green (Sharma 2007, 2009). Other colors are thought to be the consequence of color distortion or an epistatic effect. The monogenic dominance inheritance of the black (*Blt*) color has been confirmed by several authors (Vaillancourt and Slinkard 1992; Emami 1996; Emami and Sharma 2000; Sharma et al. 2004). Emami and Sharma (2000) found that its penetrance is incomplete dominance, which is influenced by the so-called dosage effect. Its expression is determined by the amount of black pigment accumulated in the seed coat during seed development. As a result, heterozygous plants (*Bltblt*) produce seeds that range in color from very dark to non-black. In contrast, the homozygous plants with two dosages (*Blt*) generate seeds with a fully black seed coat that are easy to score (Sharma 2009). On the other hand, in studies on the inheritance of brown, gray, and green seed coats in lentils, it was discovered that brown had a monogenic dominance over gray and tan (Sharma 2009). Furthermore, Vandenberg and Slinkard (1990) showed that the color of the lentil seed coat is regulated by two non-linked dominant genes and that all three seed coat colors (brown, gray, and tan) are dominant over the green color. They suggested the gene symbols Ggc for gray and Tgc for tan, as well as genotype constitutions: *GgcGgcTgcTgc* (brown), *GgcGgctgctgc* (gray), *ggcggcTgcTgc* (tan), and the recessive *ggcggctgctgc* (green). The presence of tannins (proanthocyanidins) in the lentil seed coat causes seed discoloration (browning) during storage, limiting the grain's marketability (Bezeda 1980). Vaillancourt et al. (1986) reported that zero-tannin in the lentil seed coat is controlled by a single recessive gene (*tan*) that results in a phenotype of white flowers. The *tan* gene encodes a bHLH transcription factor that is homologous to the A gene in pea (Mirali et al. 2016).

Pod dehiscence, controlled by the dominant gene *Pi*, is one of the key traits contributing to yield reductions in lentils. According to Muehlbauer and Slinkard (1981), *Pi* is one of the distinguishing factors between wild and cultivated *Lens* species. Most of the wild lentil relatives rely on it as a seed dispersal mechanism to survive. Conversely, resistance to shattering (pod-indehiscence) is controlled by a single recessive *pi* gene in lentils (Ladizinsky 1979; Vaillancourt and Slinkard 1992). Flower color is another monogenic trait that has been subjected to genetic investigation. Lentil flower color is the most variable characteristic, as it is influenced by the color of the standard wings and keel. As reviewed by Sharma (2009), the flower colors violet (Lal and Srivastava 1975), blue (Ladizinsky 1979), and purple (Kumar 2002) have all been used interchangeably. In all cases, violet/purple flowers were reported to have complete monogenic dominance over white flowers (Lal and Srivastava 1975; Gill and Marhotra 1980; Kumar 2002). However, it is impossible to rule out the possibility of multiple alleles explaining some of the intermediate colors (Sharma 2009).

Remarkably, a variety of heritable traits showing simple inheritance patterns have been studied in lentil, including leaf pigmentation (Kumar 2002; Kumar and Gupta 2019), chlorina and xantha chlorophyll mutants (Vandenberg and Slinkard 1989; Vaillancourt and Slinkard 1992), leaf shape (Kumar 2002), stipule size (Kumar 2002), plant pubescence (Kumar 2002), growth habit (Ladizinsky 1979), number of flowers per inflorescence (Khosravi et al. 2010), seed hardness (Ladizinsky et al. 1984), and seed coat spotting (Ladizinsky 1979; Vaillancourt et al. 1986).

2.4 Breeding Opportunities Offered by Development of Genomic Resources

2.4.1 Molecular Markers Repertoire

Nowadays, molecular markers play an important role in the lentil breeding pipelines, including parental line selection for crossing, ensuring purity of F_1 hybrids, linkage map construction and mapping traits in controlled-cross populations and GWA studies, and molecular profiling of breeding populations (Solanki et al. 2010; Alo et al. 2011; Kumar et al. 2014, 2021b; Verma et al. 2015). They can be categorized based on timeline (past-present-future), generation of data points per single run (high- and low-throughput), and technique used (hybridization-PCR-sequencing) (Kumar et al. 2015). The first hybridization-based RFLP (restriction fragment length polymorphism) markers were used to develop linkage maps in lentil (Havey and Muehlbauer 1989; Eujayl et al. 1998). The use of RFLP markers did not gain popularity owing to the high technical skills required for their use. The advent of molecular markers based on PCR such as RAPD (random amplified polymorphic DNA), SCAR (sequence characterized amplified region), and SSRs (simple sequence repeats) have initiated their use in lentil breeding programs (reviewed in Kumar et al. 2015), which are used till date (Gupta et al. 2016; Kumar et al. 2018a,b; Ates et al. 2018a; Polanco et al. 2019; Singh et al. 2019b). Now, the development of PCR-based genic-SSR markers has been accelerated and made more cost-effective through transcriptome analysis and subsequent NGS (Kaur et al. 2011; Singh et al. 2017, 2019a). In a recent transcriptome study, a set of 4968 SSR primers were developed in lentil (Kumar et al. 2021b).

SNP (single nucleotide polymorphism) marker systems are today's priority genotyping platforms to generate millions of data points (Morgil et al. 2020). These SNPs can be seen in high quantities across the genome of legume crops, and a variety of approaches can be used to identify and validate them (Shilpa and Lohithaswa 2021). Previously, Sanger sequencing technology was used to re-sequence unigene-derived amplicons or to discover gene-based SNPs using EST data, which facilitated the use of the kompetitive allele-specific PCR (KASP) methodology in lentil to detect SNPs from the EST data (Fedoruk et al. 2013; Sharpe et al. 2013). NGS technologies have made rapid and cost-effective mining of SNPs for mapping QTL/gene(s) in lentil possible (Kaur et al. 2011). Sharpe et al. (2013) reported that data generated using the Illumina Genome Analyzer yielded ~ 44,879 SNP markers in lentil. In another study, a set of 50,960 SNPs was discovered and used to construct a high density linkage map in lentil (Temel et al. 2015). This way, the availability of high-density SNP markers boosted the development of Illumina Golden Gate platforms with >1000 SNPs for genotyping in lentil (Sharpe et al. 2013; Kaur et al. 2014). However, transcriptome analysis using NGS platforms has opened a gateway for greater numbers of SNPs or

EST-SSRs from coding regions of the lentil genome in the recent past (Singh et al. 2017, 2019a; Kumar et al. 2021b). SNP data from heat-tolerant lentil transcriptomes were used to identify genes encoding proteins or regulating factors involved in various metabolic pathways, such as signal transduction, fatty acid biosynthesis, rRNA processing, ribosome biogenesis, gibberellin (GA) biosynthesis, and riboflavin biosynthesis (Kumar et al. 2019b), as well as the pathways regulating the seed size related traits in lentil (Dutta et al. 2022). Recently, an exome capture sequencing array targeting 85 Mb of the lentil genome's protein-coding region was developed (Ogutcen et al. 2018). This method generates a large number of high-quality and informative SNP markers, which are used in a variety of lentil molecular research activities, including the development of high-density linkage maps, GWAS, and genomic selection (Haile et al. 2020, 2021; Gela et al. 2021a).

2.4.2 Expressed Sequence Tag (EST) and Contigs/Unigenes/Transcripts

Conventional (i.e., Sanger sequencing) and advanced NGS technologies have provided nucleotide sequences of coding regions that are expressed in different tissues as well as at different plant growth stages. These sequences have opened avenues to develop functional or genic markers, including SSR, SNP, and ITP (intron-targeted polymorphism) markers (Kumar et al. 2015; Singh et al. 2020b). Previously, conventional sequencing was the way to sequence 150–400 bp cDNA clones. These correspond to mRNAs and are expressed in distinct tissues and/or at the appropriate stage of crop plants leading to the development of expressed sequence tags (ESTs). The development of ESTs has been further augmented by using serial analysis of gene expression (SAGE). For instance, ESTs were recognized for an amino oxidase gene in lentil (Rossi et al. 1992). The EST library was produced from eight cultivars with diverse seed phenotypes (Vijayan et al. 2009). Presently, 33,963 ESTs are open with respect to lentil. A study generated 5000 ESTs in lentil from leaf tissues that were infected with *Colletotrichum truncatum* (Bhadauria et al. 2011).

The use of NGS-based RNA or transcriptome sequencing has accelerated the production of ESTs and unigenes/transcripts in lentil. These were used to identify SSR and SNP markers. For instance, 1.38×10^6 ESTs were achieved from tissue specific cDNA of six lentil genotypes. It identified 15,354 contigs and 68,715 singletons through *de-novo* assembly (Kaur et al. 2011). This was utilized for recognition of 25,592 unigenes and 2,393 EST-SSR markers. Validation of a subset of 192 EST-SSR markers from this set has resulted in 47.5% polymorphism throughout a panel of 12 cultivated genotypes (Kaur et al. 2011). Moreover, diversified transcriptome resources in the case of lentil have been developed from wild and cultivated accessions using 454 pyrosequencing, which led to the development of 1.03×10^6 ESTs (Sharpe et al. 2013). This has given a base assembly with 50,146 contigs, and also produced a reference assembly of 27,921 contigs after purification by duplication, overlap, and size. These have revealed 44,879 SNPs and among them, a sub-set comprising 1536 SNPs was used to develop a high-throughput GoldenGate array platform for genotyping SNPs in lentil. A genetic map of *L. culinaris* based on SNP was constructed using the array (Sharpe et al. 2013). In a separate study, 20,009 nonredundant transcripts were developed by using 119,855,798 short reads generated by Illumina paired-end sequencing through *de novo* transcriptome assembly. These transcripts paved the way for generating 5673 SSR markers that were deployed in assessing diversity (Verma et al. 2013). By using Illumina platform for transcriptome sequencing of Precoz and WA8649041 (*L. cultivars*) 111,105,153 ESTs have been generated, which were assembled into 97,528 high quality contigs (Temel et al. 2015). These contigs were deployed in developing 50,960 SNP markers for generating a linkage map. Recently, the first lentil pan-transcriptome with 15,910 core genes encoded in all accessions and 24,226 accessory genes was published (Gutierrez-Gonzalez et al. 2022).

2.4.3 Mapping Population Development

In lentils, biparental populations, derived from crosses between two genotypes, have been widely used as genetic materials. Various types of biparental populations including F_2s, backcrosses, and recombinant inbred lines (RILs) have been developed for genetic map construction and trait mapping studies. Trait-specific RIL populations have been created for important traits such as agronomic performance (i.e., earliness, early growth vigor, and root traits), disease resistance, seed quality, drought and boron tolerance, and seed mineral concentration (i.e., Fe, Zn, Mn, and Se) in lentil (Kumar et al. 2015, 2021a). For example, RIL mapping populations developed from biparental crosses of (CDC Redberry × ILL 7502, ILL 8006 × CDC Milestone, and PI 320937 × CDC Eston) have been used to construct linkage maps as well as to identify the QTLs for absorption of Fe, Mn, and Se (Ates et al. 2016, 2018a,b; Aldemir et al. 2017). For understanding the genetic basis of Aphanomyces root rot resistance in lentil, partial resistant, and susceptible breeding lines were crossed to raise a mapping population comprising 189 RILs (Ma et al. 2020). Earlier studies of intraspecific mapping populations found low genetic diversity in general, with only a small number of markers incorporated to genetic maps (Bohra et al. 2012). This drawback of intraspecific mapping populations, in general, has been revolutionized by the use of SNPs derived from NGS platforms such as exome capture sequencing (Gela 2021; Gela et al. 2021a; Haile et al. 2021).

In addition, several interspecific RIL populations have also been developed by crossing *L. culinaris* with wild species such as *L. odemensis* and *L. ervoides* accession, which have been used in identifying functional markers associated with important agronomic traits and disease resistance (Chen 2018; Polanco et al. 2019; Gela et al. 2021b). Although incorporating beneficial alleles from wild species is important for improving the biotic and abiotic stresses of elite breeding germplasm, they may be inherited alongside deleterious genes, masking the genetic variation of the desired trait (Tanksley and Nelson 1996). Thus, backcross-derived inbred line (BIL) populations enable the efficient use of desired traits from wild parents while minimizing the presence of undesirable traits in individual introgression lines (Tanksley and Nelson 1996). Gela et al. (2021c) created a lentil BIL in the CDC Redberry cultivar background using *L. ervoides* alleles from an interspecific RIL, LR-59-81 (Fiala et al. 2009; Gela et al. 2020; Gela 2021). This population was used to map disease resistance QTL/gene(s) such as anthracnose (*Colletotrichum lentis* race 0), which was found in *L. ervoides* but not in *L. culinaris* (Gela 2021). Similarly, Kumar et al. (2019a) developed a BC_2F_5 BIL population by crossing cultivar IPL 220 with *L. orientalis* accession ILWL 118. Multi-parent advanced generation inter-cross (MAGIC) populations have been proposed to enrich allelic diversity and broaden genetic diversity (Huang et al. 2012; Scott et al. 2020). No MAGIC population has yet been reported for lentils.

2.4.4 Development of Linkage Maps

A comprehensive list of linkage maps developed using biparental inter- and intra-specific mapping populations in lentil has been detailed in our previous reviews (Kumar et al. 2015, 2021a). The first lentil's genetic linkage map using DNA markers was constructed using only 20 RFLPs in the late 1980s (Havey and Muehlbauer 1989), which was followed by the development of several genetic linkage maps using various morphological and advanced DNA markers (reviewed in Kumar et al. 2021a). Low-cost sequencing technologies have paved the way for greater number of high-density linkage maps based on SNP markers (Fedoruk et al. 2013; Gujaria-Verma et al. 2014; Sudheesh et al. 2016; Bhadauria et al. 2017; Aldemir et al. 2017; Polanco et al. 2019; Subedi et al. 2018; Ma et al. 2020; Gela et al. 2021a,b; Rajandran et al. 2022). For example, a high-density consensus map with about 10,000 SNP markers covering 977.47 cM distance was developed by using three mapping populations, which contained seven linkage groups and an average distance of 0.10 cM between two markers (Ates et al. 2018b). As NGS identifies SSR and SNP markers from coding regions of the genome, some of these are functional, which directed the development of high-density linkage map with markers generated from gene-based SNPs using an interspecific (*L. culinaris* × *L. odemensis*) mapping population. This map comprises 6153 markers grouped in 4682 unique bins and placed on 10 linkage groups with a coverage of 5782 cM length (Polanco et al. 2019). Furthermore, exome capture sequencing was used to create high-density linkage genetic maps with 21,634 SNP markers grouped into seven linkage groups that correspond to the lentil's seven haploid chromosomes (Gela et al. 2021a; Haile et al. 2021).

2.4.5 Trait Mapping

Several QTL mapping studies have been conducted over the last 20 years using various types of molecular markers ranging from low to high fidelity in lentil. A list of QTLs/genes developed from lentil mapping populations for key traits up to 2019 was presented in Kumar et al. (2021a). However, the majority of these QTL/genes have not been deployed in marker-assisted breeding (MAB) in lentil for a variety of reasons, including a lack of a large number of genome-wide molecular markers, poor linkage between markers and traits, and low phenotypic predictive values (variation explained) of the markers (Kumar et al. 2019a). Lately, advances in NGS technology and the availability of the reference genome have enabled the detection of large-scale and high-throughput SNP variations across the whole genome, resulting in significant progress in identifying QTLs for important lentil traits such as resistance to anthracnose (Gela et al. 2021a,b), Ascochyta blight (Dadu et al. 2021), Aphanomyces root rot (Ma et al. 2020), and Stemphylium blight (Adobor et al. 2022). QTLs for early plant vigor (Mane et al. 2020), flowering time (Yuan et al. 2021; Haile et al. 2021), and phenological traits (Haile et al. 2021) have also been reported. Marker-trait associations using the GWAS approach were recently reported for seed size (Khazaei et al. 2017), seed quality traits (Khazaei et al. 2018), prebiotic carbohydrates (Johnson et al. 2021), agronomic traits (Rajandran et al. 2022), phenological traits (Neupane et al. 2022), anthracnose resistance (Gela et al. 2021a), and seed protein and amino acid contents (Jiayi 2021). The SNPs linked to the QTL for desirable traits should be converted to robust breeder-friendly markers to be used in MAB. For instance, Haile et al. (2021) converted the coding sequences of SNPs within the flowering time QTL interval to KASP markers and validated them in a population with different genetic backgrounds. They indicated that the newly discovered SNP markers could be used in MAB to precisely transfer genomic regions from exotic germplasm into elite lentil cultivars while preserving adaptation. More importantly, the density of SNP markers in lentils has been significantly increased, allowing for fine mapping of QTL regions and gene identification; thus, converting target SNP markers into PCR-based KASP markers requires more attention in order to use them in lentil breeding programs (Wang et al. 2020).

2.4.6 Development of Reference Genome Sequence

A draft lentil genome sequence (*L. culinaris*) has been revised several times (Bett et al. 2016). The current *L. culinaris* reference genome (Lcu.2RBY; CDC Redberry) was assembled using 54× long reads polished with both long and short reads (Ramsay et al. 2021). This assembly is composed of scaffolded contigs that cover 3.76 Gbp of the estimated 3.92 Gbp size using K-mer distribution analysis, which is comparable to the flow cytometry-determined 4.06 Gbp for this species (Arumuganathan and Earle 1991). It also has 39,778 high-confidence annotated genes, which account for 94% of the total BUSCOs, and is available on the KnowPulse web portal (http://knowpulse.usask.ca) for facilitating in-depth genetic and genomic studies in lentil. In addition, the wild lentil reference genome, *L. ervoides* (Ler.1DRT; IG 72815), was also assembled using the same methods as *L. culinaris* (*Lcu.2RBY*), 52× long reads, and additional short reads (Ramsay et al. 2021). The contigs in this assembly were scaffolded into seven pseudomolecules, representing 96.1% of the assembly, which contains 37,045 annotated genes and covers 2.9 Gbp of the predicted size based on K-mer analysis (2.9 Gbp). The *L. ervoides* genomes are smaller in size than the *L. culinaris* genomes, with a comparable number of high-confidence annotated genes.

2.5 Comparative and Functional Genomics

According to a recent comparative mapping study, severe rearrangements occurred after *Lens* species diverged from other cool-season legumes but before *L. culinaris* and *L. ervoides* diverged (Ramsay et al. 2021). Despite the fact that several translocations and inversions exist among legume species, including the unbalanced reciprocal translocation of lentil (*L. culinaris*) chromosomes 1 and 5 relatives to the genomes of *Medicago truncatula*, pea, and *L. ervoides*, there is large-scale genome collinearity among them (Sharpe et al. 2013; Gujaria-Verma et al. 2014; Bhadauria et al. 2017; Ramsay et al. 2021). Understanding the structural rearrangements of the cultivated and wild lentil genomes could aid in accessing the favorable wild genes located in the reciprocal translocation region via genome editing (CRISPR/Cas9, Rossato et al. 2022), allowing inversion of inaccessible wild genome regions relative to the cultivated genome while reducing linkage drag (Schmidt et al. 2020). Previous research has also shown that comparative genetics can be used to develop functional markers, investigate candidate genes, and determine their roles in regulating desirable agronomic traits in lentil. Comparative mapping of *MLO* genes from lentil and other legume species, for example, has resulted in the identification of two candidate genes, *LcMLO1* and *LcMLO3*. Only SNPs and small indels in introns differentiated these genes within and between species, but they produced identical amino acid sequences. Two amino acid substitutions were found in the *MLO3* candidate gene between *L. culinaris* and *L. orientalis*, and amino acid substitutions and indels were found in the carboxyl intracellular domain of this gene among three species (*L. odemensis*, *L. ervoides*, and *L. lamottei*) (Polanco et al. 2018), and thus could be used to develop powdery mildew resistance in lentil. A number of transcriptomes based on comparative mapping have been reported for lentils (reviewed in Kumar et al. 2021a). Furthermore, the draft lentil genome sequence (Ramsay et al. 2021) will provide lentil researchers with an excellent opportunity and a wealth of resources to study the important gene functions annotated in other legumes via sequence homology.

2.6 Conclusions

The global consumption of lentils is rapidly expanding, mainly due to its contribution to green agriculture, protein, and nutritional security. Progress in lentil genomics has accelerated our efforts to make genomics-assisted breeding a reality in the near future. This involves the precise discovery and efficient deployment of traits through the advanced techniques in current lentil breeding programs. Speed-breeding tools to shorten breeding cycles have been developed and utilized in lentil (Lulsdorf and Banniza 2018). Haile et al. (2020) also suggested genomic selection as a tool in lentil breeding to make predictions within populations and across environments. Adopting genetic and genomic resources, as well as speed-breeding tools, will hasten the development of new climate-resilient, high-yielding lentil cultivars, paving the way for maximizing genetic gains. This task will be improved further with significant improvements in phenotyping by incorporating deep learning and artificial intelligence into lentil breeding programs, propelling lentil into the genomics-based breeding era.

References

Abbo, S., Ladizinsky, G., and Weeden, N.F. (1991). Genetic-analysis and linkage study of seed weight in lentil. *Euphytica* 8: 259–266.

Aboelwafa, A., Murai, K., and Shimada, T. (1995). Intra-specific and inter-specific variations in *Lens* revealed by RAPD markers. *Theoretical and Applied Genetics* 90: 335–340.

Adobor, S., Gela, T.S., Banniza, S., and Vandenberg, A. (2022). Mapping of genomic regions linked to stemphylium blight (*Stemphylium botryosum* Wallr.) resistance in lentil using linkage mapping and marker-trait association analysis. *bioRxiv*. https://doi.org/10.1101/2022.08.22.504834.

Ahmad, M. and McNeil, D.L. (1996). Comparison of crossability, RAPD, SDS-PAGE and morphological markers for revealing genetic relationships within and among *Lens* species. *Theoretical and Applied Genetics* 93: 788–793.

Aldemir, S., Ateş, D., Temel, H.Y. et al. (2017). QTLs for iron concentration in seeds of the cultivated lentil (*Lens culinaris* Medic.) via genotyping by sequencing. *Turkish Journal of Agriculture and Forestry* 41 (4): 243–255.

Alo, F., Furman, B.J., Akhunov, E. et al. (2011). Leveraging genomic resources of model species for the assessment of diversity and phylogeny in wild and domesticated lentil. *Journal of Heredity* 102 (3): 315–329.

Ates, D., Sever, T., Aldemir, S. et al. (2016). Identification QTLs controlling genes for Se uptake in lentil seeds. *PLoS One* 11 (3): e0149210.

Ates, D., Aldemir, S., Alsaleh, A. et al. (2018a). A consensus linkage map of lentil based on DArT markers from three RIL mapping populations. *PLoS One* 13 (1): e0191375.

Ates, D., Aldemir, S., Yagmur, B. et al. (2018b). QTL mapping of genome regions controlling manganese uptake in lentil seed. *G3: Genes, Genomes, Genetics* 8 (5): 1409–1416.

Bett, K., Ramsay, L., Chan, C., et al. (2016). Lentil v1.0 and beyond, *Proceedings of XXIV Plant and Animal Genome Conference*, San Diego, CA (9–22 January 2016).

Bezeda, M. (1980). Effect of environmental conditions on lentil seeds. MSc thesis. University of Ottawa, Ottawa, Canada.

Bhadauria, V., Banniza, S., Vandenberg, A. et al. (2011). EST mining identifies proteins putatively secreted by the anthracnose pathogen *Colletotrichum truncatum*. *BMC Genomics* 12 (1): 327.

Bhadauria, V., Ramsay, L., Bett, K.E., and Banniza, S. (2017). QTL mapping reveals genetic determinants of fungal disease resistance in the wild lentil species *Lens ervoides*. *Scientific Reports* 7 (1): 3231.

Bohra, A., Saxena, R.K., Gnanesh, B.N. et al. (2012). An intra-specific consensus genetic map of pigeonpea [*Cajanus cajan* (L.) Millspaugh] derived from six mapping populations. *Theoretical and Applied Genetics* 125 (6): 1325–1338.

Chauhan, M.P. and Singh, I.S. (1995). Inheritance of protein content in lentil (*Lens culinaris* Medik.). *Legume Research* 18 (1): 5–8.

Chen, L. (2018). Assessing impacts of crop-wild introgression in lentil using interspecific lens species recombinant inbred line populations. PhD thesis. University of Saskatchewan, Saskatoon, Canada.

Dadu, R.H.R., Bar, I., Ford, R. et al. (2021). *Lens orientalis* contributes quantitative trait loci and candidate genes associated with Ascochyta blight resistance in lentil. *Frontiers in Plant Science* 12: 703283.

Dissanayake, R., Braich, S., Cogan, N.O. et al. (2020). Characterization of genetic and allelic diversity amongst cultivated and wild lentil accessions for germplasm enhancement. *Frontiers in Genetics* 11: 546.

Dissanayake, R., Cogan, N.O.I., Smith, K.F., and Kaur, S. (2021). Application of genomics to understand salt tolerance in lentil. *Genes* 12 (3): 332.

Dutta, H., Mishra, G.P., Aski, M.S. et al. (2022). Comparative transcriptome analysis, unfolding the pathways regulating the seed-size trait in cultivated lentil (*Lens culinaris* Medik.). *Frontiers in Genetics* 13: 942079.

El haddad, N., Rajendran, K., Smouni, A. et al. (2020). Screening the FIGS set of lentil (*Lens culinaris* Medikus) germplasm for tolerance to terminal heat and combined drought-heat stress. *Agronomy* 10 (7): 1036.

Emami, M.K. (1996). Genetic mapping in lentil (*Lens culinaris* Medik.). PhD thesis. Division of Genetics, Indian Agricultural Research Institute, New Delhi, India.

Emami, M.K. and Sharma, B. (2000). Inheritance of black testa colour in lentil (*Lens culinaris* Medik.). *Euphytica* 115: 43–47.

Erdoğan, C. (2015). Genetic characterization and cotyledon color in lentil. *Chilean Journal of Agricultural Research* 75: 383–389.

Eujayl, I., Baum, M., Powell, W. et al. (1998). A genetic linkage map of lentil (Lens sp.) based on RAPD and AFLP markers using recombinant inbred lines. *Theoretical and Applied Genetics* 97: 83–89.

FAO (2010). *The Second Report on the State of the World's Plant Genetic Resources for Food and Agriculture*. Rome: Food and Agriculture Organization of the United Nations. https://www.fao.org/3/i1500e/i1500e00.pdf (accessed 15 December 2022).

FAOSTAT (2022). Statistical databases. Food and Agriculture Organization of the United Nations, Italy. http://www.fao.org/faostat/en/#data/QC. (accessed 12 December 2022).

Fedoruk, M.J., Vandenberg, A., and Bett, K.E. (2013). Quantitative trait loci analysis of seed quality characteristics in lentil using single nucleotide polymorphism markers. *The Plant Genome* 6 (3). https://doi.org/10.3835/plantgenome2013.05.0012.

Fiala, J.V., Tullu, A., Banniza, S. et al. (2009). Interspecies transfer of resistance to anthracnose in lentil (*Lens culinaris* Medic.). *Crop Science* 49: 825–830.

Ford, R.E.C.K., Pang, E.C.K., and Taylor, P.W.J. (1997). Diversity analysis and species identification in *Lens* using PCR generated markers. *Euphytica* 96 (2): 247–255.

Gela, T.S. (2021). Mapping and analysis of genetic loci conferring resistance to anthracnose in lentil. PhD. thesis. University of Saskatchewan, Saskatoon, Canada.

Gela, T.S., Banniza, S., and Vandenberg, A. (2020). Lack of effective resistance to the virulent race of *Colletotrichum lentis* in *Lens culinaris* Medikus subsp. *culinaris*. *Plant Genetic Resources: Characterisation and Utilisation* 18 (2): 81–87.

Gela, T.S., Ramsay, L., Haile, T.A. et al. (2021a). Identification of anthracnose race 1 resistance loci in lentil by integrating linkage mapping and genome-wide association study. *The Plant Genome* 14: e20131.

Gela, T.S., Koh, C.S., Caron, C.T. et al. (2021b). QTL mapping of lentil anthracnose (*Colletotrichum lentis*) resistance from *Lens ervoides* accession IG 72815 in an interspecific RIL population. *Euphytica* 217: 70.

Gela, T.S., Adobor, S., Khazaei, H., and Vandenberg, A. (2021c). An advanced lentil backcross population developed from a cross between *Lens culinaris* × *L. ervoides* for future disease resistance and genomic studies. *Plant Genetic Resources: Characterization and Utilization* 19: 167–173.

Gill, A.S. and Marhotra, R.S. (1980). Inheritance of flower colour and flower number per inflorescence in lentils. *Lens* 7: 93–19.

Guerra-García, A., Gioia, T., von Wettberg, E. et al. (2021). Intelligent characterization of lentil genetic resources: evolutionary history, genetic diversity of germplasm, and the need for well-represented collections. *Current Protocols* 1: e134.

Gujaria-Verma, N., Vail, S.L., Carrasquilla-Garcia, N. et al. (2014). Genetic mapping of legume orthologs reveals high conservation of synteny between lentil species and the sequenced genomes of *Medicago* and chickpea. *Frontiers in Plant Science* 5: 676.

Gupta, D.S., Cheng, P., Sablok, G. et al. (2016). Development of a panel of unigene-derived polymorphic EST–SSR markers in lentil using public database information. *The Crop Journal* 4 (5): 425–433.

Gutierrez-Gonzalez, J.J., García, P., Polanco, C. et al. (2022). Multi-species transcriptome assemblies of cultivated and wild lentils (*Lens* sp.) provide a first glimpse at the lentil pangenome. *Agronomy* 12: 1619.

Haile, T.A., Heidecker, T., Wright, D. et al. (2020). Genomic selection for lentil breeding: empirical evidence. *The Plant Genome* 13 (1): e20002.

Haile, T.A., Stonehouse, R., Weller, J.L., and Bett, K.E. (2021). Genetic basis for lentil adaptation to summer cropping in northern temperate environments. *The Plant Genome* 14 (3): e20144.

Hamwieh, A., Udupa, S.M., Sarker, A. et al. (2009). Development of new microsatellite markers and their application in the analysis of genetic diversity in lentils. *Breeding Science* 59 (1): 77–86.

Havey, M.J. and Muehlbauer, F.J. (1989). Linkages between restriction fragment length, isozyme, and morphological markers in lentil. *Theoretical and Applied Genetics* 77 (3): 395–401.

Huang, B.E., George, A.W., Forrest, K.L. et al. (2012). A multiparent advanced generation inter-cross population for genetic analysis in wheat. *Plant Biotechnology Journal* 10 (7): 826–839.

Hussain, S.A., Iqbal, M.S., Akbar, M. et al. (2022). Estimating genetic variability among diverse lentil collections through novel multivariate techniques. *PLoS One* 17 (6): e0269177.

Jiayi, H. (2021). Genome-wide association study of seed protein and amino acid contents in cultivated lentils as determined by near-infrared reflectance spectroscopy. PhD thesis. University of Manitoba, Winnipeg, Canada.

Johnson, N., Boatwright, J.L., Bridges, W. et al. (2021). Genome-wide association mapping of lentil (*Lens culinaris* Medikus) prebiotic carbohydrates toward improved human health and crop stress tolerance. *Scientific Reports* 11: 13926.

Kahraman, A., Kusmenoglu, I., Aydin, N. et al. (2004). Genetics of winter hardiness in 10 lentil recombinant inbred line population. *Crop Science* 44: 5–12.

Kaur, S., Cogan, N.O., Pembleton, L.W. et al. (2011). Transcriptome sequencing of lentil based on second-generation technology permits large-scale unigene assembly and SSR marker discovery. *BMC Genomics* 12 (1): 265.

Kaur, S., Cogan, N.O., Stephens, A. et al. (2014). EST-SNP discovery and dense genetic mapping in lentil (*Lens culinaris* Medik.) enable candidate gene selection for boron tolerance. *Theoretical and Applied Genetics* 127 (3): 703–713.

Khazaei, H., Street, K., Bari, A. et al. (2013). The FIGS (focused identification of germplasm strategy) approach identifies traits related to drought adaptation in *Vicia faba* genetic resources. *PLoS One* 8: e63107.

Khazaei, H., Caron, C.T., Fedoruk, M. et al. (2016). Genetic diversity of cultivated lentil (*Lens culinaris* Medik.) and its relation to the world's agro-ecological zones. *Frontiers in Plant Science* 7: 1093.

Khazaei, H., Podder, R., Caron, C.T. et al. (2017). Marker–trait association analysis of iron and zinc concentration in lentil (*Lens culinaris* Medik.) seeds. *The Plant Genome* 10 (2): 1–8. https://doi.org/10.3835/plantgenome2017.02.0007.

Khazaei, H., Fedoruk, M., Caron, C.T. et al. (2018). Single nucleotide polymorphism markers associated with seed quality characteristics of cultivated lentil. *The Plant Genome* 11 (1): 170051.

Khazaei, H., Subedi, M., Nickerson, M. et al. (2019). Seed protein of lentils: current status, progress, and food applications. *Food* 8: 391.

Khosravi, E., Khodambashi, M., and Mohammadi, S. (2010). A linkage group of four genes related to morphological traits in lentil (*Lens culinaris* Medik.). *International Journal of Plant Production* 4: 303–308.

Kumar, Y. (2002). Inheritance and linkage of genes for morphological traits in lentil (*Lens culinaris* Medik.). PhD thesis. Charan Singh University, Meerut, India.

Kumar, S. and Gupta, J. (2019). Inheritance of qualitative and quantitative traits in interspecific crosses of lentil. *Indian Journal of Genetics* 79 (3): 626–631.

Kumar, S., Barpete, S., Kumar, J. et al. (2013). Global lentil production: constraints and strategies. *SATSA Mukhapatra-Annual Technical* 17: 1–13.

Kumar, J., Srivastva, E., Singh, M. et al. (2014). Diversification of indigenous gene-pool by using exotic germplasm in lentil (*Lens culinaris* Medikus subsp. *culinaris*). *Physiology and Molecular Biology of Plants* 20 (1): 125–132.

Kumar, S., Rajendran, K., Kumar, J. et al. (2015). Current knowledge in lentil genomics and its application for crop improvement. *Frontiers in Plant Science* 6: 78.

Kumar, J., Gupta, D.S., Kumar, S. et al. (2016). Current knowledge on genetic biofortification in lentil. *Journal of Agricultural and Food Chemistry* 64 (33): 6383–6396.

Kumar, J., Basu, P.S., Gupta, S. et al. (2018a). Physiological and molecular characterisation for high temperature stress in *Lens culinaris*. *Functional Plant Biology* 45 (4): 474–487.

Kumar, J., Gupta, S., Biradar, R.S. et al. (2018b). Association of functional markers with flowering time in lentil. *Journal of Applied Genetics* 59 (1): 9–21.

Kumar, J., Gupta, S., Gupta, D.S., and Singh, N.P. (2018c). Identification of QTLs for agronomic traits using association mapping in lentil. *Euphytica* 214 (4): 75.

Kumar, S., Choudhary, A.K., Rana, K.S. et al. (2018d). Bio-fortification potential of global wild annual lentil core collection. *PLoS One* 13 (1): e0191122.

Kumar, J., Kumar, S., Gupta, D.S. et al. (2019a). Molecular marker assisted gene pyramiding. In: *Lentils: Potential Resources for Enhancing Genetic Gains* (ed. M. Singh), 125–139. Academic Press.

Kumar, H., Singh, A., Dikshit, H. K., Mishra, G. P., Aski, M., Meena, M. C., et al. (2019b). Genetic dissection of grain iron and zinc concentrations in lentil (Lens culinaris medik.). *Journal of Genetics* 98: 66. https://doi.org/10.1007/s12041-019-1112-3.

Kumar, J., Sen Gupta, D., Baum, M. et al. (2021a). Genomics-assisted lentil breeding: current status and future strategies. *Legume Science* 3 (3): e71.

Kumar, J., Gupta, D.S., Kesari, R. et al. (2021b). Comprehensive RNAseq analysis for identification of genes expressed under heat stress in lentil. *Physiologia Plantarum* 173: 1785–1807.

Ladizinsky, G. (1979). The genetics of several morphological markers in lentil. *Journal of Heredity* 70: 135–137.

Ladizinsky, G., Braun, D., Goshen, D., and Muehlbauer, F.J. (1984). The biological species of the genus *Lens* L. *Botanical Gazette* 145: 253–261.

Lal, S. and Srivastava, R.S. (1975). Inheritance of flower colour in lentils. *Indian Journal of Genetics and Plant Breeding* 35: 29–30.

Liber, M., Duarte, I., Maia, A.T., and Oliveira, H.R. (2021). The history of lentil (*Lens culinaris* subsp. culinaris) domestication and spread as revealed by genotyping-by-sequencing of wild and landrace accessions. *Frontiers in Plant Science* 12: 628439.

Lombardi, M., Materne, M., Cogan, N.O. et al. (2014). Assessment of genetic variation within a global collection of lentil (*Lens culinaris* Medik.) cultivars and landraces using SNP markers. *BMC Genetics* 15 (1): 150.

Lulsdorf, M.M. and Banniza, S. (2018). Rapid generation cycling of an F_2 population derived from a cross between *Lens culinaris* Medik. and *Lens ervoides* (Brign.) Grande after aphanomyces root rot selection. *Plant Breeding 37*: 486–491.

Ma, Y., Marzougui, A., Coyne, C.J. et al. (2020). Dissecting the genetic architecture of aphanomyces root rot resistance in lentil by QTL mapping and genome-wide association study. *International Journal of Molecular Sciences* 21 (6): 2129.

Malhotra, N., Panatu, S., Singh, B. et al. (2019). Genetic resources: collection, conservation, characterization and maintenance. In: *Lentils: Potential Resources for Enhancing Genetic Gains* (ed. M. Singh), 21–41. Academic Press.

Mane, R., Katoch, M., Singh, M. et al. (2020). Identification of genomic regions associated with early plant vigour in lentil (*Lens culinaris*). *Journal of Genetics* 99: 21.

Mirali, M., Purves, R.W., Stonehouse, R. et al. (2016). Genetics and biochemistry of zero-tannin lentils. *PLoS One* 11: e0164624.

Morgil, H., Gercek, Y.C., and Tulum, I. (2020). Single nucleotide polymorphisms (SNPs) in plant genetics and breeding. In: *The Recent Topics in Genetic Polymorphisms* (ed. M. Çalışkan, O. Erol, and G. Öz), 137–145. London: Intech Open.

Muehlbauer, F.J. and Slinkard, A.E. (1981). Genetic and breeding methodology. In: *Lentil* (ed. C. Webb and G. Hawtin), 69–90. Fernham, England: CAB-ICARDA.

Neupane, S., Wright, D.M., Martinez, R.O. et al. (2022). Focusing the GWAS *Lens* on days to flower using latent variable phenotypes derived from global multi-environment trials. *bioRxiv*. https://doi.org/10.1101/2022.03.10.483676.

Ogutcen, E., Ramsay, L., von Wettberg, E.B., and Bett, K.E. (2018). Capturing variation in *Lens* (Fabaceae): development and utility of an exome capture array for lentil. *Applications in Plant Sciences* 6 (7): e01165.

Pavan, S., Bardaro, N., Fanelli, V. et al. (2019). Genotyping by sequencing of cultivated lentil (*Lens culinaris* Medik.) highlights population structure in the Mediterranean gene pool associated with geographic patterns and phenotypic variables. *Frontiers in Genetics* 10: 872.

Polanco, C., Sáenz de Miera, L.E., Bett, K., and Pérez de la Vega, M. (2018). A genome-wide identification and comparative analysis of the lentil *MLO* genes. *PLoS One* 13 (3): e0194945.

Polanco, C., Sáenz de Miera, L.E., González, A.I. et al. (2019). Construction of a high-density interspecific (*Lens culinaris* x *L. odemensis*) genetic map based on functional markers for mapping morphological and agronomical traits, and QTLs affecting resistance to *Ascochyta* in lentil. *PLoS One* 14 (3): e0214409.

Rajandran, V., Ortega, R., Vander Schoor, J.K. et al. (2022). Genetic analysis of early phenology in lentil identifies distinct control of component traits. *Journal of Experimental Botany* 73: 3963–3977.

Ramsay, L., Koh, C.H., Kagale, S. et al. (2021). Genomic rearrangements have consequences for introgression breeding as revealed by genome assemblies of wild and cultivated lentil species. *bioRxiv*. https://doi.org/10.1101/2021.07.23.453237.

Rossato, M., Marcolungo, L., De Antoni, L. et al. (2022). CRISPR/Cas9-based repeat depletion for the high-throughput genotyping of complex plant genomes. *bioRxiv*. https://doi.org/10.1101/2022.11.22.517518.

Rossi, A., Petruzzelli, R., and Agrò, A.F. (1992). cDNA-derived amino-acid sequence of lentil seedlings' amine oxidase. *FEBS Letters* 301 (3): 253–257.

Sadras, V., Rosewarne, G., and Lake, L. (2022). Genetic gain in lentil yield between 1988 and 2019 has been larger under stress. *Proceedings of the 20th Agronomy Australia Conference*, Toowoomba Qld (18–22 September 2022). http://www.agronomyaustraliaproceedings.org.

Sarker, A., Erskine, W., Sharma, B., and Tyagi, M.C. (1999). Inheritance and linkage relationships of days to flowering and morphological loci in lentil (*Lens culinaris* Medikus subsp. *culinaris*). *The American Genetic Association* 90: 270–275.

Schmidt, C., Fransz, P., Rönspies, M. et al. (2020). Changing local recombination patterns in Arabidopsis by CRISPR/Cas mediated chromosome engineering. *Nature Communications* 11: 4418.

Scott, M.F., Ladejobi, O., Amer, S. et al. (2020). Multi-parent populations in crops: a toolbox integrating genomics and genetic mapping with breeding. *Heredity* 125: 396–416.

Sharma, B. (2007). Genetic of economic traits. In: *The Lentil Botany, Production, and Uses* (ed. W. Erskine, F.J. Muehlbauer, A. Sarker, and B. Sharma), 76–101. Cambridge: CABI International.

Sharma, B. (2009). Genetic of economic traits. In: *The Lentil Botany, Production, and Uses* (ed. W. Erskine, F.J. Muehlbauer, A. Sarker, and B. Sharma), 76–101. Cambridge: CABI International.

Sharma, S.K., Dawson, I.K., and Waugh, R. (1995). Relationships among cultivated and wild lentils revealed by RAPD analysis. *Theoretical and Applied Genetics* 91 (4): 647–654.

Sharma, S.K., Knox, M.R., and Ellis, T.N. (1996). AFLP analysis of the diversity and phylogeny of *Lens* and its comparison with RAPD analysis. *Theoretical and Applied Genetics* 93 (5, 6): 751–758.

Sharma, B., Tyagi, M.C., Mishra, S.K., and Kumar, Y. (2004). Confirmation of the inheritance pattern of black testa colour in lentil. *Journal of Lentil Research* 1 (1): 19–20.

Sharpe, A.G., Ramsay, L., Sanderson, L.A. et al. (2013). Ancient orphan crop joins modern era: gene-based SNP discovery and mapping in lentil. *BMC Genomics* 14 (1): 192.

Shilpa, H.B. and Lohithaswa, H.C. (2021). Discovery of SNPs in important legumes through comparative genome analysis and conversion of SNPs into PCR-based markers. *Journal of Genetics* 100: 75.

Singh, T.P. (1978). Inheritance of cotyledon colour in lentil. *Indian Journal of Genetics and Plant Breeding* 38: 11–12.

Singh, R.J. and Chung, G.H. (2016). Landmark research for pulses improvement. *Indian Journal of Genetics and Plant Breeding* 76 (4): 399–409.

Singh, M., Bisht, I.S., Kumar, S. et al. (2014). Global wild annual *Lens* collection: a potential resource for lentil genetic base broadening and yield enhancement. *PLoS One* 9 (9): e107781.

Singh, D., Singh, C.K., Taunk, J. et al. (2017). Transcriptome analysis of lentil (*Lens culinaris* Medikus) in response to seedling drought stress. *BMC Genomics* 18 (1): 206.

Singh, D., Singh, C.K., Taunk, J. et al. (2019a). Genome wide transcriptome analysis reveals vital role of heat responsive genes in regulatory mechanisms of lentil (*Lens culinaris* Medikus). *Scientific Reports* 9 (1): 12976.

Singh, A., Dikshit, H.K., Mishra, G.P. et al. (2019b). Association mapping for grain diameter and weight in lentil using SSR markers. *Plant Gene* 20: 100204.

Singh, M., Kumar, S., Basandrai, A.K. et al. (2020a). Evaluation and identification of wild lentil accessions for enhancing genetic gains of cultivated varieties. *PLoS One* 15 (3): e0229554.

Singh, D., Singh, C.K., Tribuvan, K.U. et al. (2020b). Development, characterization, and cross species/genera transferability of novel EST-SSR markers in lentil, with their molecular applications. *Plant Molecular Biology Reporter* 38 (1): 114–129.

Sinha, R.P., Chaudhary, S.K., and Sharma, R.N. (1987). Inheritance of cotyledon colour in lentil. *Lens Newsletter* 14 (1, 2): 3.

Slinkard, A.E. (1978). Inheritance of cotyledon colour in lentil. *Journal of Heredity* 69: 139–140.

Solanki, R.K., Singh, S., and Kumar, J. (2010). Molecular marker assisted testing of hybridity of F_1 plants in lentil. *Journal of Food Legumes* 23: 21–24.

Stagnari, F., Maggio, A., Galieni, A., and Pisante, M. (2017). Multiple benefits of legumes for agriculture sustainability: an overview. *Chemical and Biological Technologies in Agriculture* 4: 2.

Subedi, M., Bett, K.E., Khazaei, H., and Vandenberg, A. (2018). Genetic mapping of milling quality traits in lentil (*Lens culinaris* Medik.). *The plant Genome* 11 (2): 1–10. https://doi.org/10.3835/plantgenome2017.10.0092.

Subedi, M., Khazaei, H., Arganosa, G. et al. (2021). Genetic stability and genotype × environment interaction analysis for seed protein content and protein yield of lentil. *Crop Science* 61 (1): 342–356.

Sudheesh, S., Rodda, M.S., Davidson, J. et al. (2016). SNP-based linkage mapping for validation of QTLs for resistance to ascochyta blight in lentil. *Frontiers in Plant Science* 7: 1604.

Tahir, M., Muehlbauer, F.J., and Spaeth, S.C. (1994). Association of isozyme markers with quantitative trait loci in random single seed descent derived lines of lentil (*Lens culinaris* Mekid.). *Euphytica* 75: 111–119.

Tanksley, S.D. and Nelson, J.C. (1996). Advanced backcross QTL analysis: a method for the simultaneous discovery and transfer of valuable QTLs from unadapted germplasm into elite breeding lines. *Theoretical and Applied Genetics* 92: 191–203.

Temel, H.Y., Göl, D., Akkale, H.B.K. et al. (2015). Single nucleotide polymorphism discovery through Illumina-based transcriptome sequencing and mapping in lentil. *Turkish Journal of Agriculture and Forestry* 39 (3): 470–488.

Thavarajah, D., Abare, A., Mapa, I. et al. (2017). Selecting lentil accessions for global selenium biofortification. *Plants* 6 (3): 34.

Thomas, T. (2016). Understanding the genetic basis of carotenoid concentration in lentil (*Lens culinaris* Medik.) seeds. MSc thesis. University of Saskatchewan, Saskatoon, Canada.

Tiwari, M., Singh, B., Min, D., and Jagadish, S.V.K. (2022). Omics path to increasing productivity in less-studied crops under changing climate – lentil a case study. *Frontiers in Plant Science* 13: 813985.

Tripathi, K., Kumari, J., Gore, P.G. et al. (2022). Agro-morphological characterization of lentil germplasm of Indian national genebank and development of a core set for efficient utilization in lentil improvement programs. *Frontiers in Plant Science* 12: 751429.

Tschermak-Seysenegg, E. (1928). Lentil and field bean crosses. *Sityringsber Akademy Wissenschaft Wein Mathematics Natural Kl. I, Abt* 137 (3/4): 171–181. (cf. Muehlbauer, F. J. and A. E. Slinkard, 1987. Breeding methodology. In: *Lentil* (eds. C. Webb and G. Hawtin). CAB-ICARDA, Fernham, England, 69–90.

Tullu, A., Kusmenoglu, I., McPhee, K.E., and Muehlbauer, F.J. (2001). Characterization of core collection of lentil germplasm for phenology, morphology, seed and straw yields. *Genetic Resources and Crop Evolution* 48 (2): 143–152.

Vaillancourt, R. and Slinkard, A.E. (1992). Inheritance of new genetic markers in lentil. *Euphytica* 64: 227–236.

Vaillancourt, R., Slinkard, A.E., and Reichert, R.D. (1986). The inheritance of condensed tannin concentration in lentil. *Canadian Journal of Plant Science* 66: 241–246.

Vandenberg, A. and Slinkard, A.E. (1989). Inheritance of four new quantitative genes in lentil. *Journal of Heredity* 80: 320–322.

Vandenberg, A. and Slinkard, A.E. (1990). Genetics of seed coat colour and pattern in lentil. *Journal of Heredity* 81: 484–488.

Verma, P., Shah, N., and Bhatia, S. (2013). Development of an expressed gene catalogue and molecular markers from the de novo assembly of short sequence reads of the lentil (*Lens culinaris* Medik.) transcriptome. *Plant Biotechnology Journal* 11 (7): 894–905.

Verma, P., Goyal, R., Chahota, R.K. et al. (2015). Construction of a genetic linkage map and identification of QTLs for seed weight and seed size traits in lentil (*Lens culinaris* Medik.). *PLoS One* 10 (10): e0139666.

Vijayan, P., Vandenberg, A., and Bett, K. E. (2009). A mixed genotype lentil EST library representing the normalized transcriptome of different seed development stages. http://www.ncbi.nlm.nih.gov/nucest/?term=lens%20culinaris (accessed 13 December 2022).

Wang, D., Yang, T., Liu, R. et al. (2020). RNA-Seq analysis and development of SSR and KASP markers in lentil (*Lens culinaris* Medikus subsp. *culinaris*). *The Crop Journal* 8 (6): 953–965.

Wilson, V.E., Law, A.G., and Warner, R.L. (1970). Inheritance of cotyledon colour in *Lens culinaris* Medik. *Crop Science* 10: 205–207.

Wong, M.M., Gujaria-Verma, N., Ramsay, L. et al. (2015). Classification and characterization of species within the genus *Lens* using genotyping-by-sequencing (GBS). *PLoS One* 10 (3): e0122025.

Wright, D.M., Neupane, S., Heidecker, T. et al. (2021). Understanding photothermal interactions will help expand production range and increase genetic diversity of lentil (*Lens culinaris* Medik.). *Plants, People, Planet* 3: 171–181.

Yadav, N.K., Ghimire, S.K., Shakya, S.M. et al. (2016). Genetic diversity analysis of lentil (*Lens culinaris* L.) germplasm using DNA based SSR markers. *American Journal of Food Science and Health* 2 (3): 18–24.

Yuan, H.Y., Caron, C.T., Ramsay, L. et al. (2021). Genetic and gene expression analysis of flowering time regulation by light quality in lentil. *Annals of Botany* 128: 17.

3

Preharvest Quality Management, Postharvest Handling, and Consumption Trends of Lentils

Muhammad Siddiq[1], Rabiha B. Sulaiman[2], and Mark A. Uebersax[3]

[1] *Food Science Consultant, Windsor, ON, Canada*
[2] *Department of Food Technology, University Putra Malaysia (UPM), Serdang, Selangor, Malaysia*
[3] *Department of Food Science and Human Nutrition, Michigan State University, East Lansing, MI, USA*

3.1 Introduction

Lentils (*Lens culinaris* Medik.), like other food legumes (pulses), are a nutrient-rich staple consumed widely in many regions of the world. In recent years, lentil production has grown at a faster rate outside of traditional lentil-producing/-consuming regions, particularly in the United States, Canada, and Australia, which has also helped expand lentil consumption in those countries (Joshi et al. 2017; Kaale et al. 2022; Oduro-Yeboah et al. 2022). A Farm-to-fork quality maintenance approach is commonly practiced across the value-chain operations of lentil production, marketing, and utilization. Sinha et al. (2009) reported that genetic purity, physical purity, size uniformity, freedom from pest and viability are the important quality attributes of this legume. Lentil quality management starts at the preharvest stage with the application of desiccants. The desiccant use achieves uniform drying or maturity, aids harvest efficiency, maximizes yield, and facilitates better handling and optimum storage (Sravanthi et al. 2013; Chelladurai and Erkinbaev 2020; Yang et al. 2021).

Maintenance of lentil quality during postharvest handling and storage is critical for its economic value and marketing as well as the consumer acceptance of raw lentils and lentil-based processed products. However, similar to other grain crops (legumes and cereals), postharvest losses are also common in lentils, especially in many developing countries in Asia and Africa, which occurs due to rather poor infrastructure and/or inappropriate postharvest handling, storage, and transportations (Wang et al. 2010; Njoroge et al. 2019; Oduro-Yeboah et al. 2022). Grading, standards, and appropriate packaging are essential components of lentil value chain (Kaale et al. 2022). Dhuppar et al. (2012) noted that the more research on adaptive capabilities of current crop production technologies and innovative postharvest handling will further maximize lentil production and utilization. A two-pronged approach, i.e., maximizing lentil yields and minimizing postharvest quantitative/

Lentils: Production, Processing Technologies, Products, and Nutritional Profile, First Edition.
Edited by Jasim Ahmed, Muhammad Siddiq, and Mark A. Uebersax.

qualitative losses, can play a critical role in enhancing the economic returns for all stakeholders (from growers to wholesale and retail marketers).

Lentils are a rich source of protein (~24%) and contain 60% carbohydrates (mostly starch) and important minerals. It is noteworthy that lentil carbohydrates are slowly digested in human gut, thus contributing to a lower glycemic index of about 30 versus 100 for white bread (Bressani and Elias 1988; Singh et al. 2021). The consumption trends and utilization of lentils in different cuisines vary widely across the globe. In South Asia, lentils are frequently consumed with flat wheat bread (*Chapati*), boiled rice, or incorporated in a range of recipes (savory soups and stews) and ready-to-eat snacks (mostly roasted and fried products). Other uses of lentils include as ingredient for preparing different non-traditional food products. Lentils are a healthy food choice especially for vegetarians and/or those consumers looking for environmentally sustainable meat alternatives. Lentil-based gluten-free products (e.g., pasta) is commercially produced and marketed in North America. Lentils are also being used increasingly in a variety of cuisines in the developed countries in North America, EU, and the United Kingdom (Amin and Borchgrevink 2022).

In order to expand the utilization of legumes, including lentils, the impact and control of postharvest handling, storage, and quality maintenance must be thoroughly understood (Siddiq et al. 2022). This chapter provides an overview of lentil preharvest quality management, postharvest handling/storage, and postharvest quantitative and qualitative losses, quality grading and standards, consumption trends, and role in food security.

3.2 Preharvest Quality Management

Ensuring the postharvest quality of lentils starts with the crop management practices prior to harvesting, such as the use of desiccants. Since lentil plants have indeterminate growth habits, this particular trait typically makes it necessary for commercial growers (e.g., in Canada) to apply preharvest desiccants to lentil crop. The main objectives of treatment with desiccants are to force uniform maturity (or plant dry-down), improve harvest efficiency, maximize crop yield, and facilitate retention of postharvest quality of lentils (Ellis et al. 1988; Bruce 2008; McVicar et al. 2017; Bertholet 2019). Preharvest herbicides are generally applied after the lentil crop is mature and when the seed color is changing. The timing of harvest and harvesting aid application of herbicides can impact the end use of lentils, i.e., whether for export to other countries, local consumption, or on-farm use as seed by the growers. Furthermore, each preharvest herbicide has specific or different restrictions regarding the end use of the crop after harvest (Baig et al. 2003; Bertholet 2019).

Lentils are deemed mature when the color of the lower one-third of the pods turns yellow to brown and seeds rattle when the pod is shaken (SPG 2012). The increases in large-scale production of lentil in Canada, United States, and Australia have been achieved primarily through mechanized harvesting, while in most traditional lentil producing countries (e.g., India, Pakistan, Nepal, and Bangladesh), lentil is still harvested mainly by hand (Haddad et al. 1988; Materne and Reddy 2007). Erskine et al. (1991) reported that manual harvesting is also the main limitation to lentil production in North Africa and the Middle-East, especially in Jordan and Syria. Sarker and Erskine (2002) indicated that cultivation of lentil varieties suited for mechanical harvesting, which were introduced in the Middle-East, has significantly increased the net returns to growers.

Bruce (2008) investigated the effect of selected preharvest treatments (desiccants and swathing) on the milling quality of selected red lentil cultivars. Results of this study showed that under favorable or optimum harvest conditions, the applied treatments had minimal effect on the efficiency of milling and dehulling processes. However, under cool and wet harvest conditions, lentils of highly variable milling quality were produced. Early application of desiccants had the most negative effect on dehulling efficiency, i.e., significantly reducing milling efficiency to below 70%.

3.3 Postharvest Handling, Storage, Grading, and Packaging

Most legume crops require a careful handling and proper postharvest storage before marketing/consumption. Jones et al. (2012) reported that the main goals of postharvest handling, drying, and storage are to preserve the harvest quality and to add value. These goals can be assured by removing impurities, foreign matter, and separating lots based on quality attributes, drying (if needed), storing in appropriate facilities, monitoring storage conditions, and implementing cleanliness/hygiene protocols for the products, personnel, and physical facilities. Oduro-Yeboah et al. (2022) noted that the qualitative and quantitative losses typically occur due to poor handling and drying, and/or a lack of appropriate storage facilities. The critical quality defect of damaged seed coat is impacted at all stages of harvest, handling, and storage. Care must be taken to minimize impact damage and seed coat abrasion during handling since the damaged seed coats directly affect seed appearance and are also indicative of overall quality due to adverse effect on cooking characteristics. The seed coat quality indicator commands significant value and has economic consequences throughout the supply chain. Tabil et al. (2010) reported that the storage temperature, postharvest treatments, and storage time of lentils have significant effect on both the dehulling efficiency and the cooking quality.

3.3.1 Receiving and Handling

Lentils are recommended to be harvested at 18–20% moisture content (MC) to avoid shattering damage and loss of the ripe pods. After harvesting, lentils are dried to 13–14% MC to optimize yield of high quality, i.e., seed color uniformity and for safe long-term storage without subsequent seed breakage or damage (Tang et al. 1990; Ghosh et al. 2007). It is recommended that any green weed seeds and other high-moisture field debris must be separated without any delays to prevent respiration-induced heating (Wood et al. 1977; Wang et al. 2012; SPG 2012; Barker 2016). The commonly used unit operations for handling and cleaning of lentils prior to storage, retail sale, and processing are shown in Figure 3.1. Amin et al. (2004) noted that the size and shape are the important criteria for designing the cleaning, sizing, and grading equipment.

Upon delivery to the elevator (cleaning, sorting, and storage facility), truckloads of lentil are weighed before being dumped into pits at the receiving facility. It is advisable to avoid unnecessary handling by employing conveying equipment, which is gentler on the lentil seeds. In this regard, belt conveyors are known to cause minimal damage than metal (steel) augers. The damage can also occur when lentils are dropped into receiving bins from significant height; therefore, the use of "bean ladder" type equipment is

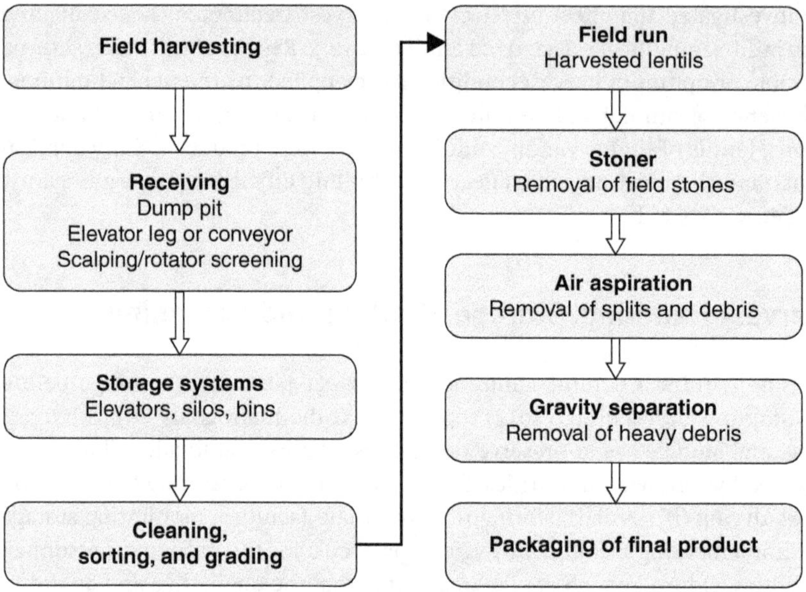

Figure 3.1 Postharvest unit operations for handling and cleaning of lentils. *Source:* Adapted from Uebersax et al. (2022b).

recommended for damage-free drop (SPG 2012). Removal of field stones, split/damaged lentils, and light and heavy debris can be achieved by stoners, air aspirators, and gravity separators. The use of variable mesh-sized screens is common to remove extraneous organic/and inorganic materials, such as immature seeds, stalks, stones, and other field debris (Joshi et al. 2017).

Uebersax et al. (2022b) reported that the final quality of pulses was directly associated with carefully controlling the critical physical, chemical, and biochemical processes during handling and storage that can ensure safe and high-quality product. Lentil quality is susceptible to numerous problems during storage (insects, molds, rodents, and fluctuations in the storage conditions – temperature and relative humidity) and requires implementation of appropriate postharvest management throughout the supply chain channels to preserve quality.

3.3.2 Drying and Optimum Moisture Content

Generally, harvested lentils require some degree of drying to attain optimum moisture level (13–14% MC) for long-term storage and quality preservation. The selection of drying method should be based on the requirements related to the end-use quality characteristics of lentils (Ghosh et al. 2007). It is recommended that, preferably, lentils should be threshed at 16–18% moisture content. Lantin et al. (1996) noted that typically the moisture content of harvested lentils is generally higher than that suitable for long-term storage, and some degree of drying (usually with heated air) is employed. Since lentils are harvested in warm and dry air conditions, it facilitates natural field drying. For example, as Bradford et al. (2018) reported, the April–June daily maximum temperatures are generally above 40 °C (with dry ambient air) in northwestern India, eastern Pakistan, and Nepal – countries where lentils are widely grown. In developing countries, some quantitative losses also occur during threshing operations, while such losses are negligible with mechanized harvesting/threshing in the developed countries that have advance agricultural production systems.

Lentil crop with 20% MC presents some challenges due to difficulty in threshing without incurring some damage to lentil seeds. Moreover, seeds with ≥18% moisture need higher energy input for longer drying times, which can also have a potential detrimental effect on the quality of lentils (McVicar 2006). It is advisable that to maintain safe storage and optimum quality, some degree of aeration may be needed to reduce grain temperature in the bins even when lentils are harvested at dry stage (Barker 2016).

For mechanical drying of lentils, to lower MC to ~14%, the hot-air temperature in the dryer should not exceed 45 °C to avoid seed shrinkage, which lowers physical quality of lentils. For red lentil, buyers and processors prefer 13% MC, which improves the efficiency of dehulling/splitting processes thereby ensuring better quality (SPG 2012). Amin et al. (2004) reported that bulk density, particle density, and porosity are major considerations in designing the drying and aeration, and storage systems, as these properties affect the resistance to air flow. For automated drying, the use of batch dryers is time consuming and drying efficiency is also relatively low. Generally, continuous air-flow dryers are used for grain (legumes) drying, which can be classified into three types based on the drying air-flow direction: (i) *Cross-flow dryers* – air is blown across the grain belt, (ii) *Counter-flow dryers* – air and the grain move in opposite directions, and (iii) *Concurrent-flow dryers* – air and grain move in the same direction (Jones et al. 2012).

3.3.3 Storage Conditions and Storage Life

The most important factors of grain quality during postharvest storage are moisture level, storage temperature, and equilibrium relative humidity (RH) (Zhang et al. 2008; Bradford et al. 2018; Sangeetha and Mohan 2020). The interrelationship between these three quality determinants is presented in Figure 3.2. Bradford et al. (2018) reported that storage life at a given moisture level increases exponentially as both the storage temperature and equilibrium RH decrease. Bello and Bradford (2016) indicated that below 95% RH, the seed respiration stops, and bacteria cannot grow below ~90% RH. Fungi are unable to be metabolically active or grow below 65% RH, which closely corresponds with the recommended maximum MC of 12–14%, 13–15%, and 6–9% for safe storage of cereals, pulses, and oilseed crops, respectively (FAO 2014; Bradford et al. 2018). In addition to quality deterioration by microbes and insects, other specific quality changes that occur during storage are associated with flavor deterioration (mustiness, sour, bitter), seed discoloration (browning, darkening), and hard shell or hard-to-cook phenomena (resulting in reduced water uptake, longer cooking time). All these postharvest defects have been reported to result in a significant loss of not only legume quality but economic value as well (Chu et al. 2020; Uebersax et al. 2022a).

Storage of lentils at 14% MC is recommended for safe, long-term storage to minimize damage to the seed coat during handling. At 14% MC and 15 °C storage temperature, lentils can be stored for up to 40 weeks (Table 3.1) (McVicar 2006; Barker 2016). Regardless of the

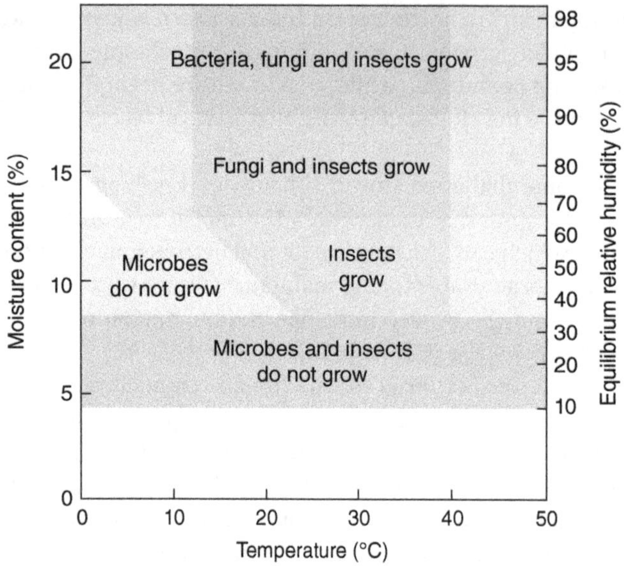

Figure 3.2 Interaction of grain moisture, storage temperature, and equilibrium relative humidity required for the growth of different organisms during storage. *Source:* Bradford et al. (2018), with permission of Elsevier.

Table 3.1 Weeks of lentils' storage at the specified grain moisture and storage temperature.

Temperature (°C)	Moisture content					
	12%	13%	14%	16%	18%	21%
25	31	16	13	7	4	2
20	55	28	23	13	7	4
15	100	50	40	20	12	6
10	200	95	80	38	20	21
5	370	175	150	70	39	20
	= Safe long-term					
	= Safe medium-term					
	= Safe short-term					
	= High risk (Unsafe or quality at risk)					

Source: Adapted from McVicar (2006); Barker (2016).

storage temperature, the storage life decreases significantly at >16% MC. When lentils are stored under appropriate conditions (i.e., 20 °C temperature and 12% RH), no significant changes in the protein quality occur even after three-year storage. Lentils are susceptible to increased chipping and peeling if handled or kept at or below a temperature of −20 °C. Ghosh et al. (2007) reported that long-term storage, especially above 25 °C, can results in the darkening of lentil seeds, due possibly to the oxidation of seed coat tannins. Such darkening or discoloration of lentils severely reduces their quality and market value.

The storage-related damage and contamination of lentils can be controlled or minimized through the use of good handling and monitoring practices. Uebersax et al. (2022b) noted that the established standards for food handling are important to sanitary control and defined lot or batch identity of legumes. In this regard, the use of Good Agricultural Practices (GAPs), Good Manufacturing Practices (GMPs), and other food safety and quality standards, including ISO 9000, HACCP (hazard analysis critical control points), and SQF (safe quality food), can be applied to handling and storage of lentils.

3.3.4 Monitoring of Storage Conditions

During postharvest storage of lentils, careful monitoring of storage temperature and RH is important for quality preservation, as any fluctuations in the storage conditions can negatively impact quality. Bradford et al. (2018) suggested cost-effective and easy-to-use temperature and relative humidity measurement tools (Figure 3.3) that can be used in developing countries. The DryCard™, shown in Figure 3.3c, was developed by researchers at the University of California, Davis, with USAID support (Thompson et al. 2017). Most of the postharvest losses of grain legumes in developing countries are due to inadequate and/or poorly maintained storage facilities, which can also affect the end-use quality characteristics. However, in the developed countries (e.g., Canada, United States, Australia), storage facilities are carefully designed and equipped with automated, continuous monitoring systems, which ensure no or minimal postharvest quantity or quality losses.

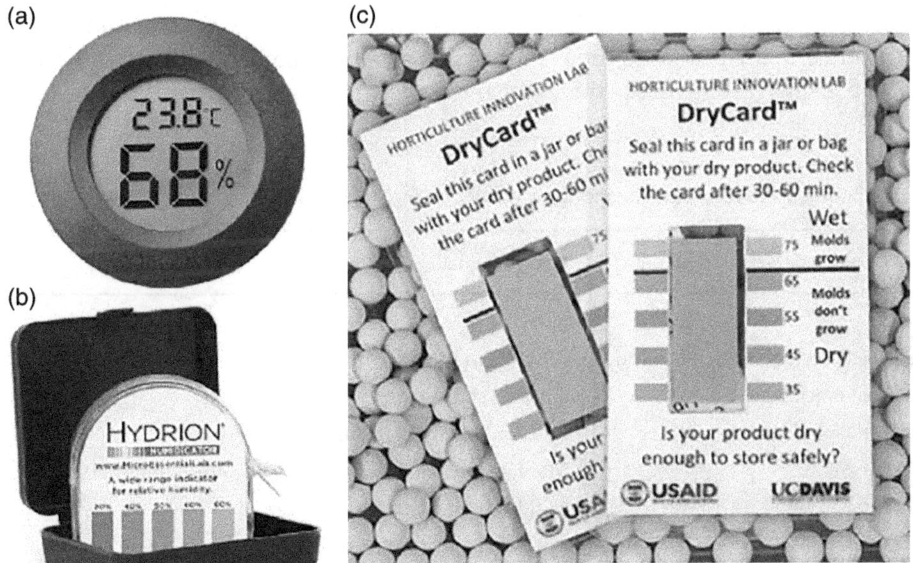

Figure 3.3 Temperature and relative humidity (RH) measurement tools used during postharvest storage. (a) Inexpensive electronic temperature and RH meter, (b) RH indicator paper that changes color in response to the ambient RH, and (c) DryCard™, with laminated RH indicator strip and adjacent RH scale – the back of the strip is exposed allowing it to equilibrate with ambient air. *Source:* Bradford et al. (2018), with permission of Elsevier.

3.3.5 Postharvest Quality Defects and Losses

Adverse storage conditions and their fluctuations are reported to induce a number of quality defects in legumes, e.g., seed hydration defects and longer cooking time and decreased digestibility and nutrient bioavailability (Paredes-Lopez et al. 1989; Uebersax et al. 2022a). Furthermore, long-term storage can produce discoloration/darkening of seeds, stemming from oxidation of phenolics/tannins in the seed coat, thereby reducing the quality and market value of lentils (Ghosh et al. 2007). Several studies have reported that under adverse storage conditions, storage defects, such as *hard-shell* and *hard-to-cook* (HTC) phenomena, can occur in legumes, including lentils, resulting in a significant loss of quality (Bhatty 1990; Cenkowski and Sosulski 1997; Joshi et al. 2010). Nozzolillo and Bezada (1984) reported that HTC may develop in lentils grown under unfavorable agroecological or growing conditions unlike in common beans (*Phaseolus* spp.) where such quality defect develop during storage at high moisture level and high temperature. Furthermore, lentil also undergo quality deterioration (i.e., HTC defect) under high storage moisture and temperature. The storage at the recommended conditions (and their careful monitoring) can retard legume seed deterioration, with seed moisture, storage temperature, RH, and time being the main controlling factors (Uebersax et al. 2022a). Storage-induced defects in legumes and their effect on quality are summarized in Table 3.2.

The postharvest losses can generally be classified as quantity loss (e.g., weight loss due to spoilage and insects/pests) and quality loss (e.g., nutritional and seed viability loss) (Boxall 2001). These losses in legumes and cereals are a serious problem, especially in developing countries due to inadequate storage facilities, pests, and poor handling and transportations protocols (Jayas and White 2003; Ikegwu et al. 2022). Sangeetha and Mohan (2020) reported that since pulse production systems, both at preharvest and postharvest phases, are highly mechanized, the postharvest losses are minimal in the developed

Table 3.2 Storage-induced defects and their impact on legume quality.

Storage/cooking defect	Effect on quality
Dry seed:	
Seed coat/cotyledon discoloration	Mold, "bin burn"
Seed coat cracking	Checks, splits
Hard-to-cook defect	Reduced hydration, longer cooking time
Off flavor/odor	Mold, chemical taints (from warehouse/containers)
Microbial growth	Mold, storage bacteria
Insect damage	Field damage, storage insects
Cooked/thermally processed:	
Discoloration	Discoloration and pigment leaching
Hard texture	Firm, hard versus soft and mushy
Off-flavor/odor	Moldy, sour, bitter, fishy, and absorbed volatiles

Source: Oduro-Yeboah et al. (2022); Uebersax et al. (2022a).

countries. However, such losses can reach as high as 30% in the developing countries due to inefficient crop harvesting and storage systems.

Roy et al. (2019) reported that in the case of lentils, losses could be due to postharvest diseases, infestation by insects/pests, and reduction in the viability of seed. These authors also noted that beetles or Bruchids (*Bruchinae*) are the primary pests of pulses but very few *Bruchid* species are known to attack lentils. However, in some cases, grain insects from cereal and other legumes crops can cross-contaminate lentils during postharvest handling and storage. Kumar and Kalita (2017) reported that reducing postharvest losses can be a sustainable solution to increase the availability of food, reduce pressure on finite natural resources, eliminate food insecurity/hunger, as well as improve livelihoods of farmers. Reddy and Reddy (2010) studied the supply-side constraints of pulses in India and reported that during postharvest handling and storage of lentils, controlling diseases alone can significantly reduce the economic losses.

3.4 Quality Grading and Packaging

3.4.1 Grading and Quality Standards

Grading of lentils for quality designation is commonly done before bulk and retail sales. Kenkel and Adam (2012) indicated that grades and standard system improves the marketing efficiency of grain crops, by communicating to both sellers and buyers the properties of a commodity, in determining the quality and value accurately. It is recommended that an efficient grading system must measure characteristics that are important to the users and that can be measured precisely and uniformly. Care should be taken that grading and standard system is easily applicable without slowing down handling and transportation through marketing channels Kenkel and Adam (2012).

According to Saskatchewan Pulse Growers guidelines, the basic quality parameters for red lentils are "*dictated by milling requirements and by color, size, condition of dehulled split seeds, or the condition of dehulled whole seeds having both cotyledons attached/intact.*" Whereas, for green (seed coat color) lentils, the basic quality parameters are "*seed diameter, seed thickness and uniformity, color uniformity, and intact green seed coats without wrinkling or staining*" (SPG 2012). It is advised that producers should not mix lentils from successive years to avoid having the entire batch downgraded. Since lentils with green seed coat are prone to discoloration during extended storage, green lentils should not be stored for more than one year (i.e., during a second summer season) to minimize/avoid excessive discoloration, which can significantly downgrade lentils marketability (SPG 2012).

In the United States, lentil grades were established relatively recently in 2017 (USDA 2017), after lentil production increased significantly from 137,393 metric tons in 2000 to three times this tonnage by early 2010s. The US lentil grades, published by the United States Department of Agriculture, are presented in Table 3.3. The four lentil grades are *U.S. #1*, *U.S. #2*, *U.S. #3*, and *U.S. Sample Grade* (USDA 2017), with the *U.S.* #1 and *U.S. Sample Grade* representing the highest and the lowest quality grades, respectively.

In the developing countries, lentil grades and standards are either not well defined as those in Canada and the United States, and/or implemented rather inconsistently. This

Table 3.3 US Grades and grade requirements of lentils (% allowable limits).

Grading factors	Grade (Based on percent defects/damage)			
	U.S. #1	U.S. #2	U.S. #3	U.S. sample grade
Defective lentils:				Lentils which: (a) Do not meet the requirements for the U.S. grades #1, #2, or #3; or (b) Contain more than 14.0 percent moisture, live weevils, or other live insects, metal fragments, broken glass, or a commercially objectionable odor; or (c) Are materially weathered, heating, or distinctly low quality
Total[1]	2.0	3.5	5.0	
Weevil-damaged lentils	0.3	0.8	0.8	
Heat-damaged lentils	0.2	0.5	1.0	
Foreign material:				
Total[2]	0.2	0.5	0.5	
Weevil-damaged lentils	0.1	0.2	0.2	
Skinned lentils	4.0	7.0	10.0	
Wrinkled lentils[3]	5.0	10.0	>10.0	
Contrasting lentils[4]	2.0	4.0	>4.0	
Inconspicuous admixture	0.5	0.8	1.0	
Minimum color requirements	Good	Fair	Poor	

[1] Defective lentils total = weevil-damaged + heat-damaged + split lentils + damaged, other
[2] Foreign material total includes stones
[3] Lentils with more than 10.0% wrinkled lentils shall grade no higher than U.S. No. 3
[4] Lentils with more than 4.0% contrasting lentils shall grade no higher than U.S. No. 3

Source: USDA (2017).

issue along with poor cleaning and sorting of harvested lentils not only impacts their economic value negatively but also affects the export potential, as importing countries require consistent quality standards.

3.4.2 Packaging and Shipping

After cleaning and grading, whole lentils are packaged and shipped based on the local or export specifications. For bulk shipments, a variety of bagging and containerization methods can be used. For example, shipping in bulk-hopper railcars is common for delivery to shipping vessels at port facilities for exports or to local processors/packagers. For some customers, primary suppliers fill lentils in heavy-duty polypropylene bulk/tote bags, which are then loaded onto standard 20- or 40-ton shipping containers. Typically, bulk shipments undergo a secondary cleaning at receiving facilities and then packaged in smaller retail or institutional packages under different brand names. The size and the quality of bags and labelling information are predetermined by the vendors/end users (Vandenberg 2009). The package sizes generally vary from 25 to 100 lb (~11–45 kg) bags for commercial or 1–5 lb (~0.45–2.27 kg) bags for the retail customers. Vandenberg (2009) reported that packages may be up to 15 kg (33 lbs) in regions with higher lentil consumption, especially for restaurant and institutional users.

The high relative humidity, which is common in humid climates, can increase the moisture content of grains stored in porous/woven bags (i.e., gunny/jute bags), which enables fungal and insect infestations (Bradford et al. 2018). In this regard, poor packaging and handling is one of the main factors contributing to postharvest losses in developing countries. Therefore, the use of polypropylene bags with adequate moisture barrier is recommended for lentils packaging and shipping.

3.5 Role in World Food Security

Food security, especially in developing and under-developed countries, continues to be a major worldwide concern. Food legumes or pulses contribute to world food supplies and food intake significantly. In this regard, lentils play a major role in the food and nutritional security of millions, mainly among low-income families in Asia, due to the high protein content (Erskine et al. 2011; Bessada et al. 2019). The functional characteristics of proteins from pulses, e.g., peas, chickpeas, lentils, and commons beans, have been extensively researched in the preparation and development of nutrient-dense and value-added products, e.g., bakery products, soups, extruded products, and ready-to-eat snacks (Boye et al. 2010). The significance of research for crop improvement and improved utilization of dry beans and pulses is evident through the scale and diversity of research programs (Siddiq et al. 2022). There are a number of research centers focusing on pulses. For example, International Center for Tropical Agriculture (CIAT, Columbia), International Crops Research Institute for the Semi-Arid Tropics (ICRISAT, India), International Center for Agricultural Research in the Dry Areas (ICARDA, Syria), and the International Institute of Tropical Agriculture (IITA, Nigeria). The improved pulses production, while strengthening household food and nutrition security, also provides women with surplus grain to sell in local markets.

The USAID (United States Agency for International Development) has long played a role in the pulses' improvement programs globally. Besides strengthening pulse value chains, research has focused on (i) increasing pulse productivity via genetic improvement and integrated crop management, and (ii) increasing pulse utilization for improved nutrition and health (USAID 2012). The USAID-funded Collaborative Research Support Program (CRSP) on pulses contributes to economic growth and food and nutrition security through knowledge and technology generation in developing countries of Africa and Latin America.

In addition, several countries through their international development agencies support pulse research programs, including lentils. Selected such agencies are Australian Agency for International Development (AusAID), Canadian International Development Agency (CIDA), International Development Research Centre (IDRC, Canada), German Agency for International Cooperation (GIZ), Japan International Cooperation Agency (JICA), Swedish International Development Cooperation Agency (SIDA), and U.K.'s Department for International Development (DFID) (Redden et al. 2007; GCDT 2008). Utilization of prepared pulses has been advocated consistently by both governmental and nongovernmental organizations (NGOs). Such nutrition programs are particularly important to infants and children. Pulses (common beans and lentils) are also frequently considered for applications in regions experiencing sustained crop failure (Siddiq et al. 2022).

3.6 Consumption Trends

Lentils are used in various cuisines worldwide and most commonly used as main dishes, side dishes, soups, stews, and as sprouted seeds in salads. Lentils can also be fried or roasted and seasoned for other uses and snacks. Lentil flour can be mixed with cereals to make breads and cakes and as food for infants. Lentils are used as a staple in the diet of many

Asian countries, especially the Indian subcontinent. Lentils are typically consumed as whole grains and as dehulled split form (*dhal*) (Sidhu et al. 2022).

In the Middle Ages, lentils were a highly consumed legume crop. For example, in Russian, Asian, North African, European, and American cuisines lentil dishes were an indispensable part of the main diet. However, with time this culture was not lost (Mamakhai and Zagoruiko 2022). Nevertheless, due to technology development and with increased health awareness globally as a result of an increase in the number of noncommunicable diseases, the demand for healthier foods such as lentil has also increased. In addition, the knowledge of food fortification which targets development of functional foods has made the dominant role of lentil dishes in the diet to become important in local and international markets. Lentils have important minerals and vitamins, fatty acids, carbohydrates, and high-quality proteins (Mamakhai and Zagoruiko 2022), which makes them to be among the best food fortificants. More interest is that the protein content in lentils has been reported to be significantly higher than that in the most cereals (such as wheat, rice, corn) and similar to that in meat (Joshi et al. 2017).

In recent years, the popularity of plant-based protein sources, as meat alternatives, has been increasing among consumers in developed countries. Among legumes, lentils, being rich in many nutrients, align well with these changing trends (Hill 2022). Lentils are finding applications in various bakery and extruded products. However, the long-term success of these expanded uses depends on evaluation and understanding lentil flours' functional properties and their effects on the sensory properties of the end-products (Sidhu et al. 2022). Similarly, lentil flour substitution levels need to be optimized in those products that are traditionally produced using 100% wheat flour. The functional/nutritional properties of lentils and other pulses make them an attractive choice for pasta manufacturers. By replacing the wheat flour with pulse flours or blending the two flours, they can produce pasta with increased protein and dietary fiber compared to traditional wheat-based pasta (Amin and Borchgrevink 2022). As summarized in Table 3.4, several companies are commercially producing lentil-based pastas.

There are wide variations in lentil consumption trends across different countries. *Red* and *Green* lentils are the most widely used lentil types in most countries where this legume is consumed regularly, whereas *Yellow* and *Spanish Brown* types of lentils are consumed

Table 3.4 Range of lentil-based pastas by the lentil type, product shape, and manufacturer.

Pulse/lentil type	Pasta shape	Manufacturer
Red lentil	Penne, rotini, spaghetti	Barilla
	Fusilli (spirals)	San Remo
	Penne, rigatoni, spaghetti	Explore Cuisine
	Penne, rotini	Tolerant Organic
Green lentil	Penne, lasagna	Explore Cuisine
	Elbows, penne, rotini	Tolerant Organic
Lentils, peas, chickpea, borlotti beans	Penne	San Remo
	Spaghetti	San Remo

Source: Adapted from Amin and Borchgrevink (2022).

Table 3.5 Common lentil market classes and consuming countries.

Lentil type	Seed size (100 seed weight)	Consuming countries
Red	Extra-small (2.9–3.2 g); small (3.3–4.5 g); large (5.5–7.3 g)	Canada, United States, Turkey, Egypt, India, Australia, Sri Lanka, Pakistan, Bangladesh, Syria, Nepal
Yellow	Extra-small (2.9–3.2 g), small (3.3–4.5 g), large (5.5–7.3 g)	Spain, England, United States, Germany
Green	Extra-small (2.9–3.2 g), small (3.3–4.5 g), medium (5.1–5.2 g)	Morocco, Greece, Italy, Egypt, Mexico, Northwestern Europe, Spain, Algeria, United States
Spanish Brown	Small (3.3–4.5 g)	Spain

Source: Adapted from Siva et al. (2017); Thavarajah et al. (2008).

only in a few countries (Table 3.5). Lentils are relatively quick and easy to prepare compared to other pulses and dry beans. Typically, green lentils are consumed as whole seeds or split seeds while red lentils are generally dehulled before cooking and consumption (Joshi et al. 2017; Oduro-Yeboah et al. 2022).

Lentils have been incorporated into different regional cuisines throughout the world. For example, besides being commonly used as curry that is consumed with wheat chapati or flat bread, lentils are commonly mixed with cereals such as rice, as in the South Asian dish *Khitchri*, whereas *Koshari* and *Mujaddara* are the common dishes in Egypt and Syria (Dagher 1991; Faris and Attlee 2017). Williams and Singh (1988) reported that lentils are also boiled and fried and seasoned before consumption. Lentil flour is typically used as a basic ingredient in stews and purees besides being mixed with cereals to make bread/cakes and as a food for infants. In Creole and Cajun cuisine in southern United States, *Cajun lentils and Rice* is prepared in a similar fashion to red beans and rice; however, red beans are replaced with red lentils and tomatoes. The intensity of aromatic heat flavor (i.e., increased Scoville units) is increased by using generous application of spices, e.g., garlic, green chilies, coriander, lemon zest, and even some curry powder, allowing for influences from those of Indian descent. Typically, mushrooms are also added to this southern dish. The rice in *Cajun Lentils and Rice* is typically a brown rice (Amin and Borchgrevink 2022). A separate chapter in this book, entitled *Global Consumption and Culinary Trends in Lentils Utilization,* covers lentil cuisines and consumption trends in more detail.

3.7 Conclusion

Lentils occupy an important place among legumes, due primarily to their production, international commerce, and consumption. For several decades, especially since 1970s, lentils have shown a consistent growth in production and cultivated area. Quality management of lentils starts at the preharvest stage, with the application of desiccants (to ensure uniform drying in the field). The use of appropriate handling and storage practices is essential for quality preservation of lentils and reducing the quantitative and qualitative losses during postharvest storage and transportation. Optimum seed moisture and storage temperature are the two most important factors in ensuring the best postharvest quality of lentils. Adverse storage conditions and their fluctuations are shown to induce a variety of quality defects, e.g., seed hydration defects and longer cooking time as well as decreased digestibility and nutrient bioavailability. Lentils are utilized in many traditional cuisines and as ingredient in various food applications. Lentils, owing to their rich protein profile and presence of many bioactive phytochemicals, are well suited for preparing a range of processed products. Careful handling and storage of lentils not only offer healthy food options to consumers but provide maximum economic return to all stakeholder across lentil value chain, i.e., growers, wholesalers, retailers, processors, and foodservice establishments. The emerging consumer preferences and trends are aligned well for expanding utilization of lentils as an important legume crop.

References

Amin, S. and Borchgrevink, C.P. (2022). A Culinology® perspective of dry beans and other pulses. In: *Dry Beans and Pulses: Production, Processing, and Nutrition*, 2e (ed. M. Siddiq and M.A. Uebersax), 453–480. Hoboken, NJ: Wiley.

Amin, M.N., Hossain, M.A., and Roy, K.C. (2004). Effects of moisture content on some physical properties of lentil seeds. *Journal of Food Engineering* 65: 83–87.

Baig, M.N., Darwent, A.L., Harker, K.N., and O'Donovan, J.T. (2003). Preharvest applications of glyphosate affect emergence and seedling growth of field pea (*Pisum sativum*). *Weed Science* 17: 655–665.

Barker, B. (2016). *Lentil Harvest Quality and Storage*. Saskatoon, Canada: Saskatchewan Pulse Growers.

Bello, P. and Bradford, K.J. (2016). Single-seed oxygen consumption measurements and population-based threshold models link respiration and germination rates under diverse conditions. *Seed Science Research* 26: 199–221.

Bertholet, E. (2019). Evaluation of efficacy of new and existing desiccants in lentil (*Lens culinaris* Medik). Doctoral dissertation. University of Saskatchewan, Saskatoon, Canada.

Bessada, S.M., Barreira, J.C., and Oliveira, M.B.P. (2019). Pulses and food security: dietary protein, digestibility, bioactive and functional properties. *Trends in Food Science & Technology* 93: 53–68.

Bhatty, R.S. (1990). Cooking quality of lentils: the role of structure and composition of cell walls. *Journal of Agricultural and Food Chemistry* 38: 376–383.

Boxall, R.A. (2001). Postharvest losses to insects – a world overview. *International Biodeterioration & Biodegradation* 48: 137–152.

Boye, J.I., Zare, F., and Pletch, A. (2010). Pulse proteins: processing, characterization, functional properties and applications in food and feed. *Food Research International* 43: 414–431.

Bradford, K.J., Dahal, P., Van Asbrouck, J. et al. (2018). The dry chain: reducing postharvest losses and improving food safety in humid climates. *Trends in Food Science & Technology* 71: 84–93.

Bressani, R. and Elias, L.G. (1988). Seed quality and nutritional goals in pea, lentil, faba bean and chickpea breeding. In: *World Crops: Cool Season Food Legumes* (ed. R.J. Summerfield), 381–404. Dordrecht, The Netherlands: Kluwer Academic Publishers.

Bruce, J.L. (2008). The effects of preharvest treatments on the milling efficiency of red lentil. MS thesis. University of Saskatchewan, Saskatoon, Canada.

Cenkowski, S. and Sosulski, F.W. (1997). Physical and cooking properties of micronized lentils. *Journal of Food Process Engineering* 20: 249–264.

Chelladurai, V. and Erkinbaev, C. (2020). Lentils. In: *Pulses* (ed. A. Manickavasag and P. Thirunathan), 129–143. Cham, Switzerland: Springer.

Chu, J., Ho, P., and Orfila, C. (2020). Growth region impacts cell wall properties and hard-to-cook phenotype of canned navy beans (*Phaseolus vulgaris*). *Food and Bioprocess Technology* 13: 818–826.

Dagher, S.M. (1991). *Traditional Foods in the Near East.* Rome, Italy: The Food and Agriculture Organization (FAO).

Dhuppar, P., Biyan, S., Chintapalli, B., and Rao, S. (2012). Lentil crop production in the context of climate change: an appraisal. *Indian Research Journal of Extension Education* 2 (Special Issue): 33–35.

Ellis, R.H., Agrawal, P.K., and Roos, E.E. (1988). Harvesting and storage factors that affect seed quality in pea, lentil, faba bean and chickpea. In: *World Crops: Cool Season Food Legumes* (ed. R.J. Summerfield), 303–329. Dordrecht, The Netherlands: Kluwer Academic Publishers.

Erskine, W., Diekmann, J., Jegatheeswaran, P. et al. (1991). Evaluation of lentil harvest systems for different sowing methods and cultivars in Syria. *Journal of Agricultural Science* 117: 333–338.

Erskine, W., Sarker, A., and Kumar, S. (2011). Crops that feed the world 3. Investing in lentil improvement toward a food secure world. *Food Security* 3: 127–139.

FAO (Food and Agriculture Organization) (2014). *Code of Practice for the Prevention and Reduction of Mycotoxin Contamination in Cereals. Publication# CAC/RCP 51–2003, Codex Alimentarius.* Rome, Italy: Food and Agriculture Organization of the United Nations.

Faris, M.A.E. and Attlee, A. (2017). Lentils (*Lens culinaris* L.): a novel functional food. In: *Exploring the Nutrition and Health Benefits of Functional Foods* (ed. H.U. Shekhar, Z.H. Howlader, and Y. Kabir), 42–72. Hershey, PA: ICI Global.

GCDT (Global Crop Diversity Trust) (2008). Global strategy for the *ex situ* conservation of lentil (*Lens* Miller). https://www.genebanks.org/resources/publications/lens-strategy-2008 (accessed 16 December 2022).

Ghosh, P.K., Jayas, D.S., Srivastava, C., and Jha, A.N. (2007). Drying and storing lentils: engineering and entomological aspects. In: *Lentil* (ed. S.S. Yadav, D. McNeil, and P.C. Stevenson), 385–414. Dordrecht, The Netherlands: Springer.

Haddad, N.I., Salkini, A.B., Jegatheeswaran, P., and Snowbar, B.A. (1988). Methods of harvesting pulse crops. In: *World Crops: Cool Season Food Legumes* (ed. R.J. Summerfield), 431–350. Dordrecht, The Netherlands: Kluwer Academic Publishers.

Hill, H. (2022). Utilization of dry beans and other pulses as ingredients in diverse food products. In: *Dry Beans and Pulses: Production, Processing, and Nutrition*, 2e (ed. M. Siddiq and M.A. Uebersax), 307–329. Hoboken, NJ: Wiley.

Ikegwu, T.M., Ezegbe, C.C., Okolo, C.A., and Ofoedu, C.E. (2022). Postharvest preservation technology of cereals and legumes. In: *Postharvest Technology-Recent Advances, New Perspectives and Applications* (ed. M. Ahiduzzaman), 25–44. London, UK: IntechOpen.

Jayas, D.S. and White, N.D.G. (2003). Storage and drying of grain in Canada: low cost approaches. *Food Control* 14: 255–261.

Jones, C., Casada, M., and Loewer, O. (2012). Drying, handling, and storage of raw commodities. In: *Stored Product Protection* (ed. D.W. Hagstrum, T.W. Phillips, and G. Cuperus), 100–120. Manhattan, KS: Kansas State University Research and Extension Service.

Joshi, M., Adhikari, B., Panozzo, J., and Aldred, P. (2010). Water uptake and its impact on the texture of lentils (*Lens culinaris*). *Journal of Food Engineering* 100: 61–69.

Joshi, M., Timilsena, Y., and Adhikari, B. (2017). Global production, processing and utilization of lentil: a review. *Journal of Integrative Agriculture* 16: 2898–2913.

Kaale, L.D., Siddiq, M., and Hooper, S. (2022). Lentil (*Lens culinaris* Medik) as nutrient-rich and versatile food legume: a review. *Legume Science* 4: e169.

Kenkel, P. and Adam, B.D. (2012). Economics of commodity grading and segregation. In: *Stored Product Protection* (ed. D.W. Hagstrum, T.W. Phillips, and G. Cuperus), 326–330. Manhattan, KS: Kansas State University Research and Extension Service.

Kumar, D. and Kalita, P. (2017). Reducing postharvest losses during storage of grain crops to strengthen food security in developing countries. *Foods* 6: 8.

Lantin, R.M., Paita, B.L., and Manaligod, H.T. (1996). Revisiting sun drying of grain: widely adopted but technologically neglected. In: *Grain Drying in Asia* (ed. B.R. Champ, E. Highley, and G.I. Johnson), 302–307. Canberra: Australian Centre for International Agricultural Research.

Mamakhai, A.K. and Zagoruiko, M.G. (2022). Territorial prospects for growing lentils. *IOP Conference Series: Earth and Environmental Science* 988: 1–6.

Materne, M. and Reddy, A. (2007). Commercial cultivation and profitability. In: *Lentil: An Ancient Crop for Modern Times* (ed. S.S. Yadav, D. McNeil, and P.C. Stevenson), 173–186. Dordrecht, The Netherlands: Springer.

McVicar, R. (2006). Pulse crop storage – 2006. In: *The Pulse Agronomy Network Partnership with Industry*. Edmonton, Canada: Pulse Alberta.

McVicar, R., McCall, P., Brenzil, C. et al. (2017). Lentils in Saskatchewan – fact sheet. http://publications.gov.sk.ca/documents/20/86381-Lentils_in_Saskatchewan.pdf (accessed 12 December 2022).

Njoroge, A.W., Baoua, I., and Baributsa, D. (2019). Postharvest management practices of grains in the Eastern region of Kenya. *Journal of Agricultural Science* 11: 33–42.

Nozzolillo, C. and Bezada, M.D. (1984). Browning of lentil seeds, concomitant loss of viability, and the possible role of soluble tannins in both phenomena. *Canadian Journal of Plant Science* 64: 815–824.

Oduro-Yeboah, C., Sulaiman, R., Uebersax, M.A., and Dolan, K.D. (2022). A review of lentil (*Lens culinaris* Medik) value-chain: postharvest handling, processing and processed products. *Legume Science* 4: e171.

Paredes-Lopez, O., Maza-Calvino, E.C., and Gonzalez-Castaneda, J. (1989). Effect of hardening phenomenon on some physico-chemical properties of common bean. *Food Chemistry* 31: 225–236.

Redden, B., Maxted, N., Furman, B., and Coyne, C. (2007). Lens biodiversity. In: *Lentil: An Ancient Crop for Modern Times* (ed. S. Yadav, D.L. McNeil, and P.C. Stevenson) 462 pp. New York: Springer.

Reddy, A. and Reddy, G.P. (2010). Supply side constrains in production of pulses in India: case study of lentils. *Agricultural Economics Research Review* 23: 129–136.

Roy, A., Mani, A., Roy, P., and Sarkar, T. (2019). Advances in postharvest management of lentil (*Lens culinaris* Medik). In: *Recent Trends and Advances in Food Science and Postharvest Technology* (ed. I. Chakraborty, R. Ilahy, B. Vikram, et al.), 43–49. Delhi, India: Satish Serial Publishing House.

Sangeetha, A. and Mohan, R.J. (2020). Pulses postharvest technology. In: *Pulse Foods: Processing, Quality and Nutraceutical Applications* (ed. B.K. Tiwari, A. Gowen, and B. McKenna), 193–212. London, UK: Academic Press.

Sarker, A. and Erskine, W. (2002). Lentil production in the traditional lentil world. In: *Proceedings of Lentil Focus 2002* (ed. J.B. Brouwer), 35–40. Victoria, Australia: Horsham.

Siddiq, M., Uebersax, M.A., and Siddiq, F. (2022). Global production, trade, processing and nutritional profile of dry beans and other pulses. In: *Dry Beans and Pulses: Production, Processing, and Nutrition* (ed. M. Siddiq and M.A. Uebersax), 1–28. Hoboken, NJ: Wiley.

Sidhu, J.S., Zafar, T., Benyathiar, P., and Nasir, M. (2022). Production, processing, and nutritional profile of chickpeas and lentils. In: *Dry Beans and Pulses: Production, Processing, and Nutrition*, 2e (ed. M. Siddiq and M.A. Uebersax), 383–407. Hoboken, NJ: Wiley.

Singh, M., Manickavasagan, A., Shobana, S., and Mohan, V. (2021). Glycemic index of pulses and pulse-based products: a review. *Critical Reviews in Food Science and Nutrition* 61: 1567–1588.

Sinha, J.P., Vishwakarma, M.K., and Sinha, S.X. (2009). Quality improvement in lentil (*Lens culinaris* Medik) seed through mechanical seed processing. *Journal of Research (SKUAST)* 8: 7–17.

Siva, N., Johnson, C.R., Duckett, S. et al. (2017). Can lentil (*Lens culinaris* Medikus) reduce the risk of obesity? *Journal of Functional Foods* 38: 706–715.

SPG (Saskatchewan Pulse Growers) (2012). *Lentil Production Manual*. Saskatoon, Canada: Saskatchewan Pulse Growers.

Sravanthi, B., Jayas, D.S., Alagusundaram, K. et al. (2013). Effect of storage conditions on red lentils. *Journal of Stored Products Research* 53: 48–53.

Tabil, L.G., Opoku, A., Fadeyi, O.A., and Zhang, Y. (2010). Effects of postharvest processes including pre-conditioning, drying and rewetting cycles and storage period on dehulling and cooking quality characteristics of red lentils. *Proceeding of American Society of Agricultural and Biological Engineers (ASABE)*, Pittsburgh, Pennsylvania (20–23 June).

Tang, J., Sokhansanj, S., Slinkard, A.E., and Sosulski, F.W. (1990). Quality of artificially dried lentil. *Journal of Food Process Engineering* 13: 229–238.

Thavarajah, D., Ruszkowski, J., and Vandenberg, A. (2008). High potential for selenium biofortification of lentils (*Lens culinaris*). *Journal of Agricultural and Food Chemistry* 56: 10747–10753.

Thompson, J.F., Reid, M.S., Felix, L. et al. (2017). DryCard™ – a low-cost dryness indicator for dried products. *AIMS Agriculture and Food* 2: 339–344.

Uebersax, M.A., Siddiq, M., and Borbi, M. (2022a). Hard-to-cook and other storage-induced quality defects in dry beans. In: *Dry Beans and Pulses: Production, Processing, and Nutrition*, 2e (ed. M. Siddiq and M.A. Uebersax), 105–127. Hoboken, NJ: Wiley.

Uebersax, M.A., Siddiq, M., Cramer, J., and Bales, S. (2022b). Harvesting, postharvest handling, distribution, and marketing of dry beans. In: *Dry Beans and Pulses: Production, Processing, and Nutrition*, 2e (ed. M. Siddiq and M.A. Uebersax), 81–104. Hoboken, NJ: Wiley.

USAID (US Agency for International Development) (2012). *Pulse CRSP (Collaborative Research Support Program)*. Washington, DC: USAID.

USDA (United States Dept of Agriculture) (2017). *United States Standards for Lentils*. https://www.ams.usda.gov/sites/default/files/media/LentilsStandards.pdf (accessed 9 June 2022).

Vandenberg, A. (2009). Postharvest processing and value addition. In: *The Lentil – Botany, Production and Uses* (ed. W. Erskine, F.J. Muehlbauer, A. Sarker, and B. Sharma), 391–407. Wallingford, Oxfordshire, UK: CABI.

Wang, S., Tiwari, G., Jiao, S. et al. (2010). Developing postharvest disinfestation treatments for legumes using radio frequency energy. *Biosystems Engineering* 105: 341–349.

Wang, L., Gruber, S., and Claupein, W. (2012). Optimizing lentil-based mixed cropping with different companion crops and plant densities in terms of crop yield and weed control. *Organic Agriculture* 2: 79–87.

Williams, P.C. and Singh, U. (1988). Quality screening and evaluation in pulse breeding. In: *World Crops: Cool Season Food Legumes* (ed. R.J. Summerfield), 445–457. Dordrecht, The Netherlands: Kluwer Academic Publishers.

Wood, D.W., Longden, P.C., and Scott, R.K. (1977). Seed size variation: its extent, source and significance in field crops. *Seed Science and Technology* 5: 331–352.

Yang, T., Liu, K., Poppy, L. et al. (2021). Minimizing lentil harvest loss through improved agronomic practices in sustainable agro-systems. *Sustainability* 13: 1896.

Zhang, Y., Tabil, L.G., Li, X., and Opoku, A. (2008). Effect of moisture content, postharvest treatment and storage time on physical properties and dehulling quality of red lentil. *Proceeding of American Society of Agricultural and Biological Engineers (ASABE)*, Providence, Rhode Island (29 June–2 July 2008).

Part II

Processing, Physical and Functional Properties, and Food and Nonfood Applications

4

Value-Added Processing of Lentils and Emerging Research Trends

Muhammad Siddiq[1], Charlotte Oduro-Yeboah[2], and George O. Abong[3]

[1] *Food Science Consultant, Windsor, ON, Canada*
[2] *Food Technology Research Division, CSIR – Food Research Institute, Accra, Ghana*
[3] *Department of Food Science, Nutrition & Technology, University of Nairobi, Kangemi, Kenya*

4.1 Introduction

Lentil (*Lens culinaris* Medik.) is an important food legume crop worldwide considering its production, trade, and consumption. Lentils originated in Turkey and their consumption dates to early human civilization (Chelladurai and Erkinbaev 2020). Yang et al. (2021) reported that lentil cultivation is commonly used to diversify the cereal-oilseed cropping systems. With respect to nutritional composition, lentils are a rich source of protein (~25%) and selected minerals and vitamins. Besides being rich in protein and important minerals, lentils contain high contents of dietary fiber, resistant starch, and a number of bioactive phytochemicals (Kamboj and Nanda 2018; Ramírez-Ojeda et al. 2018; Sidhu et al. 2022). It is noteworthy that due to a large number of lentil varieties grown commercially, considerable differences exist in the nutritional composition of lentils globally (Zia-ul-Haq et al. 2011). The consumption of lentils is growing rapidly beyond the traditional regions of its production and utilization, e.g., in Canada, United States, Australia, United Kingdom, and European Union countries (Sidhu et al. 2022). Canada is the top-most producer of lentils globally, though, as reported by McVicar et al. (2017), commercial production of this legume crop in Canada began in the 1970s. Khazaei et al. (2019) noted that among legumes, lentil is the most rapidly expanding legume crop since mid-1960s – a trend that is expected to continue due to its desirable protein source for various food applications.

Value-added processing of lentils and development of new products are helpful in expanding economic significance and consumer acceptance of this important food legume. In recent years, consumers have become increasingly aware of the negative environmental consequences of different foods in their diets and they are making their food choices based on environmental concerns and sustainability (Hill 2022; Uebersax et al. 2022). These trends offer potential opportunities for increased lentil utilization through value-added

Lentils: Production, Processing Technologies, Products, and Nutritional Profile, First Edition.
Edited by Jasim Ahmed, Muhammad Siddiq, and Mark A. Uebersax.

processing and develop new lentil-based products. Various processing methods, e.g., cooking, autoclaving, dehulling, extrusion, baking, roasting, frying, germination, and fermentation, can be employed for lentil processing. Generally, these processing methods have been shown to improve protein digestibility and bioavailability of legumes, including lentils. Lentils, like other most legumes, possess some antinutrient factors (ANFs), such as lectins, enzyme inhibitors, phytates, saponins, and oligosaccharides. Therefore, processing is also needed to inactivate or reduce the antinutrients (Sidhu et al. 2022; Wiesinger et al. 2022). Additionally, some innovative technologies have emerged as an alternative to traditional thermal processing, which can also be employed in value-added processing of legumes, especially their nutritional and sensory qualities.

In recent years, increasing emphasis on plant foods has emerged as a major consumer preference trend due to a higher carbon footprint of producing animal-based foods (Rondoni and Grasso 2021; Uebersax et al. 2022). Hill (2022) reported that legume-based food products align well with consumer concerns and emerging marketing trends to offer healthy, cost-effective, and sustainable food options. In this regard, lentil-based products/ingredients can be used in a variety of food products, such as snack foods, flour mixes/doughs, baked goods, gluten-free products, nutraceutical applications (Table 4.1). Due to these product development efforts, lentil utilization has found applications in diverse food products. Lentil consumption has been growing in recent years beyond its traditional regions of consumption since lentils are a healthy food choice for diverse and environmentally conscious consumers, including vegetarians and vegans. Gluten-free lentil products (pasta, snacks) are now commercially produced and marketed in North America. Lentils are also being used increasingly in a variety of cuisines in the developed countries (Amin and Borchgrevink 2022; Oduro-Yeboah et al. 2022).

This chapter presents an overview of lentil processing methods, lentil-based products/ingredients, and innovative technologies used for lentil processing.

Table 4.1 Quality characteristics and food product applications of green and red lentils.

Lentil type	Quality characteristics	Food product applications
Green lentils	• Yellow cotyledon, light-green seed coat • Available in large, medium, and small sizes • 25–35 min cooking without pre-soaking • Retain their shape after cooking • Nut-like flavor	• Ready-to-eat meals, soups, and salads • Vegetarian products • Gluten-free applications • Ethnic cuisine[a] • Snack products
Red lentils	• Orange-red cotyledon, dark-brown seed coat • Available whole or split (*dhal*) • 15–20 min cooking without pre-soaking • Do not retain shape after cooking • Mild, somewhat bland flavor	• Ready-to-eat meals and soups • Vegetarian products • Gluten-free applications • Ethnic cuisine[a] • Easy to add to sauces

[a] Indian, Middle Eastern, and Mediterranean.
Source: Adapted from Pulse Canada (2012); Sidhu et al. (2022).

4.2 Value-Added Lentil Processing

Lentils offer great versatility and are suited well for processing into different food products. Lentils are either used as cooked (whole or dehulled spilt seeds) for direct consumption or processed into different ingredients, e.g., flour, protein, starch, and fiber. Red and green lentils are the most commonly consumed types in most countries where lentils are consumed regularly, while yellow and Spanish brown lentils are consumed in relatively few countries. Green lentils are eaten as whole seeds or as dehulled split form (*dhal*), while red lentils are generally dehulled before cooking (Siva et al. 2017; Thavarajah et al. 2008). Joshi et al. (2017) reported that value-added lentil processing can be divided into three types: (i) *Primary processing*, which includes some basic operation, e.g., cleaning, sorting, grading, and packaging of whole lentils for domestic/export markets and for commercial processing; (ii) *secondary processing*, which involves a range of unit operations, e.g., dehulling, splitting, and polishing of the whole and split lentils; and (iii) *tertiary processing*, which involves grinding/milling of whole or dehulled lentils, separating protein- and starch-rich fractions, and extracting protein and starch isolates.

Figure 4.1 outlines ingredients and products obtained from *secondary* and *tertiary* processing of lentils. In addition, thermally processed, canned lentils are available in some developed countries, whereas traditional cooking/boiling continues to be the main method of lentil utilization in developing countries (Oduro-Yeboah et al. 2022; Sidhu et al. 2022). Siva and Thavarajah (2018) recommended that for optimization of nutritional value of lentil products, due consideration should be given to the selection of lentil market class, processing, and cooking method since all these can potentially impact the quality of

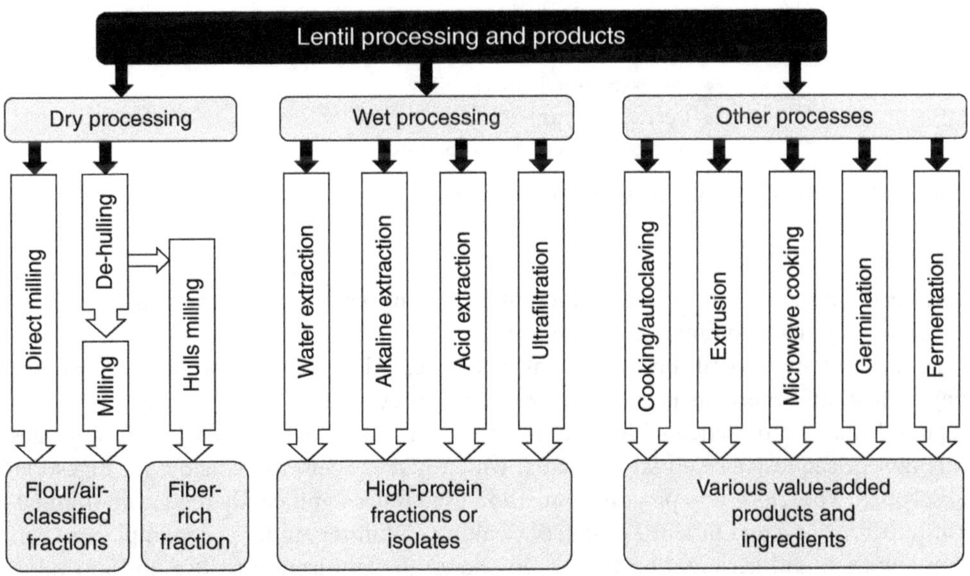

Figure 4.1 Flow diagram of lentil processing methods and end-product. *Source:* Dhull et al. (2022)/John Wiley & Sons/Public Domain CC BY 4.0.

Table 4.2 Effect of different methods on the nutritional quality of lentils.

Processing method	Nutritional effect
Dehulling	• Increases protein quality/digestibility • Increases total and resistant starch • Increases phytic acid • Decreases tannin content • Decreases trypsin inhibitor activity
Cooking	• Increases protein quality/digestibility • Increases total and resistant starch • Decreases tannin content • Decreases phytic acid • Decreases trypsin inhibitor activity
Extrusion processing	• Mixed effect on protein content • Increases total carbohydrates • Decreases tannin content • Decreases phytic acid • Decreases trypsin inhibitor activity
Germination	• Mixed effect on protein content • Increases total carbohydrates • Decreases resistant starch • Decreases total dietary fiber • Decreases flatulence-causing oligosaccharides
Fermentation	• Mixed effect on protein content • Increases resistant starch • Decreases phytic acid • Decreases trypsin inhibitor activity
Roasting	• Increases protein digestibility • Decreases trypsin inhibitor activity • Improves functional properties

Source: Adapted from Bressani and Elias (1988); Dhull et al. (2022).

lentil-based products. The effect of different processing techniques on the nutritional composition of lentils is summarized in Table 4.2.

Lentils contain several antinutritional factors, e.g., lectins, oligosaccharides, tannins, phytic acid, and saponins (Joshi et al. 2017; Sidhu et al. 2022). Nosworthy et al. (2018) reported that the presence of some antinutritional factors decreases digestive enzymes' function or sequester essential nutrients, which makes them inaccessible for digestion/absorption. Therefore, any processing method used must significantly reduce or eliminate antinutritional factors in lentil products. Soaking, dehulling, milling, cooking, extrusion, fermentation, and germination have been shown to significantly reduce antinutritional factors in lentils (Dhull et al. 2022; Hefnawy 2011; Wang et al. 2009). Removal of raffinose family oligosaccharides, which are known to cause flatulence and stomach discomfort, is especially important to improve consumer acceptability of lentil products. It has been reported that, generally, the presence of "beany" flavor and oligosaccharides is the primary reason for the plateau in consumption rates of legumes (Sharma 2021; Uebersax and Siddiq 2012).

4.2.1 Dehulling and Splitting

The seed coat (hull) of pulses is rich in tannins and imparts a bitter taste and is indigestible. The seed coat removal, referred to as "dehulling," has been shown to improve the taste, palatability, and nutritional quality of lentils and other pulses (Dhull et al. 2022). In addition to improving sensory flavor, dehulling also improves water uptake during soaking/cooking and also reduces cooking time. The longer cooking times needed for most legumes is considered a barrier to expand legume consumption, especially in developed countries. Dehulling, followed by splitting, is a common practice to satisfy consumer preference for most market classes of lentils (Joshi et al. 2017; Singh and Singh 1992). Wang et al. (2009) reported that dehulling was shown to significantly increase the content of total starch, protein, and resistant starch. Most of red lentil are consumed in the dehulled form, either split (*dhal*) or whole seed (Bruce 2008).

Afam-Mbah et al. (2018) reported that Canada has increased the export of dehulled red lentil, which has resulted in generation of more by-product (tailings) that typically goes underutilized. Tailings of lentil dehulling/splitting include a mixture of the seed coat, endosperm, broken seeds, germ (embryo), and whole shriveled seeds. Afam-Mbah and co-authors (2018) studied ways to maximize the utilization of lentil by-product through air fractionation to obtain protein- and starch-rich fractions. A lab-scale air classifier was used to fractionate the by-product samples into fine (protein-rich) and coarse (starch-rich) fractions. Significant differences in the macronutrient compositions of the two fractions were observed, and these variations affected the yield of fractions and their makeup. A correlation was noticed between the particle size and overall protein content, i.e., the smaller the particle sizes the higher the protein content.

McDonald et al. (2021) noted that consumer preference dictates that the split product, i.e., *dhal*, contain no unevenly split cotyledon or those with an attached hull or seed coat. Therefore, a high split-yield is deemed an important economic consideration for the legume processing industry since other milling fractions typically have a lower economic value. McDonald and co-workers developed a machine vision approach to classify split and dehulled fractions from multispectral images of lentils. The machine vision system was shown to classify pulse milling fractions with 88.1% validation accuracy.

4.2.2 Milling – Flour and Fractions

Milling is one of the common processes for lentil flour and protein- and starch-rich fractions, using either dry milling or wet milling (Dhull et al. 2022). The milling method used significantly affects the functional properties of lentil flours and fractions. Lentil flour in different foods can be further separated into protein- and starch-rich fractions using water or solvent extraction. These fractions offer greater opportunities for developing new lentil-based products to expand lentil utilization. Funke et al. (2022) reported that, in recent years, dry fractionation has gained increasing attention (compared to solvent extraction) for producing sustainable ingredients from legumes. Milling of lentils, their fractionation, and flours/fractions applications are discussed in more detail in a separate chapter, entitled, *Milling and Fractionation Processing of Lentils*.

Pulivarthi et al. (2021) noted that the research and development (R&D) of new food products by adding nutrient-rich lentils or lentil-based ingredients could have a significant impact on the overall utilization of lentils and other food legumes as well as the food industry's goals of manufacturing sustainable foods. Lentils and other legumes must be size-reduced for making them convenient and easy to use for incorporation into different food products. Pal et al. (2017) reported that dehulling of lentils before milling reduces tannins by over 90% and phytic acid by nearly 60%, thereby significantly increasing protein digestibility and mineral absorption. Xu et al. (2020) reported that lentil flour possesses high lipoxygenase activity that can cause off-flavors during storage, and various thermal treatments can be applied to inactivate this quality deteriorating enzyme (Jiang et al. 2016).

Ahmed et al. (2022) reported that milling operations separate the ingredients from the legume grains with definite sizes. The in-depth knowledge of microstructural properties of legume flour and ingredients, therefore, is vital in the design of processing equipment and operational planning. Additionally, the assessment of the functional properties of lentil flour and protein/starch fractions and other ingredients is critical for their successful incorporation in different food products (Jarpa-Parra 2018; Lee et al. 2021). Tyler et al. (1981) mechanically dehulled eight legumes to assess whether there were significant differences in the efficiency of separating protein from starch, which were fractionated by pin milling and air classification. The studied legumes were mung bean, green lentils, Great Northern beans, faba beans, field peas, navy beans, lima beans, and cowpeas. Results showed that the starch fractions contained 58.0–76.1% starch (highest in lentils) and 11.4–20.1% proteins (lentil, 16.5%), whereas the protein fractions contained 49.6–75.1% protein (lentil, 64.6%) and 0.0–4.6% starch (highest in lentils). It was concluded on the basis of protein separation efficiency and protein contents that mung beans and lentils were the most suitable legumes for the separation by air classifications.

Table 4.3 presents a summary of important functional properties of lentil flour and their relationship to the quality of lentil-based products.

4.2.3 Thermal Processing – Cooking/Boiling

Generally, lentils do not require soaking prior to cooking but a short-time soaking (≤15 minutes) cleans surface dust and assists in partial water uptake. Lentils, especially, dehulled and split red type, are cooked fully in about 15–20 minutes in boiling water. Whole lentils (both red and green types) require 25–30 minutes of cooking in boiling water (Sidhu et al. 2022). It is important to add optimum amount of water during cooking, without any need to discard excess water that can reduce the nutritional quality of cooked lentils. A separate chapter in this book, entitled, *Global Consumption and Culinary Trends in Lentils Utilization*, provides more details with respect to cooking lentils and lentil-based cuisines popularly consumed around the world.

Nosworthy et al. (2018) reported that the protein quality, assessed as *in vitro* digestibility level, was improved after cooking of lentils. A significant increase in resistant starch (RS)

Table 4.3 Overview of functional properties and uses of lentil flour (LF) in food products.

Functional property	Food products and % LF level	Quality attributes
Water absorption capacity	• Baked goods (<30%) • Meat products (4–15%) • Dairy products – yogurt (1–4%)	• Improvement of viscosity • Resistance to dough expansion • Better cooking quality • Shelf-life improvement
Oil absorption capacity	• Baked goods (<30%) • Meat products (4–15%)	• Improvement of texture • Better cooking quality • Improvement of sensory quality • Shelf-life extension
Solubility	• Baked goods (<30%) • Pasta (up to 100%)	• Enhancement of emulsification • Improvement of foaming and gelling properties
Emulsification	• Baked goods (<30%) • Meat products (4–15%) • Salad dressing (3–11%)	• Control of creaming • Control of coalescence • Lower flocculation and sedimentation
Gelation	• Cereal-based products (up to 100%) • Meat products (4–15%) • Dairy products – yogurt (1–4%)	• Improvement in texture • Enhancement of sensory quality

Source: Adapted from Romano et al. (2021).

was also observed after cooking (Wang et al. 2009) but fat, ash content (some minerals), and total carbohydrates were reduced in cooked lentils (Dueñas et al. 2016). Cooking also has a positive effect in reducing or eliminating a number of antinutritional factors, e.g., protease inhibitors (trypsin inhibitors), raffinose family oligosaccharides, and phytic acid contents. Pal et al. (2017) reported that trypsin inhibitors were significantly inactivated (by ~85%) after cooking. The oligosaccharides were significantly reduced by cooking of pre-soaked lentils, due primarily to leaching into soak/cook water (Wang et al. 2009). Hefnawy (2011) reported that autoclaving and microwave cooking reduced phytic acid and tannin contents significantly.

Siva and Thavarajah (2018) studied the changes in the content of prebiotic carbohydrate in red, green, and Spanish brown lentils, which were subjected to dehulling, splitting, cooking, cooling, and reheating treatments. Dehulling and splitting of lentil decreased sugar alcohols in red and green lentils; however, prebiotic carbohydrates, i.e., raffinose-family oligosaccharides (RFO) and fructo-oligosaccharides (FOS), decreased significantly in only dehulled/split-red lentil. It was observed that both dehulling and splitting of red lentil significantly decreased resistant starch (RS) content as compared to whole lentil. It was further shown that cooking followed by cooling and reheating significantly increased RS content in all three market classes of lentils. Wang et al. (2009) noted that lentils prebiotic carbohydrates (RFO, FOS) are affected by different processing treatments, including cooking, cooling, and reheating. Vidal-Valverde and Frias (1991) reported that cooking of lentil resulted in softening of soluble fibers, which partially reduces its content. On the other hand, Bressani (1993) reported that the increase in insoluble dietary fiber could be ascribed to the formation of protein–fiber complexes due possibly to chemical modification, during cooking of legumes.

Generally, consumers do not consider legumes to be rich in antioxidants as compared to fruits and vegetables but legumes do possess significant content of bioactive compounds with demonstrated antioxidant activity. Xu and Chang (2009) investigated the effect of thermal processing on phytochemical profiles and antioxidant capacities of commonly consumed food legumes (green pea, yellow pea, chickpea, and lentil). Their results showed that all four processing methods (boiling, conventional steaming, pressure boiling, and pressure steaming) had a significant reducing effect on total phenolic content and antioxidant capacities, measured as ferric reducing antioxidant power (FRAP) and peroxyl radical scavenging capacity. With the exception of conventional steaming, all other thermal processing methods resulted in a significant reduction in gallic, chlorogenic, *p*-coumaric, subtotal benzoic, and cinnamic acids as well as total phenolic content.

Aguilera et al. (2010) reported that a 16-hour pre-soaking of lentils followed by cooking was shown to decrease the antioxidant activity (measured as oxygen radical absorbing capacity, ORAC), hydroxycinnamics, catechins, procyanidins, dihydroflavonols, flavones, and flavanones. By contrast, an increase was observed in the content of hydroxybenzoics and some other phenolic compounds.

4.2.4 Extrusion Processing

Extrusion processing can be used to produce lentil flours possessing optimal functional properties by varying the extrusion parameters of feed rate, moisture content, and barrel temperature (Berrios et al. 2022; Dhull et al. 2022). Extrusion is a versatile, high-temperature short-time (HTST) process, which converts raw flours and granules into fully cooked and ready-to-eat food products or ready-to-use ingredients (Berrios et al. 2022). Extruded products are low moisture and shelf-stable, which require no additional preservatives. Extrusion has been used to prepare a variety of lentil-based products or ingredients, e.g., snacks (Ciudad-Mulero et al. 2020; Lazou and Krokida 2010), noodles (Rathod and Annapure 2017), lentil flour (Berrios et al. 2010), modified lentil starch (Gonzalez and Perez 2002), and lentil analog product (Abu-Ghoush et al. 2015). Developing new extruded products from lentil flour and utilization of lentil protein isolates and concentrates can potentially expand lentil consumption beyond traditional uses.

Extrusion has a positive effect on the nutritional and antinutritional properties of lentil-based extruded products (Ciudad-Mulero et al. 2020; Gonzalez and Perez 2002; Hernández-Nava et al. 2011; Lazou and Krokida 2010; Ochoa-Yepes et al. 2019). Rathod and Annapure (2017) reported that extrusion processing improved the *in vitro* protein and starch digestibility in lentil noodles, with no change in total protein content. Additionally, extrusion significantly reduced different antinutrients, e.g., phytic acid, tannins, and trypsin inhibitors – all by over 98%. Extrusion was shown to result in higher amino acid score and percent true protein digestibility in lentil flour (Nosworthy et al. 2018). Pasqualone et al. (2021) reported that, typically, milder or severe extrusion parameters (especially, temperature and screw speed) have no or minimal impact on the content of total protein, whereas total carbohydrates and dietary fiber concentration could be variable in flours extruded obtained under different extrusion processing parameters. Ciudad-Mulero et al. (2020) observed a significant decrease in lectins and trypsin inhibitor activity in extruded lentil flour.

4.2.5 Baking Applications

Baking application using lentil flour and fractions (e.g., protein concentrates) is one of the easiest and convenient way to boost the nutritional content of the prepared products. In this regard, lentil flours and fractions offer suitable functional properties for incorporation in a variety of baked products, such as bread, cookies, cakes, and tortillas. Lentil-based baked products offer a great potential to be commercialized due to their nutrient dense nature and increased awareness among consumers for healthy food choices.

Idris (1982) isolated proteins from lentil with 66–69% recovery. The isolated proteins were incorporated into white wheat bread formulations at 5%, 7.5%, and 10% replacements levels and the breads thus prepared were compared with the 100% wheat flour bread (control) for various quality attributes. The addition of lentil protein isolates (LPI) increased the protein content of bread samples by 11.44–14.60%, as the LPI level was increased from 5% to 10%. However, in comparison to the control bread, specific volume, crumb softness, and instrumental color (Hunter L values) of LPI-added breads decreased significantly. Although the LPI-added breads received lower sensory scores than the control bread, they were rated acceptable for various attributes (color, texture, moistness, flavor, and overall desirability).

Kohajdová et al. (2013) studied the feasibility of lentil and bean flours in baked bread dough product (rolls). Dough development and baking behavior of wheat flour, substituted with 10%, 20%, or 30% of legume flours, including lentils, were evaluated. The addition of legume flours resulted in increased water absorption capacity (from 58.5% to 74.9%) and dough development time (from 3.5 to 5.5 minutes) while dough stability was reduced (from 6.7 to 2.3 minutes). The legume flours showed a negative effect on the physical quality attributes of baked rolls, e.g., reduction in volume, specific volume, and cambering. Based on sensory evaluation, the best acceptable quality baked rolls were obtained with 90 : 10 wheat-legume flour blend. An adverse effect on the shape, crust color, and hardness of baked rolls was observed with greater than 10% legume flour addition.

Lentil flour addition in layer and sponge cakes and the resulting effect on the quality of the batter and of the final product were investigated by de la Hera et al. (2012). Lentil flour addition significantly increased batter viscosity in both layer and sponge cake formulations. The densities of layer cake and sponge cake batters with 100% lentil flour or 1 : 1 lentil-wheat flours were lower and higher, respectively, than that of the control sample (100% wheat flour). Higher firmness values were noted in the layer cake made with lentil flour than the control cake, and this effect was more pronounced in 100% lentil flour cake.

4.2.6 Roasting and Frying

Roasting applications have been traditionally investigated for their effect on lentil flours and fractions. However, in recent years, roasted whole lentils in different flavors have become commercially available in the United States, where these products are being marketed as healthy snack options for general consumers as well as nutrition source for the vegetarians and vegans. Fried lentils are commercially available in some ready-to-eat snack mixes in India and Pakistan.

Baik and Han (2012) investigated the effects of roasting, along with cooking and fermentation, on the composition and protein properties of selected legumes, including lentils, and evaluated the properties of dough and bread incorporated with legume flours. The protein solubility was shown to decrease as a result of roasting while *in vitro* protein digestibility was improved. Bread dough from roasted legume flours was less sticky than that prepared with raw or fermented flours. The roasted legume flour also exhibited a more appealing aroma and greater loaf volume of bread than the other treatments. Ma et al. (2011) studied the effect of roasting and boiling treatments on trypsin inhibitor activity, functional properties, and microstructural characteristics of lentil, chickpea, and pea flours. A significant reduction in trypsin inhibitor activity was observed by both thermal treatments. Roasted pulse flours had similar microstructure (i.e., starch granule and protein matrix) as the raw flour. Both treatments had minimal effect on the emulsifying and foaming properties of the legume flours. It was concluded that thermally treated legume flours had a good potential for use as value-added food ingredients in diverse food applications.

Bahnassey and Khan (1986) investigated the feasibility of developing spaghetti fortified with flours from common beans (navy and pinto), lentil, and protein concentrates of these three legumes. The legume flours were prepared from roasted and non-roasted (control) legumes. The effect of fortification was evaluated on various quality attributes, including rheological properties, cooking quality of spaghetti, and sensory attributes (evaluated by taste panels). Legume flour fortification levels of 5%, 10%, 15%, 20%, and 25% were used in place of durum wheat semolina in spaghetti formulations. Dough development time and the stability were generally higher with legume flour fortifications. The uncooked fortified spaghetti exhibited weaker hardness or texture, as it shattered sooner than the control spaghetti. Moreover, higher cooking loss occurs with the increasing fortification levels. Sensory evaluation showed that spaghetti with 10% added legume flours or protein concentrates was acceptable for all attributes tested (appearance, color, mouthfeel, and acceptability). Panelists preferred spaghetti fortified with roasted legume flour compared to the non-roasted samples and reported beany flavor in spaghetti prepared with non-roasted legume flours or their concentrates (especially, at 25% fortification level).

Benmeziane-Derradji et al. (2020) studied the effect of roasting on the physicochemical and functional properties of whole lentil flour. Roasting had a positive effect on the physicochemical properties but, as expected, the moisture content decreased significantly. Most of the functional properties were improved after roasting; however, oil absorption capacity, emulsion stability, and foaming activity were reduced significantly. It was further observed that roasting did not significantly affect the microstructure of lentil flour.

The uses of lentil flour are rather limited in fried products, but its feasibility has been explored in some studies in different products. Khorsand et al. (2018) prepared fried doughnut by adding soy or lentil flour (at 15% and 30%) to the formulation and studied the effect on doughnut's quality. The doughnut hardness was decreased with the addition of legume flour. Additionally, a decrease in moisture loss, fat content, volume, porosity, and lightness were also observed in doughnut made with added soy or lentil flour. Shokrolahi et al. (2019) investigated the effect of adding 10%, 25%, or 50% lentil flour on the batter formulation quality of simulated deep-fried crust. The formulation of batter and frying time exhibited significant effect on the color, moisture and oil content of crusts. The crust color of lentil-flour-added samples was significantly darker compared to that of the control samples.

4.2.7 Germination/Sprouting

Germination is used for enhancing the nutritional and sensory properties of legumes. Germination of lentils is a traditional and economic process to enhance their health benefits by improving the nutritional value, boosting antioxidant capacity, and reducing the antinutritional factors, i.e., phytic acid and tannins (Dhull et al. 2022).

Ghumman et al. (2016) reported that germination of lentil seeds resulted in significant compositional changes and nutritional quality, mainly by reducing carbohydrates and lipid contents and enhancing the crude protein content. Reduction in carbohydrate content is attributed to the activation of endogenous enzymes, such as α-amylase, β-amylase, glucosidase, and dextranase. Tárrago and Nicolás (1976) assessed starch content and the total amylolytic (α-amylase, β-amylase) and phosphorylase activities in germinated lentil seeds (13-day germination at 25 °C, in dark). The enzymic degradation of starch in higher plants occurs due mainly to the amylase and phosphorylase activities. It was observed that starch degraded relatively slowly, as the total amylolytic activity peaked on day 7 of germination. By day 13, the residual starch content was still around 50%. During the germination process, an increase in the total amylolytic activity was observed due primarily to an 80-fold increase in the α-amylase activity on the 6th day of germination. The phosphorylase activity increased rapidly during germination, as it peaked on the 3rd day.

Dueñas et al. (2016) reported that germination of lentils effected a decrease in non-starch carbohydrates, namely insoluble, soluble, and total dietary fibers. Flour produced from germinated lentil showed improved water absorption capacity, foaming capacity, and breakdown viscosity but emulsifying index and foaming stability were reduced compared to control (raw) flour.

Morad et al. (1980) investigated the physicochemical properties of flours and starches from germinated and ungerminated (control) legumes, including lentils. The α-amylase activity of flours increased by 1.5, 2.4, and 8.1 times in faba bean, lentil, and yellow pea flours, respectively, that were obtained after four days of germination. Structural analysis, assessed by scanning electron microscopy (SEM), showed some alteration of the starch granules' surfaces after germination. It was further observed that germination resulted in changes in starch pasting properties of legume flours and their fractions. Incorporation of germinated legumes starches in wheat bread (at 5%, 10%, 15%, and 20%) had a generally significant negative effect on loaf volume, crumb gran, and crust color of breads. Loaf volume, which is an important quality attribute of breads, was reduced successively with increasing substitution levels from 5% to 20%. However, off-flavor or odor was observed in breads made with germinated legume flours. The germinated cowpea flour had the highest detrimental effect on baking properties of bread than those from faba bean or lentil flours.

Hsu et al. (1982) also studied the physical and chemical properties of protein isolates from germinated and ungerminated lentils, faba beans, and cowpea. The SDS-PAGE showed different degree of protein modification among three legumes after four-day germination. Germination also resulted in varying effects on the functional properties (water absorption, viscosity, gelation, emulsification, and foam capacity/stability) of protein isolates. It was observed that germination negatively affected the baking properties of legume protein isolates, with this effect being more pronounced in faba bean starch than that from yellow peas or lentils. When the protein isolates from germinated legumes were incorporated into bread formulations (at 3%, 5%, and 8%), with the exception of 3% substitution level, a detrimental effect on loaf volume and crumb grain of the bread was observed with 5% or 8% levels.

4.2.8 Fermentation

Fermentation is a cost-effective process that can be used to improve the nutritional and functional quality of lentils. There are significant changes in the sensory quality (improved flavor) as well. The protein digestibility is enhanced by fermentation as a result of partial reduction in antinutrients. Alrosan et al. (2021) reported that protein digestibility was significantly enhanced from 76.4% in unfermented lentils to 84.1% after five days of fermentation. Bautista-Exposito et al. (2018a) reported that 30-hour fermentation at alkaline pH, using *Lactobacillus plantarum*, was reported to have a positive effect on proteins and phenolic contents, besides increase in antioxidant, hypoglycemic, hypolypidemic, anti-inflammatory, and antihypertensive properties.

Shekib (1988) investigated the effect of fermentation on the protein efficiency ratio (PER), methionine, lysine, phytic acid, total phosphorus, and inorganic phosphorus contents in lentils, rice, and 1 : 1 ratio (w/w) lentil-rice blend. The PER of raw lentils, rice, and lentil-rice blend was 0.74, 1.19, and 1.39, respectively, before fermentation, which increased significantly to 1.12 (lentil), 1.98 (rice), and 2.22 (lentil-rice blend) after three days of fermentation. Similarly, the methionine content increased significantly in fermented lentils and lentil-rice blend, whereas the lysine content increased significantly in fermented rice and lentil-rice blend. It was observed that fermentation significantly reduced phytic acid (an antinutritional factor) content by 26.7% and 45.15% in lentils and lentil-rice blend, respectively. However, no changes were noted in the total phosphorus in fermented lentils, rice, or lentil-rice blend.

4.3 Emerging Research and Development Trends

In the last two decades, a number of innovative processing technologies (high-pressure processing, ultrafiltration, ionizing radiation, ultraviolet radiation, pulsed electric field, pulsed light treatment, ohmic heating, and ultrasound treatment) have been introduced (Ahmed 2018; Ahmed et al. 2019; Gharibzahedi et al. 2018; Tokuşoğlu and Swanson 2014). Some of these innovative technologies can also be employed in value-added processing of legumes, especially their nutritional and sensory qualities. Lentils, and most legumes, possess some antinutrient factors (ANFs), such as lectins, enzyme inhibitors, phytates, saponins, and oligosaccharides (Campos-Vega et al. 2010; Sidhu et al. 2022). The presence of these ANFs is reported to decrease the activity of digestive enzymes or sequester essential nutrients, making them inaccessible for digestion and absorption (Dhull et al. 2022;

Nosworthy et al. 2018). Therefore, appropriate processing methods must be applied to significantly reduce or completely eliminate ANFs.

Ahmed et al. (2009) investigated the thermal properties of high-pressure (HP)-treated lentil flour dispersions (LFDs) as a function of pressure level (350–650 MPa) and moisture content (14–58%). An incomplete denaturation of lentil proteins by HP treatment was observed in LFDs by differential scanning calorimetry (DSC). No starch gelatinization peak was detected during thermal scanning of LFDs (either in control or HP-treated samples) regardless of moisture content or heating rate. Dynamic rheological measurements indicated that slurries from HP-treated lentils exhibited a true viscoelastic behavior. It was further shown that HP treatment of LFDs significantly reduced the retrogradation behavior when compared to that obtained from the thermally gelatinized LFDs. In another study, Ahmed et al. (2016) concluded that the HP treatment of lentil starch could be utilized to improve its functional properties and develop food products with desirable quality and enhanced consumer acceptance.

Saricaoglu (2020) studied the application of HP homogenization (HPH) to modify the functional, structural, and rheological properties of lentil proteins. The HPH treatment was shown to cause unfolding of lentil proteins between 50 and 150 MPa and decreased particle size and improved functional properties of proteins. Primozic et al. (2018) investigated HPH at 5000 and 15,000 psi (~34–103 MPa) for producing nanoemulsions from lentil protein isolate (LPI) for improved digestibility. The HPH treatment resulted in significantly reducing the particle size of LPI and its moduli and surface hydrophobicity. It was concluded that physical modification of LPI by HPH prior to emulsification could be a potential novel method to utilize lentils and other legume proteins in the formation and stabilization of nanoemulsions, which could potentially improve the LPI's digestibility.

The use of ultrafiltration (UF) and isoelectric precipitation (IEP) was investigated by Boye et al. (2010) to assess their effects on the functional properties of protein concentrates (PCs) prepared from pea, chickpea, and lentil. The initial protein content of the pulses (16.7–24.8%) was concentrated nearly fourfolds. The UF application yielded higher protein content in the PCs (69.1–88.6%) than the IEP method, which yielded 63.9–81.7% protein. It was further observed that red and green lentils' PCs had higher protein solubility of 70–77% using UF isolation.

Supercritical fluid (carbon dioxide) extrusion (SCFX) was used by Lv et al. (2018) to process a powder blend of lentil flour and pregelatinized potato starch. This blend was extruded at temperature of 95–99.5 °C using screw speeds of 40, 80, and 120 rpm. The physical properties (porosity, expansion ratio, water solubility, and rehydration rate) were enhanced as the screw speed increased from 40 to 120 rpm. SCFX resulted in an increase of the transition temperature of starch and protein in lentil flour, and increased screw speed was shown to increase starch gelatinization. In contrast to traditionally cooked lentils, no retrogradation of starch was observed for the extruded red lentil products stored at 4 °C for 60 days. The phenolics and 2,2-diphenyl-1-picrylhydrazyl (DPPH) radical scavenging activity in supercritical fluid extruded lentil products increased by 30% and 18%, respectively.

Gharibzahedi and Smith (2020) reported that the sonication or high-intensity ultrasound (HIU) treatment, as a relatively low-energy means to improve pulse protein functionality, is an efficient method to cleave disulfide bonds without the use of chemical reductants. However, the process parameters need to be optimized to improve the gel strength and its thermal stability. The use of HIU treatment, which resulted in reduced particle size, turbidity, and viscosity of legume proteins slurries, significantly increased the sulfhydryl and hydrophobic functional groups.

Najib et al. (2022) studied the combination of germination and microwave-infrared (MW-IR) drying to process lentil flour with enhanced quality attributes for use in diverse food applications. Results of this study showed that microwave power had the most impact on drying rate, followed by infrared power, whereas the effect of germination time was not significant. Based on the kinetic results, the drying rate constant (k) of the diffusion approach was shown to be significantly correlated with the MW-IR power while no correlation with germination time was observed.

Guillermic et al. (2021) employed a novel technique, X-ray microtomography imaging, to characterize and quantify the detailed microstructure of red lentil extrudates. Extrusion parameters (screw speed, moisture content, and injection pressure of a blowing agent) can be varied to achieve optimum microstructure and texture of puffed extruded products. An increase in the injected pressure of the physical blowing agent was shown to be correlated with a decrease in cell size and wall thickness, and an increase in the number of cells. Weak spots in the extrudate microstructure resulted in the reduced hardness and crunchiness. The cell wall rupture with an increased screw speed was also observed; however, this negative effect could be offset by a higher moisture content during extrusion processing.

4.4 Development of Lentil-Based Products

The traditional utilization of lentils is primarily in home-prepared cuisines. In many South Asian and African countries, lentils have been a major source of protein and part of traditional cuisines that are consumed regularly. Research and development efforts have been made in recent years to formulate lentil-based products, as shown is Figure 4.2. Selected examples of lentil-based products include pan/flat bread (Ahmed et al. 2021; Marchini et al. 2021), pasta/spaghetti (Bresciani et al. 2021; Valdelvira et al. 2022), crackers (Han et al. 2010; Polat et al. 2020), gluten-free extruded snacks (Ciudad-Mulero et al. 2020), puffed snacks (Guillermic et al. 2021; Sinaki et al. 2021), extruded snacks (Luo et al. 2020; Lv et al. 2018), lentil fortified yogurt (Zare et al. 2011), and energy bars (Gao et al. 2020; Zello et al. 2012). Besides flour, lentil protein isolates and concentrates can be used in developing new products (Bhatty and Christison 1984; Boye et al. 2010; Oduro-Yeboah et al. 2022).

The chickpea-based hummus, a relatively unknown product in Western countries just two decades ago, has become a popular healthy food choice among non-Middle-Eastern consumers. This is an area where red lentils could be readily incorporated to market chickpea-lentil hummus or hummus prepared with lentils and other pulses. Keskin et al. (2022) reported that despite a variety of nutritional benefits, the per capita consumption of legumes, including lentils, continues to be relatively low in the developed countries where a major portions of protein consumption is derived from animal-based foods. The current consumer trends show a move away from animal products and opt for sustainably produced foods (especially, plant-based choices) with significantly lower negative impact on the environment compared to raising animals for food. The superior functionality of legume-based ingredients can play an important role in expanding legume consumption

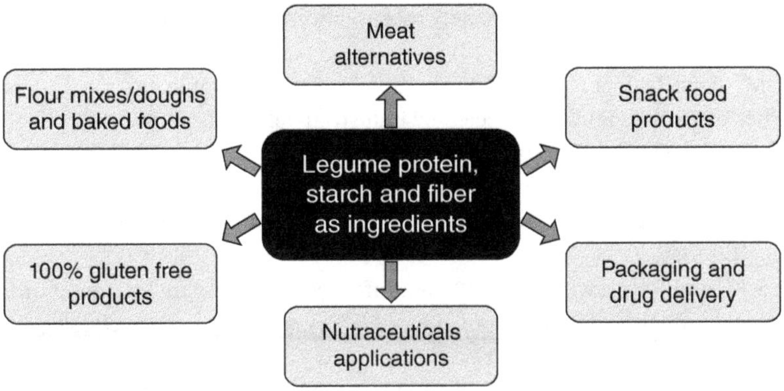

Figure 4.2 Diverse application of legume protein, starch, and fiber ingredients. *Source:* Keskin et al. (2022)/John Wiley & Sons/Public Domain CC BY 4.0.

beyond traditional products, uses, and consumers. Sidhu et al. (2022) reported that the success of newly developed lentil-based products depends on fully understanding the functionality of lentil flour/isolates, their effects on physico-chemical and consumer acceptance.

4.5 Conclusion

Lentils constitute an important part of diets in many regions of the world and are used in a wide variety of traditional cuisines and as an ingredient in various food applications. Besides traditional domestic soaking and cooking, a number of other processing techniques as discussed in this chapter can be used to improve the quality of lentils and other food legumes. In addition to improving taste/flavor, cooking and processing of lentils also significantly reduce antinutritional factors and enhance the bioavailability of nutrients and bioactive compounds. Lentils being rich in proteins and many bioactive compounds are well suited for producing a range of value-added products and functional foods. Lentil whole flours and segregated fractions (rich in starch and protein) offer a great potential for developing gluten-free products. In recent years, some innovative technologies have emerged as an alternative to traditional thermal processing, which can also be used for lentil processing and product development. Development and commercial production of lentil ingredients can not only offer healthy food options to consumers but economic benefits to the food industry and lentil growers as well. Overall, lentils offer a great potential to be incorporated into gluten-free and plant-based food products, which align well with changing consumer preferences for healthier foods that are produced in environmentally sustainable manner.

References

Abu-Ghoush, M., Alavi, S., and Al-Shathri, A. (2015). A novel cooked extruded lentils analog: physical and chemical properties. *Journal of Food Science and Technology* 52: 4216–4225.

Afam-Mbah, C., Ketabi, L., Emami, S. et al. (2018). Air classification of by-product of lentil dehulling into protein-rich and starch-rich fractions. *In 2018 ASABE Annual International*

Meeting (29 July–1 August 2018). American Society of Agricultural and Biological Engineers, St. Joseph, MI.

Aguilera, Y., Duenas, M., Estrella, I. et al. (2010). Evaluation of phenolic profile and antioxidant properties of Pardina lentil as affected by industrial dehydration. *Journal of Agricultural and Food Chemistry* 58: 10101–10108.

Ahmed, J. (2018). Minimal processing and novel technologies applied to vegetables. In: *Handbook of Vegetables and Vegetable Processing* (ed. M. Siddiq and M.A. Uebersax), 317–339. Hoboken, NJ: Wiley.

Ahmed, J., Varshney, S.K., and Ramaswamy, H.S. (2009). Effect of high pressure treatment on thermal and rheological properties of lentil flour slurry. *LWT Food Science and Technology* 42: 1538–1544.

Ahmed, J., Thomas, L., Taher, A., and Joseph, A. (2016). Impact of high pressure treatment on functional, rheological, pasting, and structural properties of lentil starch dispersions. *Carbohydrate Polymers* 152: 639–647.

Ahmed, J., Mulla, M., Al-Ruwaih, N., and Arfat, Y.A. (2019). Effect of high-pressure treatment prior to enzymatic hydrolysis on rheological, thermal, and antioxidant properties of lentil protein isolate. *Legume Science* 1: e10.

Ahmed, H.E., Yousef, H.Y., Nofa, Z.A., and Saad, Z.H. (2021). Chemical composition and mineral contents of gluten free flat bread fortified with lentil and lupin flour. *Al-Azhar Journal of Agricultural Research* 46: 219–230.

Ahmed, J., Mulla, M.Z., Siddiq, M., and Dolan, K.D. (2022). Micromeritic, thermal, dielectric, and microstructural properties of legume ingredients: a review. *Legume Science* 4: e123.

Alrosan, M., Tan, T.C., Mat Easa, A. et al. (2021). Effects of fermentation on the quality, structure, and nonnutritive contents of lentil (*Lens culinaris*) proteins. *Journal of Food Quality* 2021: 5556450.

Amin, S. and Borchgrevink, C.P. (2022). A Culinology® perspective of dry beans and other pulses. In: *Dry Beans and Pulses Production, Processing, and Nutrition* (ed. M. Siddiq and M.A. Uebersax), 453–480. Hoboken, NJ: Wiley.

Bahnassey, Y. and Khan, K. (1986). Fortification of spaghetti with edible legumes. II. Rheological, processing, and quality evaluation studies. *Cereal Chemistry* 63: 216–219.

Baik, B.K. and Han, I.H. (2012). Cooking, roasting, and fermentation of chickpeas, lentils, peas, and soybeans for fortification of leavened bread. *Cereal Chemistry* 89: 269–275.

Bautista-Exposito, S., Penas, E., Duenas, M. et al. (2018a). Individual contributions of Savinase and *Lactobacillus plantarum* to lentil functionalization during alkaline pH-controlled fermentation. *Food Chemistry* 257: 341–349.

Benmeziane-Derradji, F., Djermoune-Arkoub, L., Ayat, N.E.H., and Aoufi, D. (2020). Impact of roasting on the physicochemical, functional properties, antioxidant content and microstructure changes of Algerian lentil (*Lens culinaris*) flour. *Journal of Food Measurement and Characterization* 14: 2840–2853.

Berrios, J.D.J., Morales, P., Cámara, M., and Sánchez-Mata, M.C. (2010). Carbohydrate composition of raw and extruded pulse flours. *Food Research International* 43: 531–536.

Berrios, J.D.J., Losso, J.N., and Albertos, I. (2022). Extrusion processing of dry beans and pulses. In: *Dry Beans and Pulses Production, Processing, and Nutrition* (ed. M. Siddiq and M.A. Uebersax), 225–246. Hoboken, NJ: Wiley.

Bhatty, R.S. and Christison, G.I. (1984). Composition and nutritional quality of pea (*Pisum sativum* L.), faba bean (*Vicia faba* L. spp. minor) and lentil (*Lens culinaris* Medik.) meals, protein concentrates and isolates. *Plant Foods for Human Nutrition* 34: 41–51.

Boye, J.I., Aksay, S., Roufik, S. et al. (2010). Comparison of the functional properties of pea, chickpea and lentil protein concentrates processed using ultrafiltration and isoelectric precipitation techniques. *Food Research International* 43: 537–546.

Bresciani, A., Giuberti, G., Cervini, M., and Marti, A. (2021). Pasta from yellow lentils: how process affects starch features and pasta quality. *Food Chemistry* 364: 130387.

Bressani, T. (1993). Grain quality of common beans. *Food Review International* 9: 237–297.

Bressani, R. and Elias, L.G. (1988). Seed quality and nutritional goals in pea, lentil, faba bean and chickpea breeding. In: *World Crops: Cool Season Food Legumes* (ed. R.J. Summerfield), 381–404. Dordrecht, The Netherlands: Kluwer Academic Publishers.

Bruce, J.L. (2008). The effects of preharvest treatments on the milling efficiency of red lentil. MS thesis. University of Saskatchewan, Saskatoon, Canada.

Campos-Vega, R., Loarca-Pina, G., and Oomah, B.D. (2010). Minor components of pulses and their potential impact on human health. *Food Research International* 43: 461–482.

Chelladurai, V. and Erkinbaev, C. (2020). Lentils. In: *Pulses* (ed. A. Manickavasag and P. Thirunathan), 129–143. Cham, Switzerland: Springer.

Ciudad-Mulero, M., Fernandez-Ruiz, V., Cuadrado, C. et al. (2020). Novel gluten-free formulations from lentil flours and nutritional yeast: evaluation of extrusion effect on phytochemicals and non-nutritional factors. *Food Chemistry* 315: 126175.

Dhull, S.B., Kinabo, J., and Uebersax, M.A. (2022). Nutrient profile and effect of processing methods on the composition and functional properties of lentils (*Lens culinaris* Medik): a review. *Legume Science* 2022: e156.

Dueñas, M., Sarmento, T., Aguilera, Y. et al. (2016). Impact of cooking and germination on phenolic composition and dietary fibre fractions in dark beans (*Phaseolus vulgaris* L.) and lentils (*Lens culinaris*). *LWT Food Science and Technology* 66: 72–78.

Funke, M., Boom, R., and Weiss, J. (2022). Dry fractionation of lentils by air classification-composition, interfacial properties and behavior in concentrated O/W emulsions. *LWT Food Science and Technology* 154: 12718.

Gao, R., Shaw, K., Kaviani, M. et al. (2020). Glycemic index of lentil- and cherry-based sport nutrition products for endurance athletes. *Science & Sports* 36: 234–234.

Gharibzahedi, S.M.T. and Smith, B. (2020). The functional modification of legume proteins by ultrafiltration: a review. *Trends in Food Science & Technology* 98: 107–116.

Gharibzahedi, S.M.T., Roohinejad, S., George, S. et al. (2018). Innovative food processing technologies on the transglutaminase functionality in protein-based food products: trends, opportunities and drawbacks. *Trends in Food Science & Technology* 75: 194–205.

Ghumman, A., Kaur, A., and Singh, N. (2016). Impact of germination on flour, protein and starch characteristics of lentil (*Lens culinari*) and horsegram (*Macrotyloma uniflorum* L.) lines. *LWT Food Science and Technology* 65: 137–144.

Gonzalez, Z. and Perez, E. (2002). Evaluation of lentil starches modified by microwave irradiation and extrusion cooking. *Food Research International* 35: 415–420.

Guillermic, R.M., Aksoy, E.C., Aritan, S.E.R.D.A.R. et al. (2021). X-ray microtomography imaging of red lentil puffed snacks: processing conditions, microstructure and texture. *Food Research International* 140: 109996.

Han, J.J., Janz, J.A., and Gerlat, M. (2010). Development of gluten-free cracker snacks using pulse flours and fractions. *Food Research International* 43: 627–633.

Hefnawy, T.H. (2011). Effect of processing methods on nutritional composition and anti-nutritional factors in lentils (*Lens culinaris*). *Annals of Agricultural Science* 56: 57–61.

de la Hera, E., Ruiz-Paris, E., Oliete, B., and Gomez, M. (2012). Studies of the quality of cakes made with wheat-lentil composite flours. *LWT Food Science and Technology* 49: 48–54.

Hernández-Nava, R.G., Bello-Pérez, L.A., Martín-Martínez, S. et al. (2011). Effect of extrusion cooking on the functional properties and starch components of lentil/banana blends: response surface analysis. *Revista Mexicana de Ingenieria Química* 10: 409–419.

Hill, H. (2022). Utilization of dry beans and other pulses as ingredients in diverse food products. In: *Dry Beans and Pulses Production, Processing, and Nutrition* (ed. M. Siddiq and M.A. Uebersax), 307–329. Hoboken, NJ: Wiley.

Hsu, D.L., Leung, H.K., Morad, M.M. et al. (1982). Effect of germination on electrophoretic, functional, and bread-baking properties. *Cereal Chemistry* 59: 344–350.

Idris, N.A. (1982). Extraction of lentil proteins and their use in supplementation of bread. MS thesis. University of Oregon, USA.

Jarpa-Parra, M. (2018). Lentil protein: a review of functional properties and food application. An overview of lentil protein functionality. *International Journal of Food Science and Technology* 53: 892–903.

Jiang, Z.Q., Pulkkinen, M., Wang, Y.J. et al. (2016). Faba bean flavor and technological property improvement by thermal pre-treatments. *LWT Food Science and Technology* 68: 295–305.

Joshi, M., Timilsena, Y., and Adhikari, B. (2017). Global production, processing and utilization of lentil: a review. *Journal of Integrative Agriculture* 16: 2898–2913.

Kamboj, R. and Nanda, V. (2018). Proximate composition, nutritional profile and health benefits of legumes – a review. *Legume Research* 41: 325–332.

Keskin, S.O., Ali, T.M., Ahmed, J. et al. (2022). Physico-chemical and functional properties of legume protein, starch, and dietary fiber — a review. *Legume Science* 4: e117.

Khazaei, H., Subedi, M., Nickerson, M. et al. (2019). Seed protein of lentils: current status, progress, and food applications. *Food* 8: 391.

Khorsand, M., Aliabadi, M., and Jarrahiferiz, J. (2018). The effect of adding flour soy and lentil to doughnut on its physicochemical properties. *Research and Innovation in Food Science and Technology* 7: 197–212.

Kohajdová, Z., Karovičová, J., and Magala, M. (2013). Effect of lentil and bean flours on rheological and baking properties of wheat dough. *Chemical Papers* 67: 398–407.

Lazou, A. and Krokida, M. (2010). Structural and textural characterization of corn–lentil extruded snacks. *Journal of Food Engineering* 100: 392–408.

Lee, H.W., Lu, Y., Zhang, Y. et al. (2021). Physicochemical and functional properties of red lentil protein isolates from three origins at different pH. *Food Chemistry* 358: 129749.

Luo, S., Chan, E., Masatcioglu, M.T. et al. (2020). Effects of extrusion conditions and nitrogen injection on physical, mechanical, and microstructural properties of red lentil puffed snacks. *Food and Bioproducts Processing* 121: 143–153.

Lv, Y., Glahn, R.P., Hebb, R.L., and Rizvi, S.S.H. (2018). Physico-chemical properties, phytochemicals and DPPH radical scavenging activity of supercritical fluid extruded lentils. *LWT Food Science and Technology* 89: 315–321.

Ma, Z., Boye, J.I., Simpson, B.K. et al. (2011). Thermal processing effects on the functional properties and microstructure of lentil, chickpea, and pea flours. *Food Research International* 44: 2534–2544.

Marchini, M., Carini, E., Cataldi, N. et al. (2021). The use of red lentil flour in bakery products: how the particle size and substitution level affect rheological properties of wheat bread dough? *LWT Food Science and Technology* 136: 110299.

McDonald, L.S., Assadzadeh, S., and Panozzo, J.F. (2021). Images, features, or feature distributions? A comparison of inputs for training convolutional neural networks to classify lentil and field pea milling fractions. *Biosystems Engineering* 208: 16–27.

McVicar, R., McCall, P., Brenzil, C. et al. (2017). Lentils in Saskatchewan – fact sheet. https://pubsaskdev.blob.core.windows.net/pubsask-prod/86381/86381-Lentils_in_Saskatchewan.pdf (accessed 16 March 2023).

Morad, M.M., Leung, H.K., Hsu, D.L., and Finney, P.L. (1980). Lentil, and faba bean flours and starches. *Cereal Chemistry* 57: 390–396.

Najib, T., Heydari, M.M., and Meda, V. (2022). Combination of germination and innovative microwave-assisted infrared drying of lentils: effect of physicochemical properties of different varieties on water uptake, germination, and drying kinetics. *Applied Food Research* 2: 100040.

Nosworthy, M.G., Medina, G., Franczyk, A.J. et al. (2018). Effect of processing on the *in vitro* and *in vivo* protein quality of red and green lentils (*Lens culinaris*). *Food Chemistry* 240: 588–593.

Ochoa-Yepes, O., Di Giogio, L., Goyanes, S. et al. (2019). Influence of process (extrusion/thermo-compression, casting) and lentil protein content on physicochemical properties of starch films. *Carbohydrate Polymers* 208: 221–231.

Oduro-Yeboah, C., Sulaiman, R., Uebersax, M.A., and Dolan, K.D. (2022). A review of lentil (*Lens culinaris* Medik) value-chain: postharvest handling, processing and processed products. *Legume Science* 4: e171.

Pal, R.S., Bhartiya, A., Yadav, P. et al. (2017). Effect of dehulling, germination and cooking on nutrients, anti-nutrients, fatty acid composition and antioxidant properties in lentil (*Lens culinaris*). *Journal of Food Science and Technology* 54: 909–920.

Pasqualone, A., Costantini, M., Labarbuta, R., and Summo, C. (2021). Production of extruded-cooked lentil flours at industrial level: effect of processing conditions on starch gelatinization, dough rheological properties and techno-functional parameters. *LWT Food Science and Technology* 147: 111580.

Polat, H., Capar, T.D., Inanir, C. et al. (2020). Formulation of functional crackers enriched with germinated lentil extract: a response surface methodology Box-Behnken design. *LWT Food Science and Technology* 123: 109065.

Primozic, M., Duchek, A., Nickerson, M., and Ghosh, S. (2018). Formation, stability and *in vitro* digestibility of nanoemulsions stabilized by high-pressure homogenized lentil proteins isolate. *Food Hydrocolloids* 77: 126–141.

Pulivarthi, M.K., Nkurikiye, E., Watt, J. et al. (2021). Comprehensive understanding of roller milling on the physicochemical properties of red lentil and yellow pea flours. *Processes* 9: 1836.

Pulse Canada (2012). *Lentils (Lens culinaris)*. Winnipeg, Manitoba: Pulse Canada.

Ramírez-Ojeda, A.M., Moreno-Rojas, R., and Cámara-Martos, F. (2018). Mineral and trace element content in legumes (lentils, chickpeas and beans): bioaccessibility and probabilistic assessment of the dietary intake. *Journal of Food Composition and Analysis* 73: 17–28.

Rathod, R.P. and Annapure, U.S. (2017). Physicochemical properties, protein and starch digestibility of lentil based noodle prepared by using extrusion processing. *LWT Food Science and Technology* 80: 121–130.

Romano, A., Gallo, V., Ferranti, P., and Masi, P. (2021). Lentil flour: nutritional and technological properties, *in vitro* digestibility and perspectives for use in the food industry. *Current Opinion in Food Science* 40: 157–167.

Rondoni, A. and Grasso, S. (2021). Consumers behaviour towards carbon footprint labels on food: a review of the literature and discussion of industry implications. *Journal of Cleaner Production* 301: 127031.

Saricaoglu, F.T. (2020). Application of high-pressure homogenization (HPH) to modify functional, structural and rheological properties of lentil (*Lens culinaris*) proteins. *International Journal of Biological Macromolecules* 144: 760–769.

Sharma, A. (2021). A review on traditional technology and safety challenges with regard to antinutrients in legume foods. *Journal of Food Science and Technology* 58: 2863–2883.

Shekib, L.A.E. (1988). Evaluation of protein quality, methionine and lysine availability, and phytic acid in natural fermented lentils, rice and their blend [Egypt]. *Alexandria Journal of Agricultural Research* 33: 135–144.

Shokrolahi, B., Mohebbi, M., Varidi, M. et al. (2019). Performance of lentil and chickpea flour in deep-fried crust model (DFCM): oil barrier and crispy properties. *Journal of Food Measurement and Characterization* 13: 296–304.

Sidhu, J.S., Zafar, T., Benyathiar, P., and Nasir, M. (2022). Production, processing, and nutritional profile of chickpeas and lentils. In: *Dry Beans and Pulses Production, Processing, and Nutrition* (ed. M. Siddiq and M.A. Uebersax), 383–407. Hoboken, NJ: Wiley.

Sinaki, N.Y., Masatcioglu, M.T., Paliwal, J., and Koksel, F. (2021). Development of cellular high-protein foods: third-generation yellow pea and red lentil puffed snacks. *Food* 11: 38.

Singh, U. and Singh, B. (1992). Tropical grain legumes as important human foods. *Economic Botany* 46: 310–321.

Siva, N. and Thavarajah, P. (2018). The impact of processing and cooking on prebiotic carbohydrates in lentil. *Journal of Food Composition and Analysis* 70: 72–77.

Siva, N., Johnson, C.R., Duckett, S. et al. (2017). Can lentil (*Lens culinaris* Medikus) reduce the risk of obesity? *Journal of Functional Foods* 38: 706–715.

Tárrago, J.F. and Nicolás, G. (1976). Starch degradation in the cotyledons of germinating lentils. *Plant Physiology* 58: 618–621.

Thavarajah, D., Ruszkowski, J., and Vandenberg, A. (2008). High potential for selenium biofortification of lentils (*Lens culinaris* L.). *Journal of Agricultural and Food Chemistry* 56: 10747–10753.

Tokuşoğlu, Ö. and Swanson, B.G. (ed.) (2014). *Improving Food Quality with Novel Food Processing Technologies*. Boca Raton, FL: CRC Press.

Tyler, R., Youngs, C., and Sosulski, F. (1981). Air classification of legumes. I. Separation efficiency, yield, and composition of the starch and protein fractions. *Cereal Chemistry* 58: 144–148.

Uebersax, M.A. and Siddiq, M. (2012). Postharvest storage quality, packaging and distribution of dry beans. In: *Dry Beans and Pulses Production, Processing and Nutrition*, 1e (ed. M. Siddiq and M.A. Uebersax), 75–100. Ames, IA: Wiley.

Uebersax, M.A., Cichy, K.A., Gomez, F.E. et al. (2022). Dry beans (*Phaseolus vulgaris* L.) as a vital component of sustainable agriculture and food security – a review. *Legume Science* 4: e155.

Valdelvira, R., Garcia-Medina, G., Crespo, J.F., and Cabanillas, B. (2022). Allergenic content of new alimentary pasta made of lentils compared with lentil seeds and analysis of the impact of boiling processing. *Plant Foods for Human Nutrition* 77: 443–446.

Vidal-Valverde, C. and Frias, J. (1991). Legume processing effects on dietary fiber components. *Journal of Food Science* 56: 1350–1352.

Wang, N., Hatcher, D.W., Toews, R., and Gawalko, E.J. (2009). Influence of cooking and dehulling on nutritional composition of several varieties of lentils (*Lens culinaris*). *LWT Food Science and Technology* 42: 842–848.

Wiesinger, J.A., Marsolais, F., and Glahn, R.P. (2022). Health implications and nutrient bioavailability of bioactive compounds in dry beans and other pulses. In: *Dry Beans and Pulses Production, Processing, and Nutrition* (ed. M. Siddiq and M.A. Uebersax), 505–529. Hoboken, NJ: Wiley.

Xu, B. and Chang, S.K. (2009). Phytochemical profiles and health-promoting effects of cool-season food legumes as influenced by thermal processing. *Journal of Agricultural and Food Chemistry* 57: 10718–10731.

Xu, M., Jin, Z., Gu, Z. et al. (2020). Changes in odor characteristics of pulse protein isolates from germinated chickpea, lentil, and yellow pea: role of lipoxygenase and free radicals. *Food Chemistry* 314: 126184.

Yang, T., Liu, K., Poppy, L. et al. (2021). Minimizing lentil harvest loss through improved agronomic practices in sustainable agro-systems. *Sustainability* 13: 1896.

Zare, F., Boye, J.I., Orsat, V. et al. (2011). Microbial, physical and sensory properties of yogurt supplemented with lentil flour. *Food Research International* 44: 2482–2488.

Zello, G.A., Jochim, J.T., Rooke, J., and Chilibeck, P.D. (2012). Consuming a lentil-based sports nutrition bar affects metabolic and performance measures during endurance exercise. *FASEB Journal* 26: 820.19.

Zia-ul-Haq, M., Ahmad, S., Shad, M.A. et al. (2011). Compositional studies of lentil (*Lens culinaris* Medik.) cultivars commonly grown in Pakistan. *Pakistan Journal of Botany* 43: 1563–1567.

5

Milling and Fractionation Processing of Lentils

Gaurav Kumar[1], *Pramod K. Prabhakar*[2], *and Santanu Basu*[3]

[1] *Department of Food Engineering, National Institute of Food Technology Entrepreneurship and Management, Kundli, India*
[2] *Department of Food Science and Technology, National Institute of Food Technology Entrepreneurship and Management, Kundli, India*
[3] *Department of Molecular Sciences, Swedish University of Agricultural Sciences, Uppsala, Sweden*

5.1 Introduction

Pulses, including lentils, are dicotyledonous seeds of leguminous plants that are broadly classified into beans (like pinto bean, soybean, kidney beans, and white beans), chickpeas, lentils, and peas (yellow pea and green pea) (Tosh and Yada 2010). They are good sources of plant proteins and have high amounts of starch, dietary fibers, and polyphenols, making pulses a cost-effective nutritious food. Lentils are desirable from a sustainable agricultural point of view as well since leguminous plants have high nitrogen fixing capacity, adding to the fertility of the soil and improving biodiversity (Escobedo and Mojica 2021; Marti et al. 2022).

The size, shape, and structure of pulse grains vary according to species and geographical origins; however, almost all share some common features. In general, cotyledons are covered by a hull layer with a layer of gum, mucilage, and cementing materials like pectin present between them, which binds them together (Tiwari et al. 2010). Similar adhesive substances are also present between both the cotyledons. The gum layer majorly consists of non-starch polysaccharides, which can be water soluble and insoluble, as well as some proteinaceous substances. The pectin layer is majorly composed of 1,4-β-D-galacturonic acid, 1,3-β-D-xylogalacturonan, rhamnogalacturonan, and homogalacturonan. Apart from the size and shape of the pulse seeds, seed coat composition and composition of the adhesive layer largely influences the dehulling characteristics of pulses (Kurien and Parpia 1968; Vishwakarma et al. 2017).

Lentil milling is not yet scientifically developed like rice or wheat milling. The purpose of lentil milling is three folds. First, we need to produce either dehulled seed or split seeds for cooking purpose by removal of the seed coat. This step called dehulling can be achieved easily for certain variety of lentils but is rather complicated for example in pulses like

Lentils: Production, Processing Technologies, Products, and Nutritional Profile, First Edition.
Edited by Jasim Ahmed, Muhammad Siddiq, and Mark A. Uebersax.

pigeon pea. Therefore, pulse-specific dehulling technique development is very much necessary to have an efficient low wastage operation. Second purpose of lentil milling is to produce lentil or flour fractions for specific industrial or home cooking applications. Different sieving, micro milling, or air fractionation methods can be adopted to produce flour with different functional characteristics. Third goal of lentil milling is to achieve desired damaged starch characteristics by stress-induced milling operation. The properties of protein and fibers in lentil also gets altered due to this mechanical operation and changes the technical properties of the lentil flour. There is need to develop tests to understand and define lentil flour quality in the industry.

This chapter is broadly divided into three sections. The first section deals with milling, factors affecting pulse milling, traditional, and novel pretreatment strategies for dehulling and milling, and efficiency calculations. The second section deals with pulse flour, factors affecting flour particle size and functionality as well as the application of newer nonthermal technologies to modulate pulse flour functionality. The last section focuses on the fractionation process and individual pulse fractions, i.e., starch, proteins, and dietary fibers.

5.2 Milling Process

In general, milling can be defined as a size reduction mechanism, although the exact meaning of milling can change depending upon the grains and desired final produce. For instance, milling in rice processing is very different from milling in wheat, pulse, and maize processing. The aim of different types of pulse milling operations is based on the desired output, which is dehulling or decortication, splitting, and flour production. Dehulling is the removal of the seed coat or hull, splitting is the process of disintegrating the seed into two halves (cotyledons), and flour production is the process of grinding the whole lentil seeds or cotyledons to produce lentil flour fractions.

The milling of pulses is not new to mankind and has been there in different forms in different parts of the world. In rural areas in Asia, traditional stone mills are extensively used for dehulling and splitting, even till date. Other popular domestic methods include hand pounding, manually operated stone mills, and mortar and pestle. There are numerous problems associated with the traditional methods, like high labor requirements, low productivity, almost no scope of scalability, and high time consumption. Over the course of years, researchers have developed various machinery and equipment aimed at increasing process efficiency and minimizing processing time and losses. Modern day lentil processing mills are equipped with emery-coated rollers and carborundum rollers, which are known to have a high abrasive effect and produce excellent quality end products.

5.2.1 Dehulling

Removal of the outer layer of lentil seed known as the seed coat is known as dehulling. The outer layer on seeds is a protective biological barrier, which protects the grains from external hazards. The outer hull layer or seed coat of the pulses are rich in nondigestible polysaccharides, minerals, and antinutritional factors like trypsin inhibitors, phytic acids, and tannins, which can have a detrimental effect on human health, thus requires dehulling.

The dehulling process comprises of three major steps and several intermediary steps depending upon the local conditions and machinery requirements. The first step is conditioning by various pre milling operations, followed by drying and finally passing through the mills where actual removal of the hull takes place (Deshpande et al. 1982; Erskine et al. 1991; Goyal et al. 2008). The details of different dehulling processes are outlined in Section 5.2.5. Loss in protein and grain breakage are some disadvantages associated with the dehulling process. Wang (2008) reported that dehulling efficiency is positively correlated with starch content but negatively correlated with the protein and crude fiber content of raw seeds. Moreover, there can be grain losses in terms of brokens and powders which might affect the economic prospects. Though, the advantages of dehulling outweigh the disadvantages and dehulling (or dehulling with splitting) is widely practiced throughout pulse processing.

5.2.2 Hard to Mill and Easy to Mill Pulses

Ease of milling can differ with pulse species, cultivars, storage condition, and storage time. Pulses can be classified into two categories depending upon their milling characteristics. First category is the easy to mill pulses, in which the seed coat can be easily removed by single or double milling cycle. These constitute pulses like lentils, chickpeas, and dry peas. Second category is the hard to mill pulses, which require multiple passes in the milling machine. There are two major influencing factors responsible for the differences in milling behavior of pulses: the presence of soluble fibers like gums and mucilage and the physical interlocking mechanism in the adjacent surfaces at the junction of the seed coat and cotyledons. Hard to mill pulse varieties have strong cohesive forces between the hulls and cotyledons due to the presence of high concentrations of gums and mucilage. The soluble dietary fibers like pectin play a role in adhesion by cementing cell walls together in the middle lamella in the seeds of difficult-to-mill genotypes and hull and making the dehulling process difficult (Dabhi et al. 2019; Vishwakarma et al. 2017; Wood et al. 2022). Major hard to mill varieties are pigeon peas, black grams, green grams, mung bean, and urd bean. Whereas higher proportion of insoluble non starch polysaccharides like lignin or cellulosic components make the seed coat part slightly brittle and dehulling becomes easier for the easy to mill variety of pulses. Wood et al. (2016) showed that varietal differences observed in the ease of milling are linked to seed structure and the surface topography at the junctions between seed coat and cotyledons. Wood et al. (2016) also reported that differences in the milling performance of pulse seeds can be explained by the seed shape with diverse patterns of epicuticular wax at both the abaxial and adaxial cotyledon junctions of their seeds.

5.2.3 Splitting

Annually, about 67 MT of pulse is produced globally. India is the largest producer, with a contribution of about 17% of global production. About 75% of total pulse consumption in India is in the form of dhals. The intact dehulled lentil seed is called *Gota*, while the split lentil cotyledons are called Dhal, which is widely consumed in the Indian subcontinent and south east Asian countries. Splitting is the process of disintegrating the cotyledons into two halves which reduces its cooking time and increases palatability. It is one of the most common unit operations performed on lentils. Dhal yield can vary depending on a variety of factors like the adhesive force between cotyledons, moisture content of the seeds prior to milling, and the milling machine parameters.

5.2.4 Factors Affecting Milling Efficiency

Several factors like variety, species, agronomical and climatic variables, geometric attributes, mechanical properties of the grains and the machine operating parameters have a significant effect on pulse milling and composition of pulse thereafter. The major factors are outlined in this section.

5.2.4.1 Species and Variety

Genotypical variations and agronomical practice can alter the composition and nutritional profiles of the lentils. Different lentil varieties will have different hull thicknesses, geometries, seed hardness, chemical composition, gum, and mucilage composition. These varietal differences have a pronounced effects on the milling efficiency as they can affect the ability of grains to withstand abrasive forces. In addition, the amount of rainfall, irrigation and conservation practices, climatic conditions, and soil type influence the composition which in turn influences the dehulling efficiency of the seeds (Erskine et al. 1985, 1991). It is interesting to note that two varieties of the pulse grain can have opposite milling behavior as observed for lentils, faba beans, peas, and chickpeas by Jeganathan et al. (2022). Apart from lentils, Sefa-Dedeh and Stanley (1979) observed that the dehulling efficiency of eight different varieties of cowpeas was very different owing to the differences in hull thickness and microstructural properties. Likewise, Reichert et al. (1984) demonstrated that brown cowpeas had lower dehulling efficiency compared to black eyed cowpeas, which had a higher seed coat content.

5.2.4.2 Seed Shape and Size

Roundness, which is a measure of the shape of seeds, increases their machinability and milling efficiency compared to seeds with rough edges. The heat transfer within the grains is greatly dependent on the size parameters affecting the milling efficiency and composition. Regular shaped seeds have a more uniform temperature distribution profile and better dehulling. However, nonuniform seeds might have localized hot spots having a negative effect on the grain quality as well as milling performance (Sreerama et al. 2009). Similarly, larger seeds have higher dehulling efficiency than smaller seeds (Wang 2008). The grains having thicker seed coat has higher dehulling efficiency compared to grains with thinner seed coats. Uniformity in size distribution of seeds also plays a vital role in determining hulling efficiency. Reichert et al. (1984) examined the factors affecting the efficiency of various varieties of pulses and found that the dehulling efficiency is highly dependent on the anatomy of seeds. The physical position of the hilum plays a very important role in this regard. Some seeds like fava beans have the hilum at an end while other seeds like kidney beans have it in the reentrant curves, causing the difference in dehulling efficiency of these seeds.

5.2.4.3 Seed Moisture Content

Research has shown that the moisture content of seed is one of the most dominant factors affecting its milling performance. High moisture content can promote scouring losses and can considerably affect the hardness and shear resistance of the grain. But conditioning the lentil seeds to optimal moisture content level facilitates the separation of the hull or seed coat from the cotyledons during the dehulling process. However, at high moisture content hull might lose its brittleness making it difficult to remove. Whereas, low moisture content can make the grains highly brittle and result in high powder and brokens. Goyal et al. (2010) studied the effect of the moisture content of pigeon pea grains on its milling characteristics and observed that an increase in moisture content corresponds with an increase in dehulling efficiency. However, beyond 10% moisture content the yield starts to decrease due to reduced friction between the hulls and the rollers of the milling machine. Therefore, finding the optimum moisture content of pulse seeds before dehulling is very important for an efficient milling operation.

5.2.4.4 Mechanical Properties

Mechanical properties are very important from an engineering standpoint because it affects the functional and physical properties of the grain. Milling performance is also significantly affected by the mechanical properties like hardness, shear resistance, and brittleness of the seeds. As outlined in Section 5.2.4.3, moisture content of the grain has the most dominant effect on all the mechanical properties. Moreover, heat treatments like roasting and drying also have a significant effect on the mechanical properties of the grains (Mcdaniel et al. 2012). The major mechanical properties are as follows.

Hardness is a measure of the grain's resistance against plastic deformation upon application of an external shearing force. Abrasive hardness index (AHI) is a measure of grain hardness and is defined as the time to scratch 1% of seed kernel as fines (Reichert et al. 1984). The hardness of grain depends on its chemical composition, moisture content, and cultivar. Oomah et al. (2010) evaluated the hulling efficiency of Canadian dry beans using the tangential abrasive dehulling device (TADD) and observed hardness as one of the main factors affecting the hulling efficiency.

Brittleness is the property of a material to develop cracks when deforming stress is applied. Grains with low moisture content are more brittle compared to the ones with high moisture content. Highly brittle cotyledons can produce a larger amount of powder and brokens when subjected to milling and thus are undesirable. However, a brittle seed coat can be advantageous as it will facilitate dehulling.

5.2.5 Milling Processes/Pretreatments

Lentils are processed in a variety of ways throughout the globe depending on the pulse variety, geography, tradition, and available manpower and machinery. Traditionally, pulse milling pretreatments are categorized as wet pretreatments and dry pretreatments. Later, several researchers studied other modern milling pretreatments like chemical, enzymatic, ultrasonication, microwave, etc. Milling pretreatment assumes critical importance since the pretreatment greatly influences the lentil quality and compositional characteristics. Green lentils pretreatment with moisture and mechanical scouring exhibited very different particle sizes and flour characteristics post milling (Choo et al. 2022). Sections 5.2.5.1–5.2.5.6 outlines the mechanism and process of different pretreatments used prior to pulse milling operation in the industry.

5.2.5.1 Wet Milling

As shown in Figure 5.1, in this method, first the pulse seeds are cleaned of all extraneous matters. The next step is grading that is optional and is not practiced in domestic settings. Following this, the seeds are soaked in water for about 4–12 hours depending upon the processor's preference and traditional knowledge. The seeds are then mixed with about 2–3% red earth and sun dried for about two to four days. During the nights, the seeds are heaped for conditioning. Once optimum dried, the seeds are then milled in emery coated mills. The addition of red earth is an optional step, and the motive behind is to increase the drying rate and loosen the husk. The red earth particles also help in generating high abrasion and increase dehulling efficiency. Seed soaking is an important step in wet milling of the pulse. After soaking, the seeds hydrate and swell. Subsequently, upon drying the seeds shrink disproportionately compared to the hull creating a cavity between the hull and cotyledons, aiding in dehulling and splitting of the pulses. In wet milling, the yield is superior compared to dry milling, but the quality of dhal is inferior. The major disadvantages of wet milling include high labor requirements, weather dependency, and high soaking time which can cause undesirable changes in the dhal quality.

5.2.5.2 Dry Milling

Figure 5.2 depicts the dry milling process and similar to the previous one, dry milling also starts with cleaning and grading of the seeds. These seeds are then subjected to an operation called pitting, in which the seed surface is scratched. Following pitting, edible oil is added to the seeds at a rate of 0.1–1% (w.b.). The amount of oil to be added depends on the genotypic variant and weather conditions. The oiled seeds are then sun dried for next two to four days. In nights, the seeds are heaped and again spread in sun the next day. Lastly, about 2–5% of water is sprinkled on the seeds and it is left for tempering overnight. The seeds are then dried to desired moisture content, usually 10–12% (w.b.) prior to milling. Industrialized countries replace sun drying with mechanical dryers and manual mixing with screw conveyors. However, the overall process flow remains the same. Easy to mill pulses can be properly milled in a single pass; however, hard to mill pulses may require multiple passes for desirable yield. Pitting increases the surface area of contact and accelerates the absorption of water and oil into the seed. Many edible oils like linseed, mustard, palm, sunflower, sesame, and soyabean oil have been used for dry milling operations. Oils penetrate into the hull and reach the cotyledon-hull interface. It causes hydrolysis of gum and mucilage present in the interface and loosens the bond, making dehulling easy. The dhal produced by this method cooks faster and has better quality. However, this method may produce a high amount of brokens and powder.

5.2.5.3 Chemical Pretreatment

Numerous chemicals like $NaHCO_3$, Na_2CO_3, NaOH, acetic acid, and urea (CH_4N_2O) have been studied for their effects on milling efficiency. Among all these chemicals, $NaHCO_3$ is by far the most widely used chemical owing to its low cost and non-toxic nature. The chemicals hydrolyze the gums and mucilage present between cotyledons and the hull

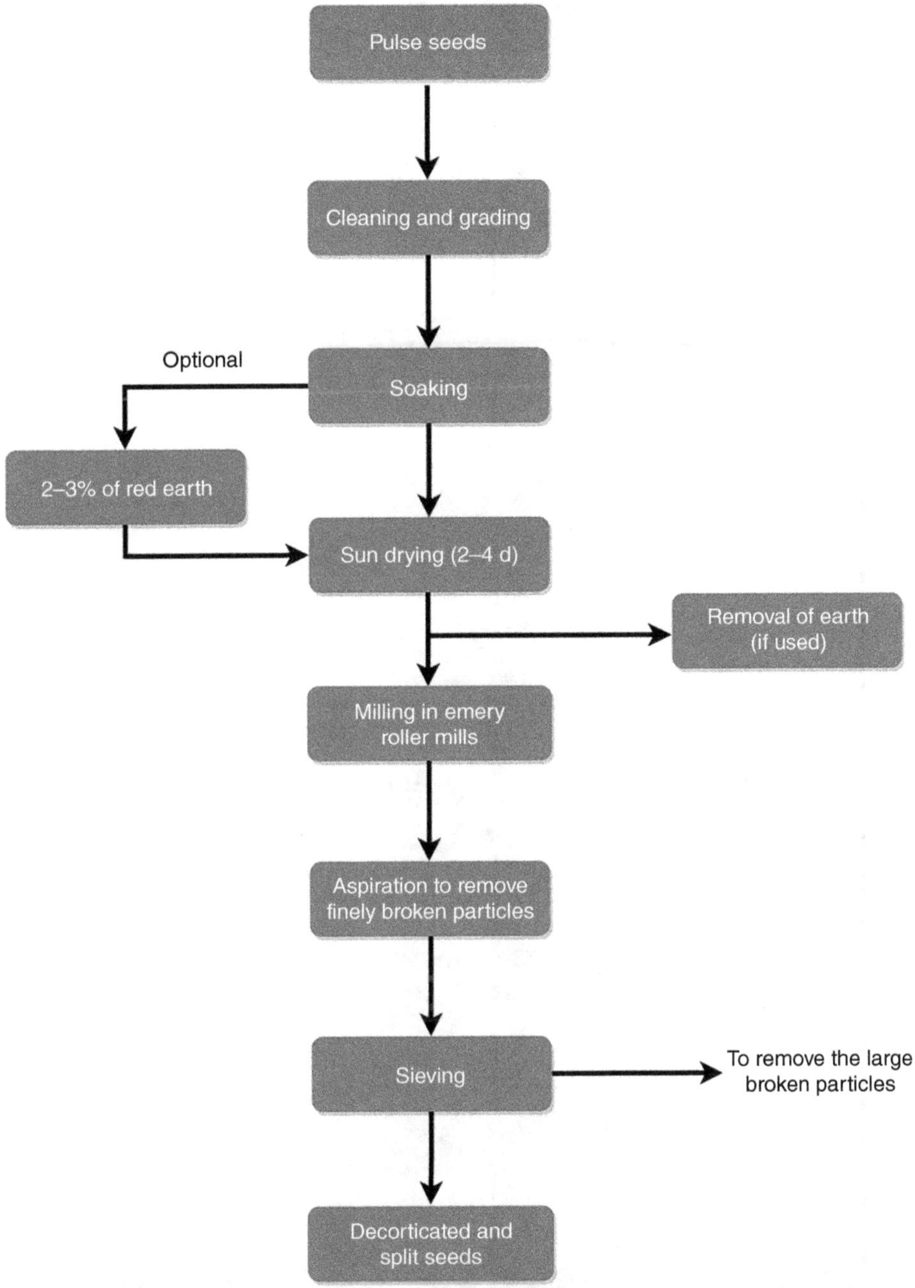

Figure 5.1 Schematic representation of wet fractionation process. *Source:* Adapted and modified from Wood and Malcolmson (2011).

resulting in the loosening of the hulls. Pitting operation has a positive effect on milling efficiency, as it increases the surface area of grain and promotes the absorption of chemicals in between the cotyledon and hulls of the grain. Saxena et al. (1988) observed that $NaHCO_3$ pretreatment reduced the gum and pectinase content of pigeon peas and increases

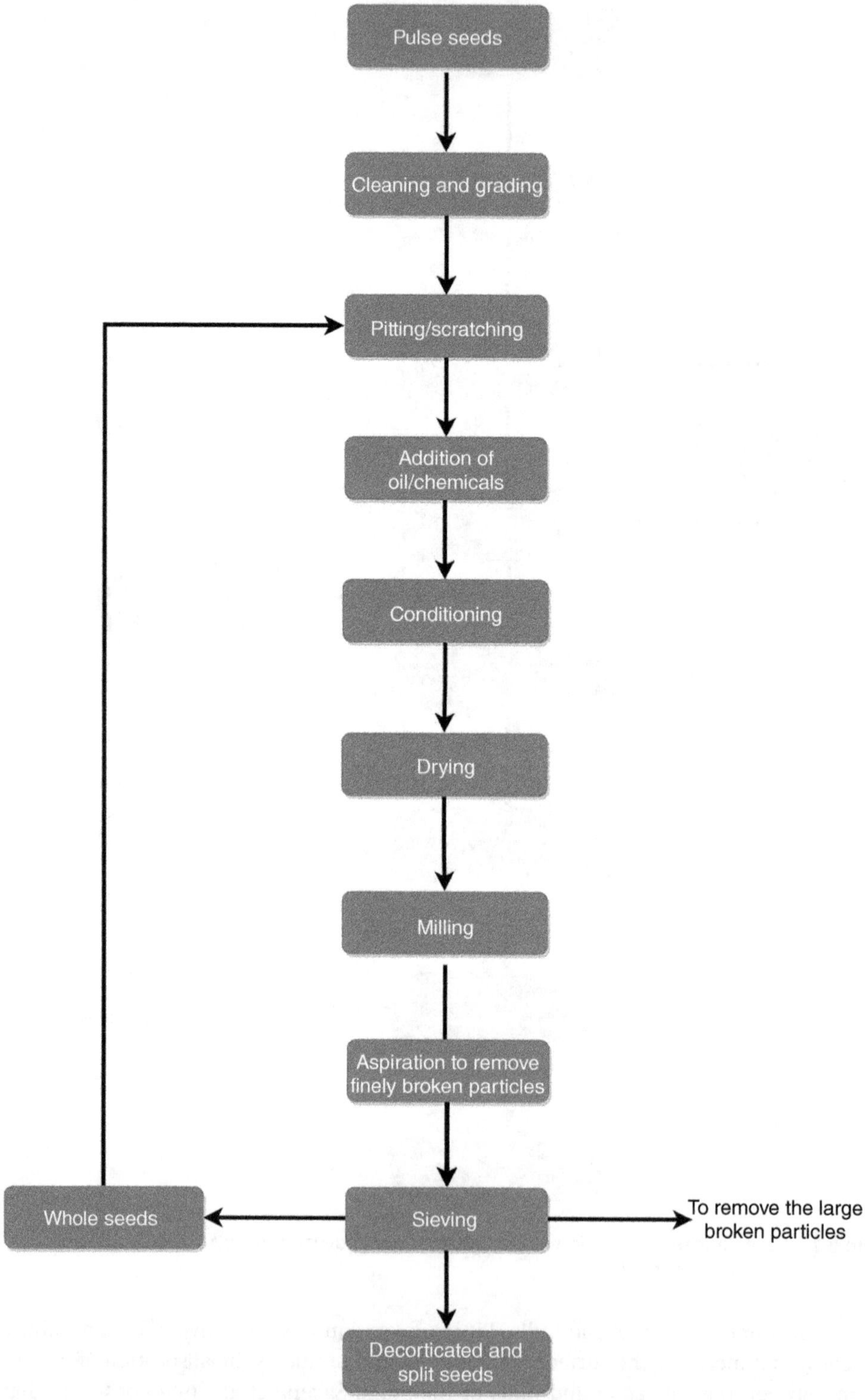

Figure 5.2 Schematic representation of dry fractionation process. *Source:* Adapted and modified from Wood and Malcolmson (2011).

the dehulling efficiency of the grains. Phirke et al. (1995) reported urea pretreatment causes a reduction in the shear force requirement for dehulling of pigeon peas. Other hard to mill pulse variety like black gram also showed a significant increase in hulling efficiency following chemical pretreatment (Tiwari et al. 2008b). Special attention and care should be taken while selecting the chemicals and only those chemicals should be used which do not have any detrimental effect on human health as well as environmental safety.

5.2.5.4 Enzymatic Pretreatment

Enzymes are proteins which accelerate metabolic and chemical reactions. Enzymatic pretreatment has a marked effect on the dehulling efficiency of pulses. It enhances the functional and nutritional properties of pulses by reducing the antinutritional factors present in it. Enzymatic pretreatment causes degradation of nonstarch polysaccharides in the seed coat layers leading to softening of the seed coat attachment to cotyledon and an overall increase in the geometric dimension of the seed. Additionally, it disrupts the gum and mucilage layer present in the hull–cotyledon interface, thus facilitating efficient dehulling (Murumkar et al. 2016; Sreerama et al. 2009). The pulse grains are soaked in an aqueous solution of enzymes and incubated at different times and treatment combinations. Subsequently, the grains are dried and passed through the mills for dehulling. The concentration of enzymes plays a crucial role in dehulling efficiency. The optimum concentration has to be determined in order to mitigate the adverse effects of the enzyme on the treated seeds. The adverse effects include a reduction in dehulling efficiency and seed hardening. Similarly, incubation time is also an important factor that affects the dehulling efficiency and dhal yield. Too high incubation temperature or time might cause denaturation of the enzymes thus reducing the process yield. Some commonly used enzymes for the treatment on hard to mill cultivars are xylanase, pectinase, cellulase, and protease. Dabhi et al. (2019) demonstrated that enzymatic pretreatment of pigeon peas increased the protein content and reduced the energy requirements when compared to conventionally oil pretreated grains. Wood et al. (2022) concluded from their chickpea enzymatic pretreatments prior to milling that three enzymes (*endo*-polygalacturonase, α-galactosidase, and cellulase) have immense scope for industrial milling applications and could help pulse millers to achieve high quality dhal with minimal breakage.

5.2.5.5 Hydrothermal Pretreatment

Pitting and oil pretreatment are the two widely prevalent steps used for the dry milling of pulses. These steps not only increase the processing time but also the overall production cost. Wet milling of pulses also suffers from a similar disadvantage where soaking of pulses for long hours can produce fermented flavor in the grains making it unacceptable in the market. Hydrothermal pretreatment is a novel pulse milling pretreatment, which aims at the elimination of these steps so that the oil can be used for edible purposes and the overall efficiency and cost-effectiveness of the operation can improve (Veena et al. 1995; Tiwari et al. 2008a). In hydrothermal treatment, first, the grains are kept in contact with high pressure steam for a specified period of time, followed by convective drying in hot air (Tiwari et al. 2010). Joyner and Yadav (2015a) observed that hydrothermally pretreated black grams had low losses and significantly higher dehulling efficiency. After steaming, the grains absorb water and swell. After drying the cotyledons shrink and the hull does not create a gap between the hull and cotyledon and can be the reason for increased hulling efficiency.

5.2.5.6 Others Pretreatments

Research has shown that the application of novel food processing operations like gamma radiation, ultrasound, and microwave pretreatment can significantly reduce the soaking time and improve the milling efficiency for hard to mill pulse varieties. Sunil et al. (2018) examined the effect of ultrasound pretreatment on the milling efficiency of black gram seeds and found that the pretreatment causes denaturation of the gums and proteins present between the hulls and cotyledons. This resulted in an increase in the milling efficiency and a reduction in soaking time, hulling loss, and cooking times caused due to microstructural changes triggered by acoustic cavitation. Benmeziane-Derradji et al. (2020) explored the potential of roasting pretreatment Algerian lentils and suggested that the flavonol and flavonoids increased for roasted samples. Also, a remarkable increase in the functional properties like emulsification and foaming properties was recorded. Although, the phenolic content and antioxidant potential had diminutive effects. This study highlight the need for optimization, which if conducted properly can provide excellent nutritional profiles while enhancing the functional properties. Recently, Nickhil et al. (2021) demonstrated that the gaseous ozone pretreatment of chickpeas increased the hulling efficiency and reduced the cooking time. Ozone reacts with the non-starchy polysaccharides and silica present in the hulls, resulting in the development of fissures and loosening of the hull. Similar results were reported for microwave pretreatment of a black gram by Joyner and Yadav (2015b), who observed that an increase in the microwave dosage decreased the time requirement for hulling. Although very high microwave dosage can produce brokens and facilitate browning in seeds due to sudden thermal shock and erratic microwave distribution. On the contrary, pretreatment of pigeon pea with gamma radiation did not show any significant effect on the hulling efficiency. Since gamma radiation is a cold treatment and does not generate thermal energy, it could not facilitate denaturation and hydrolysis in the gums and mucilage layer and hence was largely ineffective (Pawar et al. 2021). Kumar et al. (2022) demonstrated that infrared heat treatment prior to dehulling helps in lowering down the husk adhesion strength, thus improving the dehulling efficiency. They also demonstrated that infrared heat treatment of pulses equilibrated at suitable moisture level also improves the ease of splitting of the pulses during milling. Vidal-Valverde et al. (2002) investigated the efficiency of short time soaking (STS) process for red lentils and observed very promising results as the flour obtained from STS-treated lentil flour exhibited significantly lower antinutritional factor and very high vitamin contractions. Further development in this direction is needed to develop a sustainable and green pretreatment process.

5.2.6 Milling Efficiency and Calculation

Determination of the milling parameters is of paramount importance for the development of milling equipment and evaluating process efficiencies. Various researchers have devised several ways of calculating different milling attributes.

Milling efficiency I: is given by the following equation

$$E = \mu_{\mathrm{d}} \times \mu_{\mathrm{kw}} \times 100$$

where μ_d is the coefficient of dehulling and μ_{kw} is the coefficient of kernel wholeness, respectively.

- *Coefficient of dehulling (μ_d):*

$$\mu_d = 1 - \left(\frac{M_{uh}}{M_i}\right)$$

  M_{uh} is mass of dehulled pulse grains and M_i is the initial mass of the grains before dehulling.
- *Coefficient of kernel wholeness (μ_{kw}):*

$$\mu_{kw} = \frac{M_d}{M_d + M_{br} + M_p}$$

  M_d is mass of intact dehulled grain kernel, M_{br} is the mass of brokens, and M_p is the mass of powder produced during the milling operation.
- *Dehulling efficiency (μ):*
  The formula for milling efficiency does not consider pulse splitting, which occurs simultaneously with decortication. Reichert et al. (1984) proposed the following equation for the calculation of dehulling index taking the amount of split seeds produced during decortication into account. The equation is as follows

$$\mu = \left(1 - \frac{M_{uh}}{M_i}\right) \times \left(\frac{M_f}{M_f + M_{br} + M_p}\right)$$

$$M_f = M_d + M_s$$

  M_f is mass of final grain output after milling and M_s is mass of dehulled split pulses after milling.
  Black et al. (1998) suggested the following equation for the calculation of the dehulling efficiency

$$\mu = \frac{M_i - (M_{uh} + M_h + M_{br} + M_p)}{M_i}$$

  M_h is the mass of the hull removed after milling.
- *Dehulling Index (D.I.):* The value for D.I. ranges between maximum +1 and minimum −1. The maximum value +1 indicates complete removal of hull with no broken or powder. Minimum value −1 indicates complete disintegration of grain and no removal of hulls.

$$\text{D.I.} = \frac{(M_f + M_h) - (M_{uh} + M_{br} + M_p)}{M_i}$$

- *Dhal Yield* (Y_d): It is the ratio of dehulled pulse, including both whole and split grains, to the total mass of dehulled pulses. Higher dhal yield is desirable as it indicates a better milling performance of the machine.

$$Y_{\mathrm{d}} = \frac{M_{\mathrm{dp}}}{M_{\mathrm{i}}}$$

M_{dp} is the mass of dehulled pulse.

- *Dehulling Loss* (L_d): It is the ratio of broken and powdered pulse fractions to the total mass of dehulled pulses. Low value of dehulling loss is desirable.

$$L_{\mathrm{d}} = \frac{M_{\mathrm{p}} + M_{\mathrm{br}}}{\mathrm{M}_i}$$

5.3 Pulse Flour

Pulse flour is of particular interest because of its superior nutritive value, relatively less glycemic index and gut microbiota modulating capabilities (Borsuk et al. 2012; Xiong et al. 2022). These are not only a rich source of proteins but also contain high quantities of dietary fibers, minerals, and polyphenolic compounds. In a recent study, Chizhikova et al. (2017) observed that addition of red lentils into wheat flour not only increased the protein and dietary fiber contents but also the mineral contents. Researchers recommended fortifying cereal-based snacks with pulse flour to improve their bio-functional and therapeutic value. From an economic perspective, pulse flour has the versatility to be incorporated into a plethora of food products and thus has the high potential to increase consumption (Marti et al. 2022). These reports make a strong case for pulse flour utilization in the food industry.

In contrast to wheat flour, the term "Pulse flour" lacks universal definition and has varied region-specific interpretations. Based on literary evidences, an intuitive definition of pulse flour can be developed as a fine, medium, or coarse powder obtained after milling/grinding the pulse grains. Flour can be obtained from both whole and dehulled grains (Scanlon et al. 2018; Thakur et al. 2019; Wood and Malcolmson 2011). As highlighted in Section 5.1, the major factors affecting the particle size of flours are species, agronomic practices, geographical parameters, pretreatment, type of milling machinery, and number of passes. Among these, the latter three are of more interest to the food processors. It is well known that machineries like needle mills and hammer mill produce smaller particle sizes while attrition mills and stone chakkis produce coarser particles. Similarly, higher number of passes also have negligible effects on the particle size. Particle size also greatly depends on the milling pretreatments; for instance, flour obtained from dehulled lentils might have a coarser texture and high water absorption (Sopiwnyk et al. 2020; Wood and Malcolmson 2011).

5.3.1 Functional Properties of Pulse Flour

Apart from chemical composition, particle size distribution is the major factor influencing the flour functionality. Since the term pulse flour is not well defined in terms of particle sizes, comparison between the functional properties of different pulse flour becomes uncertain and dubious. However, the individual effect of lentil flour can still be studied with application of proper particle size criterion (Thakur et al. 2019). The importance of particle size of lentil flours was demonstrated by Ahmed et al. (2016), who reported that the difference in flour particle size manifested in significantly different physical, chemical, as well as thermal properties. Borsuk et al. (2012) investigated the effect of incorporation of pulse in wheat flour for pita bread development and observed that at high level of substitution, pulse flour decreased the breakability of the bread and induced stretchiness. The findings revealed the critical role of glutens present in wheat flour for crust texture development in products like breads and pizzas. Glutens provide the structural backbone for texturization, thus an absence of glutens in pulse flour poses a critical challenge in product development. It is well known that environmental and genotypic variations influence the mill ability of lentils; however, Lagassé et al. (2022) showed that these variations also influence the flour characteristics and most importantly the product quality. It was observed that the bread incorporated with different lentil varieties had significantly different baking performances. Similarly, Ek et al. (2021) studied the effects of lentil varieties on final extrudate characteristics and found that the physical properties as well as the specific mechanical energy requirement varied across different varieties. The varied compositional difference is the primary causes of these observations.

Despite being a good source of protein, dietary fibers, and bioactive substances, the consumer acceptability of pulse flour is constrained by the grassy flavor produced due to the presence of various ketone and aldehyde groups (Shi et al. 2021). To increase the consumer preference, strategies for reducing off flavor are imperative. Ma et al. (2011) found that thermal pretreatment of lentils caused significant reduction in the antinutritional factors like trypsin inhibitor activity while having positive effect on lentil flour functionality. Young et al. (2020) studied the effect of thermal pretreatment, roasting, on yellow pea and reported a significant reduction in undesirable aroma and aftertaste. Moreover, roasting pretreatment partially gelatinized the starch molecules and caused dietary fibers to swell resulting in increased water absorption capacity of the yellow bean flour. The bread baked from this flour retained higher moisture, had lower crumb firmness, and maintained freshness for a longer time period. Stone et al. (2021) suggested that wet roasting affects pulses flour functionality differently, i.e., different species of pulses may exhibit differential effects even after being subjected to the same pretreatment. Authors outlined that generalized pretreatment may not suit all the pulses, and to obtain an optimum response, processing conditions have to be pulse specific.

Owing to the presence of a high amount of protein and dietary fibers, the good gelling property is observed for lentil flour. Thus, rather than conventional products, pulse flour can also be used for the development of unconventional and new products like cheese from fava bean and yellow bean flour (Ferawati et al. 2021a,b) and yogurt/tofu analogues from fava bean flour (Jiang et al. 2020). Enriching cereal flours with lentil flours or developing blended products from cereal-lentil flour or lentil flour-based new products have several nutritional advantages like better amino acid profile and higher dietary fiber intake. But the major challenges remain in creating the product with the right texture and microstructure. A recent study outlined the potential of lentil flour to be used for preparation of electrospun nanofibers. It was found that at optimal concentrations, lentil flour can positively modulate the rheological and flow characteristics of the solution and impart desired mechanical and chemical properties (Tam et al. 2017). Readers are advised to read the work of Monnet et al. (2019a,b) to understand that pulses like pea have a higher chance for fracture but less tendency to produce damaged starch than wheat flour. Therefore, there is a need to understand the functional properties of lentil flour to develop products with targeted texture (Monnet et al. 2019a).

5.3.2 Effect of Novel Nonthermal Processing Technologies on Pulse Flour

Recent times have witnessed an increased interest in nonthermal technologies because of their application in modulating the food functionality with low thermal stress to the food matrix. Miano et al. (2016) demonstrated that high-intensity ultrasound treatment significantly increased the apparent viscosity of the mung bean flour pastes. Ultrasound waves alter the protein size and aggregation patterns and increase the protein solubility. Increased solubility of protein augments the potential binding sites and imparts viscosity to the product. It was interesting to note that sonication did not have much effect on the mung bean starch. For fava bean flour, Martínez-Velasco et al. (2018) observed similar results. Ultrasound pretreatment enhanced the emulsification and foaming properties of fava bean protein due to reduced aggregate size and increased solubility, ultimately leading to lower interfacial tension and lower zeta potential.

Likewise, the application of high hydrostatic pressure promotes structural changes in the pulse flours. The degree of change in a structure depends on the pressure level and holding time. Higher pressure promotes the disintegration of cementing materials like pectin and cellulose and reduces the insoluble dietary fiber content. Similarly, high pressure and time combination results in an unfolding of the starch granules and increased hydrolytic degradation, resulting in reduced resistant starch content (Lin et al. 2020). Ahmed et al. (2009) found that high-pressure processing can reduce the retrograding tendency of lentil flour slurry while having insignificant effect in the amide bands. After gelling the processed lentil slurry had significantly higher viscoelastic (G' and G'') properties compared to the control samples suggesting denaturation of the protein and melting of the crystalline region in starch granules. The effect of other technologies like pulse electric field (PEF), ozone processing, and cold plasma on pulse flour presents many exciting research opportunities and not yet explored much.

Just like lentil flour, individual fractions (for example fine fraction and coarse fraction, starch rich fraction, protein rich fraction, fiber rich fraction, etc.) obtained from lentils too are finding application in the food industry, especially with an increase in demand of plant-based products. Pulse fractionation methods and individual fractions including pulse starch, proteins, and dietary fibers are discussed hereafter in Sections 5.4 and 5.5.

5.4 Fractionation Methods

Lentils are consumed as either whole grains or split dals all around the globe. However, recent times have seen a change in the production and consumption pattern of food products where lentil flour with different functionalities (starch, fiber, or protein rich fractions) is widely used for product development. Therefore, recent times have seen increased interest toward pulse fractionation for valuable ingredient manufacturing. Fractionation process is the separation of ground lentil flour into different macromolecular fractions, i.e., starch, protein, and dietary fibers. Different macromolecule rich fractions are used in the industry for specific purposes for enhancing the quality, taste, and texture of the food products. The special functions of all these fractions are elucidated in Section 5.5. Milling of pulse into powder is a prerequisite step to increase the effective surface area and facilitate the separation of all the different components. Particle size of pulse powder is also an important consideration as it greatly influences the separation efficiency. In general, the starch particle size ranges from 15 to 40 μm and protein particle size is approximately 5 μm. The final target particle size after milling should be in appropriate range as too fine mill powder can negatively affect the protein separation. Finer powder will have large quantity of damaged starch molecules, which will be almost same in size with protein molecules, making separation difficult. Too coarse powder might not disentangle the protein starch matrix and contribute to inefficient separation (Pelgrom et al. 2013; Schutyser and van der Goot 2011).

Pulses are targeted for pulse protein extraction in the industry. In comparison to pulse proteins, pulse starch and fiber have less commercial importance and rather neglected. Interestingly lentils have much higher starch than total protein (15–30%) along with other components like lipid and fibers (Figure 5.3). Protein bodies are embedded in the starch matrix, thus needs different sequential separation processes for extraction, purification, and concentration of the protein fractions. There are broadly two main ways the pulse fractionation for protein extraction is achieved in the industry: dry and wet fractionation. Sieving, air classification, and electrostatic separation have become popular techniques in the dry fractionation method in pulse fractionation process. Whereas pH shift leading to isoelectric precipitation and membrane separation are the common methods of wet extraction for protein recovery from the pulses (Figure 5.4). Dry fractionation process

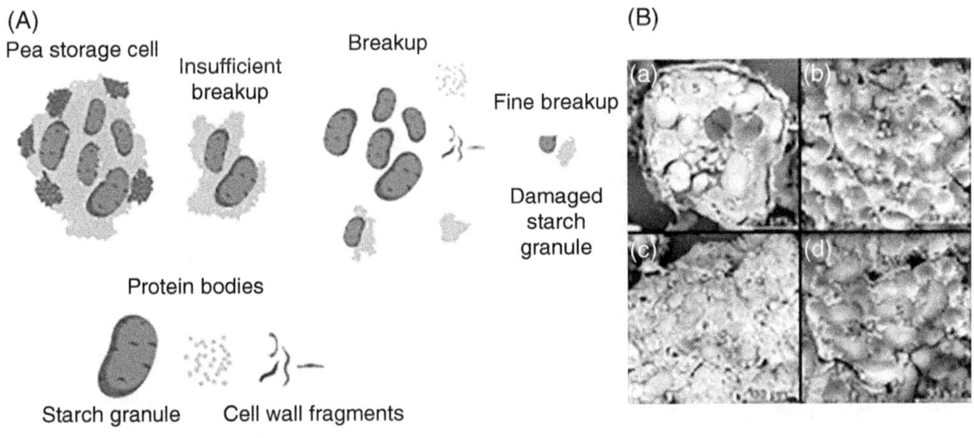

Figure 5.3 (A) Schematic drawing of cells of pea and the fragments after milling (representative figure, not in scale). *Source*: Adapted from Pelgrom et al. (2015). (B) Morphology of pea (a), bean (b), chickpea (c), and lentil (d). Starch granules, protein bodies, and cell wall can be distinguished. *Source:* Pelgrom et al. (2015), with permission of Springer Nature.

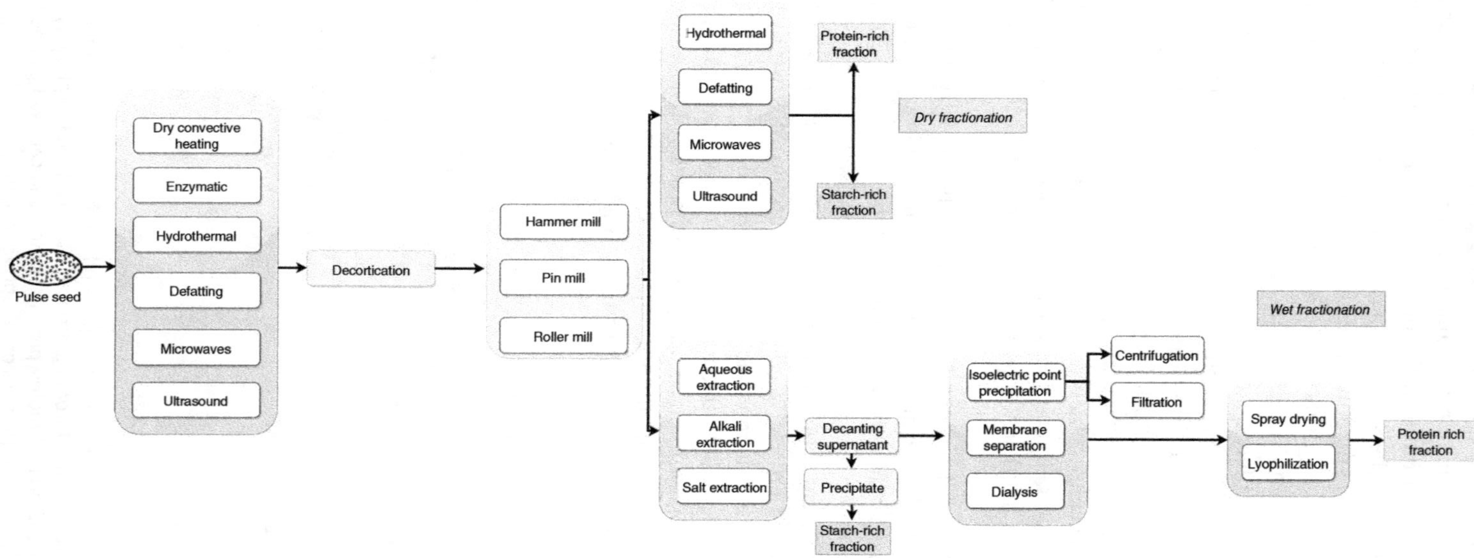

Figure 5.4 The dry and wet fractionation methods for pulse–protein extraction. *Source:* Adapted from Allotey et al. (2022).

leads to ingredient fractions with excellent techno-functional properties, but having very low purities in comparison to wet fractionation. However, dry fractionation process has low energy and water footprint due to the absence of a solvent extraction step and a drying step, and considered as a more sustainable process. A pictorial diagram is presented in Figure 5.4 for the dry and wet fractionation process for pulse protein extraction with starch byproduct.

The two primary methods for fractionation of pulses, dry and wet fractionation, are discussed below:

5.4.1 Dry Fractionation

In dry fractionation, pulse grains are subjected to milling followed by separation into different fractions. The major advantage of this technique is that it limits the usage of water and chemicals and withholds the functional attributes of the individual fractions as opposed to the wet fractionation. Dry fractionation is desirable when the complete separation of different components is not intended, as suggested by (Schutyser and van der Goot 2011). The fractionation starts with milling the pulse into desired particle size based on the desired end product (Ren et al. 2021; Thakur et al. 2019). The mills commonly used for this purpose is the hammer mills, pin mills, roller mills, or the jet mills (Pelgrom et al. 2013). After milling, the method of separation can be sieving, air classification, or electrostatic separation (Figure 5.5). In sieving, the separation is based on the difference in size of the granules, in which starch is the heavier fraction and protein is the lighter one. For production of enriched powder, this method may be suitable; however, for production of ultrapure fractions, air classification is employed. In air classification, the separation is based on the difference in aerodynamic properties of the starch, protein, and fibers (Fernando 2021; Ren et al. 2021). After milling at 4000 rpm, Pelgrom et al. (2013) reported a protein content of as high as 77% for from fractionation of dry pea flour and demonstrated that the machine parameters have greater effect on separation process than the flour composition.

When the difference in particle size and density is very small, air classification doesn't perform well. A novel fractionation strategy, electrostatic separation, can be useful in such cases. This technique is based on the surface charge difference of the various macromolecules (Wang et al. 2014a, 2016). For triboelectric separation of protein and starch-rich fractions from navy beans, Tabtabaei et al. (2016) observed that protein molecules get highly charged due to the presence of several highly ionizable groups in their amino acid side chains and thus this method is suitable for separation of protein-rich fraction. Triboelectric separation may not perform well for starch separation alone (Fernando 2021).

5.4.2 Wet Fractionation

Wet fraction is a method where the separation of components takes place in an aqueous medium. Generally, pulse seeds are soaked in the aqueous media, usually water. Some studies have also reported use of chemicals like sodium hydroxide, sodium bisulfite, and sodium metabisulfite, which aids in the separation of proteins from starch by dissociating the disulfide bonds present in protein subunits (Ren et al. 2021). Soaking is an important step because it aids the loosening of hull around cotyledons, helps in easy dissolution of

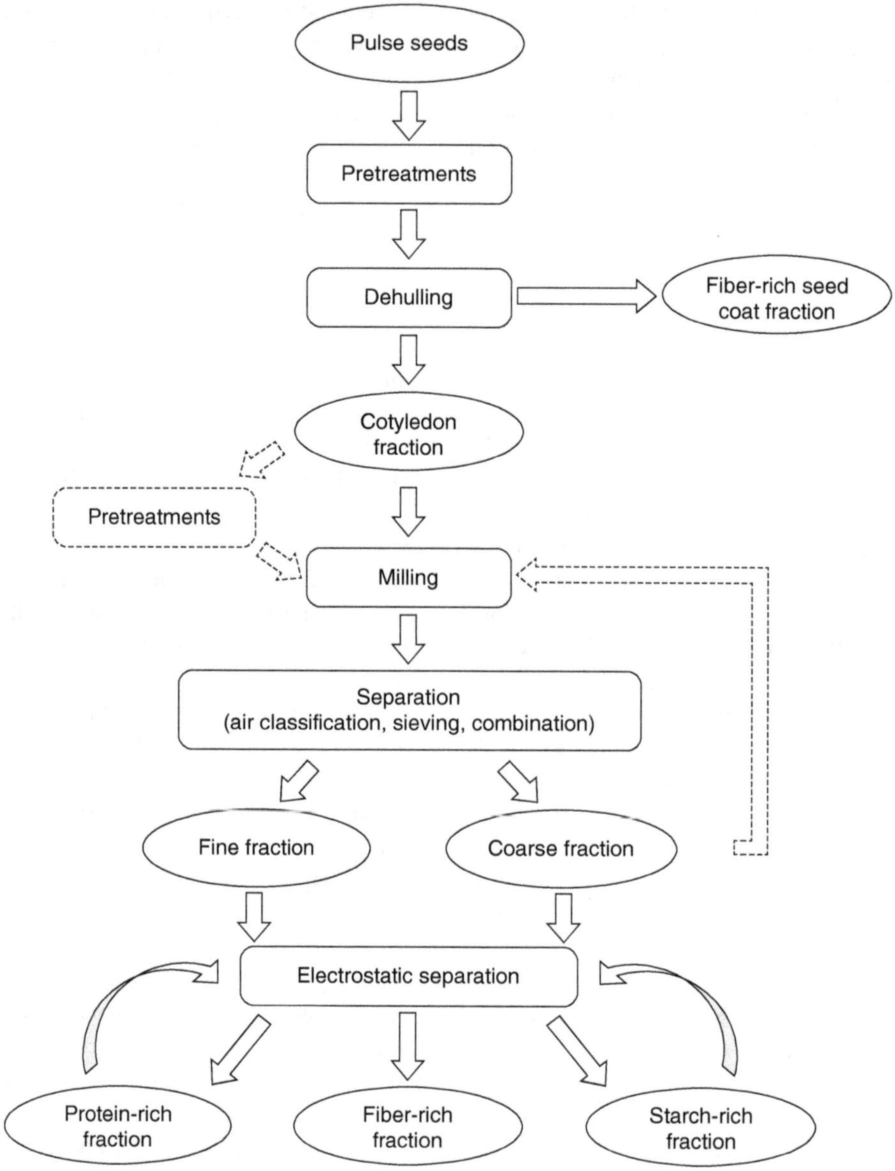

Figure 5.5 Dry fractionation process of pulses.

starch in the media, and partially solubilizes the protein fractions. After soaking, pulses are ground to form a slurry and the slurry is then allowed to stand causing sedimentation. Starch being the heaviest component starts settling at the bottom of the container and can be separated. For high purity, the slurry can be centrifuged (Figure 5.6). For protein isolation, precipitation at the isoelectric point is the widely employed method, as discussed

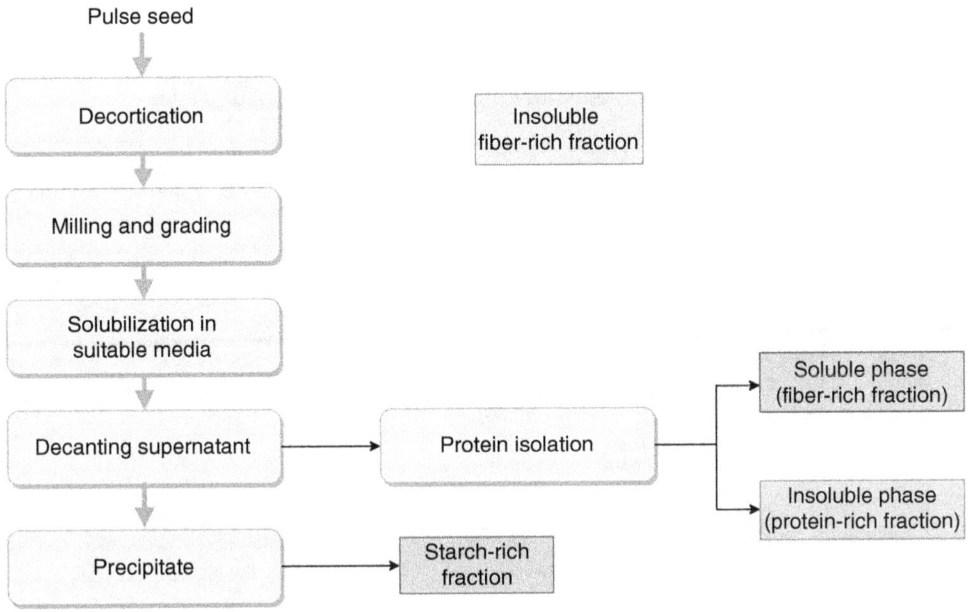

Figure 5.6 Simplified biorefinery process flow diagram for simultaneous recovery of fiber, protein, and starch.

earlier. In this method, the slurry is firstly adjusted at an alkaline pH of 8.5–10, thus solubilizing the protein fraction. The supernatant is transferred into a separate container and the pH is readjusted to 4–4.5. At this pH, the protein starts to precipitate and the supernatant is centrifuged to obtain the protein fraction (Espinosa-Ramírez and Serna-Saldívar 2019; Sánchez-Vioque et al. 1999). The dietary fibers are isolated using similar wet fractionation approaches. Although superior to dry fractionation, high capital requirements and consumption of a large amount of water during processing put severe constraints on the usage of this technique.

An important point to note here is that the usage of dry and wet fractionation methods is not mutually exclusive and thus both methods can be used in combination to produce pulse fractions with reduced impurities, processing cost, and minimal effort. According to Li et al. (2019), pea and fava bean starch produced after washing of the air-classified enriched starch fraction has significantly lesser damaged starch content. Similar hybrid techniques for fractionation have the potential to reduce the processing cost and efforts even further.

A new hybrid process has been developed, which combines dry and aqueous fractionation processes, and has been found to give higher protein yield with much less water and energy requirement compared to the wet separation process only (Pelgrom et al. 2015). In the hybrid process, pulse flour suspension is manufactured at lower dilution without any acid or alkali addition compared to the traditional wet fractionation process. The energy and water use of this combined process are graphically represented per kilogram product in the Figure 5.7.

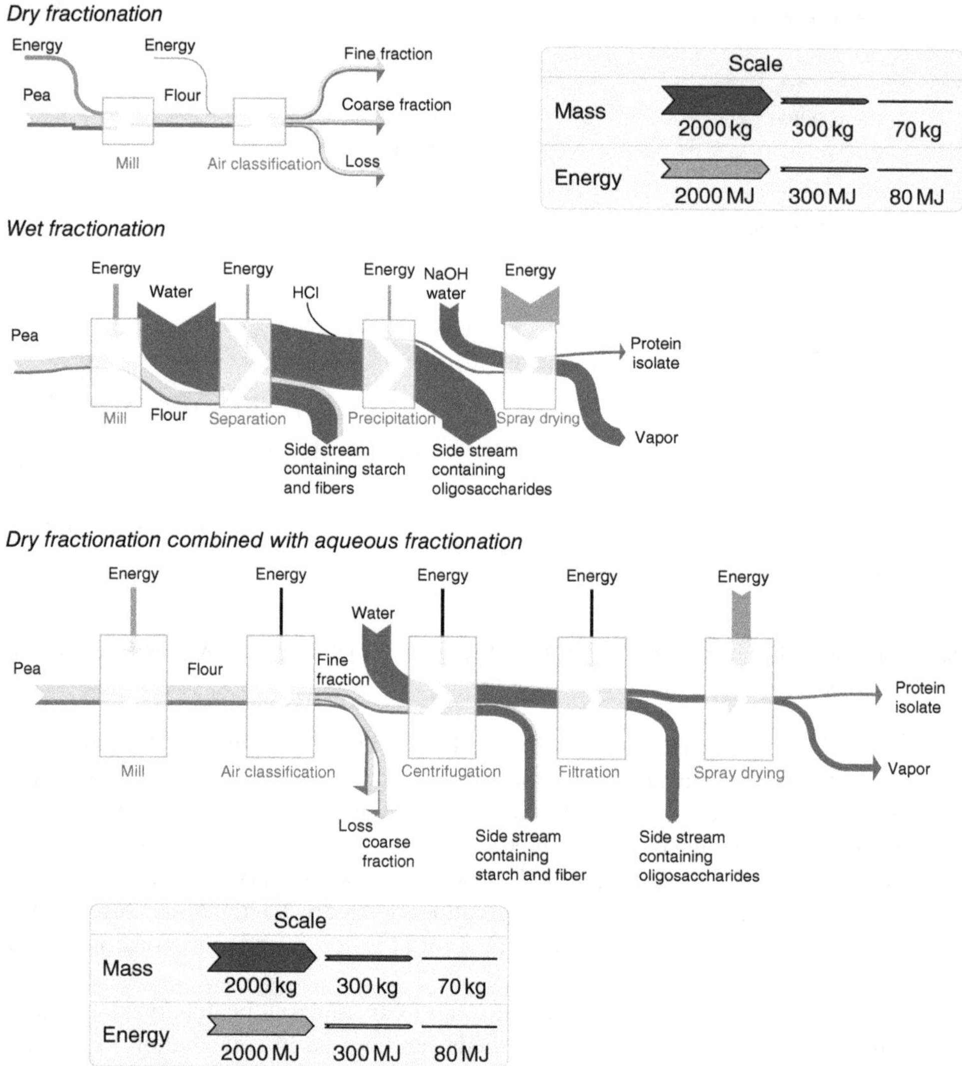

Figure 5.7 Sankey diagrams of dry fractionation, wet fractionation and hybrid fractionation process. Process streams containing water, protein, starch, fiber, oil and ash, and energy are depicted. For simplicity reasons mass flows and energy are given in the same diagram. *Source:* Adapted from Schutyser et al. (2015).

5.5 Pulse Fractions

Pulses including lentils are a rich source of dietary macronutrients which includes carbohydrates (starch and fibers), proteins and micronutrients, i.e., vitamin and minerals. Moreover, a wide range of secondary metabolites (flavonoids, alkaloids, polyphenols, tannins, etc.) that have high nutraceutical properties make pulse an excellent substitute. The main lentil fractions, i.e., protein, starch, and fiber, are discussed in the following paragraphs.

5.5.1 Protein

Lentils contain a significant amount of protein (about 21–31%) comparable to other commercially available legumes like mung bean, peas, and soybeans. Lentil proteins are rich in globulin which amounts for about 70% of total protein content, followed by albumin (16%), glutelin (11%) and prolamin (3%). Among globulins, legumin (11S) and vicilin (7S) like protein fractions are present in abundance in lentil proteins (Shrestha et al. 2022). The concentration of antinutritional factors like trypsin inhibitors is very low in lentil seeds. In a comparative study between chickpea, fava bean, and Lentil proteins, it was observed that the lentil protein had highest emulsion capacity (Joehnke et al. 2021; Karaca et al. 2011). In this regard, the lentil proteins are largely underutilized.

Several studies have also suggested that the method of protein extraction has significant effect on its functionality. Karaca et al. (2011) extracted lentil proteins using salt extraction and isoelectric precipitation and observed that the mean droplet size, creaming stability, and emulsion stability between both the methods vary significantly. In other similar study, it was observed that during isoelectric precipitation, the pH of extraction media plays dominant role in determining the functionality of isolated proteins. A study on two lentil varieties found that an increase in pH of extraction media had detrimental effect on foam expansion, however, increased the foam stability (Lee et al. 2007).

Although lentils have a low trypsin inhibition activity (TIA) compared to other legume seeds, lentil protein isolates have even lower TIA compared to that of lentil flour. Also, lentil isolates have very low galacto-oligosaccharide content, which reduces the chances of causing flatulence upon consumption (Joehnke et al. 2021). Barbana and Boye (2013) reported that protein digestibility for isolates is much higher as opposed to the flour, mainly because the protein present in flour is complexed with other macromolecules like starch and lipids, which reduces its bioavailability. Also, in the process of protein extraction, protein molecules may undergo partial denaturation, exposing more sites for enzymatic activity.

5.5.2 Starch

Lentil starch shows unimodal granular size distribution with mean granular diameter of 13–34 μm (Wang et al. 2014b). It is also characterized by a significantly high amylose content compared to other traditional starch sources like potato and maize (Joshi et al. 2013). Owing to its high amylose to amylopectin ratios, the relative crystallinity for this starch is lower, which suggests that the density of crystalline lamella is relatively low in lentil starch. Consequently, the susceptibility to undergo swelling is higher in lentil starch. A comparative study between pea starch and lentil starch reported that an increase in temperature causes many folds higher starch swelling in lentils due to higher susceptibility for melting of the crystalline region. Lentil starches also have strong gelling ability. Joshi et al. 2013 found that the elastic moduli of gels prepared from lentil starch was about eight times of potato starch gels. High amylose content in lentil starch caused an increase in water binding sites resulting in a firmer solid-like viscoelastic gel.

Chung et al. (2008) found that among pea, lentils, and chickpea starches, lentil starch had the least expected glycemic index. Although the processing condition and genotypical variation affect the digestibility of starch, but in general, lentil starch has significant amount of slowly digestible starch. Lu et al. (2018) reported that digestibility of starch varied significantly within eight different Canadian lentil cultivars. Further, it was observed that cooking has pronounced effect on the resistant starch content as cooked lentils had much higher resistant starch content than raw flour and isolated starch. Lentil starch exhibited significantly lower retrogradation tendency and noodles prepared from it had superior textural attributes than noodles prepared with pea starch (Wang et al. 2014b). Thus, lentil starch can find wide application in gluten-free food manufacturing industries targeting the diabetic or health conscious consumers.

5.5.3 Fiber

Dry lentils are rich source of slowly digestible resistant starch and galacto-oligosaccharides, which are hard to digest in the small intestine and pass on to the large intestine. Fiber consumption is associated with low glycemic response, probiotic effects, and better gut microbiota profiles (Clarke et al. 2022). Total fiber content, insoluble and soluble fiber content, and composition may vary according to genetic variations (Singh et al. 2016). Dalgetty and Baik (2003) found that isolation of dietary fiber in itself is a very challenging process as starch molecules get entrapped into the fibers and make purification very difficult. Moreover, the soluble and insoluble fibers need to be isolated separately further adding to the complexity of process.

Traditional cooking of lentils can have detrimental effect on the soluble and insoluble dietary fiber content. High temperature can cause depolymerization in the fiber structure and consequently can cause its hydrolysis (Dueñas et al. 2016). Bioavailability of dietary fibers can be increased by fractionating and incorporating in food products directly. For instance, Dalgetty and Baik (2006) developed bread fortified with dietary fibers isolated from lentils and found that bread loaf with added fibers develops a favorable crumb color without any deteriorative effect on its hardness and moisture profiles. Fibers can also be used for the development of biodegradable and edible packaging materials. Ochoa-Yepes et al. (2018) reported that the packaging film developed with lentil dietary fibers mixed in starch composite had high storage modulus, shear resistance, folding strength and significantly lower water vapor permeability in addition to high nutritional properties.

5.6 Conclusions

In comparison to wheat or rice milling, lentil milling, and fractionation research is still not a well-developed area. There is a major scientific interest to develop innovative fractionation processes and dehulling operations. As lentils vary widely in size, shape, hardness, and material properties, therefore fundamental physical grain properties must be studied for developing milling technologies for a specific pulse. Different lentil fractions are finding their way into food product development in the industry. We do need more fundamental research to relate lentil microstructure with milling operation and its link to flour functionalities to develop different food and nonfood products in the future.

References

Ahmed, J., Varshney, S.K., and Ramaswamy, H.S. (2009). Effect of high pressure treatment on thermal and rheological properties of lentil flour slurry. *LWT Food Science and Technology* 42 (9): 1538–1544.

Ahmed, J., Taher, A., Mulla, M.Z. et al. (2016). Effect of sieve particle size on functional, thermal, rheological and pasting properties of Indian and Turkish lentil flour. *Journal of Food Engineering* 186: 34–41. https://doi.org/10.1016/j.jfoodeng.2016.04.008.

Allotey, D.K., Kwofie, E.M., Adewale, P. et al. (2022). A meta-analysis of pulse-protein extraction technologies: impact on recovery and purity. *Journal of Food Engineering* 327: 111048.

Barbana, C. and Boye, J.I. (2013). In vitro protein digestibility and physico-chemical properties of flours and protein concentrates from two varieties of lentil (*Lens culinaris*). *Food & Function* 4 (2): 310–321.

Benmeziane-Derradji, F., Djermoune-Arkoub, L., Ayat, N.E.H., and Aoufi, D. (2020). Impact of roasting on the physicochemical, functional properties, antioxidant content and microstructure changes of Algerian lentil (*Lens culinaris*) flour. *Journal of Food Measurement and Characterization* 14 (5): 2840–2853.

Black, R.G., Singh, U., and Meares, C. (1998). Effect of genotype and pretreatment of field peas (Pisum sativum) on their dehulling and cooking quality. *Journal of the Science of Food and Agriculture* 77: 251–258.

Borsuk, Y., Arntfield, S., Lukow, O.M. et al. (2012). Incorporation of pulse flours of different particle size in relation to pita bread quality. *Journal of the Science of Food and Agriculture* 92 (10): 2055–2061.

Chizhikova, O.G., Korshenko, L.O., and Pavlova, M.A. (2017). Development of composite flour mixes with the use of milled lentil seeds. *Food Processing: Techniques and Technology* 46 (3): 89–95.

Choo, K., Ramachandran, R.P., Spiwnyk, E., and Paliwal, J. (2022). Effects of Seed Moisture Conditioning and Mechanical Scouring Pre-Treatments on Roller-Milled Green Lentil (Lens culinaris) and Chickpea (Cicer arietinum) Flours. *Food and Bioprocess Technology* 1–16.

Chung, H.J., Liu, Q., Hoover, R. et al. (2008). In vitro starch digestibility, expected glycemic index, and thermal and pasting properties of flours from pea, lentil and chickpea cultivars. *Food Chemistry* 111 (2): 316–321.

Clarke, S.T., Sarfaraz, S., Qi, X. et al. (2022). A review of the relationship between lentil serving and acute postprandial blood glucose response: effects of dietary fibre, protein and carbohydrates. *Nutrients* 14 (4): 849.

Dabhi, M.N., Sangani, V.P., and Rathod, P.J. (2019). Effect of enzyme pretreatment on dehulling, cooking time and protein content of pigeon pea (variety BDN2). *Journal of Food Science and Technology* 56 (10): 4552–4564.

Dalgetty, D.D. and Baik, B.-K. (2003). Isolation and characterization of cotyledon fibers from peas, lentils, and chickpeas. *Cereal Chemistry* 80 (3): 310–315.

Dalgetty, D.D. and Baik, B.-K. (2006). Fortification of bread with hulls and cotyledon fibers isolated from peas, lentils, and chickpeas. *Cereal Chemistry* 83 (3): 269–274.

Deshpande, S.S., Sathe, S.K., Salunkhe, D.K., and Cornforth, D.P. (1982). Effects of dehulling on phytic acid, polyphenols, and enzyme inhibitors of dry beans (*Phaseolus vulgaris* L.). *Journal of Food Science* 47 (6): 1846–1850.

Dueñas, M., Sarmento, T., Aguilera, Y. et al. (2016). Impact of cooking and germination on phenolic composition and dietary fibre fractions in dark beans (*Phaseolus vulgaris* L.) and lentils (*Lens culinaris* L.). *LWT Food Science and Technology* 66: 72–78.

Ek, P., Gu, B.J., and Ganjyal, G.M. (2021). Whole seed lentil flours from different varieties (Brewer, Crimson, and Richlea) demonstrated significant variations in their expansion characteristics during extrusion. *Journal of Food Science* 86 (3): 942–951.

Erskine, W., Williams, P.C., and Nakkoul, H. (1985). Genetic and environmental variation in seed yield, seed size and cooking quality of lentil. *Field Crops Research* 12: 153–161.

Erskine, W., Williams, P.C., and Nakkoul, H. (1991). Splitting and dehulling lentil (*Lens culinaris*): effects of genotype and location. *Journal of the Science of Food and Agriculture* 57 (1): 85–92.

Escobedo, A. and Mojica, L. (2021). Pulse-based snacks as functional foods: processing challenges and biological potential. *Comprehensive Reviews in Food Science and Food Safety* 20 (5): 4678–4702.

Espinosa-Ramírez, J. and Serna-Saldívar, S.O. (2019). Wet-milled chickpea coproduct as an alternative to obtain protein isolates. *LWT Food Science and Technology* 115: 108468.

Ferawati, F., Hefni, M., Östbring, K., and Witthöft, C. (2021a). The application of pulse flours in the development of plant-based cheese analogues: proximate composition, color, and texture properties. *Food* 10 (9): 2208.

Ferawati, F., Zahari, I., Barman, M. et al. (2021b). High-moisture meat analogues produced from yellow pea and faba bean protein isolates/concentrate: effect of raw material composition and extrusion parameters on texture properties. *Food* 110 (4): 843.

Fernando, S. (2021). Production of protein-rich pulse ingredients through dry fractionation: a review. *LWT Food Science and Technology* 141: 110961.

Goyal, R.K., Vishwakarma, R.K., and Wanjari, O.D. (2008). Optimisation of the pigeon pea dehulling process. *Biosystems Engineering* 99 (1): 56–61.

Goyal, R.K., Vishwakarma, R.K., and Wanjari, O.D. (2010). Effect of moisture content on pitting and milling efficiency of pigeon pea grain. *Food and Bioprocess Technology* 3 (1): 146–149.

Jeganathan, B., Temelli, F., and Vasanthan, T. (2022). Micromorphological and elemental characteristics of chickpea, faba bean, field pea, and lentil cotyledon topographies. *Cereal Chemistry* 99 (2): 380–392.

Jiang, Z.Q., Wang, J., Stoddard, F. et al. (2020). Preparation and characterization of emulsion gels from whole faba bean flour. *Food* 9 (6): 755.

Joehnke, M.S., Jeske, S., Ispiryan, L. et al. (2021). Nutritional and anti-nutritional properties of lentil (*Lens culinaris*) protein isolates prepared by pilot-scale processing. *Food Chemistry: X* 9: 100112.

Joshi, M., Aldred, P., McKnight, S. et al. (2013). Physicochemical and functional characteristics of lentil starch. *Carbohydrate Polymers* 92 (2): 1484–1496.

Joyner, J.J. and Yadav, B.K. (2015a). Optimization of continuous hydrothermal treatment for improving the dehulling of black gram (*Vigna mungo* L.). *Journal of Food Science and Technology* 52 (12): 7817–7827.

Joyner, J.J. and Yadav, B.K. (2015b). Microwave assisted dehulling of black gram (*Vigna mungo* L.). *Journal of Food Science and Technology* 52: 2003–2012.

Karaca, A.C., Low, N., and Nickerson, M. (2011). Emulsifying properties of chickpea, faba bean, lentil and pea proteins produced by isoelectric precipitation and salt extraction. *Food Research International* 44 (9): 2742–2750.

Kumar, P., Chakraborty, S.K., and Kate, A. (2022). Influence of infrared (IR) heating parameters upon the hull adherence and cotyledon integrity of whole pigeon pea (*Cajanus cajan* L.) grain. *LWT Food Science and Technology* 154: 112792.

Kurien, P.P. and Parpia, H.A.B. (1968). Pulse milling in India. I – processing and milling of Tur, Arhar (*Cajanus cajan* Linn). *Journal of Food Science and Technology* 5 (4): 203–207.

Lagassé, S., Boyd, L., Young, G. et al. (2022). Influence of genotype and environment on the flour and bread baking properties of peas and lentils. *Cereal Chemistry* 99 (2): 325–342.

Lee, H.C., Htoon, A.K., and Paterson, J.L. (2007). Alkaline extraction of starch from Australian lentil cultivars Matilda and Digger optimised for starch yield and starch and protein quality. *Food Chemistry* 102 (3): 551–559.

Li, L., Yuan, T.Z., Setia, R. et al. (2019). Characteristics of pea, lentil and faba bean starches isolated from air-classified flours in comparison with commercial starches. *Food Chemistry* 276: 599–607.

Lin, T., O'Keefe, S., Duncan, S., and Fernández-Fraguas, C. (2020). Manipulation of the dry bean (*Phaseolus vulgaris* L.) matrix by hydrothermal and high-pressure treatments: impact on in vitro bile salt-binding ability. *Food Chemistry* 310: 125699.

Lu, Z.H., Donner, E., Tsao, R. et al. (2018). Physicochemical and digestion characteristics of flour and starch from eight Canadian red and green lentils. *International Journal of Food Science & Technology* 53 (3): 735–746.

Ma, Z., Boye, J.I., Simpson, B.K. et al. (2011). Thermal processing effects on the functional properties and microstructure of lentil, chickpea, and pea flours. *Food Research International* 44 (8): 2534–2544.

Marti, A., Sadohara, R., Winham, D.M., and Cichy, K.A. (2022). Food industry views on pulse flour – perceived intrinsic and extrinsic challenges for product utilization. *Food* 11 (14): 2146.

Martínez-Velasco, A., Lobato-Calleros, C., Hernández-Rodríguez, B.E. et al. (2018). High intensity ultrasound treatment of faba bean (*Vicia faba* L.) protein: effect on surface properties, foaming ability and structural changes. *Ultrasonics Sonochemistry* 44: 97–105.

Mcdaniel, K.A., White, B.L., Dean, L.L. et al. (2012). Compositional and mechanical properties of peanuts roasted to equivalent colors using different time/temperature combinations. *Journal of Food Science* 77 (12): C1293–C1299.

Miano, A.C., Pereira, J.D.C., Castanha, N. et al. (2016). Enhancing mung bean hydration using the ultrasound technology: description of mechanisms and impact on its germination and main components. *Scientific Reports* 6 (1): 1–14.

Monnet, A.-F., Laleg, K., Michon, C., and Micard, V. (2019a). Legume enriched cereal products: a generic approach derived from material science to predict their structuring by the process and their final properties. *Trends in Food Science & Technology* 86: 131–143.

Monnet, A.F., Eurieult, A., Berland, S. et al. (2019b). Damaged starch in pea versus wheat flours: fragmentation behavior and contribution of fine and coarse fractions. *Cereal Chemistry* 96 (3): 465–477.

Murumkar, R.P., Borkar, P.A., Munje, S.S. et al. (2016). Effect of enzyme pre-treatments on milling of pigeonpea. *International Journal of Science, Environment and Technology* 5 (6): 4029–4051.

Nickhil, C., Mohapatra, D., Kar, A. et al. (2021). Delineating the effect of gaseous ozone on disinfestation efficacy, protein quality, dehulling efficiency, cooking time and surface morphology of chickpea grains during storage. *Journal of Stored Products Research* 93: 101823.

Ochoa-Yepes, O., Medina-Jaramillo, C., Guz, L., and Famá, L. (2018). Biodegradable and edible starch composites with fiber-rich lentil flour to use as food packaging. *Starch* 70 (7, 8): 1700222.

Oomah, B.D., Ward, S., and Balasubramanian, P. (2010). Dehulling and selected physical characteristics of Canadian dry bean (*Phaseolus vulgaris* L.) cultivars. *Food Research International* 43 (5): 1410–1415.

Pawar, D.A., Joshi, D.C., and Sharma, A.K. (2021). Effect of gamma irradiation and microwave energy on milling characteristics of pigeon pea. *Journal of Agricultural Engineering* 58 (1): 29–39.

Pelgrom, P.J.M., Vissers, A.M., Boom, R.M., and Schutyser, M.A.I. (2013). Dry fractionation for production of functional pea protein concentrates. *Food Research International* 53 (1): 232–239.

Pelgrom, P.J.M., Boom, R.M., and Schutyser, M.A.I. (2015). Method development to increase protein enrichment during dry fractionation of starch-rich legumes. *Food and Bioprocess Technology* 8: 1495–1502.

Phirke, P.S., Bhole, N.G., and Adhaoo, S.H. (1995). Shear forces for dehulling, splitting and breaking raw and pretreated pigeonpeas. *International Journal of Food Science & Technology* 30 (4): 485–491.

Reichert, R.D., Oomah, B.D., and Youngs, C.G. (1984). Factors affecting the efficiency of abrasive-type dehulling of grain legumes investigated with a new intermediate-sized, batch dehuller. *Journal of Food Science* 49 (1): 267–272.

Ren, Y., Yuan, T.Z., Chigwedere, C.M., and Ai, Y. (2021). A current review of structure, functional properties, and industrial applications of pulse starches for value-added utilization. *Comprehensive Reviews in Food Science and Food Safety* 20 (3): 3061–3092.

Sánchez-Vioque, R., Clemente, A., Vioque, J. et al. (1999). Protein isolates from chickpea (*Cicer arietinum* L.): chemical composition, functional properties and protein characterization. *Food Chemistry* 64 (2): 237–243.

Saxena, R.P., Chand, L., Garg, G.K., and Singh, B.P.N. (1988). Effect of bicarbonate soaking on dehusking efficiency and composition of pigeon pea (*Cajanus cajan* L.) seeds and dhal. *International Journal of Food Science & Technology* 24 (3): 237–241.

Scanlon, M.G., Thakur, S., Tyler, R.T. et al. (2018). The critical role of milling in pulse ingredient functionality. *Cereal Foods World* 63 (5): 201–206.

Schutyser, M.A.I. and van der Goot, A.J. (2011). The potential of dry fractionation processes for sustainable plant protein production. *Trends in Food Science & Technology* 22 (4): 154–164.

Schutyser, M.A.I., Pelgrom, P.J.M., van der Goot, A.J., and Boom, R.M. (2015). Dry fractionation for sustainable production of functional legume protein concentrates. *Trends in Food Science & Technology* 45 (2): 327–335.

Sefa-Dedeh, S. and Stanley, D.W. (1979). The relationship of microstructure of cowpeas to water absorption and dehulling properties. *Cereal Chemistry* 56: 379–385.

Shi, Y., Singh, A., Kitts, D., and Pratap-Singh, A. (2021). Lactic acid fermentation: a novel approach to eliminate unpleasant aroma in pea protein isolates. *LWT Food Science and Technology* 150: 111927.

Shrestha, S., van't Hag, L., Haritos, V.S., and Dhital, S. (2022). Lentil and Mungbean protein isolates: processing, functional properties, and potential food applications. *Food Hydrocolloids* 135: 108142.

Singh, J., Srivastava, R., Gupta, S. et al. (2016). Genetic variability for vitamin B9 and total dietary fiber in lentil (*Lens culinaris* L.) cultivars. *International Journal of Food Properties* 19 (4): 936–943.

Sopiwnyk, E., Young, G., Frohlich, P. et al. (2020). Effect of pulse flour storage on flour and bread baking properties. *LWT Food Science and Technology* 121: 108971.

Sreerama, Y.N., Sashikala, V.B., and Pratape, V.M. (2009). Expansion properties and ultrastructure of legumes: effect of chemical and enzyme pre-treatments. *LWT Food Science and Technology* 42 (1): 44–49.

Stone, A.K., Parolia, S., House, J.D. et al. (2021). Effect of roasting pulse seeds at different tempering moisture on the flour functional properties and nutritional quality. *Food Research International* 147: 110489.

Sunil, C.K., Chidanand, D.V., Manoj, D. et al. (2018). Effect of ultrasound treatment on dehulling efficiency of blackgram. *Journal of Food Science and Technology* 55 (7): 2504–2513.

Tabtabaei, S., Jafari, M., Rajabzadeh, A.R., and Legge, R.L. (2016). Development and optimization of a triboelectrification bioseparation process for dry fractionation of legume flours. *Separation and Purification Technology* 163: 48–58.

Tam, N., Oguz, S., Aydogdu, A. et al. (2017). Influence of solution properties and pH on the fabrication of electrospun lentil flour/HPMC blend nanofibers. *Food Research International* 102: 616–624.

Thakur, S., Scanlon, M.G., Tyler, R.T. et al. (2019). Pulse flour characteristics from a wheat flour Miller's perspective: a comprehensive review. *Comprehensive Reviews in Food Science and Food Safety* 18 (3): 775–797.

Tiwari, B.K., JaganMohan, R., and Vasan, B.S. (2008a). Effect of several pre-treatments on the physical characteristics of dehulled fraction of pigeon pea (*Cajanus cajan* L). *International Journal of Food Science and Technology* 43 (8): 1458–1463.

Tiwari, B.K., JaganMohan, R., and Vasan, B.S. (2008b). Effect of different premilling treatments on dehulling of black gram (*Phaseolus mungo* L.). *Journal of Food Processing and Preservation* 32 (4): 610–620.

Tiwari, B.K., JaganMohan, R., Venkatachalapathy, N. et al. (2010). Optimisation of hydrothermal treatment for dehulling pigeon pea. *Food Research International* 43 (2): 496–500.

Tosh, S.M. and Yada, S. (2010). Dietary fibres in pulse seeds and fractions: characterization, functional attributes, and applications. *Food Research International* 43 (2): 450–460.

Veena, A., Urooj, A., and Puttaraj, S. (1995). Effect of processing on the composition of dietary fibre and starch in some legumes. *Food/Nahrung* 39 (2): 132–138.

Vidal-Valverde, C., Sierra, I., Frias, J. et al. (2002). Nutritional evaluation of lentil flours obtained after short-time soaking processes. *European Food Research and Technology* 215 (2): 138–144.

Vishwakarma, R.K., Shivhare, U.S., Gupta, R.K. et al. (2017). Status of pulse milling processes and technologies: a review. *Critical Reviews in Food Science and Nutrition* 58 (10): 1615–1628.

Wang, N. (2008). Effect of variety and crude protein content on dehulling quality and on the resulting chemical composition of red lentil (*Lens culinaris*). *Journal of the Science of Food and Agriculture* 88 (5): 885–890.

Wang, J., De Wit, M., Schutyser, M.A.I., and Boom, R.M. (2014a). Analysis of electrostatic powder charging for fractionation of foods. *Innovative Food Science & Emerging Technologies* 26: 360–365.

Wang, N., Warkentin, T.D., Vandenberg, B., and Bing, D.J. (2014b). Physicochemical properties of starches from various pea and lentil varieties, and characteristics of their noodles prepared by high temperature extrusion. *Food Research International* 55: 119–127.

Wang, J., Zhao, J., De Wit, M. et al. (2016). Lupine protein enrichment by milling and electrostatic separation. *Innovative Food Science & Emerging Technologies* 33: 596–602.

Wood, J.A. and Malcolmson, L.J. (2011). Pulse milling technologies. In: *Pulse Foods* (ed. B.K. Tiwari, A. Gowen, and B. Mckenna), 193–221. Oxford: Academic Press.

Wood, J.A., Knights, E.J., and Choct, M. (2016). Topography of the cotyledon surfaces and adjoining seed coat of chickpea (*Cicer arietinum* L.) genotypes differing in milling performance. *Cereal Chemistry* 94: 104–109.

Wood, J.A., Knights, E.J., Campbell, G.M. et al. (2022). Enzyme pre-milling treatments improved milling performance of chickpeas by targeting mechanisms of seed coat and cotyledon adhesion with various effects on dhal quality. *Journal of the Science of Food and Agriculture* 102 (1): 62–72.

Xiong, W., Devkota, L., Zhang, B. et al. (2022). Intact cells: "nutritional capsules" in plant foods. *Comprehensive Reviews in Food Science and Food Safety* 21 (2): 1198–1217.

Young, G., Bourré, L., Frohlich, P. et al. (2020). Effect of roasting as a premilling treatment on the functional and bread baking properties of whole yellow pea flour. *Cereal Chemistry* 97 (2): 183–195.

6

Functional Properties of Lentils and Its Ingredients in Natural or Processed Form

Semin O. Keskin[1] *and Gulum Sumnu*[2]

[1] *Department of Food Processing, Kocaeli University, Kocaeli, Turkey*
[2] *Department of Food Engineering, Middle East Technical University, Ankara, Turkey*

6.1 Introduction

Lentil (*Lens culinaris* Medik.) seed is a nutrient-rich food legume due to its considerable content of essential macro- and micronutrients, e.g., proteins, resistant starches, dietary fibers, vitamins and minerals, and bioactive phytochemicals, e.g., polyphenols, phytosterols, bioactive peptides, saponins, phytates (Dhull et al. 2022; Faris et al. 2013). These constituents of lentil offer significant health promoting effects attributed to their antioxidant, antibiotic, anti-inflammatory, anticarcinogenic, hypoglycemic, hypocholesterolemic, antihypertensive, and laxative properties, and bifidogenic and prebiotic activity (Didinger and Thompson 2022; Faris et al. 2013; Faris and Attlee 2017). For example, vitamin E, β-carotene, and polyphenols are considered as the responsible constituents related to antioxidant activity. Additionally, flavonoids, Bowman-Birk type trypsin–chymotrypsin inhibitor, phytic acid, phytosterols, defensin, lectins, resistant starch, saponins, and β-carotene are regarded as anticarcinogenic constituents. Moreover, defensins have antibiotic, phytosterols has anti-inflammatory and hypocholesterolemic, and resistant starch has hypoglycemic, laxative, and bifidogenic functions (Faris et al. 2013).

Lentils, in addition to their contribution to a healthy, balanced diet and their vital role in management and prevention of several human chronic illnesses, such as diabetes, obesity, cancer, and cardiovascular diseases, have potential to be used in food applications (e.g., as a functional food ingredient or an additive for encapsulation or a component of edible/biodegradable packaging materials). Lentil ingredients have been added in food formulations as nutri- and techno-functional constituents, such as plant-based protein sources, or novel vegan edible foaming, gelling, and emulsifying agents, or antimicrobial and antioxidant agents (Jarpa-Parra et al. 2017; Jeske et al. 2019; Klopsch et al. 2019; Ma et al. 2016; Polat et al. 2020; Portman et al. 2018, 2019; Rizzello et al. 2017; Shariati-levari et al. 2016; Stantiall et al. 2018; Turfani et al. 2016). Furthermore, development of value-added food products, such as gluten-free functional products with low glycemic index and high dietary fiber

Lentils: Production, Processing Technologies, Products, and Nutritional Profile, First Edition.
Edited by Jasim Ahmed, Muhammad Siddiq, and Mark A. Uebersax.

(Gularte et al. 2012; Han et al. 2010; Morales et al. 2015), or using them as an encapsulation adjunct for edible/biodegradable packaging materials are attractive, novel food applications in recent years (Aydogdu et al. 2018, 2019; Bamdad et al. 2006; Boyaci and Yemenicioglu 2018; Chang et al. 2016; Chang and Nickerson 2018; Karaca et al. 2013a,b; Tam et al. 2017).

Besides the valuable nutritional and technological advantages, lentils also have environmental benefits that were reported by Chaudhary and Tremorin (2020). From the environmental and economical perspectives, replacing foods from animal sources with food from plant sources in terms of food sustainability and reducing carbon, water, and land footprints is an important issue for the governments (Chaudhary and Tremorin 2020). The authors reformulated beef burgers with whole cooked lentil puree and compared lentil-reformulated beef burgers with the traditional ones. Nutrient balance score (considering 27 essential macro and micronutrients) and environmental footprints (carbon, bluewater, water scarcity, land use, and biodiversity) was determined for the samples. It was observed that replacing a portion of beef with lentils improved the nutrient value (by >20% compared with traditional beef burger), as well as dietary fiber, total folate, manganese, and selenium contents. In terms of the environmental and economical indices, partial replacement of a lean beef burger with cooked lentil puree have been found to decrease the life cycle environmental footprint by ~33% and to reduce the cost by 26% (Chaudhary and Tremorin 2020).

In recent years, various value-addition studies have been to increase lentil consumption by improving the bioavailability of lentil proteins and enhancing the nutri- and techno-functional characteristics of lentil products and ingredients. In this regard, different conventional (e.g., germination, fermentation, hydrothermal processing, and extrusion) or emerging technologies (microwave, infrared, high hydrostatic pressure [HP], and ultrasound) have been researched to reach these targets (Ahmed et al. 2019; Asensio-Grau et al. 2020; Fouad and Rehab 2015; Ghumman et al. 2016a,b; Hefnawy 2011; Montemurro et al. 2019; Pasqualone et al. 2021; Quintero-Quiroz et al. 2021).

This chapter provides comprehensive overview about not only the nutri- and techno-functional properties of lentil ingredients in natural or processed form and the effects of different processing techniques on their nutri- and techno-functional characteristics but also their usage in various food applications.

6.2 Nutri-Functional and Health-Promoting Properties

Lentil is a carbohydrate-, protein-, and fiber-rich functional dietary ingredient with valuable nutrients, such as minerals (especially iron, zinc, potassium), vitamins, and so on, and notable antioxidant activity (Dhull et al. 2022; Grela et al. 2017; Khazaei et al. 2019). The carbohydrate, protein, and total dietary fiber contents of lentil seeds vary between the average values of 60–64.44 g/100 g edible portion (USDA 2016), 25.6–28.9 and 16.70–31.05 g/100 g (on dry basis), respectively (Keskin et al. 2022). Besides considerable amount of protein content, lentils have bioactive peptides, such as protease inhibitor proteins, angiotensin 1-converting enzyme (ACE) inhibitor, and defensins, providing health benefits with their anti-inflammatory, anti-hypertensive, and immune system supporting properties, respectively (El-Adawy et al. 2003). Dietary fibers in lentil has potential hypocholesterolemic (related to soluble dietary fiber fraction and saponins), antidiabetic, anticancer (related to lectins or hemagglutinins), antitumor, antimicrobial (related to lectins), and hypoglycemic (related to saponins) effects (Ganesan and Xu 2017). The resistant starches in lentils (at 9.1–13.2%) (Keskin et al. 2022), acting as prebiotics, contribute significantly to the gastrointestinal health and gut microbiota (Ganesan and Xu 2017).

Lentil seeds have been considered as one of the excellent vegetable sources of iron (Ganesan and Xu 2017) with the content varying between 6.51 and 7.71 mg/100 g edible portion (USDA 2016). Since studies have shown that iron is an important mineral in the prevention of anemia (Soltan 2013), its daily intake at required amounts is significant. This requirement could be met by incorporation of lentil in daily diet. The zinc and potassium contents of lentil seeds vary between 3.27–5.89 mg and 677–943 mg per 100 g edible portion, respectively (USDA 2016). Zinc is an essential micronutrient for human metabolism that catalyzes enzymes and has function on protein synthesis (Aslani et al. 2015). Potassium is another functional mineral in lentils. The high potassium (677–943 mg/100 g edible portion), low sodium (3–6 mg/100 g edible portion), and low total fat content (0.92–1.06 g/100 g edible portion) make lentil a suitable dietary ingredient for patients with cardiovascular diseases and obesity by lowering the blood pressure and providing low calorie intake (Ganesan and Xu 2017).

The antioxidant activity is another nutri-functional property of lentils. The notable antioxidant activity of lentils (38.9–82.0 μmol Trolox equivalent [TE]/g) (Xu and Chang 2010) is related to its naturally available polyphenolic compounds. Phenolic acids (hydroxybenzoic acids, hydroxycinnamic acids) and flavonoids (flavonols, flavanols, flavones, flavanones, antocyanins, proanthocyanidins) are the main functional polyphenols in lentils (Ganesan and Xu 2017). These phenolic compounds, which have the ability to act as antioxidants with chemopreventive potential, are capable of inhibiting oxidation reactions and reduce free radicals in the body, thus resulting in the protection of cell membranes against damage mediated by oxygen radicals and exhibition of anticancer activity (Ganesan and Xu 2017; Kang 2015). Moreover, phenolic compounds have anti-inflammatory, anti-obesity, antidiabetic, and antimicrobial effects, which promote human health (Ganesan and Xu 2017; Zhang et al. 2018). Inflammation is a basic response of immune system to infection, burn, toxic chemicals, etc. and inflammatory reactions often include the formation of oxidation products that may lead to tissue damage. Since lentils have phenolic compounds with high antioxidant activity, they have potential to reduce inflammation in the body (Zhang et al. 2018). Furthermore, studies have shown that flavonoids and fibers in lentils help to reduce the amount of food intake and flavonoids decrease the glucose and fat digestion (by inhibiting the α-glucosidase and pancreatic lipase activity) and absorption in the intestine, resulting in anti-obesity and antidiabetic activity (Ganesan and Xu 2017). Antimicrobial activity of lentils mainly come from bioactive peptides, the defensins, as *in vitro* studies have demonstrated that defensins with phenolic compounds can act as potential inhibitors of microbial growth (Ganesan and Xu 2017).

6.3 Techno-Functional Properties

Lentil ingredients, such as flours, protein isolates (PIs), lentil starch, and fiber, have been used in food applications because of their versatile techno-functional properties. The binding, emulsifying, foaming, and gelling properties of lentil-based ingredients promote the physical, textural, and sensorial characteristics of various foods.

Use of lentil flour in food products concerning their techno-functional properties have been recently overviewed by Romano et al. (2021). Water and oil absorption capabilities, emulsification, foaming, and gelation properties of lentil flour provided improvements in techno-functional and sensory quality of baked goods, meat, and dairy products (yogurt), related to the interactions between constituents of lentil (protein, starch, dietary fiber, etc.) and the food medium. De Angelis et al. (2020) determined some functional properties, such as water (WAC) and oil absorption capacities (OAC) of 15 lentil accessions. The WAC and OAC of lentil flours were found to be 0.89–1.37 (g water/g flour) and 0.49–0.86 (g oil/g flour), respectively. Lentil protein isolate (LPI) is one of the most important lentil ingredients with desirable techno-functional properties for application in various food systems. Previously, Jarpa-Parra (2018) reviewed the studies done between years 2002 and 2017, and reported techno-functional properties of LPIs. Recent studies (since 2018) have been summarized in Table 6.1.

The most considered techno-functional properties of any food protein can be listed as solubility; water binding/absorption/holding capacity; fat and oil binding/absorption/holding capacity; emulsifying, foaming, and gelation properties (Boye et al. 2010). These properties are dependent on the extraction method and conditions, which affect their amino acid composition, protein structure, and conformation, along with their surface hydrophobicity/hydrophilicity (Aryee and Boye 2017). Furthermore, processing conditions and the composition of the food matrix, such as pH, temperature of the medium and the interaction of proteins with other food components, also influence aforementioned properties and the behavior of proteins in food matrices (Boye et al. 2010; Farooq and Boye 2011).

Lee et al. (2021) investigated the techno-functional properties of LPI from red lentils (from United States, Nepal, and Turkey) over a wide range of pH values from 2 to 10. It was observed that protein solubility, foaming, and emulsifying properties of isolates were dependent on pH and the lowest (i.e., poorest) properties were obtained at the isoelectric point, pH 5.2. At high alkaline and acidic conditions, gradual improvements in functional properties of LPIs, such as solubility, foaming capacity (FC), and emulsifying stability index (ESI), were observed, all of which make them suitable for use as potential emulsifying and foaming agents in high-acidic or high-alkaline food medium (Lee et al. 2021). In another study, Osemwota et al. (2021) investigated various isolation protocols to obtain LPIs from lentil (coded as ISO [isoelectric isolate], i.e., MEM_NaOH [alkaline-soluble protein], and MEM_NaCl [salt-soluble protein]) with improved functional properties. At alkaline pH values, solubility of protein isolates were found to be higher as compared to neutral and acidic pH values. The highest solubility was 100% for ISO and MEM_NaOH at pH 9 and 92% for MEM_NaCl at pH 8. Water and oil holding capacities of LPIs were found to be significantly affected by isolation protocol, whereas the gelling ability of LPIs obtained by different methods were found to be insignificant (Osemwota et al. 2021) (Table 6.1).

Tang et al. (2021) compared the techno-functional properties of proteins isolated from common pulses, including lentil. WAC, OAC, and the least gelling concentration of yellow and white lentils were found as 1.2 and 4.9 g water/g protein, 1.78 and 1.80 g oil/g protein, and 140 and 110 g/l, respectively. Lentil and pigeon pea isolates were found to have more desirable functional properties compared to soy and yellow pea proteins (Tang et al. 2021). Gunes and Karaca (2022) investigated the techno-functional properties of five proteins isolated from local lentil landraces of Anatolia. The solubility of proteins was found to be the

Table 6.1 Techno-functional properties of lentil protein isolates (LPIs).

Lentil/lentil ingredient	Solubility (%)	Water absorption capacity, WAC (g water/g protein)	Oil absorption capacity, OAC (g oil/g protein)	Emulsifying activity index, EAI (m^2/g)	Emulsifying stability index, ESI (min)	Foaming capacity, FC (%)	Foaming stability, FS (%)	Least gelling concentration, LGC (g/L)	References
LPI (from red lentils and three different origins and isolated by ISO)	—	3.1–3.5[a]	5.8–7.3	9.6–25.1 (pH 2–10)	9.1–25.9 (pH 2–10)	85.8–96.7 (highest at pH 2) 37.1–40.8 (lowest at pH = 5)	70–90[b] (pH 2–10)	—	Lee et al. (2021)
LPI (from dehulled green lentil isolated by ISO, MEM_NaOH, and MEM_NaCl)	100 (for ISO and MEM_NaOH at pH 9) 92 (for MEM_NaCl at pH 8)	0.15[a] (ISO) 0.47[a] (MEM_NaOH) 0.10[a] (MEM_NaCl)	1.63[c] (ISO) 1.07[c] (MEM_NaOH) 0.92[c] (MEM_NaCl)	—	—	—	75–100 (at pH 7)	11.50 (ISO) 11.00 (MEM_NaOH) 10.75 (MEM_NaCl)	Osemwota et al. (2021)
LPI (from white and yellow lentils isolated by ISO)	—	1.2 (Yellow lentil) 4.9 (White lentil)	1.78 (Yellow lentil) 1.80 (White lentil)	—	—	—	—	140 (Yellow lentil) 110 (White lentil)	Tang et al. (2021)
LPI (from green, red, and brown lentils isolated by ISO)	6–9 (at pH 5) 93–100 (at pH 9)	1.4–1.5[a]	2.7–4.2[c]	18.8–24.2 (at pH 7)	43.4–125.5 (at pH 7)	71–91 (at pH 7)	74–77[b] (at pH 7)	110–150	Gunes and Karaca (2022)

ISO: isoelectric precipitation; MEM_NaOH: membrane protein isolates from alkaline solution extracts; MEM_NaCl: membrane protein isolates from salt solution extracts.
[a] Water holding capacity (g/g).
[b] Foam stability after 30 min.
[c] Oil holding capacity (g/g).

lowest at pH 5, which was close to their isoelectric points (pH 4.25–4.74), and the highest at pH 9. Notable varietal differences in the functional properties of those proteins were observed, especially in the ESI values, changing in the range of 43.4–125.5 minutes (Table 6.1).

Starch is the most prominent carbohydrate in lentil seeds, whose structural and functional characteristics influence its functionality in food systems (Keskin et al. 2022). Joshi et al. (2013) measured physicochemical (composition, crystallinity, gelatinization behavior) and functional (swelling, pasting, and gel formation) properties of lentil starch and compared with those of cereal (corn starch) and tuber (potato starch) starches. Lentil starch exhibited the highest amylose content (32.52%) and the lowest crystallinity (9.12%) and gelatinization enthalpy (11.80 J/g) among three starches. The gelatinization and pasting temperatures of lentil starch were found to be 69.3 °C (peak temp.) and 71.9 °C, which were between those of corn and potato starches. The final elastic modulus at the end of cooling period of lentil starch (11,073 Pa) was eight times and about three times higher, respectively, than that of potato starch (1362 Pa) and corn starch (3828 Pa). Lentil starch gel exhibited higher viscoelastic solid properties and gel strength and final viscosity (6516 mPa·s), making it suitable ingredient for food formulations where higher gel strength and pasting viscosity required/desired.

Majeed et al. (2017) determined swelling, pasting, and gelation properties of lentil starch. Swelling and solubility index is known to give information about the interaction of crystalline and amorphous regions of starch chains with water molecules, exhibiting the hydration ability of starch. Swelling and solubility indices of native lentil starch were observed to increase significantly from 2.32 to 16.77 g/g and 0.02 to 0.24, respectively, with the increase in temperature from 50 to 90 °C. This temperature-dependent effect was attributed to the weakening of intermolecular interactions like hydrogen bonding and higher amylose leaching, respectively (Majeed et al. 2017). The emulsion, water, and oil absorption capacities of lentil starch have been reported as 5.01%, 5.49 g water/g sample, and 4.83 g oil/g sample, respectively. The peak, trough, breakdown, final, and setback viscosities of lentil starch have been observed as 3898.3, 2424.0, 2353.0, 4912.0, and 2740.5 cP, respectively. The pasting temperature and peak time were reported as 71.1 °C and 4.2 minutes.

Lentil is a good source of dietary fiber, one of the important bioactive components of lentil, which has been incorporated into processed foods to increase their dietary fiber content and/or as a functional ingredient (Tosh and Yada 2010). The functional properties of dietary fiber fractions affect the preparation, processing, and storage behavior of lentil in food systems and influence the sensory, nutritional, and textural attributes of end products (Wang and Toews 2011). Fat absorption, water hydration, swelling, and water retention capacities of lentil (dehulled) fibers from three different varieties were reported to be 1.10–1.12 g oil/g, 5.0–5.7 g water/g, 22.7–27.1 ml/g, and 8.9–11.4 g water/g, respectively (on dry weight [dw] basis) (Wang and Toews 2011). The swelling, water, and oil binding capacities of lentil seed coat (hull) have been determined by Dalgetty and Baik (2003) as 2.38 ml/g, 3.60 ml water/g, and 1.63 ml oil/g (dw basis), respectively. Authors compared the functional properties of insoluble or soluble fibers from lentil seed coat and cotyledon fibers. The swelling, water, and oil binding capacities of insoluble cotyledon fiber were found to be higher than that of lentil seed coat (hull) and soluble cotyledon fiber, which make it suitable functional ingredient for various foods, such as meat and cereal-based products (Dalgetty and Baik 2003).

6.4 Effect of Processing on Quality Characteristics of Lentil Ingredients

Both food processing techniques and processing conditions provide remarkable changes in the functionality of lentil ingredients. The effects of processing on nutri- and techno-functional characteristics of lentil ingredients, as reported in the literature, are discussed in Sections 6.4.1.2–6.4.1.4 and related results are summarized in Tables 6.2–6.4.

6.4.1 Conventional Processing Techniques

6.4.1.1 Germination

One of the technologies to improve lentil functionality is seed germination or the sprouting process. It is an inexpensive but effective conventional bioprocessing technique, which can positively impact the quality of lentil-based products and ingredients. Germination provides significant improvements in nutritional profile, phenolic content, and bioactivity, and reduction in antinutritional factors (tannins and phytic acid) (Fouad and Rehab 2015). From the techno-functional point of view, germination was reported to increase lentil flour's foaming capacity (FC) and water absorption capacity (WAC), and the foaming stability (FS) and emulsifying activity index (EAI) were decreased (Ghumman et al. 2016a). The efficacy of a germination process is a function of type of pulse seed and processing conditions, e.g., temperature, quantity of water added, germination time, presence or absence of light (Acquah et al. 2021).

Germination process improves the nutritional and functional quality of legumes, e.g., lentils, and reduces their antinutrient content, such as trypsin inhibitor, phytates, tannins, and raffinose family oligosaccharides (Fouad and Rehab 2015; Pal et al. 2017). Fouad and Rehab (2015) studied the effect of germination time (3, 4, 5, and 6 days) on the proximate composition, bioactive content, and antioxidant capacity in lentil sprouts. They observed increase in water, protein, ash, crude fiber, total, reducing and nonreducing sugar, and free amino acid contents and decrease in crude oil, total carbohydrates, and antinutritional factors (phytic acid and tannins) after germination process. Furthermore, phenolic content of sprouted lentils were found to increase from 1341.13 mg/100 g dry weight (in raw lentil seeds) to 1411.50, 1463.00, 1630.20, and 1510.10 mg/100 g dry weight for the samples germinated for 3, 4, 5, and 6 days, respectively. Duenas et al. (2016) investigated the effects of germination on dietary fiber fractions of lentils. Germination process was found to decrease the total, insoluble, and soluble dietary fiber contents of lentils by 25%, 23%, and 47%, respectively. Moreover, germination improves protein solubility and digestibility and antioxidant capacity and increases soluble sugar, free amino acid contents, and foaming and water absorption capacities of lentils (Bubelova et al. 2018; Fouad and Rehab 2015; Ghumman et al. 2016a; Swieca et al. 2012). The antioxidant activity of lentil ingredients is significantly affected by germination though there are mixed findings reported in the literature. Bubelova et al. (2018) reported that, compared to raw lentils, germinated lentils exhibited higher total phenolic and flavonoid contents and antioxidant activity, which increased from 1089.9 to 1599.8 mg (gallic acid equivalent [GAE]/kg dry matter), from 3.4 to 38.5 (mg rutin equivalent/kg dry matter), and 329 to 1209.7 (mg TEAC) in red lentils, respectively. Fouad and

Table 6.2 Effect of fermentation on nutri- and techno-functional characteristics of lentil ingredients.

Lentil ingredient	Process conditions	Effects of processing	References
Flour	Microorganisms: *Lactobacillus plantarum* C48, *Lactobacillus brevis AM7* Process temperature and time: 30 °C, 24 h	• Increase in the nutritional and functional values (increased antioxidant and phytase activities) of flour • Decrease in the levels of antinutritional factors	Curiel et al. (2015)
Flour	Microorganism: *L. plantarum* CECT 748 Hydrolysis with Savinase	• Improved beneficial health effects in terms of antioxidant, antihypertensive, and hypoglycemic properties	Bautista-Expósitoa et al. (2018)
Flour	Soaking I: in 1.25% sodium hypochlorite (NaClO) at ~25 °C for 30 min Soaking II: in tap water, at ~10 °C for 20–30 min Process temperature and time: 16.5 °C, 24 h (in dark) Microorganisms: mixed starter (ratio 1 : 1 : 1; *Lactobacillus rossiae* LB5, *L. plantarum* 1A7 and *Lactobacillus sanfranciscensis* DE9) Process temperature and time: 30 °C, 24 h	• Improvement of functional and nutritional quality by combining germination with sourdough fermentation • Increase in peptides, free amino acids, and γ-aminobutyric acid concentrations • Decrease in phytic acid, condensed tannins, raffinose, and trypsin inhibitory activity	Montemurro et al. (2019)
Seed	Microorganism: an edible fungus (*Pleurotus ostreatus*) Process temperature and time: 28 °C, 14 d	• Enhanced nutritional (increase of protein, resistant starch, and polyphenols) and digestibility (higher fraction of digested protein) profile	Asensio-Grau et al. (2020)
Protein	Microorganisms: from Kefir seeds Process temperature and time: 25 °C, 5 d	• Increase in protein digestibility from 76.4% to 84.1% over five days of fermentation • Increase in total phenolic content from 443.4 to 792.6 mg GAE/100 g after two days of fermentation • Reduction in α-helix: β-sheet ratio with the increase in protein digestibility	Alrosan et al. (2021)

GAE, gallic acid equivalent.

Table 6.3 Effect of hydrothermal processing (boiling) on nutri- and techno-functional characteristics of lentil seeds.

Process conditions	Effects of processing	References
Soaking: in water for 18 h, at room temperature; process conditions: atmospheric pressure for 20 min	• Decrease in the concentration of insoluble and soluble fiber fractions • Identification of hydroxybenzoic and hydroxycinnamic compounds, flavan-3-ols, procyanidins, flavonols, and flavones in dietary fiber fractions	Duenas et al. (2016)
Soaking: in water; process temperature: 95 °C for 30 min	• Higher water holding capacity and fat absorption capacity compared to raw lentil flour • Decrease in trypsin inhibitors • Reduction in the content of sulfur-containing amino acids • Decrease in polyphenolic compounds	Aryee and Boye (2017)
Soaking: in distilled water 1 : 4 (seed: water, w/w) for 12 h at room temperature; process conditions: cooking in a boiling water bath	• Increase in total soluble sugar • Decrease in protein and total lipid contents • Reduction in trypsin inhibitors • Decrease in the content of antioxidant metabolites (gallic acid, catechin, and quercetin) and antioxidant activities	Pal et al. (2017)
Process time: 25 min	• Disintegration of the cell wall matrix • Reduction in the content of insoluble-bound phenolics • Loss of phenolic compounds • Structural and positional changes of bound phenolics and possible alteration of chemical bonding	Yeo and Shahidi (2017)
Soaking: in cold distilled water for 2 h (non-dehulled samples); process time: 15 min (applied to brown, red, and black lentils) and 20 min (applied to both green lentils); Dehulled samples were not soaked prior to cooking and were boiled for 5 min	• Increase in digestibility • Reduction in total phenolic and flavonoid contents, antioxidant activity, and both crude and neutral-detergent fiber contents	Bubelova et al. (2018)

Rehab (2015) also reported that sprouted lentil seeds exhibited higher DPPH (2,2-diphenyl-1-picrylhydrazyl) radical scavenging and reducing power activities. However, according to the Pal et al. (2017), germination process decreased the content of antioxidant metabolites, e.g., gallic acid from 214.45 to 160.07 μg/g, catechin from 144.56 to 70.80 μg/g, and quercetin from 78.78 to 59.28 μg/g and total antioxidant activities of lentils (reported as average of five varieties). The differences in lentil's origin and analysis methods could be the most likely reason for showing mixed results for the antioxidant

Table 6.4 Effect of extrusion processing on nutri- and techno-functional characteristics of lentil ingredients.

Lentil ingredient	Process conditions	Effects of processing	References
Flour	Feed moisture: 15, 20, 25 g/100 g; feed rate: 20 kg/h; extrusion temperature: 100, 125, 150 °C; screw speed: 300 rpm	• Increase in oil absorption capacity with an increase in feed moisture • Decrease in water absorption capacity with an increase in both extrusion temperature and feed moisture • Increase in foaming capacity and decrease in viscosity and *in vitro* protein digestibility with an increase in feed moisture and a decrease in extrusion temperature	Ghumman et al. (2016b)
Flour	Feed moisture: 14–22%; feed rate: 16 rpm (340 g/min); die temperature: 140–180 °C; screw speed: 150–250 rpm	• Decrease in trypsin inhibitors (99.54%), phytic acid (99.30%) and tannin (98.83%) without altering the protein content • Decrease in total polyphenols • Improved protein and starch digestibility, up to 89% and 96%, respectively	Rathod and Annapure (2016)
Flour	Feed rate: 36 kg/h; barrel temperature: 30–50, 70–90, and 100–120 °C; screw speed: 650 rpm	• Increase in the protein digestibility-corrected amino acid score of extruded lentil flour from 55.0 and 50.8 to 63.01 and 57.09 for red and green lentils, respectively • Increase in the average digestible indispensable amino acid score of extruded lentil flours from 0.54 and 0.49 to 1.30 and 1.34 for red and green lentils, respectively	Nosworthy et al. (2017)
Flour	Barrel temperature: 140, 160 °C; die head temperature: 140, 160 °C; screw speed: 500 rpm	• Significant increase, up to 85% in total α-galactosides content • Significant decrease in inositol phosphates content • Reduction in trypsin inhibitor activity as well as lectins content higher than 90% • Increase in desirable prebiotic compounds • Decrease in non-nutritional factors	Ciudad-Mulero et al. (2020)

Flour	Barrel temperature: 110, 125, and 140 °C; screw speed: 150, 200, and 250 rpm	• Significant impact on the extrusion process responses (die back pressure, torque, specific mechanical energy) and the expansion ratio, density, water absorption index, and water solubility index of the extrudates • Highest expansion at low extrusion temperature (110 °C)	Ek et al. (2021)
Lentil flour (extruded at two different conditions; milder [EF1] and more severe [EF2])	Feed moisture content: 28.8 g/100 g for milder condition (E1), 27.1 g/100 g for more severe condition (E2); barrel temperature: 80–90 °C for E1, 100–105 °C for E2; die head temperature: 100 °C for E1, 115 °C for E2; screw speed: 220 rpm for E1 and 230 rpm for E2	• Increase in the viscoamylograph initial viscosity with a significantly higher value in milder extrusion conditions • Increase in the Mixolab maximum consistency • Decrease in starch retrogradation levels • Increase in water absorption index (higher in EF1 than in EF2) and oil absorption capacity (higher in EF2 than in EF1) • Decrease in bulk density	Pasqualone et al. (2021)
Flour	Feed rate: 36 kg/h; barrel temperature: 30–50, 70–90, and 100–120 °C; screw speed: 650 rpm	• Decrease in amino acid score, essential amino acid/total amino acid ratio • Increase in *in vitro* protein digestibility and protein digestibility corrected amino acid score • Improvement in antioxidant capacity • Decrease in phytic acid content	Sanchez-Velazquez et al. (2021)

activity. In an earlier study, Swieca et al. (2012) determined the antioxidant compounds and antioxidant capacity of lentil sprouts as a function of germination time and type of illumination. Both germination and illumination conditions were found to modify the polyphenolic profiles and biological activity of lentils. The authors suggested germination of lentils under continuous light were helpful to stimulate the phenolic phytochemicals (*p*-hydroxybenzoic, benzoic, caffeic acids).

Lentil flours obtained from germinated seeds from three lines were analyzed for foaming capacity and stability (FC, FS) and emulsifying activity index (EAI) and water absorption capacity (WAC) by Ghumman et al. (2016a). It was shown that germination increased the WAC and FC and decreased FS and EAI of lentil flours. FC of germinated and ungerminated lentil flours were 17.7–23.3% and 16.7–17.2%, respectively. Although FC values were higher, the germinated flours exhibited lower FS, which was related to the hydrolysis of large peptides into small ones, accompanied by the release of amino acids during germination, wherein the protein–protein interaction is lowered. The increase in WAC of germinated flours from 1.2–1.9 to 1.1–2.7 g/g was attributed to the formation of low–molecular weight proteins having large number of polar groups post-germination.

6.4.1.2 Fermentation

Fermentation is also a conventional bioprocessing technique, in which microorganisms are used to breakdown a material via enzymatic activity. The activation of endogenous enyzmes during fermentation causes hydrolytic reactions, which leads to processing of pulses with improved palatability and nutritional value and reduced antinutritional factors, all of which enhance health benefits of lentils (Asensio-Grau et al. 2020; Dhull et al. 2022; Montemurro et al. 2019). It has been reported that the antinutritional factors (tannins, phytic acid, trypsin inhibitors) have been decreased, whereas protein digestibility, total phenolic content, antioxidant, and phytase activities of lentils have been increased by means of fermentation (Alrosan et al. 2021; Asensio-Grau et al. 2020; Curiel et al. 2015). Additionally, beneficial health effects of lentil in terms of antioxidant, antihypertensive, and hypoglycemic properties have been improved with the fermentation process (Bautista-Expósitoa et al. 2018).

Asensio-Grau et al. (2020) reported that the solid-state fermentation of lentils with an edible fungus (*Pleurotus ostreatus*) improved protein digestibility and enhanced nutritional profile of lentil flours with an increase in protein, resistant starch, and polyphenol contents. In another study (Montemurro et al. 2019), sourdough fermentation of sprouted lentil flours with *Lactobacillus rossiae LB5*, *Lactobacillus plantarum 1A7*, and *Lactobacillus sanfranciscensis DE9* enhanced the nutritional and functional features by means of increased peptide, free amino acid, and γ-aminobutyric acid concentrations, and decreased phytic acid, condensed tannins, raffinose, and trypsin inhibitory activity. It is noteworthy that the effects of fermentation on pulse properties depend on microbial culture, enzyme activity, fermentation method (solid or liquid state fermentation), and processing conditions (temperature, time) (Acquah et al. 2021). A summary of studies related to the effects of fermentation with different microorganisms and under different processing conditions on the functional properties of lentils is presented in Table 6.2.

6.4.1.3 Boiling

Boiling is the most common conventional thermal treatment to cook pulses. Structural changes in pulse constituents, such as protein denaturation, starch gelatinization, and retrogradation, may be observed during thermal processing, affecting the functionality of pulses (Acquah et al. 2021). Boiling leads to reduction in trypsin inhibitors (Aryee and Boye 2017; Pal et al. 2017), decrease in the concentration of insoluble and soluble fiber fractions (Duenas et al. 2016), the content of total phenolics and flavonoids and antioxidant metabolites (gallic acid, catechin, and quercetin), and the antioxidant activity (Aryee and Boye 2017; Bubelova et al. 2018; Pal et al. 2017; Yeo and Shahidi 2017) in boiled lentils. The effects of boiling on nutri- and techno-functional characteristics of lentil ingredients have been summarized in Table 6.3.

Yeo and Shahidi (2017) monitored the changes in the insoluble-bound phenolics of lentil cultivars upon boiling by HPLC analysis and the antioxidant capacity determination. It was observed that the content of insoluble-bound phenolics decreased in boiled samples, indicating their release from the disintegrated or loosened cell wall matrix exposed to hydrothermal energy. However, it was shown that the released bound phenolics did not remain as soluble/free phenolics and total amount of phenolics decreased, suggesting the reason of this result as the interaction of phenolic compounds with other food constituents, such as protein, starch, and cellulose (Yeo and Shahidi 2017). From the techno-functional point of view, it was reported that lentil flours obtained from boiled seeds exhibited higher water holding and fat absorption capacities compared with raw flours (Aryee and Boye 2017). Higher water holding capacity (WHC) of cooked flours have been attributed to protein denaturation and starch gelatinization during cooking, which may cause entrapment and retainment of more water compared with the raw flour. It was suggested that about 15% increase in fat absorption capacity may increase the potential of cooked flours being used as a functional ingredient in foods (Aryee and Boye 2017).

6.4.1.4 Extrusion

Extrusion processing is one of the thermal treatments that have remarkable impacts on functionality of lentil ingredients via modifying the physicochemical properties of food components, such as starch and proteins, via exposing them to high temperature and mechanical shearing (Gomez and Martínez 2016). Typically, extrusion results in an increase in both water and oil absorption capacities and a decrease in bulk density of the extrudates (Pasqualone et al. 2021). Extrusion temperatures (barrel, die head temperatures), screw speed, feed moisture, and feed rate are the key parameters for optimization of the extrusion process. The decrease in antinutritional factors (trypsin inhibitors, phytic acid, and tannin contents) (Ciudad-Mulero et al. 2020; Rathod and Annapure 2016; Sanchez-Velazquez et al. 2021) and improvement in protein and starch digestibility (Nosworthy et al. 2017; Rathod and Annapure 2016) have been observed in extruded lentil flours. It has been reported in the literature that extrusion process conditions significantly affect both the nutri- and techno-functional properties of lentil flours, as summarized in Table 6.4 (Ek et al. 2021; Ghumman et al. 2016b; Pasqualone et al. 2021).

6.4.2 Emerging Processing Techniques

6.4.2.1 Microwave

Microwave heating is a rapid heating technology, which is used to thaw, temper, cook, dry, pasteurize, or preserve foods and have ability to modify food constituents and change their functionality in food systems. Microwave power level and exposure time are the critical parameters in this process (Acquah et al. 2021). Rapid volumetric heating via microwaves provides structural changes mainly in starch and protein components of food products in a short time, changing the nutritional and technological functionalities of them (González and Pérez 2002; Hefnawy 2011). Microwave heating, as a fast cooking method, can be used for lentil processing, which is known to improve *in vitro* protein digestibility (IVPD) and protein efficiency ratio (PER) and significantly reduces antinutritional factors (phytic acid, tannins, and trypsin inhibition activity), but it results in small losses in minerals content (Hefnawy 2011).

González and Pérez (2002) modified lentil starch by microwave heating, which caused significant changes in its functional and rheological properties. The decrease in water absorption, solubility, and swelling power values of microwaved lentil starches was attributed to the promotional effect of microwaves on intragranular starch–starch interactions and molecular rearrangement. Furthermore, all amylographic viscosities, including setback viscosity, of lentil starch were found to decrease after microwave modification. The decrease in setback viscosity have meant reduction in retrogradation tendency of lentil starches, which is a desirable property expected from the starches in most of the food applications (González and Pérez 2002). Hefnawy (2011) reported that, besides its short cooking time advantage, microwave heating has the ability to decrease antinutritional factors in lentil, e.g., by 39.17%, 34.38%, and 93.29% reduction in phytic acid, tannins, and trypsin inhibition activity, respectively.

6.4.2.2 Infrared

Infrared or micronization heating is one of the emerging technologies used in the processing of pulses, which provides significant changes in lentil constituents, mainly in starch and protein fractions, resulting in modification of functional characteristics of lentils (Fasina et al. 2001; Liu et al. 2020; Pathiratne et al. 2015). Moreover, this technology has been used to inactivate endogenous enzymes, lipoxygenase, and peroxidase to give positive features to lentil flours in food applications in terms of improved shelf stability and sensory characteristics, respectively (Pathiratne et al. 2015). Efficient and rapid heating by infrared energy in a short time provides significant changes in lentil constituents, mainly in starch and protein fractions, resulting in the modification of functional characteristics of lentils (Fasina et al. 2001; Liu et al. 2020; Pathiratne et al. 2015). Seed size, moisture level, and surface temperature upon heating have been stated as the critical parameters of micronization process, which affects physical, chemical, and functional properties of lentil flours (Liu et al. 2020; Pathiratne et al. 2015).

Liu et al. (2020) studied the effect of seed size and tempering (hydration) duration (24, 48, and 96 hours) in altering the lentil flour functionality. Authors explained the significance of seed size in modifying functional properties of lentil flours with its ability to determine not only the rate of moisture migration into the interior of seeds during tempering but also the relative surface area exposed to infrared heat (Liu et al. 2020). Tempering is known as a pretreatment (primarily hydration in water) prior to infrared processing conducted to achieve target seed moisture level. Since a relatively significant time is needed for the water migration through seed coat and into the inner part of seed, tempering duration is classified as one of the critical parameters that can have an effect on the functional properties of tempered and micronized material (Liu et al. 2020). Greater degrees of starch gelatinization and protein denaturation have been observed when the tempering period has been prolonged and a smaller size seed has been used. The increase in emulsifying activity of modified lentil flours with the increase in tempering duration (from 24 to 48–96 hours) has been attributed to denaturation of proteins (Liu et al. 2020).

The combined treatment of tempering and infrared heating has been also reported to significantly enhance the WHC of modified lentil flours as a result of starch gelatinization induced by infrared heating. The modified flours with a longer tempering time, a lower surface temperature, and a smaller seed size have been found to exhibit higher water holding capacities related to their higher damaged/gelatinized starch contents (Liu et al. 2020). In another study by Pathiratne et al. (2015), an increase in the WHC of lentil flours have been reported at the processing conditions of 16% and 23% seed moisture and micronizing temperatures above 130 °C. Furthermore, higher oil absorption capacity have been observed for micronized flours tempered to 16% and 23% seed moisture levels than at 8% moisture level. The resistant starch content and protein dispersibility index of flours decreased with all micronizing temperatures (115, 130, 150, and 165 °C) and tempering levels (8%, 16%, and 23%) (Pathiratne et al. 2015). Fasina et al. (2001) utilized infrared heating to modify lentil flours by reaching the target surface temperature of 140 °C. The results showed a significant reduction in trypsin inhibitor levels (~31%) without any alteration in the protein and starch contents, and superior water hydration capacities for lentil flours compared to raw ones (Fasina et al. 2001). Moreover, micronization technology have been used to inactivate endogenous enzymes, lipoxygenase, and peroxidase to give positive features to lentil flours in food applications in terms of improved shelf stability and sensory characteristics (Pathiratne et al. 2015).

6.4.2.3 High Hydrostatic Pressure

High pressure (HP) is an emerging nonthermal technology, which has been mainly used for food preservation and novel food development (Ahmed et al. 2019; Garcia-Mora et al. 2015). In recent years, HP treatment has been used alone or in combination with other technologies to modify the structure of starches and proteins from lentil (Ahmed et al. 2009, 2016; Han et al. 2007) or to prepare LPIs (lentil protein isolates) or lentil protein hydrolysates (LPH) with improved *in vitro* protein digestibility (IVPD) and antioxidant activity (AA) and techno-functional properties (water holding capacity (WHC), foaming capacity (FC) and stability (FS), emulsifying activity index (EAI), and emulsion stability index (ESI)) (Ahmed et al. 2019; Garcia-Mora et al. 2015; Han et al. 2007).

Han et al. (2007) reported that the HP treatment during soaking significantly decreased the soaking time from 12 hours to less than 1 hour and increased the IVPD of lentils. This improvement in IVPD was related to the structural loosening and unfolding of proteins, leading to an increase in the proteolytic susceptibility of proteins (Gharibzahedi and Smith 2020). Thermal characteristics of HP-treated lentil flour dispersions have been studied by Ahmed et al. (2009) as a function of pressure level (350–650 MPa) and moisture content (14–58%). Based on differential scanning calorimetry (DSC) and Fourier-transform infrared spectroscopy (FTIR) results, authors stated that at the pressure and moisture levels tested, the native structure of lentil protein had not been lost and the protein fraction (globulin) had not been denatured completely in HP treatment of lentil flour slurries. The increase in pressure level led to an increase in viscoelasticity. It was further observed that HP-treated lentil flour dispersion converted to a viscous gel by gelatinized and/or denatured starch and protein components (Ahmed et al. 2009).

In another study by Ahmed et al. (2016), lentil starch dispersions have been exposed to HP treatment at 400, 500, and 600 MPa for 10 minutes to evaluate the pressure-induced changes in the functional, rheological, thermal, and structural properties of lentil starch. The results showed that WHC and damaged resistant starch content increased after pH treatment at 600 MPa. Furthermore, the pasting properties were significantly influenced by the HP treatment, especially at 600 MPa, resulting in a considerable decrease in peak viscosity, breakdown, and final viscosity, and an increase in peak time. It was concluded that a threshold pressure of 600 MPa is required to gelatinize the lentil starch completely and to obtain significant changes in the aforementioned properties for HP-treated lentil starches.

The use of HP treatment in combination with enzymolysis have been found to enhance protein susceptibility to digestion and provides release of bioactive peptides with increased ACE-inhibitory and antioxidant activities and increased IVPD (Garcia-Mora et al. 2015). Garcia-Mora et al. (2015) have shown that pressure treatment at 300 MPa significantly improved proteolytic efficiency of four food-grade proteases (Alcalase™, Protamex™, Savinase™, and Corolase 7089™), eliciting production of peptides with molecular weight lower than 3 kDa with increased ACE-inhibitory activity (with the exception of Alcalase). Among the proteases studied, Savinase (69.46%) and Corolase®-7089 (70.77%) provided the maximum ACE-inhibitory and Savinase elicited the highest antioxidant activities of 403.86 μmoles Trolox equivalent (TE)/g lentil protein concentrate (LPC) to the lentil protein hydrolysate (LPH) treated with HP at 300 MPa. Another approach in the production of protein hydrolysates is the application of HP before enzymatic hydrolysis in the case of pressure-sensitive enzymes (Garcia-Mora et al. 2015).

Ahmed et al. (2019) used HP treatment (300, 450, and 600 MPa for 15 minutes) before enzymolysis (with Alcalase at 0.5% and 1.0% w/w) of LPI to improve functional and bioactive properties of LPH. The WHC of LPIs increased with the HP pretreatment, due likely to the increase in hydrophilic groups after pressurization. A reduction in FC and a significant increase in FS were also observed for LPIs treated by HP at 300 MPa for 15 minutes. It was also found that FC and FS of LPHs were reduced by HP treatment, which was related to the lower surface activity of smaller polypeptide chains (Ahmed et al. 2019). It was found that the pressurization increased the EAI of proteins, whereas the enzymatic hydrolysis decreased the EAI and ESI values of proteins, significantly. The decrease in both EAI and ESI values of hydrolysates was associated with the creation of more hydrophilic peptides or due to the weak viscoelastic film formation at the interface by small peptides (Ahmed et al. 2019). The DPPH radical scavenging activity (DRSA) of LPI and LPH has been found to increase from 15.6% to 19.4% with pressurization at 300 MPa and from 15.6% to 25.3% with Alcalase (0.5%) digestion, respectively. Furthermore, the results related to the DRSA of hydrolysates after enzymatic digestion of LPI with and without pressurization have confirmed that HP treatment has a significant effect on the release of the bioactive compounds from LPH. The highest DPPH scavenging effect has been observed for the LPHs produced by application of pressure at 300 MPa for 15 minutes before Alcalase (0.5%) treatment. These effect were associated with pressure-induced enhancement in susceptibility of cleavage sites to proteases in LPI, providing formation of low–molecular weight bioactive peptides with remarkable functional activities (Ahmed et al. 2019; Gharibzahedi and Smith 2020).

6.4.2.4 Ultrasound

Ultrasound, also called as sonication, is a simple, inexpensive, and environmentally friendly emerging technology used to extract various components, such as protein, dietary fiber, and bioactives, from lentils (Kaya et al. 2017; Quintero-Quiroz et al. 2021) or to modify the functional properties of lentil starch (Majeed et al. 2017). Use of ultrasound treatment in protein extraction provided maintenance of LPIs with higher WHC, fat absorption capability (FAC), and emulsifying activity (EA) and stability (ES) (Quintero-Quiroz et al. 2021). Sound waves with a frequency of high intensity (equal to or greater than 20 kHz) are used to agitate particles in a solution. Cavitation bubbles are formed due to pressure changes and collapse violently as the sound wave propagates, resulting into regions of high temperature and pressure due to dissipated energy (Majeed et al. 2017). This effect leads to extraction of target components from the materials efficiently or induces changes in the structure of starches and proteins, resulting in noticeable modifications in their functional properties (Majeed et al. 2017; Quintero-Quiroz et al. 2021).

Lentil-water suspensions have been sonicated at different processing conditions to extract the hull (bran) fraction from green and red lentils and determine the effect of ultrasound treatment on some functional and bioactive properties of those lentil hulls (Kaya et al. 2017). It was reported that sonication did not significantly affect the soluble dietary fiber and phytic acid contents and WAC and OAC of lentil hulls. Additionally, decrease in total phenolic content and TEAC (Trolox equivalent antioxidant capacity) values for sonicated lentil hulls is also observed. The decrease in antioxidant capacity of lentil hulls was associated with the soaking and washing procedures, which may induce loss of polar antioxidants (Kaya et al. 2017). In another study, lentil starch was sonicated to modify its functional characteristics. Ultrasound application provided maintenance of sonicated starch having similar physicochemical and functional properties with the native starch and only a decrease in amylose content of sonicated starch have been observed (Majeed et al. 2017).

Quintero-Quiroz et al. (2021) applied ultrasound to efficiently extract proteins from lentils and compared the functional properties of LPIs extracted with ultrasound-assisted or conventional method (leaching). It was reported that sonication increased protein yield from 23.19% to 57.57%. Moreover, LPIs obtained by ultrasound-assisted extraction were found to exhibit higher WAC, FAC, EAI, ESI values, and lower gelation temperatures compared to conventionally extracted isolates, exhibiting desirable protein emulsifier properties, which are related to the secondary structure of the protein. However, it was indicated that application of ultrasound in the extraction process did not statistically affect the properties such as molecular weight, isoelectric point, particle size, and SH group, which are related to the primary structure. Authors explained all these changes in properties of LPIs related to secondary and primary structures, with partial protein denaturation during ultrasound-assisted extraction (Quintero-Quiroz et al. 2021).

6.5 Food Applications of Lentil-Based Ingredients

Lentil ingredients have potential to be used as a desirable protein source, a functional food ingredient, or an additive, an encapsulation, or an edible/biodegradable material in food applications related to their nutritional and technological properties. Lentil flours, LPCs, and LPIs have been used in food applications as functional food ingredients (as a protein, dietary fiber, or a phytochemical source) or an additive, such as binding, foaming, emulsifying, or antifungal agent, or a foam and emulsion stabilizer. The functionality of lentil ingredients in food systems have been summarized in Sections 6.5.1–6.5.4.

6.5.1 Lentil Flour

Lentil flour and other ingredients can be used in many food applications, including development of gluten-free products. Incorporation of lentil flour in gluten-free cracker and cake formulations resulted in improved nutritional and technological characteristics (Gularte et al. 2012; Han et al. 2010). Gluten-free crackers including red lentil flour had desirable dark color and good crispiness (Han et al. 2010). Adding lentil flour to rice flour (50 : 50) in gluten-free cakes showed higher specific volume, similar crumb hardness, and higher springiness values than the 100% rice flour (control) cake. Additionally, lentil flour provided protein- and fiber-rich gluten-free cakes with increased protein digestibility and low starch hydrolysis rate and lower estimated glycemic index compared to the control cake (Gularte et al. 2012). Ozkahraman et al. (2016) partially replaced wheat flour with 10% lentil flour to prepare legume-based cakes. The weight loss, specific volume, and hardness values of legume cakes were found to be similar with that of control or wheat flour cakes.

Morales et al. (2015) incorporated red lentil flour into a snack-type, gluten-free formulation, which resulted in the production of gluten-free extruded product with balanced nutritional and phytochemical composition. Inclusion of 3% and 7% lentil flour decreased hardness and cooking loss values of beef patties during 12 days storage, demonstrating good textural properties (Baugreet et al. 2016). Shariati-levari et al. (2016) added micronized green lentil flour as a binding agent in low-fat beef burgers. Cook loss, drip loss, shrinkage, and shear force of low-fat beef burgers were reduced with the incorporation of non-micronized or micronized lentil flours. Authors suggested addition of micronized lentil flour at 6% level as a gluten-free binder for low-fat burgers, based on the favorable results for its physicochemical properties and consumer acceptability. The addition of whole green lentil flour (at 6%, 12%, and 24%) and lentil cotyledon flour (at 5%, 10%, 15%, 20%, and 40%) to wheat bread formulation provided improvement in nutritional quality of wheat breads, such as increase in lysine-rich proteins, dietary fiber, phenolic compounds, and lignans (Turfani et al. 2016), and increase in protein and ash contents (Portman et al. 2018).

Klopsch et al. (2019) fortified wheat bread with lentil flour to produce functional wheat bread. Lentil flour led to increase in antioxidant activity of breads. In another study by Portman et al. (2019), premium and frost-damaged lentil flours were added to cookie formulation to assess the nutritional and functional properties of cookies. Lentil flour addition resulted in a significant increase in total protein, insoluble fiber and oligosaccharides contents, and antioxidant activity of cookies. Even premium or down-graded lentil flour inclusion at 25% level significantly enhanced the nutri- and techno-quality of cookies, i.e., enhanced antioxidant activity, acceptable hardness, and color characteristics (Portman et al. 2019). Benmeziane et al. (2021) incorporated 4% lentil flour from either roasted or unroasted seeds to yogurt to develop a new probiotic and prebiotic functional food. Addition of both types of lentil flours to yogurts exhibited lower syneresis values compared to control yogurt due to the water absorbing ability of lentil constituents acting as a stabilizer in food medium. Furthermore, yogurts formulated with lentil flour exhibited acceptable sensory quality (Benmeziane et al. 2021).

6.5.2 Lentil Protein Concentrates

Jarpa-Parra et al. (2017) and Eckert et al. (2018) explored LPCs as efficient egg replacers (foam and emulsion stabilizer) in cake/muffin and doughnut formulations, respectively. The use of lentil protein concentrate as egg/milk replacer (50–100%) in angel cake and muffins formulations were investigated by Jarpa-Parra et al. (2017). Baking losses for both cake and muffins were significantly reduced with the inclusion of lentil protein in the formulation. The hardness and chewiness values of lentil-protein-added cakes and muffins were found to increase during storage, resulting in inferior quality. According to the sensory evaluation results, authors suggested that replacing 50% and 100% of egg/milk protein by lentil protein was achievable in angel cake and muffins, respectively (Jarpa-Parra et al. 2017). Incorporation of lentil protein concentrates with barley protein concentrates in doughnut formulations resulted in better cooking and nutritional characteristics, such as reduced cooking loss and diameter reduction (serving as an efficient foam stabilizer as well as a water and oil binder in doughnut formulations), and higher protein and β-glucan contents, respectively (Eckert et al. 2018). However, based on the storage-stability tests, the doughnuts containing lentil and barley protein concentrates were found to exhibit higher values of hardness, chewiness, and gumminess compared to the control (Eckert et al. 2018).

6.5.3 Lentil Protein Isolates

Ma et al. (2016) and Jeske et al. (2019) added LPI as emulsifying agents to produce salad-dressing and pasteurized milk alternatives, respectively. Lentil, pea, and chickpea PIs have been used as egg-yolk replacers in salad dressings (Ma et al. 2016). Authors demonstrated that pulse-protein-supplemented salad dressings, obtained by varying levels of pulse protein, egg yolk, and oil, had comparable physical properties with commercial dressings, providing low-cholesterol, hypo-allergenic salad dressings. Highly stable nanoemulsions were obtained with high-pressure homogenization (900 bar at 85 °C) of lentil proteins. Those emulsions exhibited good colloidal stability, appearance, and viscosity similar to cow milk, proving that lentil proteins could be used as commercial plant-based milk substitute (Jeske et al. 2019).

6.5.4 Other Lentil Ingredients

A hydrolysate obtained from a mixture of pea, lentil, and faba bean flours was added to bread formulation as a natural biopreservative (antifungal agent) to extend the shelf-life of wheat bread (Rizzello et al. 2017). Three different wheat flour breads containing legume flour hydrolysate, only wheat flour (control bread) or calcium-propionate-added wheat flour were prepared. One half of the bread slices were inoculated with a suspension of *Penicillium roqueforti* DPPMAF1 (10^2 conidia/ml) and all slices (inoculated and not inoculated) were packaged in polyethylene film and stored for 21 days at room temperature to observe fungal contamination. Wheat breads formulated with legume flour hydrolysate (natural antifungal agent) or synthetic antifungal agent (calcium propionate) delayed appearance of fungal mycelium for seven days compared to control bread, without adversely affecting the textural and sensory properties of wheat breads (Rizzello et al. 2017). In another study by Stantiall et al. (2018), pulse (haricot beans, garbanzo chickpeas, whole green lentils and split yellow peas) cooking waters have been investigated and compared with egg white in terms of foaming and gelling abilities in the production of meringues. The results did show that all pulse cooking waters exerted foaming and gelling abilities. Among the pulses, cooking waters of whole green lentils exhibited the highest foaming ability (97%); however, it was lower than that of egg white (400%). The bitter flavor of whole green lentils, related to their saponin and phenolics contents, in meringues resulted in lower consumer acceptance compared to the meringues made from egg white (Stantiall et al. 2018).

Polat et al. (2020) added 5–15% germinated lentil extracts to the crackers formulation. Incorporation of germinated lentil extract, at all levels, increased the ash and protein contents, hardness (except 5% level, similar hardness with control cracker), flexibility, instrumental color values a* and b*, total phenolic contents and TEAC (Trolox equivalent antioxidant capacity), and decreased the color L* value of crackers. Crackers with 5% germinated lentil extract were found to be the most acceptable with their similar hardness values compared to control crackers and high overall acceptability based on sensory panel results (Polat et al. 2020).

6.6 Encapsulation or Edible/Biodegradable Material

One of the prominent application areas of lentil ingredients is their utilization as encapsulation or edible/biodegradable materials in food applications. In literature, lentil flours, LPIs, and LPCs have been used as encapsulation materials (Karaca et al. 2013a,b; Chang et al. 2016; Chang and Nickerson 2018; Aydogdu et al. 2019; Kabakci et al. 2021) or film materials to produce edible films (Aydogdu et al. 2018; Bamdad et al. 2006), antimicrobial composite films (Boyaci and Yemenicioglu 2018), or electrospun nanofibers (Tam et al. 2017).

6.6.1 Lentil Flour

Lentil flour-based, polyethylene oxide, electrospun nanofibers prepared at alkaline pH have been used to encapsulate gallic acid, one of the natural antioxidants. This active packaging material have been directly electrospun on polylactic acid (PLA) sheets to package walnuts and enhance oxidative stability of them (Aydogdu et al. 2019). The loading efficiency, total phenolic content, and antioxidant activities of gallic-acid-loaded lentil-flour-based electrospun nanofibers (prepared at different pH values) were found to be 20.6%, 7.01 mg GAE/g dry matter, and 1.63 mg DPPH (2,2-diphenyl-1-picrylhydrazyl)/g, respectively. The reduction in peroxide (from 2.3 to 1.3 meqO_2/kg walnut oil), *p*-anisidine (from

2.101 to 1.075 AnV), and Totox (from 6.701 to 3.675) values was observed for walnuts packaged with gallic-acid loaded lentil-flour-based electrospun nanofibers. The results showed that the oxidative stability of walnuts during 21 days of storage at 40 °C was enhanced significantly (Aydogdu et al. 2019). In another study by Kabakci et al. (2021), lentil flour at different concentrations (15%, 20%, 25%, 30%) have been used as a hydrophilic surfactant to prepare double emulsions to entrap magnesium, a vital mineral for humans, and incorporated them into cake batter. It was shown that stability of double emulsions increased from 67.6% to 76.0% when lentil four concentration was increased from 15% to 30%. Authors recommended double emulsions produced with lentil flour to encapsulate minerals like magnesium in the production of functional foods (Kabakci et al. 2021).

Aydogdu et al. (2018) utilized lentil flour as a potential biodegradable material in edible film production. Glycerol was used as a plasticizer at different concentrations (1%, 1.5%, and 2% w/v) and the process temperatures (70 and 90 °C). Lentil flour films containing 1% glycerol and obtained at 90 °C process temperature were found to have water solubility of 30.7%, tensile strength of 7.46 MPa, Young's modulus of 249.7 MPa, 20% elongation at break, and 0.25 ng m/s/Pa water vapor permeability values, which were comparable with banana flour-glycerol films in literature (Aydogdu et al. 2018). Authors suggested lentil flour as a good biopolymer source and a potential alternative to lentil protein in food packaging applications. Lentil flour (at 1% and 2% [w/v] levels) and hydroxypropyl methylcellulose (HPMC, 0.25%, 0.5%, and 1% [w/v] levels) blend has been used in electrospinning process to produce nanofibers for packaging applications (Tam et al. 2017). Authors indicated that homogeneous nanofibers could be obtained from solutions containing polyethylene oxide, lentil flour, and HPMC at different combinations and at different pH values (pH 7, 10, and 12) by using electrospinning process. It was suggested that a composition of 1% lentil flour, 0.25% HPMC at pH 10 was a good candidate for packaging material because of its ability to form uniform, homogeneous nanofibers with the desired smallest diameter (198 nm).

6.6.2 Lentil Protein Concentrate

Bamdad et al. (2006) prepared edible films from lentil protein concentrate (5 g/100 ml water) and glycerine (50%, w/w of lentil protein concentrate). Lentil protein-based edible films were found to have comparable characteristics, in terms of tensile strength, elongation, moisture barrier property, and water solubility, with other edible protein (soy, whey, and pea proteins) films. These films were designated as strong, elastic films with good moisture barrier property and acceptable physical integrity (Bamdad et al. 2006).

6.6.3 Lentil Protein Isolate

Karaca et al. (2013a,b) employed LPI and maltodextrin as wall materials of microcapsules obtained by freeze-drying and spray-drying, respectively, to protect flaxseed oil against oxidation. The results showed that microcapsules (including LPI and maltodextrin) obtained by either freeze-drying or spray-drying protected flaxseed oil against oxidation over a 25 days storage period at room temperature by a lack of significant change in initial peroxide value and TBARS (2-thiobarbituric acid-reactive substances) values. In another study by Chang et al. (2016), canola oil (20% and 30%) has been encapsulated by spray-drying with a wall material containing a mixture of LPI (2–8%) and maltodextrin (9.5–18%) with/without lecithin (1% and 3%) and/or sodium alginate (1%). Properties reached minimum surface oil (~2.4%) and maximum entrapment efficiency (~88%) for canola oil (20%), employing the combination of LPI (8%), maltodextrin (68%), and sodium alginate (4%) as wall material, which was considered the combination, the best wall material compared to other combinations for microcapsulation of canola oil.

The oxidative stability of LP–maltodextrin–alginate microcapsules were found to exhibit better oxidative stability over a 30-day storage at 22 °C than the LP-maltodextrin microcapsules (Chang et al. 2016). Omega fatty-acid-rich oils (canola, fish, and flaxseed oils) have been encapsulated with LPI-based wall material, a combination of 8% LPI, 68% maltodextrin, and 4% sodium alginate obtained by spray-drying (Chang and Nickerson 2018). This strong wall material was found to provide great protection to the omega fatty-acid-rich oils, against hydrolytic rancidity and oxidative degradation over a 30-day storage at 22 °C. The microcapsules containing fish oil were found to exhibit greater antioxidative capacity than the synthetic antioxidant, tert-butylhydroquinone (TBHQ), which has been related to the presence of eicosapentaenoic acid (EPA) and docosahexaenoic acid (DHA) in the fish oil, adversely affecting antioxidant capacity of TBHQ (Chang and Nickerson 2018).

Boyaci and Yemenicioglu (2018) prepared composite films with potential antimicrobial activity against *Listeria innocua*, with active and activate-at-home type (consumer-controlled and pH-triggered release mechanism for lysozyme) properties, from zein soy and LPIs and lysozyme. It was shown that the active property and pH-triggered release mechanism of films were obtained by maintaining 30–50% of total lysozyme in free form and providing charge–charge interactions between lysozyme and soy and LPIs, resulting in binding of remaining lysozyme in film matrix, respectively. Sliced beef, lamb meat, and smoked salmon were selected as food samples to assess active packaging's ability of composite films. The zein-LPI films activated by acidification were found to release lysozyme in packed cold-stored beef, lamb, and smoked salmon, exhibiting antimicrobial activity on *L. innocua* (Boyaci and Yemenicioglu 2018).

6.7 Conclusion

Lentil is one of the important food legume, standing out with its multiple nutritional and health benefits. It has the potential to prevent malnutrition and chronic diseases related to its functional constituents, such as bioactive peptides, proteins, dietary fibers, and polyphenols, and so on. Furthermore, it positively contributes to the reduction of environmental (carbon, water, and land) footprints. Further, lentil ingredients have the potential to be used as functional ingredients in food applications, such as binding, emulsifying, gelling, antioxidant, or antimicrobial agents, as a vegan meat analogue or an edible film material. Various conventional (e.g., germination, fermentation, hydrothermal processing, extrusion) or emerging technologies (microwave, infrared, high hydrostatic pressure [HP], ultrasound) have been utilized in literature to facilitate the lentil consumption and to improve the bioavailability of lentil proteins and to change/improve the nutri- and techno-functional characteristics of lentil ingredients. A review of published work demonstrated that the success of the process utilized to improve nutri- and techno-functional properties of lentil ingredients depends on processing conditions, availability of pre-treatments or combined treatments, and the properties of the processing or application medium. A thorough understanding of techno-functional properties of lentil-based ingredients can help expand their utilization in diverse food applications and encapsulation processes.

References

Acquah, C., Ohemeng-Boahen, G., Power, K.A., and Tosh, S.M. (2021). The effect of processing on bioactive compounds and nutritional qualities of pulses in meeting the sustainable development goal 2. *Frontiers in Sustainable Food Systems* 5: 1–16.

Ahmed, J., Varshney, S., and Ramaswamy, H. (2009). Effect of high pressure treatment on thermal and rheological properties of lentil flour slurry. *LWT Food Science and Technology* 42: 1538–1544.

Ahmed, J., Thomas, L., Taher, A., and Joseph, A. (2016). Impact of high pressure treatment on functional, rheological, pasting, and structural properties of lentil starch dispersions. *Carbohydrate Polymers* 152: 639–647.

Ahmed, J., Mulla, M., Al-Ruwaih, N., and Arfat, Y.A. (2019). Effect of high-pressure treatment prior to enzymatic hydrolysis on rheological, thermal, and antioxidant properties of lentil protein isolate. *Legume Science* 1: e10.

Alrosan, M., Tan, T.C., Easa, A. et al. (2021). Effects of fermentation on the quality, structure, and nonnutritive contents of lentil (*Lens culinaris*) proteins. *Journal of Food Quality* 2021: 1–7.

Aryee, A.N. and Boye, J.I. (2017). Comparative study of the effects of processing on the nutritional, physicochemical and functional properties of lentil. *Journal of Food Processing and Preservation* 41: e12824.

Asensio-Grau, A., Calvo-Lerma, J., Heredia, A., and Andrés, A. (2020). Enhancing the nutritional profile and digestibility of lentil flour by solid state fermentation with *Pleurotus ostreatus*. *Food & Function* 11: 7905–7912.

Aslani, Z., Alipour, B., Mirmiran, P., and Bahadoran, Z. (2015). Lentil's (*Lens culinaris* L.) functional properties in prevention and treatment of non-communicable chronic diseases: a review. *International Journal of Nutrition and Food Sciences* 4: 15–20.

Aydogdu, A., Kirtil, E., Sumnu, G. et al. (2018). Utilization of lentil flour as a biopolymer source for the development of edible films. *Journal of Applied Polymer Science* 135: 46356.

Aydogdu, A., Yildiz, E., Aydogdu, Y. et al. (2019). Enhancing oxidative stability of walnuts by using gallic acid loaded lentil flour based electrospun nanofibers as active packaging material. *Food Hydrocolloids* 95: 245–255.

Bamdad, F., Goli, A.H., and Kadivar, M. (2006). Preparation and characterization of proteinous film from lentil (*Lens culinaris*): edible film from lentil (*Lens culinaris*). *Food Research International* 39: 106–111.

Baugreet, S., Kerry, J.P., Botineştean, C. et al. (2016). Development of novel fortified beef patties with added functional protein ingredients for the elderly. *Meat Science* 122: 40–47.

Bautista-Expósitoa, S., Martínez-Villaluengaa, C., Dueñasb, M. et al. (2018). Combination of pH-controlled fermentation in mild acidic conditions and enzymatic hydrolysis by Savinase to improve metabolic health-promoting properties of lentil. *Journal of Functional Foods* 48: 9–18.

Benmeziane, F., Raigar, R.K., Ayat, N.E.H. et al. (2021). Lentil (*Lens culinaris*) flour addition to yogurt: impact on physicochemical, microbiological and sensory attributes during refrigeration storage and microstructure changes. *LWT Food Science and Technology* 140: 110793.

Boyaci, D. and Yemenicioglu, A. (2018). Expanding horizons of active packaging: design of consumer-controlled release systems helps risk management of susceptible individuals. *Food Hydrocolloids* 79: 291–300.

Boye, J., Zare, F., and Pletch, A. (2010). Pulse proteins: processing, characterization, functional properties and applications in food and feed. *Food Research International* 43 (2): 414–431.

Bubelova, Z., Sumczynski, D., and Salek, R.N. (2018). Effect of cooking and germination on antioxidant activity, total polyphenols and flavonoids, fiber content, and digestibility of lentils (*Lens culinaris* L.). *Journal of Food Processing and Preservation* 42: e13388.

Chang, C. and Nickerson, M.T. (2018). Stability and in vitro release behaviour of encapsulated omega fatty acid-rich oils in lentil protein isolate-based microcapsules. *International Journal of Food Science and Nutrition* 69: 12–23.

Chang, C., Varankovich, N., and Nickerson, M.T. (2016). Microencapsulation of canola oil by lentil protein isolate-based wall materials. *Food Chemistry* 212: 264–273.

Chaudhary, A. and Tremorin, D. (2020). Nutritional and environmental sustainability of lentil reformulated beef burger. *Sustainability* 12: 6712.

Ciudad-Mulero, M., Fernandez-Ruiz, V., Cuadrado, C. et al. (2020). Novel gluten-free formulations from lentil flours and nutritional yeast: evaluation of extrusion effect on phytochemicals and non-nutritional factors. *Food Chemistry* 315: 126175.

Curiel, J.A., Coda, R., Centomani, I. et al. (2015). Exploitation of the nutritional and functional characteristics of traditional Italian legumes: the potential of sourdough fermentation. *International Journal of Food Microbiology* 196: 51–61.

Dalgetty, D.D. and Baik, B.K. (2003). Isolation and characterization of cotyledon fibers from peas, lentils, and chickpeas. *Cereal Chemistry* 80: 310–315.

De Angelis, D., Pasqualone, A., Costantini, M. et al. (2020). Data on the proximate composition, bioactive compounds, physicochemical and functional properties of a collection of faba beans (*Vicia faba* L.) and lentils (*Lens culinaris* Medik.). *Data in Brief* 34: 106660.

Dhull, S.B., Kinabo, J., and Uebersax, M.A. (2022). Nutrient profile and effect of processing methods on the composition and functional properties of lentils (*Lens culinaris* Medik): a review. *Legume Science* 2022: e156.

Didinger, C. and Thompson, H.J. (2022). The role of pulses in improving human health: a review. *Legume Science* 4: e147.

Duenas, M., Sarmento, T., Aguilera, Y. et al. (2016). Impact of cooking and germination on phenolic composition and dietary fibre fractions in dark beans (*Phaseolus vulgaris* L.) and lentils (*Lens culinaris* L.). *LWT Food Science and Technology* 66: 72–78.

Eckert, E., Wismer, W., Waduthanthri, K. et al. (2018). Application of barley- and lentil-protein concentrates in the production of protein-enriched doughnuts. *Journal of the American Oil Chemists' Society* 95: 1027–1040.

Ek, P., Gu, B.-J., and Ganjyal, G.M. (2021). Whole seed lentil flours from different varieties (Brewer, Crimson, and Richlea) demonstrated significant variations in their expansion characteristics during extrusion. *Journal of Food Science* 86: 942–951.

El-Adawy, T., Rahma, E., El-Bedawey, A., and El-Beltagy, A.E. (2003). Nutritional potential and functional properties of germinated mung bean, pea and lentil seeds. *Plant Foods for Human Nutrition* 58: 1–13.

Faris, M.A.E. and Attlee, A. (2017). Chapter 3: Lentils (*Lens culinaris*, L.): a novel functional food. In: *Exploring the Nutrition and Health Benefits of Functional Foods* (ed. H.U. Shekhar, Z.H. Howlader, and Y. Kabir), 42–72. IGI Global.

Faris, M.A.E., Takruri, H.R., and Issa, A.Y. (2013). Role of lentils (*Lens culinaris* L.) in human health and nutrition: a review. *Mediterranean Journal of Nutrition and Metabolism* 6 (1): 3–16.

Farooq, Z. and Boye, J. (2011). Chapter 11: Novel food and industrial applications of pulse flours and fractions. In: *Pulse Foods Processing, Quality and Nutraceutical Applications* (ed. B.K. Teiwari, A. Gowen, and B. McKenna), 283–323. Oxford: Academic Press.

Fasina, O., Tyler, B., Pickard, M. et al. (2001). Effect of infrared heating on the properties of legume seeds. *International Journal of Food Science and Technology* 36: 79–90.

Fouad, A.A. and Rehab, F.M.A. (2015). Effect of germination time on proximate analysis, bioactive compounds and antioxidant activity of lentil (*Lens culinaris* Medik.) sprouts. *Acta Scientiarum Polonorum. Technologia Alimentaria* 14 (3): 233–246.

Ganesan, K. and Xu, B. (2017). Polyphenol-rich lentils and their health promoting effects. *International Journal of Molecular Sciences* 18 (11): 2390.

Garcia-Mora, P., Peñas, E., Frias, J. et al. (2015). High-pressure improves enzymatic proteolysis and the release of peptides with angiotensin I converting enzyme inhibitory and antioxidant activities from lentil proteins. *Food Chemistry* 171: 224–232.

Gharibzahedi, S.M. and Smith, B.M. (2020). Effects of high hydrostatic pressure on the quality and functionality of protein isolates, concentrates, and hydrolysates derived from pulse legumes: a review. *Trends in Food Science and Technology* 107: 466–479.

Ghumman, A., Kaur, A., and Singh, N. (2016a). Impact of germination on flour, protein and starch characteristics of lentil (*Lens culinaris*) and horsegram (*Macrotyloma uniflorum* L.) lines. *LWT Food Science and Technology* 65: 137–144.

Ghumman, A., Kaur, A., Singh, N., and Singh, B. (2016b). Effect of feed moisture and extrusion temperature on protein digestibility and extrusion behavior of lentil and horsegram. *LWT Food Science and Technology* 70: 349–357.

Gomez, M. and Martínez, M.M. (2016). Changing flour functionality through physical treatments for the production of gluten-free baking goods. *Journal of Cereal Science* 67: 68–74.

González, Z. and Pérez, E. (2002). Evaluation of lentil starches modified by microwave irradiation and extrusion cooking. *Food Research International* 35: 415–420.

Grela, E.R., Kiczorowska, B., Samolińska, W. et al. (2017). Chemical composition of leguminous: part I-content of basic nutrients, amino acids, phytochemical compounds and antioxidant activity. *European Food Research and Technology* 243: 1385–1395.

Gularte, M.A., Gómez, M., and Rosell, C.M. (2012). Impact of legume flours on quality and in vitro digestibility of starch and protein from gluten-free cakes. *Food and Bioprocess Technology* 5: 3142–3150.

Gunes, Z.S. and Karaca, A. (2022). Examining the amino acid composition, secondary structure, and physicochemical and functional properties of proteins isolated from local lentil landraces of Anatolia. *Cereal Chemistry* 99: 78–89.

Han, I.H., Swanson, B.G., and Baik, B.-K. (2007). Protein digestibility of selected legumes treated with ultrasound and high hydrostatic pressure during soaking. *Cereal Chemistry* 84: 518–521.

Han, J., Janz, J.A.M., and Gerlat, M. (2010). Development of gluten-free cracker snacks using pulse flours and fractions. *Food Research International* 43: 627–633.

Hefnawy, T.H. (2011). Effect of processing methods on nutritional composition and anti-nutritional factors in lentils (*Lens culinaris*). *Annals of Agricultural Science* 56 (2): 57–61.

Jarpa-Parra, M. (2018). Lentil protein: a review of functional properties and food application. An overview of lentil protein functionality. *International Journal of Food Science and Technology* 53: 892–903.

Jarpa-Parra, M., Wong, L., Wismer, W. et al. (2017). Quality characteristics of angel food cake and muffin using lentil protein as egg/milk replacer. *International Journal of Food Science and Technology* 52: 1604–1613.

Jeske, S., Bez, J., Arendt, E.K., and Zannini, E. (2019). Formation, stability, and sensory characteristics of a lentil-based milk substitute as affected by homogenisation and pasteurisation. *European Food Research and Technology* 245: 1519–1531.

Joshi, M., Aldred, P.L., Mcknight, S. et al. (2013). Physicochemical and functional characteristics of lentil starch. *Carbohydrate Polymers* 92: 1484–1496.

Kabakci, C., Sumnu, G., Sahin, S., and Oztop, M.H. (2021). Encapsulation of magnesium with lentil flour by using double emulsion to produce magnesium enriched cakes. *Food and Bioprocess Technology* 14: 1773–1790.

Kang, H.W. (2015). Antioxidant activity of ethanol and water extracts from lentil (*Lens culinaris*). *Journal of Food and Nutrition Research* 3 (10): 667–669.

Karaca, A., Low, N.H., and Nickerson, M. (2013a). Encapsulation of flaxseed oil using a benchtop spray dryer for legume protein-maltodextrin microcapsule preparation. *Journal of Agricultural and Food Chemistry* 61: 5148–5155.

Karaca, A., Nickerson, M., and Low, N.H. (2013b). Microcapsule production employing chickpea or lentil protein isolates and maltodextrin: physicochemical properties and oxidative protection of encapsulated flaxseed oil. *Food Chemistry* 139: 448–457.

Kaya, E., Tuncel, N.B., and Yılmaz & Tuncel N. (2017). The effect of ultrasound on some properties of pulse hulls. *Journal of Food Science and Technology* 54: 2779–2788.

Keskin, S.O., Ali, T.M., Ahmed, J. et al. (2022). Physico-chemical and functional properties of legume protein, starch, and dietary fiber – a review. *Legume Science* 4 (1): e117. (Early view).

Khazaei, H., Subedi, M., Nickerson, M. et al. (2019). Seed protein of lentils: current status, progress, and food applications. *Food* 8: 391.

Klopsch, R., Baldermann, S., Voss, A. et al. (2019). Narrow-banded UVB affects the stability of secondary plant metabolites in kale (*Brassica oleracea* var. *sabellica*) and pea (*Pisum sativum*) leaves being added to lentil flour fortified bread: a novel approach for producing functional foods. *Food* 8: 427.

Lee, H.W., Lu, Y., Zhang, Y. et al. (2021). Physicochemical and functional properties of red lentil protein isolates from three origins at different pH. *Food Chemistry* 358: 129749.

Liu, S., Yin, H., Pickard, M., and Ai, Y. (2020). Influence of infrared heating on the functional properties of processed lentil flours: a study focusing on tempering period and seed size. *Food Research International* 136: 109568.

Ma, Z., Boye, J.I., and Simpson, B.K. (2016). Preparation of salad dressing emulsions using lentil, chickpea and pea protein isolates: a response surface methodology study. *Journal of Food Quality* 39: 274–291.

Majeed, T., Wani, I.A., and Hussain, P.R. (2017). Effect of dual modification of sonication and γ-irradiation on physicochemical and functional properties of lentil (*Lens culinaris* L.) starch. *International Journal of Biological Macromolecules* 101: 358–365.

Montemurro, M., Pontonio, E., Gobbetti, M., and Rizzello, C.G. (2019). Investigation of the nutritional, functional and technological effects of the sourdough fermentation of sprouted flours. *International Journal of Food Microbiology* 302: 47–58.

Morales, P., Berrios, J.J., Varela, A. et al. (2015). Novel fiber-rich lentil flours as snack-type functional foods: an extrusion cooking effect on bioactive compounds. *Food & Function* 6: 3135–3143.

Nosworthy, M., Medina, G., Franczyk, A. et al. (2017). Effect of processing on the in vitro and in vivo protein quality of red and green lentils (*Lens culinaris*). *Food Chemistry* 240: 588–593.

Osemwota, E.C., Alashi, A.M., and Aluko, R.E. (2021). Comparative study of the structural and functional properties of membrane-isolated and isoelectric pH precipitated green lentil seed protein isolates. *Membranes (Basel)* 11: 694.

Ozkahraman, B.C., Sumnu, G., and Sahin, S. (2016). Effect of different flours on quality of legume cakes to be baked in microwave-infrared combination oven and conventional oven. *Journal of Food Science and Technology* 53: 1567–1575.

Pal, R.S., Bhartiya, A., Yadav, P. et al. (2017). Effect of dehulling, germination and cooking on nutrients, anti-nutrients, fatty acid composition and antioxidant properties in lentil (*Lens culinaris*). *Journal of Food Science and Technology* 54: 909–920.

Pasqualone, A., Costantini, M., Labarbuta, R., and Summo, C. (2021). Production of extruded-cooked lentil flours at industrial level: effect of processing conditions on starch gelatinization, dough rheological properties and techno-functional parameters. *LWT Food Science and Technology* 147: 111580.

Pathiratne, S., Shand, P.J., Pickard, M., and Wanasundara, J.P.D. (2015). Generating functional property variation in lentil (*Lens culinaris*) flour by seed micronization: effects of seed moisture level and surface temperature. *Food Research International* 76: 122–131.

Polat, H., Dursun Capar, T., Inanir, C. et al. (2020). Formulation of functional crackers enriched with germinated lentil extract: a response surface methodology box-behnken design. *LWT Food Science and Technology* 123: 109065.

Portman, D., Blanchard, C., Maharjan, P. et al. (2018). Blending studies using wheat and lentil cotyledon flour-effects on rheology and bread quality. *Cereal Chemistry* 95: 849–860.

Portman, D., Maharjan, P., McDonald, L.S. et al. (2019). Nutritional and functional properties of cookies made using down-graded lentil- a candidate for novel food production and crop utilisation. *Cereal Chemistry* 97: 95–103.

Quintero-Quiroz, J., Celis-Torres, A., Ciro-Gómez, G. et al. (2021). Physicochemical properties and functional characteristics of ultrasound-assisted legume-protein isolates: a comparative study. *Journal of Food Science and Technology* 59: 1665–1676.

Rathod, R.P. and Annapure, U.S. (2016). Effect of extrusion process on antinutritional factors and protein and starch digestibility of lentil splits. *LWT Food Science and Technology* 66: 114–123. https://doi.org/10.1016/J.LWT.2015.10.028.

Rizzello, C.G., Verni, M., Bordignon, S. et al. (2017). Hydrolysate from a mixture of legume flours with antifungal activity as an ingredient for prolonging the shelf-life of wheat bread. *Food Microbiology* 64: 72–82.

Romano, A., Gallo, V., Ferranti, P., and Masi, P. (2021). Lentil flour: nutritional and technological properties, in vitro digestibility and perspectives for use in the food industry. *Current Opinion in Food Science* 40: 157–167.

Sanchez-Velazquez, O.A., Ribereau, S., Mondor, M. et al. (2021). Impact of processing on the in vitro protein quality, bioactive compounds, and antioxidant potential of 10 selected pulses. *Legume Science* 3: –e88.

Shariati-levari, S., Ryland, D., Edel, A. et al. (2016). Sensory and physicochemical studies of thermally micronized chickpea (*Cicer arietinum*) and green lentil (*Lens culinaris*) flours as binders in low-fat beef burgers. *Journal of Food Science* 81: S1230–S1242.

Soltan, S.S.A. (2013). The protective effect of soybean, sesame, lentils, pumpkin seeds and molasses on iron deficiency anemia in rats. *World Applied Sciences Journal* 23: 795–807.

Stantiall, S.E., Dale, K.J., Calizo, F.S., and Serventi, L. (2018). Application of pulses cooking water as functional ingredients: the foaming and gelling abilities. *European Food Research and Technology* 244: 97–104.

Swieca, M., Gawlik-Dziki, U., Kowalczyk, D., and Złotek, U. (2012). Impact of germination time and type of illumination on the antioxidant compounds and antioxidant capacity of *Lens culinaris* sprout. *Scientia Horticulturae* 140: 87–95.

Tam, N., Oguz, S., Aydogdu, A. et al. (2017). Influence of solution properties and pH on the fabrication of electrospun lentil flour/HPMC blend nanofibers. *Food Research International* 102: 616–624.

Tang, X., Shen, Y., Zhang, Y. et al. (2021). Parallel comparison of functional and physicochemical properties of common pulse proteins. *LWT Food Science and Technology* 146: 111594.

Tosh, S.M. and Yada, S. (2010). Dietary fibres in pulse seeds and fractions: characterization, functional attributes, and applications. *Food Research International* 43: 450–460.

Turfani, V., Narducci, V., Durazzo, A. et al. (2016). Technological, nutritional and functional properties of wheat bread enriched with lentil or carob flours. *LWT Food Science and Technology* 78: 361–366.

United States Department of Agriculture (USDA) (2016). Agricultural research service, national nutrient database for standard reference release 28. Nutrient Database Laboratory Home Page. https://ndb.nal.usda.gov/ndb/search/list (accessed 14 August 2016).

Wang, N. and Toews, R. (2011). Certain physicochemical and functional properties of fibre fractions from pulses. *Food Research International* 44: 2515–2523.

Xu, B. and Chang, S.K. (2010). Phenolic substance characterization and chemical and cell-based antioxidant activities of 11 lentils grown in the northern United States. *Journal of Agricultural and Food Chemistry* 58 (3): 1509–1517.

Yeo, J. and Shahidi, F. (2017). Effect of hydrothermal processing on changes of insoluble-bound phenolics of lentils. *Journal of Functional Foods* 38: 716–722.

Zhang, B., Peng, H., Deng, Z., and Tsao, R. (2018). Phytochemicals of lentil (*Lens culinaris*) and their antioxidant and anti-inflammatory effects. *Journal of Food Bioactives* 1 (1): 93–103.

7

Rheological Properties of Lentil Protein and Starch

Jasim Ahmed

Food and Nutrition Program, Environment and Life Sciences Research Center, Kuwait Institute for Scientific Research, Safat, Kuwait

7.1 Introduction

The food legumes have gained increasing importance in the global commodities market recent years due to a shifting trend in the consumption pattern of proteins from animal to plant sources. Lentil (*Lens culinaris*) is one of the leading legumes globally because of its popularity among consumers. Lentil is a highly nutritious and sustainable food source with a high content of protein, dietary fiber, and other micronutrients including iron and vitamin B and is low in fat and cholesterol-free (Jo et al. 2020; Keskin et al. 2022). Many reports are available on the health benefits of consuming lentils, in particular, for reducing hypertension, cardiovascular diseases, diabetes mellitus, and cancer (Barbana et al. 2011; Verni et al. 2020). In addition to their nutritional attributes, a growing interest in developing value-added processing of lentils into various fractional components, including protein and starch isolates, has created an interest in lentils in the food industry for improving functionality. The superior functionality of lentil-based ingredients can play an important role in expanding lentil consumption beyond traditional products and uses. Typically, lentils are distributed as seeds or flour in the supply chain.

Lentil contains 21–32% protein, and its protein has good nutritive value with a high lysine content but is limited in sulfur-containing amino acids (Wynnychuk et al. 2021). The lentil proteins have the potential to be further fractionated into higher protein concentrations and can be marketed as protein concentrates (60–85% protein) or protein isolates (>85% protein) (Aryee and Boye 2017; Khazaei et al. 2019). Globulins are the major proteins in lentils, which consist of legumin (11S; molecular mass: 320–380 kDa) and vicilin (7S; molecular mass: 50 and 70 kDa)-like proteins. Plant proteins, e.g., from lentil, can produce excellent emulsifiers because of their amphiphilic characteristics that enable them to become surface active to align and aggregate at the oil–water interface to form stabilizing films (Primozic et al. 2017). Proteins, being natural products, are in high demand for their use in emulsion stabilization for food applications. Numerous studies have been conducted

Lentils: Production, Processing Technologies, Products, and Nutritional Profile, First Edition.
Edited by Jasim Ahmed, Muhammad Siddiq, and Mark A. Uebersax.

on lentil protein-based oil-in-water emulsion and nanoemulsion (Ma et al. 2013; Primozic et al. 2017; Pang et al. 2020). Furthermore, lentil protein-based aggregates have been successfully used as a stabilizer to form oil-in-water emulsions (Wynnychuk et al. 2021).

The extraction of large quantities of proteins generates a large quantity of lentil starch during processing. To utilize such voluminous amounts of starch in industrial applications, the properties of lentil ingredients should be well understood (Ahmed et al. 2022b). Lentil seeds comprise about 69% carbohydrates, which are mostly in the form of starch (Adsule et al. 1989). Lentil starch granules (3–30 μm diameter) differ in shape from ovoid to spherical and are the main components of the flour (Sotomayor et al. 1999; Hoover et al. 2010; Ahmed et al. 2016). Starch consists of amylopectin and amylose, which are α-D-glucoses linked together in two different configurations (Hoover et al. 2010). The structural and functional characteristics of these glucan polymers influence the functionality and end uses of starch. While starch granules undergo thermal treatment in the presence of excess water, an endothermic transition occurs where semi-crystalline starch granules swell and starch crystallites melt (Ahmed et al. 2022a). Gelatinization modifies the structure and texture of starchy food materials and improves digestibility. Investigations on the starch gelatinization and gel formation are, therefore, highly recommended for the efficient design of foods with desirable properties.

The rheological characterization of lentil flour, starch, and proteins during processing and storage provides a better understanding of those materials for food applications. Both starch and protein influence texture in many food products because of their ability to form gel. A thorough rheological investigation can provide characterization of each component of lentil (e.g., starch, proteins) individually in a composite blend (e.g., starch–protein) or in an emulsion. Changes during non-isothermal heating (changes occurring simultaneously in temperature and time) of lentil protein or starch can be quantified by evaluating the reaction kinetics of the process by monitoring the evolution of the dynamic moduli (elastic modulus, G' and viscous modulus G''), and their associated enthalpies. Starch and protein gels can be well characterized by measuring gel strength and rigidity using oscillatory or creep measurements. When both starch and protein components are used to form a gel, the resulting systems are usually described as composites. The starch–protein interaction of lentils is a relatively recent subject of interest (Joshi et al. 2014). The steady-flow measurements would provide shear-dependency (e.g., thinning or thickening) of lentil ingredients or products or the presence of thixotropy. Similarly, the lentil-enriched emulsions (protein-polysaccharides) can be rheologically characterized by measuring surface properties and represented by interfacial rheology.

This chapter elucidates rheological methods adopted for the measurement and characterization of lentil seed flour, isolated starch and proteins, starch–protein interaction, and lentil protein-based emulsions. Readers can find detailed methodologies and mathematical modeling of rheology elsewhere (Ahmed 2023), as this chapter does not cover them.

7.2 Rheological Measurement Techniques Related to Lentils

The rheological properties of flour, starch, and proteins are described in this section using various rheological measurement techniques. The most common measurement techniques of rheology are steady flow, oscillatory (small amplitude and large amplitude), creep, and recovery, which lead to flow characterization and structural analysis. Conversely, the food industry still prefers to use macroscopic measurements (e.g., pasting) for maintaining quality for day-to-day activities and for production. Therefore, both microscopic and macroscopic techniques are being studied for the characterization of lentil flour and ingredients. It is to be noted that only the microscopic part has been discussed in this chapter, while a separate chapter is devoted to macroscopic measurements (Chapter 9).

7.2.1 Steady Flow

The flow behavior of aqueous dispersions of lentil flour, starch, or proteins can be performed using steady-shear measurements. The obtained experimental shear stress (τ) or the apparent viscosity (η) data can be plotted against the shear rate ($\dot{\gamma}$) to obtain the flow behavior indices of the test specimens. Many flow models have been employed to describe the flow behavior of lentil dispersions and gels (Ahmed 2023) as no single model can describe the flow behavior adequately or characterize the rheological behavior of a material over an extended range of shear rate. However, the power law model (Eqs. (7.1) and (7.2)) is commonly used to characterize lentils or their components (starch or proteins) in dispersions. A logarithmic plot of the log shear stress/apparent viscosity versus the log shear rate yields a straight line and the resultant slope (n) indicates the nature of the fluid, whether it is shear-thinning or shear-thickening. Some fluids, however, exhibit a minimum stress prior to flow, which is termed the yield stress (τ_0). An extension of the power law model can measure the yield using the Herschel–Bulkley model (Eq. (7.3)).

$$\tau = K\dot{\gamma}^n \tag{7.1}$$

$$\eta = K\dot{\gamma}^{n-1} \tag{7.2}$$

$$\tau = \tau_0 + K\dot{\gamma}^n \tag{7.3}$$

where K is the consistency index (Pa·s^n) and τ_0 is the yield stress (Pa).

7.2.2 Oscillatory Rheology

Oscillatory shear measurement provides insights into structural changes of legume flour, proteins, or starch in dispersion or a formulation in the linear viscoelastic range. When a specimen is subjected to oscillatory stress, and determining the response, both elastic and viscous or damping characteristics can be obtained. Viscoelastic materials show both the elastic (G') and viscous (G'') behavior. The measurement can be made by detecting the gelatinization of starch, denaturation of protein, starch–lipid complex, or their interaction during heating. The entire process can further yield the reaction kinetics of the process, which is useful in process design applications (Ahmed 2014).

7.2.3 Creep and Recovery

The creep and recovery tests consist of measuring the extension or deformation of a viscoelastic specimen subjected to constant shear stress (load) for a set period (creep), followed by the release of the stress to assess the ability of the samples to recover after a maximum strain is attained (recovery). Starch gels show partial recovery of structure because of their viscoelastic characteristics (Ma et al. 2013). To determine irreversible deformations, creep and recovery of legume starch gels are conducted at selected loads and temperatures. Creep-recovery measurements may give some insight into the macro-structure of starch-rich dough of lentil. A small recovery reflects the presence of small structures, while a large recovery illustrates the presence of larger structures (Weegels et al. 1996). The creep and recovery experimental data are mostly analyzed by fitting them into the four-parameter Burger's model, which is basically a combination of a Maxwell element and a Kelvin–Voigt element in series. However, the presence of a large number of parameters makes it complex to represent a dough system (Steffe 1996), and it is difficult to interpret the data.

Creep data are represented using creep compliance, J_c, which is defined as the strain divided by the applied stress. The experimental data from creep tests can be described with precision by a discrete retardation spectrum $\left\{ J_k / \tau_k \right\}$, which is a set of $2n$ positive constants J_k and τ_k, using the following equation (Kaschta and Schwarzl 1994):

$$J_c(t) = J_0 + \sum_{k=1}^{n} J_k \left[1 - e^{-t/\tau_k} \right] + \frac{t}{\eta_0} \tag{7.4}$$

where each retardation time τ_k is related to a spectral compliance magnitude (Voigt compliance) J_k, J_0 is the instantaneous compliance (Maxwell compliance), and η_0 is the steady-state viscosity (Maxwell dashpot viscosity).

For the recovery compliance $J_r(t)$, similar equations are followed. Since there is no viscous flow in the recovery phase, the Eq. (7.4) can be modified into the following form that consists only of parameters describing the elastic response after release of the shear stress. The data from creep tests can be described by thc Burgers model:

$$J_r(t) = J_0 + \sum_{k=2}^{n} J_k \left[1 - e^{-t/\tau_k} \right] \tag{7.5}$$

7.3 Rheological Properties of Lentil Constituents

7.3.1 Lentil Flour

7.3.1.1 Steady-Flow Rheology

The steady-flow rheology of lentil flour dough at selected moisture contents (14–22%) and shear rates (0.01–100 s^{-1}) at 110 °C displayed that the η decreased with increasing shear rate and the dough behaved as a shear-thinning fluid (Figure 7.1) (Rathod et al. 2017). The shear stress–shear rate data fitted the Herschel–Bulkley model well, and the rheological parameters are shown in Table 7.1. Furthermore, it was observed that the η and the τ_0 of the lentil dough decreased with increasing the moisture content in the flour as practiced during the extrusion. High moisture conditioning and extrusion temperatures of lentils resulted in higher τ_0, K, and n. An increase in τ_0 and K indicates an increment of viscosity with increasing moisture content. The changes observed in the viscosity and flow behavior properties could be attributed to the chemical changes occurring in the lentil due to the varied conditioning moisture, affecting the intensity of the extrusion process. Decreasing

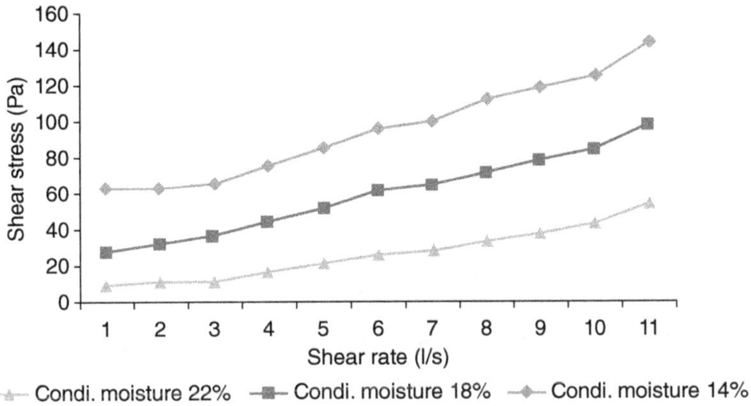

Figure 7.1 Shear stress–shear rate rheograms for the lentil dough at 110 °C. *Source:* Rathod et al. (2017), with permission of Springer Nature.

Table 7.1 Herschel-Bulkley model parameters for lentil doughs at 110 °C.

Moisture content at conditioning (%)	Yield stress (Pa)	Consistency index (*K*), Pa·s^n	Flow behavior index (*n*)
14	1.78 ± 0.03	0.94 ± 0.01	0.84 ± 0.06
18	2.64 ± 0.05	1.89 ± 0.02	0.77 ± 0.04
22	6.92 ± 0.05	2.92 ± 0.04	0.72 ± 0.03

Source: Adapted from Rathod et al. (2017).

the conditioning moisture of the lentil and increasing extrusion temperature as well as screw speed favored the extrusion process, converting more of the polymeric structure of carbohydrates into reducing sugars.

7.3.1.2 Oscillatory Rheology

The influence of particle size (74–210 μm) on the oscillatory rheology of lentil flours (LF) from two geographical locations (Indian and Turkish) was investigated using isothermal and non-isothermal heating (Ahmed et al. 2016). The evolution of the complex viscosity (η^*) with increasing heating time was demonstrated by non-isothermal heating (25–95 °C at 5 °C/min) of LF particle dispersions (flour to water = 1 : 2) at 1 Hz. The increase in the η^* was attributed to the interplay of the gradual swelling of starch granules that eventually develop into a close-packed network (Eliasson 1986). Each particle fraction in the dispersion exhibited a distinct η^* peak at a defined temperature, denoted by T_p. The peak η^* value for Turkish LF decreased as particle size decreased; however, this trend did not hold true for Indian lentils (Figure 7.2). The η^* dropped significantly for the dispersions containing the finest particles, and it could be associated with the lower porosity that resulted in their being unable to imbibe much water. T_p values were detected in the Indian and Turkish samples at 80 and 83 °C, respectively. Such differences could be attributed to the variations

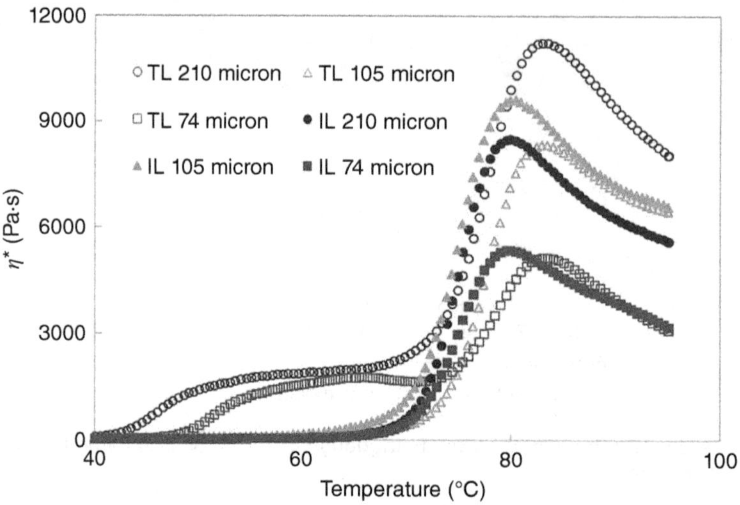

Figure 7.2 Influence of particle size on the peak complex viscosity of lentil flour dispersion during non-isothermal heating at a heating rate of 5 °C/min (TL-Turkish lentil and IL-Indian lentil flour samples). *Source:* Ahmed et al. (2016), with permission of Elsevier.

in water absorption capability, swelling behavior of individual flour particles, amylose content, amylose/amylopectin ratio, and amylopectin crystal melting behavior of the two specimens.

7.3.2 Lentil Starch

7.3.2.1 Steady-Flow Rheology

The rheological characteristics of the heat-treated lentil starch dispersions (2.6% w/w) from three cultivars were studied at 90 °C with varying heating times (2–44 minutes) by Lee (2007). Typical rheograms of heat-treated starch dispersions exhibited the thixotropic behavior of the gel (Figure 7.3). With the heating time, the steady flow data fitted well to the Power law ($R^2 > 0.98$) and the Hershel–Bulkley ($R^2 > 0.98$) flow models. Rheological model parameters are presented in Table 7.2. All three starches exhibited shear-thinning behavior up to the shear rate of $800\,s^{-1}$. The n values decreased with prolonged heating. The increasing trend of the consistency coefficient, K, for both models with heating time indicated the development of significantly higher viscosity during the heating. A relatively high yield stress of Australian lentil starches is attributed to the presence of a significantly high level of hydrogen bonds in the amylose helix structure, and therefore, forms a stable configuration that is resistant to flow when sheared. Lee (2007) further found a good correlation between all calculated K and n values from the Power law model and the starch amylose content. Conversely, the K and n obtained from the Hershel–Bulkley model did not show any correlation with amylose content. A good correlation was established between the yield stress and the amylose content. This was probably due to the formation of hydrogen bonds in the amylose helix structure, leading to the formation of a stable configuration in the starches, which increases resistance to flow when sheared, thus resulting in higher yield stress (Harrison et al. 1999).

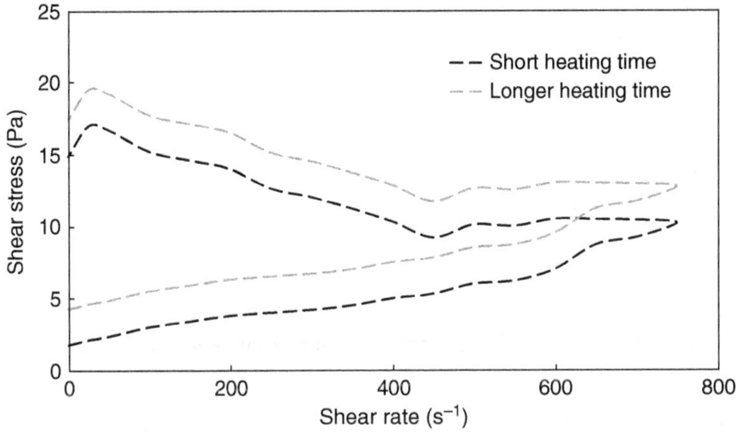

Figure 7.3 Shear stress–shear rate plot of typical lentil starch dispersions heated at two selected times between 0 and 750 s^{-1} shear rate.

Table 7.2 Rheological model parameters for three Australian lentil starch dispersions as a function of heating time at 90 °C.

		Power law model		**Herschel–Bulkley model**		
Lentil cultivar	**Heating time (min)**	**K (Pa·s^n)**	**n(−)**	**τ_0 (Pa)**	**K (Pa·s^n)**	**n(−)**
Matilda	2	0.31	0.58	2.20	0.062	0.80
	15	0.56	0.53	3.36	0.098	0.77
	30	0.53	0.53	3.33	0.085	0.79
	44	0.97	0.48	4.75	0.141	0.74
Digger	2	0.35	0.56	2.44	0.059	0.81
	15	0.34	0.58	2.73	0.049	0.85
	30	0.73	0.50	4.34	0.083	0.80
	44	0.99	0.48	5.12	0.122	0.76
Cobber	2	0.61	0.52	3.09	0.111	0.73
	15	0.82	0.47	3.52	0.146	0.70
	30	0.75	0.48	3.45	0.139	0.71
	44	1.52	0.42	5.81	0.197	0.69

Source: Adapted from Lee (2007).

Byars and Singh (2016) studied the steady flow properties of lentil starch dispersions at higher concentrations (6%, 8%, and 10% w/w) at 50 °C after the pasting test in a pasting cell. The results showed that the data fitted the power-law model in the entire shear rate range. The n did not change in the concentration range of 6–8% while the K increased threefold from 5.4 to 18.5 Pa·s^n, indicating an increase in the mechanical rigidity of the developed gel. The gel was so rigid at 10% that the viscosity of the gel was not measurable at 50 °C.

The effect of heat-moisture treatment (30% moisture, 100 °C, 16 hours) and annealing (75% moisture, 50 °C, 72 hours) on the flow behavior of 6% gelatinized lentil starch pastes demonstrated that the apparent viscosity-shear rate data fitted the power law model (Hoover and Vasanthan 1994). The native starch exhibited a non-Newtonian shear thinning behavior. The n and the K values of the native lentil starch were 0.46 and 5.53 Pa·s^n, respectively. Heat-moisture treatment and annealing decreased the K values and increased the n values significantly. However, the modification to the flow behavior was more obvious during heat-moisture treatment than from annealing. The n displays the shear thinning behavior of a starch paste subjected to increasing shear rate. Shear thinning could be attributed to deformation and subsequent disintegration (more pronounced at very high shear rates) of the swollen granules under the influence of the shear field.

7.3.2.2 Oscillatory Rheology

Gelatinization Reaction Kinetics Starch rheology is an important area of research, with a significant influence on food product development and non-food industrial applications. Starch granules gelatinize during heating in excess water at a defined temperature (55–80 °C), which is known as the gelatinization temperature (T_p). During the process, the structure of the starch granules is disrupted and swells up to several times their original volume. The T_p depends on the amount of water present in the starch matrix and the pH of the sample. T_p values for lentil starch ranged from 63 to 70 °C (Lee 2007; Ahmed and Auras 2011). The starch sample develops a three-dimensional gel when passed through the gelatinization process. The starch gels are subjected to small and large amplitude oscillatory shear, steady flow, or creep measurements during rheological measurements (Ahmed et al. 2016; Acevedo et al. 2017; Doublier 1987; Phrukwiwattanakul et al. 2014). The developed gel can be quantified by measuring the elastic modulus (G') using a rheometer. The gel rigidity is primarily a function of the starch concentration, water content, pH, and some other factors.

Since starch gelatinization occurs in a nonequilibrium state, the gelatinization reaction kinetics provide a set of process parameters (temperature, time, viscosity, and mechanical strength) for a specific starch. Such data is very useful for process and equipment design for a specific starch or a blend. The differential scanning calorimetry (DSC) measurement has limitations and is not able to detect gelatinization temperature by providing the required endothermic curve. Conversely, rheometric measurement can more precisely detect the gelatinization temperature and reaction kinetics during the non-isothermal heating of starch.

Ahmed and Auras (2011) evaluated the gelatinization reaction kinetics of native and hydrolyzed lentil starch were evaluated using oscillatory rheology. The authors used a time–temperature profile and an Arrhenius relationship to evaluate the reaction kinetics of the gelatinization using changes in the elastic modulus (G') with time and temperature. An increase in the G' with heating temperature and time represents a change in concentration in reaction kinetics. The final equation for the process is shown below. Interested readers can consult elsewhere for more details (Ahmed 2023).

$$\ln\left(\frac{1}{G'^n}\cdot\frac{\mathrm{d}G'}{\mathrm{d}t}\right) = \ln k_0 - \left(\frac{E_a}{R}\right)\left(\frac{1}{T}\right) \tag{7.6}$$

$$\ln\left(\frac{\mathrm{d}G'}{\mathrm{d}t}\right) = \ln k_0 + n\ln G' - \left(\frac{E_a}{R}\right)\left(\frac{1}{T}\right) \tag{7.7}$$

where k_0 is the pre-exponential or frequency factor, E_a the activation energy (J/mol), T the absolute temperature (K), and R the universal gas constant (8.314 J/mol/K).

The derivatives dG' and dt represent changes in the elastic modulus and time. The negative sign of the original equation changes to positive because of the development of rigidity with increasing G' during heating. The kinetic parameters, E_a and k_0, are estimated from an Arrhenius-type plot.

The kinetic data set, i.e., order of the reaction and activation energy, of lentil starch was analyzed using a multiple linear regression technique, and the results showed that the reaction orders for native and hydrolyzed lentil starch dispersions were 1.11 and 1.20, respectively (Ahmed and Auras 2011). The calculated value of n was greater than unity, thus the reaction order of each sample was considered to be second-order kinetics. Figure 7.4 illustrates an Arrhenius-type plot for starch gelatinization with data generated by linearly increasing temperature. The activation energies (E_a) were 241 ($R^2 = 0.96$) and 434 kJ/mol ($R^2 = 0.82$) for hydrolyzed and native lentil starch dispersions, respectively. The lowering of E_a in hydrolyzed starch could have been attributed to the presence of amylose only in the microstructure when compared to a native starch with a firm gel that was contributed by both amorphous amylose and crystalline amylopectin (Virtanen et al. 1993).

The evolution of complex viscosity (η^*) of lentil starch when blended with nanoclay in dispersion during non-isothermal heating from 25 to 95 °C at a heating rate of 2.5 °C/min showed a similar rheogram as observed for lentil flour (Ahmed 2014). The peak gelatinization temperature for the starch dispersion was detected at about 70 °C. However, the T_p was not distinct when clay was incorporated into starch dispersions. The η^*_{max} value marginally decreased from 81.6 to 80.4 °C with an increase in clay concentration, indicating nanoclays did not have much impact on the reinforcement at elevated temperatures for short contact times.

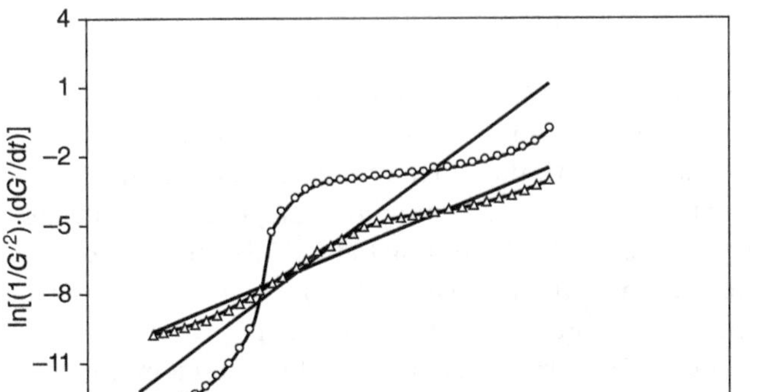

Figure 7.4 The 2nd-order reaction kinetics for lentil starch dispersion (25 g starch per 100 g water) during thermal gelatinization (B: unhydrolyzed and 6: hydrolyzed). *Source:* Ahmed and Auras (2011), with permission of Elsevier.

Isothermal Starch Gel Ahmed and Auras (2011) prepared gel by isothermal heating of native and hydrolyzed lentil starch dispersions at 95 °C for 15 minutes, followed by cooling to 25 °C, and thereafter comparing their mechanical rigidity. The G' increased with increasing frequency, and the native starch exhibited a stronger gel with a higher G' than the hydrolyzed starch. The G'' of both starches, however, was superimposed on each other. The higher mechanical rigidity of the native starch gel has been attributed to higher inherent amylose content that leached out during heating and improved the rigidity of the starch granular structure (Lii et al. 1996). Conversely, the acid hydrolysis preferentially attacks the amorphous regions in the starch granule (Manelius et al. 1974); whereas the crystallites are decoupled from and no longer destabilized by the amorphous parts. So, acid-treated starch gel exhibited lower rigidity during heating and subsequent cooling.

The viscoelastic behavior of the thermally treated gels is characterized by calculating slopes of the linear regression of the power-type relation (Eqs. (7.8) and (7.9)).

$$G' = A\omega^n \tag{7.8}$$

$$\ln G' = \ln A + n \ln \omega \tag{7.9}$$

where n is the slope (dimensionless) and A is the intercept (Pa·s^n). The magnitudes of the slope values (0.06–0.10) indicate a gradual transformation of liquid-like characteristics to solid-like characteristics following the gel formation. The viscoelasticity of gels was further verified by plotting a graph between the complex viscosity (η^*) and phase angle (δ) against frequency, which displayed a systematic decrease in the η^* with increasing ω for both starch samples, and a lower value of δ confirmed more solid-like behavior of native starch gels (Ahmed and Auras 2011). The oscillatory rheology of lentil starch gels (6%, 8%, and 10% w/w) measured at 25 °C displayed that the G' was nearly independent of frequency and 10 times larger than the loss modulus, G'', indicating the formation of a gel network (Byars and Singh 2016). Lentil starch, however, formed weaker gels at all concentrations and had the lowest rate of increase in the G' with concentration due to lower amount of amylose leaching from the starch during sample pasting.

Applicability of Time–Temperature Superposition Principle (TTSP) In oscillatory measurement, a series of isothermal curves correlating dynamic moduli (elastic modulus, G', viscous modulus, G'', and complex viscosity, η^*) with the range of measurement frequency (ω) can be brought together on a single master curve at a reference temperature by means of *time–temperature superposition* (Ahmed 2022). According to the TTSP, the frequency (or response time) function of rheological modulus at a particular temperature is very similar in shape to the same functions at adjacent temperatures. Following the TTSP rule, a series of frequency sweep (0.1 and 10 Hz) experiments of lentil starch (starch to water ratio of 1 : 3) were conducted at selected temperatures (70, 75, 80, 85, and 90 °C), and rheograms were brought together to a reference temperature of 80 °C (Ahmed 2014). The temperature range was selected based on the gelatinization temperature of lentil starch (68 °C).

The log–log plots of the isotherms of the dynamic moduli (e.g., elastic modulus (G') and complex viscosity (η^*)) were superimposed to a reference temperature (T_{ref}) of 80 °C by horizontal shifts log (a_T), along the log (ω) axis, and vertical shifts given by log (b_T). The relevant equations are shown below:

$$b_T G'(\omega; T) = G'(a_T\omega; T_{\text{ref}}) \tag{7.10}$$

$$|\eta^*|(\omega; T) = (b_T / a_T)|\eta^*|(a_T\omega; T_{\text{ref}}) \tag{7.11}$$

Figure 7.5 displays the reduced angular frequency, $a_T\omega$ dependence of G', and η^* for the lentil starch at selected temperatures. The G'-ω data was not superimposed at 70 and 90 °C, indicating the rule was not followed. Calculated values of the a_T coefficient are not close to unity, indicating that scaling has a significant effect on extending the frequency window.

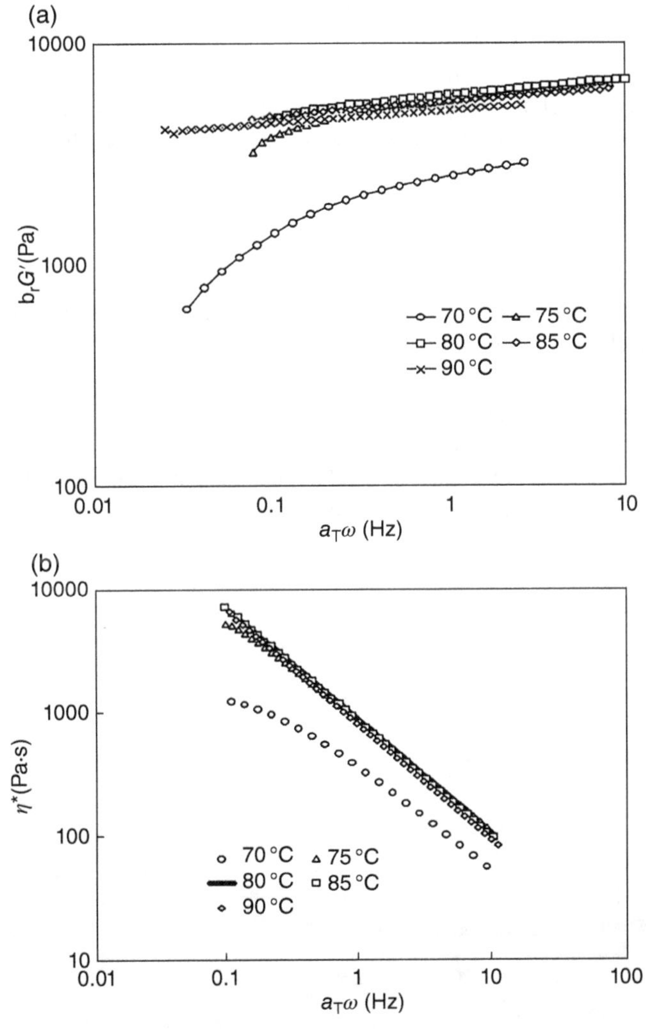

Figure 7.5 Applicability of time–temperature superposition (TTS) for neat starch sample: (a) elastic modulus and (b) complex viscosity. *Source:* Ahmed (2014), with permission of Taylor & Francis.

The TTS, on the other hand, was well fitted for the complex viscosity except for the temperature at 70 °C. The possible reason for the non-applicability of TTS at 70 °C could be the proximity of the applied temperature to the gelatinization temperature of starch where semi-crystalline starch granules were likely converted completely to an amorphous region.

7.3.2.3 Creep and Recovery

A comparative study on creep and recovery between native and modified (acid-hydrolyzed) lentil starch (25–33.3%) gel was conducted at selected temperatures by Ahmed and Auras (2011). Both gels were produced by holding starch dispersions at 95 °C for 15 minutes *in situ*, followed by cooling to 25 °C. A considerable variation in the creep behavior was observed between the two test samples (Figure 7.6). At a constant stress of 80 Pa, it was observed that the increment of strain was considerably higher for the hydrolyzed starch gel, indicating that it was less resistant to the stress and produced a weaker gel when compared with the native one. However, after removal of the applied stress, the elastic strain showed a permanent deformation with no significant difference found between the two starches.

The Burgers model was used to analyze the creep curves using the creep compliance (Eq. (7.12)) and model parameters were estimated following the method described by Steffe (1996).

$$J = f(t) = J_0 + J_1\left(1 - \exp\left(\frac{-t}{\lambda_{ret}}\right)\right) + \frac{t}{\mu_0} \tag{7.12}$$

where J_0 is the instantaneous compliance, J_1 is the retarded compliance, λ_{ret} is the retardation time (μ/G_1) of the Kelvin component, and μ_0 is the Newtonian viscosity of the free dashpot.

The Burgers model for both lentil starch gels indicate that the thermally assisted gel was significantly influenced by starch modification (e.g., hydrolysis). Furthermore, the retardation time (λ_{ret}) remains independent of the individual gel. For a perfect Hookian solid, the λ_{ret} becomes zero (Steffe 1996) and it increases with increasing viscous component.

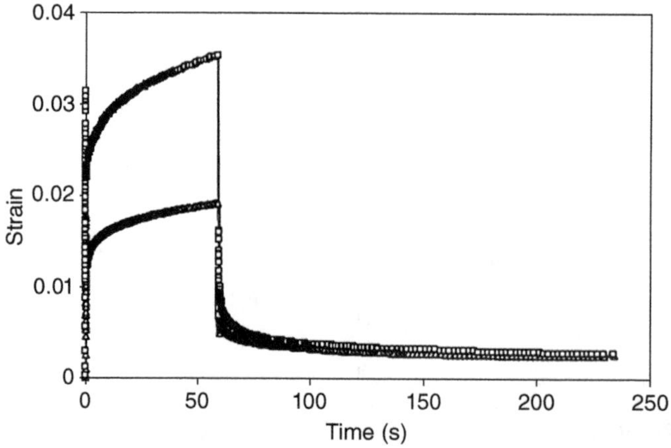

Figure 7.6 Creep and recovery curves for lentil starch gels (triangle: whole starch and, square: hydrolyzed starch). *Source:* Ahmed and Auras (2011), with permission of Elsevier.

Table 7.3 Creep properties of lentil starch gels at selected concentration.

Starch Concentration (%, w/w)	J_0 (Pa^{-1})	J_e^0 (Pa^{-1})	τ_{avg} (s)	η (Pa·s)
6	1.55×10^{-7}	3.73×10^{-4}	0.03	1.18×10^{6}
8	1.01×10^{-6}	1.24×10^{-4}	0.02	2.73×10^{6}
10	1.34×10^{-7}	2.95×10^{-5}	5.76	1.07×10^{7}

Source: Adapted from Byars and Singh (2016).

Therefore, the λ_{ret} time for hydrolyzed gel corresponds to a more viscous product. Overall, the results revealed that the hydrolyzed lentil starch had a lower ability to form gels with rigidity and stability when compared to the native starch.

The creep measurements of lentil starch gels (6%, 8%, and 10% w/w) were performed for the samples stored at 4 °C for 24 hours, and the results were fitted to a six-parameter Burgers model (Byars and Singh 2016). Values of the instantaneous compliance (J_0) and the steady-state compliance $\left(J_e^0\right)$ of lentil starch gels are presented in Table 7.3. The J_0 showed no correlation with lentil starch concentration. On the other hand, J_e^0 decreased with increasing concentration, indicating more solid-like behavior. Although a six-parameter model was used to fit the data, only two or three parameters contributed significantly to the fit for all samples. The average retardation time (τ_{avg}) was marginal (0.02–03 second) in the concentration range of 6–8% and increased significantly to 5.76 second at 10%. It was shown that the viscosity increased with the concentration, and the highest viscosity was observed for the gel containing 10% starch concentration. The results further showed that there were no significant changes in the viscosity between 6% and 8% concentrations for any of the starches.

Ma et al. (2013) conducted creep and recovery tests for lentil-based salad dressings prepared with various pre-treatments of lentil flour. The samples without lentil flour (control) had the highest J values for both the creep and recovery tests, while the dressings supplemented with pre-cooked freeze- or spray-dried lentils had the lowest values. At 0.5 Pa, the dimensional changes in the dressings with time (strain) were more for the control and relatively less for the pre-cooked lentils supplemented dressings. For the dressing supplemented with pre-cooked lentils, a stronger network results in less deformation, and for the control, which had the weakest viscoelastic properties, a large change in its dimensions would be observed owing to the weaker emulsion structure. The dressings supplemented with precooked freeze- or spray-dried lentils exhibited significantly higher ($p < 0.05$) recoverable strain ($Q(t)$%) indicative of greater elasticity. As these samples displayed the highest K, η_{app}, G', and plateau modulus ($G^0{}_N$) values, the results together indicate that this type of dressing had the highest viscoelasticity and compact structure.

7.3.3 Lentil Proteins

7.3.3.1 Steady-Flow Rheology

The effectiveness of high-pressure (HP)-assisted enzymatic hydrolysis of lentil protein isolate (LPI) was monitored by measuring steady-shear rheology (Ahmed et al. 2019). As shown in Figure 7.7, the shear stress increased as the shear rate was increased from 0.1 to

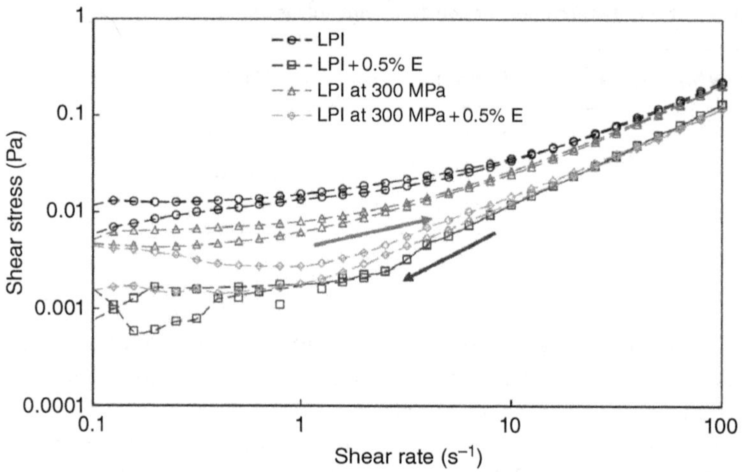

Figure 7.7 Rheograms of lentil proteins as influenced by high-pressure and enzymatic treatment Shear stress–shear rate data with thixotropic loop (LPI, lentil protein isolate; E, Alcalase enzyme). *Source:* Ahmed et al. (2019)/John Wiley & Sons/Public Domain CC BY 4.0.

$100\,s^{-1}$. The $\tau - \dot{\gamma}$ rheograms decreased in the order of LPI (control) + 0.5% Alcalase < LPI at 300 MPa + 0.5% Alcalase < LPI at 300 MPa after pressurization and Alcalase digestion. The presence of thixotropy with structural breakdown of the test samples was evident during the forward ($0–100\,s^{-1}$) and backward ($100–0\,s^{-1}$) runs of the samples. The structural breakdown of the pressurized samples was more pronounced for both enzyme and combined HP/enzyme hydrolysis. The Alcalase digestion of LPI at 0.101 and 300 MPa might have produced the largest number of low MW peptides (<3 kDa) that resulted in the lowering of the viscosity at its bottom. Additionally, the increase in the shear stress of the HP-assisted enzymatic hydrolysates could be related to the agglomeration of the smaller hydrolyzed proteins.

7.3.3.2 Oscillatory Rheology

Boye et al. (2010) investigated the gelling properties of thermally treated (100 °C for one hour) lentil protein concentrates (80% protein), and reported a firm gel requires a minimum of 8–12 wt% protein. In addition, this effect depends on the type of lentil (green or red) and the extraction process employed. For gel formation, a wide range of lentil protein concentrations (1–13% w/w), heating times (0.5–16 hours), and pH values (2–9) were investigated by Jo et al. (2020). The time–temperature rheological profiles of LPI suspensions during thermally-induced gelation at pH 2 and pH 7 are illustrated in Figure 7.8a. At both pH values, the G' decreased significantly during heating, increased marginally during 90 °C holding and then increased rapidly during the cooling and 5 °C holding. The increase in G' upon cooling was probably caused by the decreased mobility of aggregates that facilitated their interactions to form gels during the cooling step. As far as the influence of pH is concerned, the LPI suspension at pH 7 had a higher G' than that at pH 2 during the heating and heated-holding phases; however, the G' at pH 2 crossed the value at pH 7 during the cooling phase. The significantly higher G' at pH 2 indicated a much stronger network

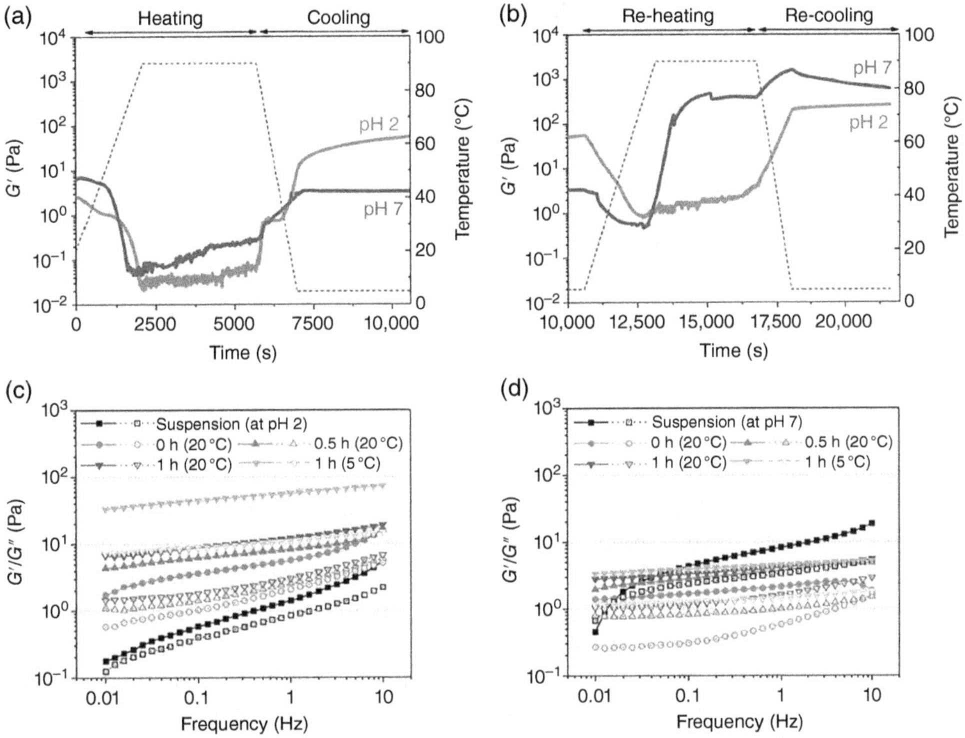

Figure 7.8 Rheological properties of LPI suspension (10%, w/v) for the formation of LPI gels at selected pH levels. (a) A time–temperature gelation profile at different pHs (pH 2, pH 7); (b) a re-heating and re-cooling profile at different pHs; (c and d) frequency dependence of storage modulus (G') and loss modulus (G'') at different pHs (pH 2: C, pH 7: D) as a function of cooling time (0, 0.5, 1 hours). The temperature is depicted in the dashed line. Solid symbols indicate storage modulus (G') and the open symbols indicate loss modulus (G''). *Source:* Jo et al. (2020), with permission of Royal Society of Chemistry.

formed during cooling from fibrillar aggregates rather than particulate aggregates, which was most likely due to their linear and fine structures, which promote aggregate interactions and the formation of highly interconnected networks.

The thermo-reversibility of LPI gels was traced during a re-heating and re-cooling process (Figure 7.8b). Upon reheating, the G' value increased with temperature, resulting in a much higher G' value compared to the level achieved during the initial cooling phase. It demonstrated that the change in gel structure at pH 2 was more thermo-reversible than that at pH 7 during re-heating and re-cooling, suggesting that physical bonds were likely responsible for the formation of fibrillar gel.

Frequency sweep tests were performed for developed gels with cooling times (0, 0.5, and 1 hours) during gel formation at 5 or 20 °C (Figure 7.8c,d). The G' of LPI gels increased with cooling time at 20 °C due to the formation of LPI gels from fibrillar and particulate aggregates, which further built up gel matrices during cooling. The LPI gels had higher G' values at pH 2 than at pH 7, indicating that fibrillar aggregates formed stronger gel networks than particulate aggregates. The G' value of LPI gels formed at pH 2 increased significantly at

5 °C, but there were no significant changes in the G' value of LPI gels formed at pH 7 (Figure 7.8c,d). Such observations indicated that fibrillar aggregates stabilized by hydrogen bonding act more predominantly in gel strengthening than particulate aggregates because hydrogen bonds are strengthened at low temperatures. It was hypothesized that the high aspect ratio allowed fibrillar aggregates to build a higher level of structures with positive characteristics along with other attractive interactions, including hydrophobic interactions and disulfide bonds, to build strong gels.

The effect of ethanol (30–80% v/v) addition on gelation properties of lentil protein powder dispersions (10–20%) revealed that the G' was dominant from the beginning of the experiment, indicating that gelation took place immediately upon contact with ethanol due to protein–protein aggregation (Yavuz 2019). Increasing protein concentration increased the gel strength and, generally, the strongest gels were observed at 20% protein when the ethanol content ranged between 30% and 40%. The gel produced by 10% lentil protein in 40% ethanol was soft (16 Pa), whereas 15% lentil protein in 50% ethanol produced a higher G' (22 Pa). Gels at 20% lentil protein in 40% ethanol showed strong gel properties. Increasing protein concentration had a significant effect on G' for time 0 minutes, but ethanol content had no effect ($p > 0.05$). In contrast to time 0, the G' for ethanol content and protein concentration was significant at 60 minutes. The statistical analysis indicated that changing ethanol content and protein concentration had a significant effect on G'' lentil protein values for 0- and 60-minute ($p < 0.05$). Increasing ethanol content beyond 60% rendered the gel structure softer for all protein types. The high amount of ethanol could have changed the balance between protein–solvent interactions and cause dissociation of protein aggregates. It was concluded that ethanol-induced aggregation of protein might be reversible.

Boeck et al. (2021) developed a lentil-based yogurt alternative by producing an emulsion of lentil protein isolate and fermenting it with lactic acid bacteria. The final lentil emulsion contained 3.5% protein. Oscillatory measurements displayed the G' of imitated yogurt, which was 4–4.5 times larger than G'' in the frequency range of 0.01–15 Hz at 6 °C. A similar trend of dynamic moduli was observed for soy drinks. The damping factor ($\tan\delta$) values ranged between 0.22 and 0.25, which indicated that tested samples had more gel-like behavior.

The three interval thixotropy test proposed by Klost and Drusch (2019) was employed for lentil-based yogurt alternatives (Boeck et al. 2021). At the first interval, oscillatory measurement was conducted at a strain of 0.1% and a frequency of 1 Hz for 120 s, followed by a second interval of rotational deformation at a shear rate of 200 s^{-1} for 120 s. The last interval was followed by the same conditions as the first interval, but lasted for 300 seconds. Structure damage resulting from this intense shearing was calculated, Eqs. (7.13) and (7.14):

$$\Delta G'(\%) = 100\% - \frac{\left(G'_{\text{end}} \times 100\%\right)}{G'_{\text{start}}} \tag{7.13}$$

$$\Delta G''(\%) = 100\% - \frac{\left(G''_{\text{end}} \times 100\%\right)}{G''_{\text{start}}} \tag{7.14}$$

The $\Delta G'$, representing the elastic portion of the viscoelastic behavior, and the $\Delta G''$, representing the viscous portion, are between the end and beginning of the thixotropy test. It has been found that the loss of structure ($\Delta G'\%$ and $\Delta G''\%$) for imitated yogurt (lentil-based) was about 55% versus to 70% for the control dairy sample.

7.4 Lentil Protein-Based Emulsion Rheology

The protein aggregates have received increasing attention in recent years in the food industry because of its functionality improvement properties. In addition, protein-based stabilizers have advantageous to be adsorbed at the oil–water or air–water interfaces because of their amphoteric characteristics (Calabrese et al. 2018). A combination of heat treatment and pH adjustments on proteins results in aggregates with distinct morphologies, such as fibrils and particles (Gao et al. 2017). For fibril formation, protein hydrolysis is performed by heating at low pH, which makes peptides assemble into proto-filaments, which are then packed into long fibrils (Kroes-Nijboer et al. 2012). It has been observed that the lentil protein aggregates, especially fibrillar type, exhibited excellent gelling properties compared with protein molecules. Therefore, understanding the design of protein aggregates may provide a new strategy for food structuration and texturization for improved sensory quality of plant protein-based food products (Jo et al. 2020). In addition, fibrillar protein aggregates provide better emulsifying effects to stabilize emulsions due to a larger aspect ratio and a high degree of flexibility at the oil–water interface compared to rigid-bound spherical particles (Kroes-Nijboer et al. 2012; Pang et al. 2020).

7.4.1 Steady-Flow Rheology

The shear-viscosity of oil in water (O/W) emulsions of lentil protein aggregate suspension as the continuous phase and canola oil as the dispersed phase was examined by Wynnychuk et al. (2021). O/W emulsions were prepared with different LPI concentrations (1.5%, 3%, and 5%, w/v) and oil fractions (0.3 and 0.5). The resulting emulsions were stored at 25 °C for 21 days. The emulsions stabilized by both particulate aggregates result in lower viscosities (7.14–16.41 mPa·s). At 0.3 of oil fraction and pH 2, emulsions stabilized by 1.5% and 3% LPI fibrillar aggregates had lower viscosity than those stabilized by non-heated LPI solution at pH 2. The viscosity of the emulsion stabilized by 5% LPI fibrillar aggregate, however, was much higher than that stabilized by non-heated LPI, which indicates the fibrillar aggregates at high concentration might interact with the fibers on neighboring oil droplets, or the high aspect ratio of fibril structures allowed them to form inter-chain networks in the continuous phase. The emulsions with a higher oil fraction of 0.5 produced higher viscosities than the emulsions with an oil fraction of 0.3. In spite of the significantly increased emulsion viscosity stabilized by fibrillar aggregates at 5.0% LP, the viscosity values (10 ~ 500 mPa·s at 10 s^{-1} of shear rate) are typical for an emulsion system with good flowability for formulating beverages, salad dressing, and sauces (Alavi et al. 2021).

Primozic et al. (2017) examined the steady-flow behavior of nanoemulsions containing lentil proteins. The nanoemulsions were prepared by adding 5% canola oil to 95% of the protein solutions (0.5–5 wt%) at pH 3, followed by high-pressure homogenization (20,000 psi for six cycles). Both the nanoemulsions and lentil protein solutions exhibited shear-thinning behavior, indicating structure formation in a quiescent state that broke down during the application of shear. It was observed that the viscosity of protein solutions did not change significantly with the concentration until it reached 1.5 wt%, thereafter it increased significantly at the higher concentrations. It is hypothesized that the significant increment in the viscosity occurred due to the protein-droplet network formation, resulting in higher shear thinning behavior.

Rheological properties of lentil-supplemented salad dressings (% [w/w]: lentil flour [7%], canola oil [35%], egg yolk [5%], vinegar [containing 5% (w/v) acetic acid] [7.0%], lemon juice [5%], salt [1.0%], sugar [3.5%], and xanthan gum [0.25%]) were characterized using flow measurements (Ma et al. 2013). Both the power law and Herschel–Bulkley models fit shear stress–shear rate to the experimental data. All the samples showed pseudoplastic, non-Newtonian behavior ($n < 1$) with apparent yield stress. The model parameters of the lentil-supplemented dressings, K, n, and τ_0 varied significantly between lentil treatments and with storage time. The τ_0 and K of the salad dressing decreased with storage for all samples, indicating a decrease in the pseudoplasticity and fewer interactions and entanglements in the emulsion system.

The structure formation of lentil protein emulsion was evaluated without disruption of the structure during the fermentation of the milk (Boeck et al. 2021). After inoculation, the samples were transferred to the geometry of the rheometer, covered with a thin layer of vegetable oil (to avoid evaporation), and fermented for six hours at 42 °C during oscillatory measurement at a frequency of 1 Hz and a strain of 0.1%. The G' of the lentil sample increased exponentially after 70 minutes and the corresponding tanδ decreased, indicating gel formation.

Tam et al. (2017) investigated the rheological properties of the lentil flour and hydroxypropyl methylcellulose (HPMC)-blended nanofiber solution prior to electrospinning as a function of pH (7, 10, and 12), lentil flour concentration (1% and 2% [w/v]), and HPMC concentration (0.25%, 0.5% and 1% [w/v]). The shear stress–shear rate data of the solution fitted well to the power law model. The flow behavior index (n) varied between 0.88 and 0.94, confirming shear-thinning behavior. The consistency index of solutions decreased significantly when the pH increased for both of the solutions composed of 1% and 2% lentil flour. Most likely, the variation in the isoelectric point of lentil protein (pH 4.5) influenced the viscosity of the solution.

7.4.2 Oscillatory Rheology

The oscillatory rheological measurements of the emulsion system produced by lentil protein-derived fibrillar and particulate aggregates with canola oil confirmed the solid-like behavior ($G' > G''$) (Wynnychuk et al. 2021). The G' of the emulsions increased with the corresponding increases in protein concentration from 1.5% to 5%. The emulsion with fibrillar aggregates (pH 2) showed frequency dependency and an increase in the G' when the frequency was lower than 0.1 Hz. However, the G' of the emulsion was less dependent on the frequency at >0.1, which suggested a stronger structure of the emulsions when fibrillar aggregates with higher LPI concentrations were incorporated into the emulsion system. The G' of the emulsion improved further when oil content was increased to 50%, and there was less dependence on frequency, which is indicative of a predominantly elastic material. Such an increase in G at higher oil content was generally correlated with a decrease in droplet size (Ge et al. 2017); thus, a more densely packed droplet was formed, which can increase emulsion's texture. The emulsion formed by fibrillar aggregates at pH 7 had a lower G' value than that stabilized by fibrillar aggregates at pH 2. However, it was less dependent on the frequency.

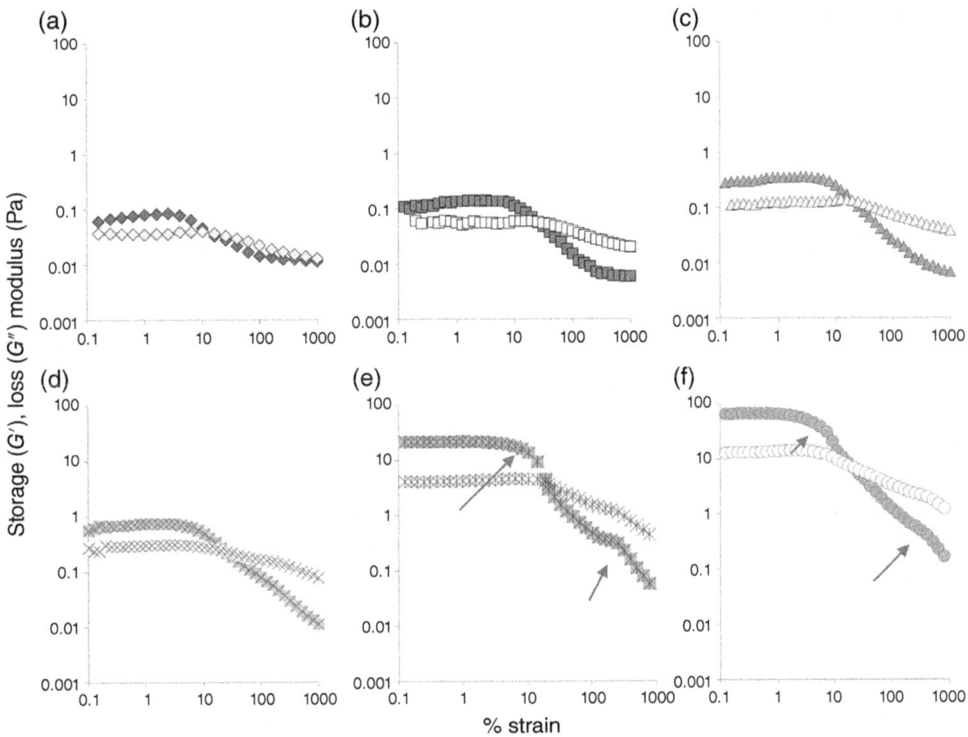

Figure 7.9 Change in G' and G'' as a function of % strain for nanoemulsions stabilized with (a) 0.5%, (b) 1%, (c) 1.5%, (d) 2%, (e) 3%, and (f) 5% LPI. Arrows indicate inflection points on the G' curves. *Source:* Primozic et al. (2017), with permission of Elsevier.

The oscillatory strain sweep measurements of the lentil protein-based nanoemulsions displayed a solid-like behavior ($G' > G''$) in the low-strain regime, and a crossover occurred between G' and G'' indicating breakdown of the gel network (Primozic et al. 2017). Changes in dynamic moduli (G' and G'') as a function of % strain for nanoemulsions stabilized with different percentages of LPI are displayed in Figure 7.9. At lower protein concentration (0.5–2 wt%), nanoemulsion behaved as a weak gel and the G' remained less than 1 Pa. However, the G' improved significantly in the LPI concentration range of 3–5%. Overall, gel strength increased with increasing protein concentration. Furthermore, the hydrophobic nature of the LPI may also promote protein aggregation in the continuous phase, leading to a stronger gel formation. The nanoemulsion gels broke down ($G' < G''$) in the high-strain regime (>10% strain) because of the shear-induced disintegration of protein and oil droplet aggregates. The crossover strain required to break the nanoemulsion gels varied significantly at different LPI percentages. For example, it increased from 10% strain for 0.5% LPI to 25% strain for 1–2% LPI nanoemulsions. On the other hand, the crossover strain decreased to 20% in the higher LPI concentrations (3% and 5%), indicating a more fragile gel structure compared to the ones with lower LPI. The formation of a strong gel with just 5% dispersed phase and 3–5% LPI could be used for food product development with a low-fat content.

The lentil-supplemented salad dressings showed viscoelastic behavior with predominating solid-like behavior ($G' > G''$) within the studied frequency range (Ma et al. 2013). A gradual increase in both dynamic moduli was observed with increasing frequency. The low phase angle values indicated a weak gel behavior of lentil-based salad dressings, which is typical of commercial dressing and mayonnaise emulsions.

7.5 Interfacial Rheology of Lentil Protein-Based Emulsion

Most macromolecules (e.g., proteins and polysaccharides) are immiscible with each other in solution due to their high molecular weights and high interfacial tension. Two immiscible macromolecules create a surface with a thermodynamic origin, which is termed an "interface." The interface has a positive free enthalpy of mixing related to the structural properties of the macromolecules. The developed interface can be considered as being the third phase and is characterized not only by its interfacial tension but also by its intrinsic rheological properties that can potentially affect the resulting bulk properties (Omari et al. 2021). The interfacial rheology of two immiscible macromolecular systems is typically assessed by investigating the response of the interface to an applied stress or strain following the same formalism and techniques developed in the field of bulk rheology. Interfacial properties have been demonstrated to control the morphology of polymer blends and contribute usefully to food emulsions. Interfacial shear rheological measurements are performed using a rheometer and an interfacial geometry (e.g., bicone) in order to shear the interface without modifying its area.

Protein–polysaccharide bilayers have recently been studied in the design of emulsion stabilizers. Development of protein–polysaccharide complexes can be accomplished by chemical or physical association, and their functionality depends on many parameters, including charge, nature of the biopolymer, ionic strength, temperature, pH, and so on. Funke et al. (2022) studied a variety of lentil fractions with different protein contents for their ability to emulsify and stabilize concentrated oil-in-water emulsions. It was observed that the differences in starch and non-starch fiber content did not contribute to the emulsification abilities of protein-rich fractions; however, differences in starch granule size and quantity did influence the bulk rheology of the concentrated emulsions. The bulk rheology differed significantly, with emulsions exhibiting a more solid-like behavior when formed with coarser fractions. Higher amounts of nonprotein components were correlated with larger shear moduli, apparent viscosities, and the consistency coefficient. The presence of granules possibly led to a higher packing of oil droplets since their presence requires volume as well, resulting in a higher shear modulus and viscosity of the concentrated emulsion. In addition, the more rigid, less deformable particles act as fillers in the network, providing additional resistance against deformation.

The interfacial rheology at a planar interface between the aqueous dispersions of the air-classified lentil fractions and Miglyol was measured in a shear rheometer using time sweeps for four hours at 1% strain followed by amplitude sweeps from 0.1% to 1000% strain (Funke et al. 2022). The results showed that G_i' exceeded G_i'' already after a few minutes, indicating the formation of a cohesive and viscoelastic interfacial layer. G_i' and G_i'' values increased over time, and the G_i'' values reached their maximum after a four hours time sweep. After the four-hour process of interfacial layer formation, the amplitude sweeps were evaluated for the stability of the developed interface by determining the interfacial crossover strains, which provide additional information on structural breakdown behavior under stress. The dynamics of the interfacial rheology comprises the adsorption, rearrangement, and intra-membrane crosslinking events of inter-facially active components. Because of the sample protein concentration in all solutions, some of these interfacial processes were probably influenced by the presence of nonprotein components or particles.

The interfacial adsorption and interfacial shear viscoelasticity of lentil protein isolate (LPI) and fenugreek gum (FG) complexes with canola oil at pH 2 and 7 demonstrated that the G_i' of LPI-FG complexes was higher at pH 2 than at pH 7 (Gadkari et al. 2019). The biopolymer interface at pH 7 was, however, able to withstand higher strain before breaking compared to that at pH 2. Both FG and LPI had opposite charges, at pH 2, indicating a superior association between biopolymers and providing lower surface activity and higher G_i'. On the other hand, at pH 7, FG and LPI had similar charges, leading to weaker complexation, higher surface activity, and lower interfacial strength. The crossover strain between G' and G'' was significantly higher at pH 7 compared to pH 2, indicating that the stronger interface that formed at pH 2 could be brittle due to strong intermolecular complexation between the protein and polysaccharide. When compared to LPI alone, the strain required to disrupt the interfacial film was increased for the LPI-FG complex at pH 2. The contrasting behavior of the interface at two different pH values was explained using protein and protein–polysaccharide interactions. The presence of 25 wt% LPI in the LPI-FG complex significantly reduced the interfacial tension and increased the interfacial rigidity compared to FG alone, which showed its potential to be used as an effective emulsifier in stabilizing emulsions and foams.

Jarpa-Parra et al. (2015) isolated and purified the legumin-like protein from lentil and used it as a plant protein model to understand how protein molecular structure impacts its surface properties by measuring interfacial rheology. Protein solutions of three different concentrations (10, 1, and 0.1 mg/ml) were prepared at pH 3, 5, and 7. Interfacial rheology of protein solutions at the air–water interface was measured using a rheometer. It was observed that the G' was 10-fold higher than the G'' indicating legumin-like lentil protein was able to form an elastic network at the water–air interface at all pH levels. It plays an important role in contributing to the foam stability as an elastic film and can better resist deformation and recover its natural shape when the deforming force is removed. The G' value increased more rapidly at pH 3 and 7, especially at 10 mg/ml due to the fast protein diffusion and adsorption because of the higher solubility and smaller hydrodynamic diameter. In addition, their higher surface hydrophobicity could facilitate protein adsorption at the air/water interface. Moreover, it was observed that the G' values at pH 5 showed an increasing trend even after six hours at all the pH levels. This effect was generally related to protein multilayer building at the interface, which might in part explain the high stability of the foam.

7.6 Rheological Properties of Lentil Protein–Starch Composites

The influence of starch to protein ratios of lentils on the formation of heat-processed composite pastes and gels was investigated by Joshi et al. (2014), by selecting the starch fraction (φ_S) in the mixtures in the range of 0–1 and 25% (w/w) total solid content. The temperature of the pre-heated starch–protein suspension (60 °C) followed three cycles of heating

(70–90 °C), holding (90 °C), and cooling (60 °C). Initially, the G'' predominated over the G', however, with elapse of time, the $G' \gg G''$, indicating a sol–gel transition. The starch, protein, and protein–starch composite gels exhibited viscoelastic behavior during the gel formation. In the protein-rich suspensions ($\varphi_S = 0.1$ and 0.2), the G' was significantly high, which decreased significantly on heating, resulting in a drop in the rigidity. Upon further heating, the G' increased and achieved a pseudo-equilibrium G' value, indicating reorganization of the network. Conversely, starch-rich suspensions ($\varphi_S = 0.5$) displayed an intense increase in G' within a short time. The G' of starch-rich composite gels leveled off much more quickly than the protein-rich composite gels. Overall, the temperature sweep experiments demonstrated that the gel formation in the protein-rich composites was much slower when compared with the starch-rich composites. Additionally, the magnitudes of G' in the starch-rich composite gels were higher compared to the magnitudes of G' in protein and protein-rich composite gels, which indicated the more solid-like behaviors in the starch and starch-rich composite gels.

The effect of temperature (60 and 10 °C) on the frequency dependence of starch, protein, and starch–protein composite is demonstrated in Figure 7.10 (Joshi et al. 2014). The moduli showed very weak to high frequency dependency, depending on the composition of the gels. While comparing the rheograms of protein and starch gels ($\varphi_S = 0$ and 1) at 60 and 10 °C, it can be seen that the dynamic moduli of protein gels had more frequency dependency than starch gels. The difference between the magnitudes of the moduli in starch gels showed much higher $G' \gg G''$ than in protein gels. The frequency sweep data further confirmed that the former is a strong gel while the latter is a weak gel (Clark and Ross-Murphy 1987).

A comparison of gel rigidity between the gels formed during the cooling–heating cycle of protein and starch gels showed that the magnitudes of G' and G'' increased with decreasing temperature from 60 to 10 °C for protein gels; however, reheating to 60 °C affected the gel network and dropped the dynamic moduli (Figure 7.10). Such a decrease in the dynamic moduli with increasing temperature is a typical behavior of gels. The hydrogen bonding potentially played an important role in stabilizing the gel network (Ferry 1980), which was destabilized during reheating and resulted in a drop in the gel rigidity. The G' of the starch gel was reduced by one log cycle at 10 °C compared to its original value at 60 °C, and it recovered during reheating to 60 °C. Such reversibility in viscoelastic parameters in starch gels can be attributed to the dominance of hydrophobic interactions, which is favored by high temperature. The difference in dynamic moduli in starch and protein gels during the heating–cooling cycle indicated that the stability in the lentil protein and lentil starch gels is brought about by different bonds and physicochemical interactions.

The frequency curves for the starch–protein composite gels at selected starch fractions (φ_S) demonstrated that both dynamic moduli of composite gels exhibited similar behavior at φ_S of 0.3 and 0.5 with respect to temperature (Figure 7.10), which indicated protein-like behavior was dominant in low-starch composite gels ($\varphi_S \leq 0.5$). The frequency dependence of dynamic moduli in low-starch composite gels ($\varphi_S = 0.5$) was less than in protein gels. For the high-starch composite gels ($\varphi_S = 0.7$ and 0.9), the magnitude of both dynamic moduli was higher at 60 °C than the gel at 10 °C, which implies the dominance of starch in these gels. However, reheating these gels to 60 °C did not restore their original dynamic modulus values. The rheological behavior of the mixed gel mostly depends on the distribution of the components in the gel network structure.

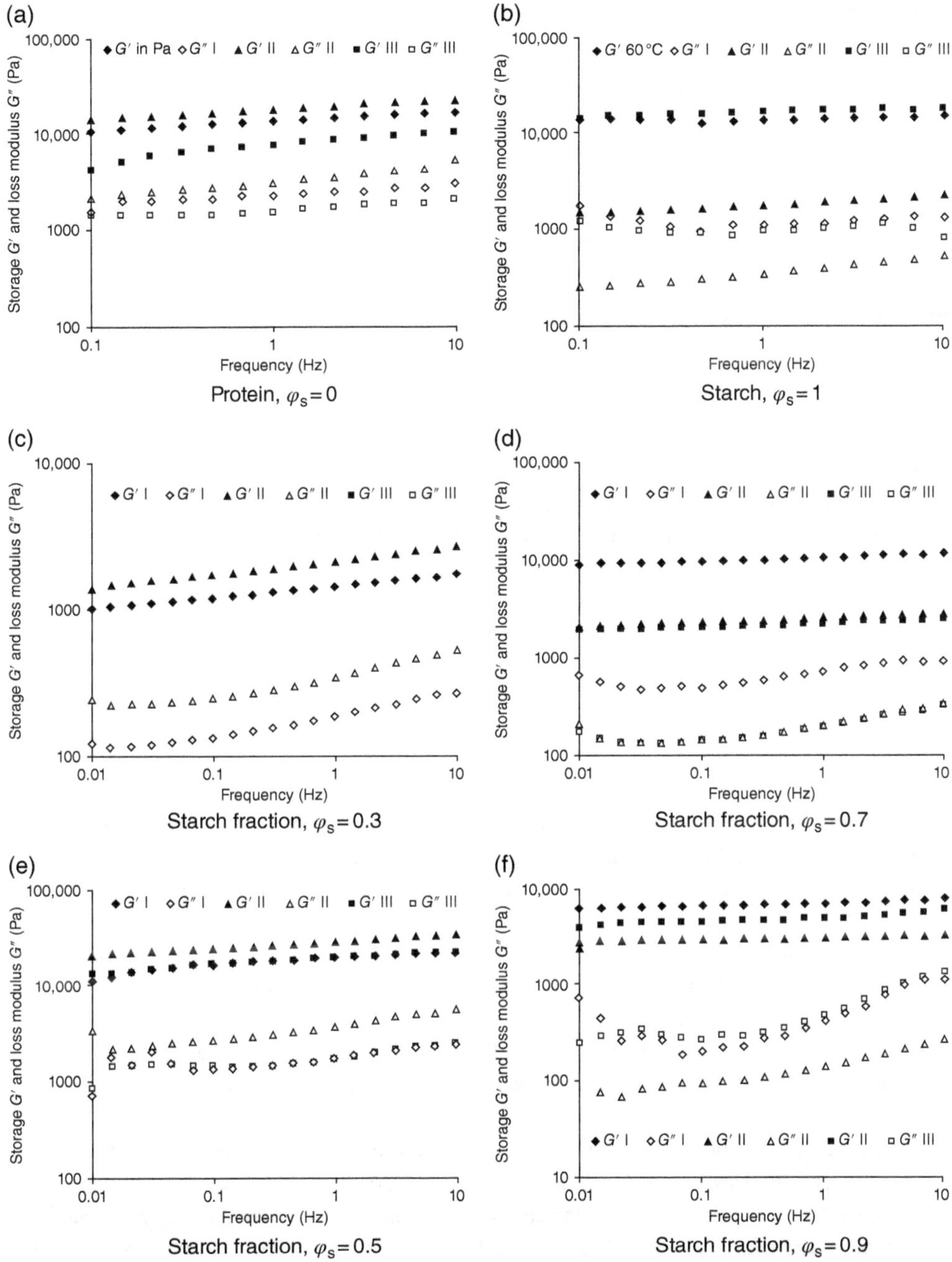

Figure 7.10 Effect of temperature (60° and 10 °C) on mechanical spectra of LP (a), LS (b), and LS-LP composite gels (c–f), where G' I/G'' I at 60 °C; G' II/G'' II at 10 °C; and G' III/G'' III at 60 °C (reheated). LP = lentil protein, LS = lentil starch. *Source:* Joshi et al. (2014), with permission of Elsevier.

7.7 Conclusions

The assessment of rheological properties of lentil ingredients provide an in-depth analysis of flow behavior and microstructure, and their roles in food product development. The starch gelatinization or protein denaturation during non-isothermal heating can be used to predict the reaction kinetics and thermodynamic properties of lentil-based ingredients and products. The functionality of lentil starch and protein during processing can be correlated to rheological data, and those properties are strongly influenced by the starch to protein ratios. The rheological measurements of lentils and derived products are not limited to flow measurements only by viscographs; they have been extended to large amplitude oscillatory shear (LAOS) and creep measurements to imitate industrial shearing and processing. Lentil proteins are characterized by their emulsifying and interfacial properties, and their understanding at the fluid/fluid interface during processing conditions is important for the application of lentil proteins as an emulsifying agent. Therefore, understanding the interfacial rheology of lentil-protein-based micro- or nanoemulsions at various processing conditions could help in developing functional foods or drug delivery systems. Rheological data on lentil starch and protein gels and emulsions could have potential for the development of bioink so that it could be used in additive manufacturing (3D printing) and biomedical applications. Tribology of lentil ingredients can be correlated with sensory evaluation to advance food product development in the food industry.

References

Acevedo, B.A., Thompson, C.M.B., González Foutel, N.S. et al. (2017). Effect of different treatments on the microstructure and functional and pasting properties of pigeon pea (*Cajanus cajan* L.), dolichos bean (*Dolichos lablab* L.) and jack bean (*Canavalia ensiformis*) flours from the north-east Argentina. *International Journal of Food Science and Technology* 52 (1): 222–230.

Adsule, R.N., Kadam, S.S., and Leung, H.K. (1989). Lentil. In: *CRS Hand Book of World Food Legumes. Nutritional Chemistry, Processing Technology and Utilization*, vol. 11 (ed. D.K. Salunkhe and S.S. Kadam), 131. Boca Raton: CRC Press.

Ahmed, J. (2014). Oscillatory rheology of lentil starch/nanoclay blend suspension: applicability of time-temperature superposition. *International Journal of Food Properties* 17: 125–138.

Ahmed, J. (2022). Application of time-temperature superposition principle in food and biopolymers. In: *Advances in Food Rheology, 2e* (ed. J. Ahmed), 221–260. UK: Elsevier.

Ahmed, J. (2023). Rheology and rheological measurements. In: *Kirk-Othmer Encyclopedia of Chemical Technology*. Wiley-VCH Verlag GmbH & Co. https://doi.org/10.1002/0471238961.1808051519030815.a01.pub3.

Ahmed, J. and Auras, R. (2011). Effect of acid hydrolysis on rheological and thermal characteristics of lentil starch slurry. *LWT Food Science and Technology* 44: 976–983.

Ahmed, J., Taher, A., Mulla, M.Z. et al. (2016). Effect of sieve particle size on functional, thermal, rheological and pasting properties of Indian and Turkish lentil flour. *Journal of Food Engineering* 186: 34–41.

Ahmed, J., Mulla, M., Al-Ruwaih, N., and Arfat, Y.A. (2019). Effect of high-pressure treatment prior to enzymatic hydrolysis on rheological, thermal, and antioxidant properties of lentil protein isolate. *Legume Science* 1: e10.

Ahmed, J., Dhull, S., and Chandok, A. (2022a). Rheological measurements of starch properties. In: *Advances in Food Rheology and Its Applications*, 2e (ed. J. Ahmed), 521–556. Woodhead Publishers.

Ahmed, J., Mulla, M.Z., Siddiq, M., and Dolan, K.D. (2022b). Micromeritic, thermal, dielectric, and microstructural properties of legume ingredients: a review. *Legume Science* 4 (1): e123.

Alavi, F., Chen, L., and Emam-Djomeh, Z. (2021). Structuring of acidic oil-in-water emulsions by controlled aggregation of nanofibrillated egg white protein in the aqueous phase using sodium hexametaphosphate. *Food Hydrocolloids* 112: 106359.

Aryee, A.N. and Boye, J.I. (2017). Comparative study of the effects of processing on the nutritional, physicochemical and functional properties of lentil. *Journal of Food Processing and Preservation* 41: e12824.

Barbana, C., Boucher, A.C., and Boye, J.I. (2011). in vitro binding of bile salts by lentil flours, lentil protein concentrates and lentil protein hydrolysates. *Food Research International* 44 (1): 174–180. https://doi.org/10.1016/j.foodres.2010.10.045.

Boeck, T., Zannini, E., Sahin, A.W. et al. (2021). Nutritional and rheological features of lentil protein isolate for yoghurt-like application. *Food* 10: 1692.

Boye, J., Zare, F., and Pletch, A. (2010). Pulse proteins: processing, characterization, functional properties and applications in food and feed. *Food Research International* 43 (2): 414–431.

Byars, J.A. and Singh, M. (2016). Rheological and textural properties of pulse starch gels. *Starch/Stärke* 68: 778–784.

Calabrese, V., Courtenay, J.C., Edler, K.J., and Scott, J.L. (2018). Pickering emulsions stabilized by naturally derived or biodegradable particles. *Current Opinion in Green and Sustainable Chemistry* 12: 83–90.

Clark, A.H. and Ross-Murphy, S.B. (1987). Structural and mechanical properties of biopolymer gels. *Advanced Polymer Science* 83: 57–192.

Doublier, J.L. (1987). A rheological comparison of wheat, maize, faba bean and smooth pea starches. *Journal of Cereal Science* 5: 247–262.

Eliasson, A.-C. (1986). Viscoelastic behavior during the gelatinization of starch. I. Comparison of wheat, maize, potato, and waxy-barley starches. *Journal of Texture Studies* 17: 253–265.

Ferry, J.D. (1980). *Viscoelastic Properties of Polymers*, 3e. New York, USA: Wiley.

Funke, M., Boom, R., and Weiss, J. (2022). Dry fractionation of lentils by air classification: composition, interfacial properties and behavior in concentrated O/W emulsions. *LWT Food Science and Technology* 154: 112718.

Gadkari, P.V., Longmore, N., Reaney, M., and Ghosh, S. (2019). Interfacial rheological behaviour of lentil protein isolate-fenugreek gum complex. *Colloid and Interface Science Communications* 30: 100180.

Gao, Z., Zhao, J., Huang, Y. et al. (2017). Edible Pickering emulsion stabilized by protein fibrils. Part 1: effects of pH and fibrils concentration. *LWT Food Science and Technology* 76 (Part A): 1–8.

Ge, S., Xiong, L., Li, M. et al. (2017). Characterizations of Pickering emulsions stabilized by starch nanoparticles: influence of starch variety and particle size. *Food Chemistry* 234: 339–347.

Harrison, G., Franks, G.V., Tirtaatmadja, V., and Boger, D.V. (1999). Suspensions and polymers – common links in rheology. *Korea-Australian Rheology Journal* 3 (11): 197–218.

Hoover, R. and Vasanthan, T. (1994). The flow properties of native, heat-moisture treated, and annealed starches from wheat, oat, potato and lentil. *Journal of Food Biochemistry* 18: 67–82.

Hoover, R., Hughes, T., Chung, H.J., and Liu, Q. (2010). Composition, molecular structure, properties, and modification of pulse starches: a review. *Food Research International* 43 (2): 399–413.

Jarpa-Parra, M., Bamdada, F., Tian, Z. et al. (2015). Impact of pH on molecular structure and surface properties of lentil legumin-like protein and its application as foam stabilizer. *Colloids and Surfaces B: Biointerfaces* 132 (2015): 45–53.

Jo, Y.-J., Huang, W., and Chen, L. (2020). Fabrication and characterization of lentil protein gels from fibrillar aggregates and the gelling mechanism study. *Food & Function* 11: 10114.

Joshi, M., Aldred, P., Panozzo, J.F. et al. (2014). Rheological and microstructural characteristics of lentil starch–lentil protein composite pastes and gels. *Food Hydrocolloids* 35: 226–237.

Kaschta, J. and Schwarzl, R.R. (1994). Calculation of discrete retardation spectra from creep data: I. method. *Rheologica Acta* 33: 517–529.

Keskin, S.O., Ali, T.M., Ahmed, J. et al. (2022). Physico-chemical and functional properties of legume protein, starch, and dietary fiber: a review. *Legume Science* 4 (1): e117.

Khazaei, H., Subedi, M., Nickerson, M. et al. (2019). Seed protein of lentils: current status, progress, and food applications. *Food* 4 (9): 391.

Klost, M. and Drusch, S. (2019). Structure formation and rheological properties of pea protein-based gels. *Food Hydrocolloids* 94: 622–630.

Kroes-Nijboer, A., Venema, P., and van der Linden, E. (2012). Fibrillar structures in food. *Food & Function* 3 (3): 221–227.

Lee, H.C. (2007). Physio-chemistry and rheology of Australian lentil flour and starch, and their implications for extrusion. PhD thesis. The University of New South Wales School of Chemical Sciences and Engineering, Australia.

Lii, C.-Y., Tsai, M.L., and Tseng, K.-H. (1996). Effect of amylose content on the rheological property of rice starch. *Cereal Chemistry* 73 (4): 415–420.

Ma, Z., Boye, J.I., Fortin, J. et al. (2013). Rheological, physical stability, microstructural and sensory properties of salad dressings supplemented with raw and thermally treated lentil flours. *Journal of Food Engineering* 116: 862–872.

Manelius, R., Nurmi, K., and Bertoft, E. (1974). Enzymatic and acidic hydrolysis of cationized waxy maize starch granules. *Cereal Chemistry* 77 (3): 345–353.

Omari, Y.E., Yousfi, M., Duchet-Rumeau, J., and Maazouz, A. (2021). Interfacial rheology testing of molten polymer systems: effect of molecular weight and temperature on the interfacial properties. *Polymer Testing* 101: 107280.

Pang, S., Shao, P., Sun, Q. et al. (2020). Relationship between the emulsifying properties and formation time of rice bran protein fibrils. *LWT* 122: 108985.

Phrukwiwattanakul, P., Wichienchotand, S., and Sirivongpaisal, P. (2014). Comparative studies on physico-chemical properties of starches from jackfruit seed and mung bean. *International Journal of Food Properties* 17 (9): 1965–1976.

Primozic, M., Duchek, A., Nickerson, M., and Ghosh, S. (2017). Effect of lentil proteins isolate concentration on the formation, stability and rheological behavior of oil-in-water nanoemulsions. *Food Chemistry* 237: 65–74.

Rathod, R.P., Gaikwad, S.B., Mhaske, S.T. et al. (2017). Rheological analysis of dough prepared with lentil flour at different moistures for production of extruded products. *Food Measure* 11: 164–172.

Sotomayor, C., Frias, J., Vidal-Valverde, C. et al. (1999). Lentil starch content and its microscopical structure as influenced by natural fermentation. *Starch/Stärke* 51: 152–156.

Steffe, J.F. (1996). *Rheological Methods in Food Process Engineering*, 2e. East Lansing: Freeman Press.

Tam, N., Oguza, S., Aydogdua, A. et al. (2017). Influence of solution properties and pH on the fabrication of electrospun lentil flour/HPMC blend nanofibers. *Food Research International* 102: 616–624.

Verni, M., Demarinis, C., Rizzello, C.G., and Baruzzi, F. (2020). Design and characterization of a novel fermented beverage from lentil grains. *Food* 9: 893.

Virtanen, T., Autio, K., Suortti, T., and Poutanen, K. (1993). Heat-induced changes in native and acid-modified oat starch pastes. *Journal of Cereal Science* 17: 137–145.

Weegels, P.L., Hamer, R.J., and Schofield, J.D. (1996). Critical review: functional properties of wheat glutenin. *Journal of Cereal Science* 23: 1–18.

Wynnychuk, C., Jo, Y.-J., Chu, Y., and Chen, L. (2021). Long-term stable emulsions prepared from lentil protein fibrillar aggregates. *Food Structure* 29: 100204.

Yavuz, H.G. (2019). Rheological and microstructural characterization of pea, lentil and almons protein gels. MS thesis. Food Science in the Graduate College of the University of Illinois, Urbana-Champaign.

8

Pasting, Thermal, and Structural Properties of Lentils

Jasim Ahmed[1] *and Najmun Nahar*[2]

[1] *Food and Nutrition Program, Environment and Life Sciences Research Center, Kuwait Institute for Scientific Research, Safat, Kuwait*
[2] *Department of Food Science, Maulana Abul Kalam Azad University of Technology, Haringhata, West Bengal, India*

8.1 Introduction

Lentils (*Lens culinaris* Medik.) are excellent sources of protein, carbohydrates (fiber, resistant starch, and oligosaccharides), dietary fibers, water-soluble vitamins, and minerals, and make a significant contribution to the human diet (Barbana and Boye 2011). Starch is the most abundant carbohydrate (22–50%) in the lentil seeds and remains the major source of carbohydrate in human nutrition (Chavan and Kadam 1989). In addition, lentils are rich in proteins (24.4–31.5%) (Li and Ganjyal 2017). Regular consumption of lentils has demonstrated several health benefits, including reducing cardiovascular diseases, lowering the glycemic index, managing obesity, and lowering the risk of cancer (Asif et al. 2013; Hall et al. 2017; Rebello et al. 2014).

Lentils are commercially available in the market either in canned or dry-packaged form, whole or split, and as flour. Lentil soup and extruded snacks are the best examples of lentil-based products. In recent times, the interest in lentils has grown to develop as a meat analog due to their high protein content. In addition to nutritional novelty, the functional, structural, thermal, and pasting properties of lentil ingredients are very important for their food and nonfood industrial applications. These properties elucidate changes in each component at the molecular level (e.g., starch gelatinization and protein denaturation) during isolation or processing. Structural changes in lentil starch and protein can be minutely measured using various analytical tools. The pasting, thermal, and structural properties of lentil ingredients have been reported in the literature. In recent times, in addition to conventional scanning electron microscopy and X-ray diffraction, atomic force microscopy and microtomography have been included in the structural analysis of lentil ingredients. The pasting, thermal, and microstructural properties of lentil flour, starch, and protein are described in this chapter.

Lentils: Production, Processing Technologies, Products, and Nutritional Profile, First Edition.
Edited by Jasim Ahmed, Muhammad Siddiq, and Mark A. Uebersax.

8.2 Pasting Properties of Lentils

Pasting is one of the important properties of starch related to a structural change in aqueous suspensions under controlled heating/cooling steps. The starch content in lentil is quite high and that warrants a thorough investigation into its pasting properties so it can be included in food product development and also nonfood applications. The composition and structure of starch have influenced the pasting properties. Starch is primarily composed of amylose and amylopectin, with the amylose to amylopectin ratio varying depending on botanical origin, cultivar, and growing season. In starch granules, amylose is present in an amorphous region and is interspersed among amylopectin molecules (Jenkins and Donald 1995). Legume starches impart distinctive properties to food systems over cereal starches because of their presence of higher amylose content (31–78%), resulting in high gelation temperature, resistance to shear thinning, high resistant starch, fast retrogradation, and high elasticity of gel (Ratnayake et al. 2002; Ren et al. 2021). The lentil contains about 32–39% amylose with a number-average degree of polymerization (DP) of 1400 (Biliaderis et al. 1986; Chung et al. 2008; Li et al. 2019), and this high amount of amylose content imparts distinct functional properties to lentil starch.

The pasting properties of lentil starch are measured using various equipment, including the Amylograph, Rapid Visco-Analyzer (RVA), and rheometer (Ahmed et al. 2016; Joshi et al. 2013). Several researchers (Ahmed 2012; Ahmed et al. 2008; Tsutsui et al. 2005) reported that rheometers are more useful for assessing starch properties than traditional viscometers because they do not break the gel structure during the measurement at a significantly lower shear strain/stress. The pasting properties of lentil starch are reported in this chapter purely on viscoamylographic data.

During heating/cooling and shearing of aqueous starch suspensions in a pasting cell, the granules swell and disrupt the structure, causing amylose to diffuse into the suspension (e.g., gelatinization) and increasing the viscosity. On prolonged heating, the viscosity drops significantly due to the breakdown of amylopectin crystallites. In the next step, the hot suspensions cool to a lower temperature when starch granules tend to reorganize dissociated chains into an ordered structure and the hardness and viscoelasticity of the starch gel increase gradually; the process is known as retrogradation (Hoover et al. 2010). This series of sequential changes (heating/holding/cooling) in apparent viscosity is being recorded and used to calculate the pasting properties of starch. The beginning of the starch gelatinization is recorded in terms of the pasting temperature (PT); the viscosity increases thereafter and attains the peak viscosity (PV) at a peak temperature, which corresponds to the close packing of the starch granules (Ahmed et al. 2015). These pasting temperatures are comparable to the onset (T_0) and peak temperature (T_p) of the glass transition of starch gelatinization.

8.2.1 Flour

The pasting properties of lentil flours as a function of the particle size in dispersions were reported by (Ahmed et al. 2016). It demonstrates that the pasting parameters varied significantly between two geographical sources of lentils (India and Turkey). The PV of Indian lentil flour increased significantly (from 544 to 714 BU) by reducing the particle size, resulting in the highest PV attributed to the finest particles (74-μm). On the other

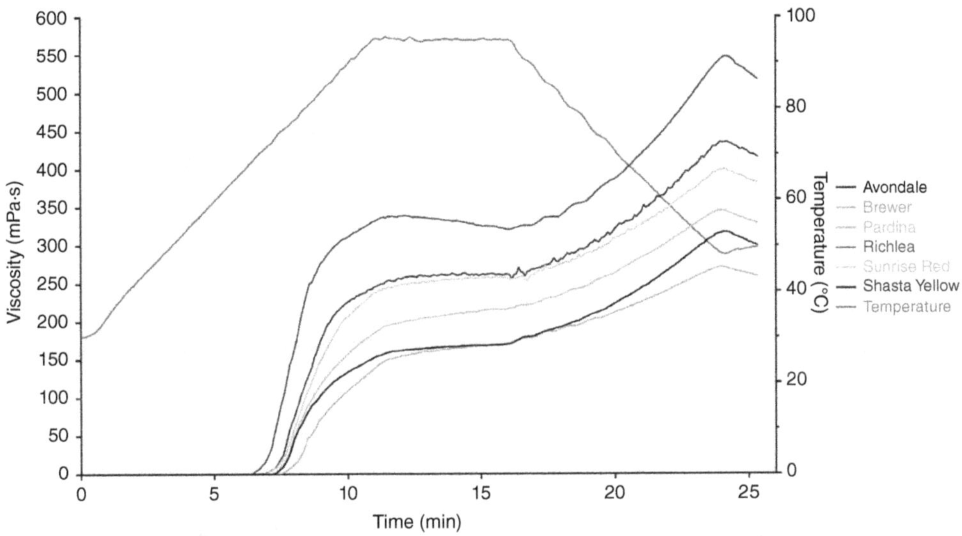

Figure 8.1 Pasting properties of lentil flours of different cultivars. *Source:* Li and Ganjyal (2017), with permission of John Wiley & Sons.

hand, the PV values of the Turkish flour decreased from 682 to 544 BU by reducing the particle size. The hot paste viscosity (HPV) and cold paste viscosity (CPV) of Indian lentil fractions increased nonlinearly with reduced particle size. The final viscosity (FV), break down viscosity (BDV), and setback viscosity (SBV) of lentil fractions increased significantly with decreasing particle size. A higher BDV of the finest flour fractions suggested that the coarser flour exhibited good paste stability and strong shearing resistance. Furthermore, an increase in the SBV of finer flours indicated more recrystallization of the gelatinized starch during cooling.

The pasting properties of six varieties of lentil flour demonstrated that the amylograph patterns were C-type, except for one (Figure 8.1) (Li and Ganjyal 2017). The gelatinization temperature of flour was measured in terms of the glass transition temperature, T_g, which ranged from 70.1 to 76.7 °C. Two peak viscosities of flour pastes were reported as PV1 (when the temperature reached 95 °C) and PV2 (the highest viscosity attained during the cooling to 50 °C and holding) that ranged from 136.0 to 331.0 and 271.3 to 547.0 mPa·s, respectively. The viscoamylogram patterns of good-and poor-cooking Canadian lentil cultivars showed consistent differences in their pasting parameters (Bhatty 1984) because of the presence of nonstarchy substances such as amylases, proteins, salts, pentosans, and sugars. Furthermore, good-cooking lentils with higher peak viscosities may have a less crystalline starch structure than poor-cooking lentils.

8.2.2 Starch

The pasting properties of lentil starch (LS) are reported in Table 8.1. The amylose content of lentil starch was significantly higher, and it contributes to the pasting properties of lentil starch by the formation of a firm gel. Pasting properties of several varieties of lentil starch

Table 8.1 Pasting properties of lentil starch suspensions of selected cultivars on Rapid Visco-Analyzer (RVA).

Sample /cultivar	Pasting /peak temperature (°C)	Peak viscosity (mPa·s)	Peak time (min)	Breakdown (mPa·s)	Setback (mPa·s)	Final viscosity (mPa·s)	References
Aldinga	71.9	7308	4.40	2244	1440	6516	Joshi et al. (2013)
CDC Plato	71.9	4625	—	1503	2323	5445	Wang et al. (2014)
CDC Impress	69.5	5244	—	1886	1593	4951	
CDC Viceroy	71.6	4861	—	1665	2880	6076	
CDC Blaze	71.6	5003	—	1790	2955	6168	
Matilda	70.3	3900	—	1062	3158s	5996	Lee (2007)
Digger	70.2	3410	—	819	2899	5490	
Cobber	73.2	3495	—	896	2974	5393	
SM-1 (native)	71.05	3898	4.20	2353	2741	4912	Majeed et al. (2017)
SM-1 (sonicated)	71.02	3788	4.23	1645	2488	4884	
SM-1 (Sonicated/IR)	70.5	3329	3.80	1474	710	1686	
Native starch (Unknown cultivar)	73.5	1150		28	472		Remya et al. (2018)
Modified starch (3%OSA)	76.5	1100		80	462		
Modified starch (5%OSA)	75.6	1070		33	390		
Modified starch (10%OSA)	78.9	1043		61	380		

OSA: octenyl succinic anhydride.

(LS) revealed that the pasting temperature ranged between 70 and 73 °C (Hoover and Ratnayake 2002; Joshi et al. 2013; Lee 2007; Li et al. 2019) and the values varied insignificantly among cultivars (Wang et al. 2014). The PV of the starches varied from 4625 to 7308 mPa·s, indicating the water-binding capacity of the starch (Joshi et al. 2013; Wang et al. 2014). The higher thermal stability of lentil starch can be related to its lower swelling capability. A higher FV of LS has been attributed to the leached amylose, which goes into the aqueous phase, reassociates, and forms a gel. The BDV indicates paste stability and is the result of the rupture of swollen starch granules. The amylose content of LS was inversely correlated with the PV and BDV, while the SBV and FV showed a positive correlation (Li et al. 2019; Wang et al. 2014). Differences in BDV among starches were attributed to the granules of these starches being disintegrated by mutual crowding and shear that resulted in more amylose and amylopectin leaching. These results show that lentil starches can be potentially used where strong gelling properties are required in the final product.

8.2.3 Starch–Protein Blend

The pasting properties of lentil starch and lentil protein composites at selected protein-to-starch ratios demonstrated that the PV of the protein solution ($\varphi_s = 0$) was significantly lower than that of the pure starch ($\varphi_s = 1$) (Table 8.2) (Joshi et al. 2014). The pasting temperature of starch, however, increased significantly from 73.5 to 82.5 °C when $\varphi_s = 0.5$ in the blend due to the high denaturation temperature (118 °C) of the lentil protein isolate. The PV decreased with increasing the protein fraction in the blend. The gelatinized amylopectin exhibited a synergistic effect on the viscosity and gel strength of starch–protein pastes (Chedid and Kokini 1992). The breakdown of starch has been severely affected when the proportion of protein in the mixture was higher ($\varphi_s < 0.9$), which indicates that the protein gel network contributed to the increase in the degree of resistance to mechanical shearing of the starch–protein composite pastes at hot conditions.

During the cooling step, the viscosity curve for starch and starch–protein composite ($\varphi_s = 0.9$) was broken down due to the rupture of the gel network.

Table 8.2 Pasting properties of lentil starch (LS), lentil protein (LP), and their composites gel with different starch fractions (φ_s = 0–1), and the effect of ionic strength and pH on the LS–LP composite (φ_s = 0.5).

Starch fraction (φ_s)	Peak viscosity (mPa·s)	Breakdown (mPa·s)	Final viscosity (mPa·s)	Set back (mPa·s)	Peak time (min)	Pasting temperature (°C)
0	32	7	189	164	6.93	—
0.1	58	13	298	253	7	—
0.3	322	128	1215	1021	7	—
0.5	1116	244	3660	2788	7	82.4
0.7	2415	90	6194	3869	7	78.45
0.9	6500	871	8859	3230	4.87	75.25
1	12,140	5391	8065	1316	4.4	73.5
NaCl ($\varphi_s = 0.5$)						
0 M	1116	244	3660	2788	7.0	82.4
0.1 M	3771	363	5659	2251	6.1	81.7
0.25 M	3263	206	4771	1714	7.0	83.5
0.5 M	2769	291	4021	1543	7.0	84.0
1.0 M	2320	595	3582	1857	7.0	85.8
pH ($\varphi_s = 0.5$)						
pH 3	3350	701	3842	1193	5.4	76.8
pH 4	2869	189	3754	1074	7.0	77.5
pH 5	4525	720	5636	1831	6.1	79.2
pH 6	3810	229	5296	1715	6.5	80.8
pH 7	4406	625	6717	2936	6.0	82.5
pH 8	4344	515	7830	4001	5.8	81.6

Source: Joshi et al. (2014), with permission of Elsevier.

The PV of the lentil starch–lentil protein paste was increased by increasing the salt concentration (ionic strength) from 0.1 to 0.25 M NaCl (Table 8.2) (Joshi et al. 2014). At low concentrations, the salt ions preferentially interact with protein molecules, and this interaction interferes with the protein–protein interaction. Conversely, the PV of composite pastes at acidic pH (3–4) was significantly lower when compared to the viscosity values at higher pH values. Such an acidic condition might affect the amorphous region of starch granules. Maximum peak PV was observed for the composite paste at pH 5 (4525 mPa·s) followed by pH 7 and pH 8. The pasting temperature increased with an increase in pH, and the highest pasting temperature was observed at pH 7.0 (82.5 °C).

The pasting properties of modified starch synthesized by octenyl succinylation of lentil starches demonstrated that PT, PV, BDV, and SBV values were significantly different from the native starch (Remya et al. 2018). The succinylation of starch significantly increased the PT from 73.5 to 78.9 °C. The setback viscosity, however, decreased and an increase in breakdown was also observed as a result of modification. A lower SBV suggests a lower rate of syneresis and more stability of starch gels at low temperatures.

8.2.4 Influence of Processing on Pasting Properties of Lentil

The pasting properties (PT, PV, SBV, and FV) of flours from three lentil varieties (Brewer, Crimson, and Richlea) prior to extrusion showed significant differences in the properties of their extrudates because of the differences in starch content in the flours (Ek et al. 2021). The effects of conventional extrusion (CV) and extrusion-cooking (EC) on the pasting behavior of 100% yellow lentils suggested that the EC promoted a more compact structure that required a higher melting temperature than the CV (66.49 versus 63.16 °C) (Bresciani et al. 2021). The test samples exhibited significant differences in PT, PV, and FV, and those differences could have been attributed to both shear and thermal stress that occur during the extrusion-cooking process, which involves starch gelatinization and higher temperatures to start gelatinizing. The influence of nitrogen gas on an extruding system of red lentil flour demonstrated that the cold viscosity of extrudates produced by N_2 gas injection was higher than those produced by conventional extrusion, which indicates the gas-assisted extrusion does affect the end-product pasting properties (Masatcioglu and Koksel 2019).

Thermally treated red lentil starch exhibited a significantly higher pasting temperature than the raw starch (72.9 versus 70.6 °C) because of the modification of starch or starch granule reorganization (Bresciani et al. 2022). Furthermore, the thermally treated flour produced a significantly higher PV (202.5 BU) and FV (394 BU) than the raw flour (PV: 164.5 BU and FV: 320 BU), indicating the thermal treatment made the starch granules more capable of absorbing water during the test, resulting in a higher viscosity due to the higher gelatinization capacity and a greater reorganization as a consequence of greater gelatinization with no impact on the retrogradation capacity.

The synergistic effects of micronization temperature (115–165 °C) and seed moisture contents (8–23% d.b.) on the pasting properties of green lentil flours resulted in significant changes in pasting parameters (Table 8.3) (Pathiratne et al. 2015). The PV values for flour samples micronized to 115 and 130 °C without seed moisture adjustment were significantly higher than for flour of non-micronized seeds (104.3 RVU versus 85.9 RVU). A significant

Table 8.3 Pasting properties of micronized lentil seed flours.

Seed treatment		Pasting parameters			
Moisture content (%)	Temperature (°C)	Pasting temperature (°C)	Peak time (min)	Peak viscosity (RVU)	Final viscosity (RVU)
8	—	72.3	5.2	85.9	129.1
8	115	71.2	4.7	104.2	139.0
	130	70.8	5.1	104.3	147.5
	150	72.6	6.2	89.3	127.0
	165	74.6	7.4	57.8	89.3
	115	71.0	5.9	91.8	139.5
16	130	71.3	6.5	79.6	117.4
	150	72.1	7.8	50.0	82.3
	165	80.3	6.6	37.3	71.1
	115	69.6	6.6	91.9	136.0
23	130	73.2	6.8	88.8	137.3
	150	76.1	6.9	37.4	59.6
	165	77.5	6.9	26.4	48.1

Source: Adapted from Pathiratne et al. (2015).

decrease in PV of flour was recorded for non-tempered seeds micronized to 165 °C, and 16% and 23% for tempered seeds micronized to 150 and 165 °C (26.4–57.8 RVU) compared to non-micronized lentil flour. The flours from micronization temperatures of 115 and 130 °C resulted in highly viscous pastes with a high FV due to the incorporation of gelatinized starch (23% tempered seeds). The results of the study conclude that different seed moisture and micronization temperatures result in lentil flours having a wide range of viscosities and pasting properties and could be uniquely utilized in the food industry.

Sonication and a dual treatment of sonication and irradiation (5 kGy) on the pasting properties of native lentil starches displayed that the modification by sonication had little effect on the PV and trough viscosity (TV). However, dual treatment resulted in a significant drop in both PV, TV, BDV, and SBV and that might have occurred because of the substantial reduction in the swelling index of starch due to its degradation upon irradiation (Majeed et al. 2017). Application of high-pressure treatment (300–600 MPa) on lentil starch dispersions displayed a significant change in pasting parameters (Ahmed et al. 2016a). The PV value increased significantly (958–981 BU) at 400 MPa but decreased at higher pressures (500–600 MPa). The significant drop of the PV at 600 MPa might have been attributed to the complete gelatinization of starch granules. The HPV, the CPV, and the BDV of LS dropped considerably at 600 MPa. A significant drop in those parameters suggests that the pressure-modified gel network is more heat-stable compared to the untreated starch.

8.3 Thermal Analysis of Lentils

The thermal analysis measures the thermal events and the associated heat involved in the process during the thermal scanning of a sample sealed in a metallic pan at a programmed heating rate on a differential scanning calorimetry (DSC) (Ahmed and Rahman 2014). Upon heating on a DSC, the expected thermal events in lentils in aqueous dispersions are starch gelatinization, protein denaturation, and amylose–lipid or protein–lipid complexes. The lentil starch dispersions exhibit melting of starch crystallites at a particular temperature (e.g., gelatinization temperature), and thereafter, form a gel. Based on the concept of glass transition (T_g), three temperatures have been associated with the thermal transition: the onset (T_o), the peak (T_p), and the conclusion (T_c). The midpoint of the thermal transition in the first scan of aqueous lentil starch is considered the T_g (Al-Attar et al. 2022). On the other hand, the globular lentil proteins denature during thermal scanning as observed through an exotherm.

8.3.1 Flour and Starch

The thermal properties (T_o, T_p, and T_c) of gelatinization and the corresponding enthalpies (ΔH) of lentil flour obtained from DSC are reported in Tables 8.4 and 8.5. There were significant differences ($P<0.05$) in T_o, T_p, and T_c of lentil flour and isolated starch among cultivars of lentil. The T_p varied significantly among cultivars, moisture contents, isolation techniques, and other factors. The T_p of lentil flour varied from 62.8 to 75 °C and the corresponding enthalpies ranged between 0.55 and 28.5 J/g. T_p and ΔH of isolated starch were 62.4–70.8 °C and 13.7–23.9 J/g, respectively. The difference between flour and the isolated starch is evident because the whole endotherm takes into account not only the starch gelatinization but also protein denaturation and melting of the amylose–lipid complex (Lu et al. 2018). Overall, lentil flours had significantly higher thermal temperatures and process enthalpy than those of their corresponding isolated starch.

The influence of moisture content (20% and 80% [w.b.]) on the thermal properties of the lentil flours revealed that the T_p at 20% moisture content ranged from 121.34 °C (Richlea) to 126.57 °C (Brewer) (Ek et al. 2021); however, these values differed significantly with other reports for the same cultivars (109 and 111 °C) (Li and Ganjyal 2017). Such differences between two different studies could be attributed to different agronomical conditions resulting in different chemical compositions. The other reason for the difference that could be contributing to the differences in thermal properties could be the interaction between starch, protein, and lipids (Rodríguez-Damian et al. 2013).

While comparing the thermal properties of lentil flour as a function of particle size, Ahmed et al. (2016) reported two DSC peaks: the first peak demonstrated a smaller peak at a lower temperature and the enthalpy that corresponds to starch gelatinization, and the second peak detected at a higher temperature ascribed to the melting of the amylose–lipid complex with higher ΔH. The T_{p1} for LF fractions varied significantly between the two samples and also with the particle size. T_{p1} values for Indian and Turkish LF fractions were 62.8–68.4 °C and 67.4–73.8 °C, respectively, with corresponding ΔH values ranging from 0.55 to 1.52 J/g and 0.65 to 2.90 J/g. The observed ΔH indicates the loss of double-helical order in lentil starch and the melting of overall amylopectin crystals rather than the

Table 8.4 Thermal properties of lentil flours and their modifications.

	Glass transition temperature (°C)				
Lentil variety	**Onset (T_o)**	**Peak (T_p)**	**Conclusion (T_c)**	**ΔH (J/g)**	**References**
Red lentil flour (RLF):					
Redberry	60.6	71.6	112.2	24.7	Lu et al. (2018)
Redcliff	61.6	71.5	112.6	26.5	
Redbow	63.6	75.0	113.4	28.5	
Blaze	62.11	68.37	—	11.27	Barbana and Boye (2013)
RLF (India):					Ahmed et al. (2016)
210-μm	66.9	62.8	68.1	0.55	
140-μm	63.4	68.4	69.2	0.56	
105-μm	61.4	64.4	67.7	1.52	
74-μm	62.6	64.4	67.8	1.14	
RLF (Turkey):					
210-μm	67.2	67.6	70.5	0.65	
140-μm	66.2	66.8	68.9	0.76	
105-μm	63.5	73.8	76.9	2.90	
74-μm	62.0	67.3	69.6	2.56	
Lentil flour:	65.11	67.26	70.43		
Crimson (20% m.c.)	113.6	120.9	130.9	1.62	Ek et al. (2021)
Crimson (80% m.c.)	66.03	74.60	81.96	2.12	
Unhydrolyzed flour	65.9	67.5	72.5	12.4	Ahmed and Auras (2011)
Hydrolyzed flour	67.3	69.2	72.7	12.2	
Green lentil flour:					
Asterix	62.9	73.0	113.8	24.9	Lu et al. (2018)
Imigreen	61.1	71.1	113.9	25.2	Lu et al. (2018)
Greenland	65.0	74.2	112.9	26.6	
Improve	63.3	73.3	113.4	25.5	
Impower	61.5	70.8	114.1	24.6	
Brewer (20% m.c.)	119.7	126.6	135.1	1.58	Ek et al. (2021)
Brewer (80% m.c.)	70.30	78.13	91.29	3.69	
Richlea (20% m.c.)	114.7	121.3	129.9	1.52	
Richlea (80% m.c.)	67.05	74.09	80.68	2.72	
Laird	63.29	69.70	—	9.53	Barbana and Boye (2013)

Table 8.5 Thermal properties of isolated lentil starches and modified starches.

	Gelatinization/Glass transition temperature (°C)				
Starch-type	**Onset (T_o)**	**Peak (T_p)**	**Conclusion (T_c)**	**ΔH (J/g)**	**References**
Native starch	56	62.4	69		Hoover and Manuel (1995)
Red lentil starch					
Redberry	58.7	66.9	84.3	14.8	Lu et al. (2018)
Redcliff	62.7	70.2	89.8	14.8	Lu et al. (2018)
Redbow	60.6	67.6	85.2	16.0	Lu et al. (2018)
Green lentil starch					
Asterix	60.3	66.9	82.8	15.7	Lu et al. (2018)
Greenland	61.6	68.1	84.0	16.2	
Imigreen	58.3	65.4	80.8	15.0	
Impower	59.0	66.2	81.8	13.7	
Improve	60.1	69.7	83.8	14.2	
Red lentil starch	55.71	63.72	72.3	8.8	Ahmed et al. (2016)
Native starch	64.0	70.0	64.0	20.8	Chung et al. (2010)
ANN-modified	71.5	71.5	71.5	17.8	
HMT-modified	67.2	67.2	67.2	23.9	
ANN-HMT-modified	69.9	69.9	69.9	20.2	
HMT-ANN-modified	70.8	70.8	70.8	21.6	
Native starch	61.6	66.3	72.3	—	Remya et al. (2018)
3% OSA-MS	62.9	69.1	77.7	—	
5% OSA-MS	65.7	70.3	75.4	—	
10% OSA-MS	66.1	70.8	76.3	—	

separation of double helices within starch crystallites (Cooke and Gidley 1992; Lopez-Rubio et al. 2009). The observed T_{p2} at a higher temperature has been related to the melting of the amylose–lipid complex or protein denaturation, and those values were not significantly different among particle sizes. The T_{p2} increased significantly for the finest particles due to the formation of a strong complex between starch and lipid, and it required more energy to break down (Table 8.4).

Chung et al. (2010) found that annealing (70% moisture content at 50 °C for 24 hour) and heat-moisture treatment (HMT) (30% moisture content/120 °C for 24 hour) of lentil starch increased the T_p and process enthalpy. The addition of 3–10% OSA to lentil starch increased the T_p from 66.3 to 70.8 °C (Remya et al. 2018). Regardless of germination time, germination of lentil flours within six days exhibited no differences in thermal properties as raw flours (Xu et al. 2019), indicating that the hydrogen bond of starch in germinated pulses can be broken down at a temperature similar to that of raw pulse flours.

8.3.2 Proteins

Protein denaturation of lentils is characterized by a distinct DSC thermal peak. The protein denaturation endotherm peak temperature (T_d) indicates the thermal stability of the protein, which is mostly globulin. A higher T_d value suggests higher thermal stability of the globular proteins. The thermal denaturation temperatures (T_d) of lentil proteins are summarized in (Table 8.6). The T_d values for the native proteins varied between 79.62 and 135.5 °C and the corresponding enthalpies were in the range of 6.1–12.57 J/g (Ahmed et al. 2009; Aryee and Boye 2017; Cai et al. 2002; Hall and Moraru 2021; Joshi et al. 2011). The influence of moisture content (6.5–57.6% wet basis) on the denaturation of lentil proteins indicated that the T_d decreased from 135.5 to 111 °C with increasing moisture content (Ahmed et al. 2009).

While protein was isolated from lentil flour, the T_d of the isolated protein was higher than the flour, which might be attributed to the higher thermo-resistance of the proteins in the isolates. Because of the presence of multiple protein fractions in lentil proteins (Shand et al. 2007), thermal analysis of lentil protein concentrates (LPC) exhibited broad transition peaks (Shand et al. 2007). Lentil protein isolate (LPI) produced using different drying methods (spray, freeze-, and vacuum) showed a single endothermic peak with T_d ranging from 118 to 123 °C, and these values were independent of the drying method (Joshi et al. 2011). The higher enthalpy of the isolated sample is indicative of a higher proportion of undenatured proteins compared with flour (Aryee and Boye 2017). However, Cai et al. (2002) observed two T_ds at 97.8 and 110.5 °C for lentil proteins. These differences in peak denaturation temperatures might be due to differences in cultivars, method of preparation, or concentration of protein used for testing.

Table 8.6 Protein denaturation temperatures of lentils.

		Protein denaturation temperature (T_d)			
Lentil protein type	**Description**	**Onset (°C)**	**Peak (°C)**	**ΔH (J/g)**	**References**
Protein concentrate	Native	78.7	83.7	6.1	Hall and Moraru (2021)
	Heat	78.2	81.5	0.4	
	Pressure	79.4	82.8	0.6	
Red lentil protein concentrate		75.05	87.82	10.87	Barbana and Boye (2013)
Green lentil protein concentrate		70.35	79.62	12.57	
Red lentil flour		79.48	83.44	0.02	Aryee and Boye (2017)
Lentil protein isolate		74.62	85.67	0.80	
Protein in flour	6.5% m.c.		135.5		Ahmed et al. (2009)
	24.6% m.c.		125.0		
	57.6% m.c.		111.0		

Untreated LPC had the following protein denaturation temperatures (T_d): T_{onset} ~79 °C and T_{peak} ~84 °C (Hall and Moraru 2021). LPC samples were subjected to treatments (heat and pressure) that resulted in subtle transitions, suggesting near complete denaturation of the proteins (Table 8.6). Pressure and heat treatments resulted in a significant decrease in the enthalpy of protein denaturation compared to untreated samples for all protein types, indicating that both treatments induced considerable protein denaturation. A decrease in residual enthalpy after the pressurization suggests that some of the protein structure has been lost. The shift in the denaturation temperature at higher pressure is imposed by specific volume change (ΔV) (Ahmed et al. 2009).

Thermal properties of lentil protein concentrate and enzyme-assisted hydrolyzed samples exhibited a sharp endothermic peak that corresponded to protein denaturation (T_d) and ranged between 93.95 and 94.79 °C (Rezvankhah et al. 2021). The process enthalpies (ΔH) for the hydrolyzed and untreated samples were in the range of 10.09–10.75 J/g.

8.4 Scanning Electron Microscopy (SEM)

Scanning electron microscopy (SEM) provides information on the microstructure and morphology of lentil seeds, flour, starch, and proteins. The microstructures of red and green lentils under compression loading (moisture content: 9.5–21.1%, w.b.) demonstrated that the cotyledon cells of green lentils create a regular structure containing ellipsoidal and smooth-surfaced starch granules within a honeycomb-like structure embedded in the protein network (Gharibzahedi et al. 2014). The shape of the starch grain of red lentils appeared to be more elongated and larger than those of green lentils. Additionally, the loaded samples suggested lower fracture magnitude in green lentils than in red lentils for both vertical and horizontal orientations.

8.4.1 Flour and Starch

Ahmed et al. (2016) compared the microstructure of lentil flour particle fractionates from India and Turkey. Overall, the shapes of the lentil starch granules varied from round to oval to anomalously shaped. The starch granule size ranged from 9 to 38 μm in length, with an average length of 25 μm, and the corresponding values for Turkish lentils were 12–31 μm in length, with an average length of 22.2 μm (Figure 8.2a–d). The starch became agglomerated or linked with either fiber or other components for the Turkish sample, whereas for its Indian counterpart, the granules were distinct. Furthermore, the average starch granule size of Indian cultivars showed a little change with size reduction (27–18 μm).

For isolated starch, the granules were spherical-oval to elliptical to irregular in shape, with characteristic dimensions in the range 2.5–25 μm (Aguilera et al. 2009; Ahmed et al. 2016; Hoover and Manuel 1995). The starch granules were characterized by a smooth surface. The starches isolated from the air-classified flours, however, showed some defects in the granular structure due to harsh dry milling of the flours (Li et al. 2019). The granules were broken, and scratches were on the surface of some granules. However, these damages appeared to be minimal to the structure of starch granules because the isolated pulse starches showed damaged-starch content of about 1%.

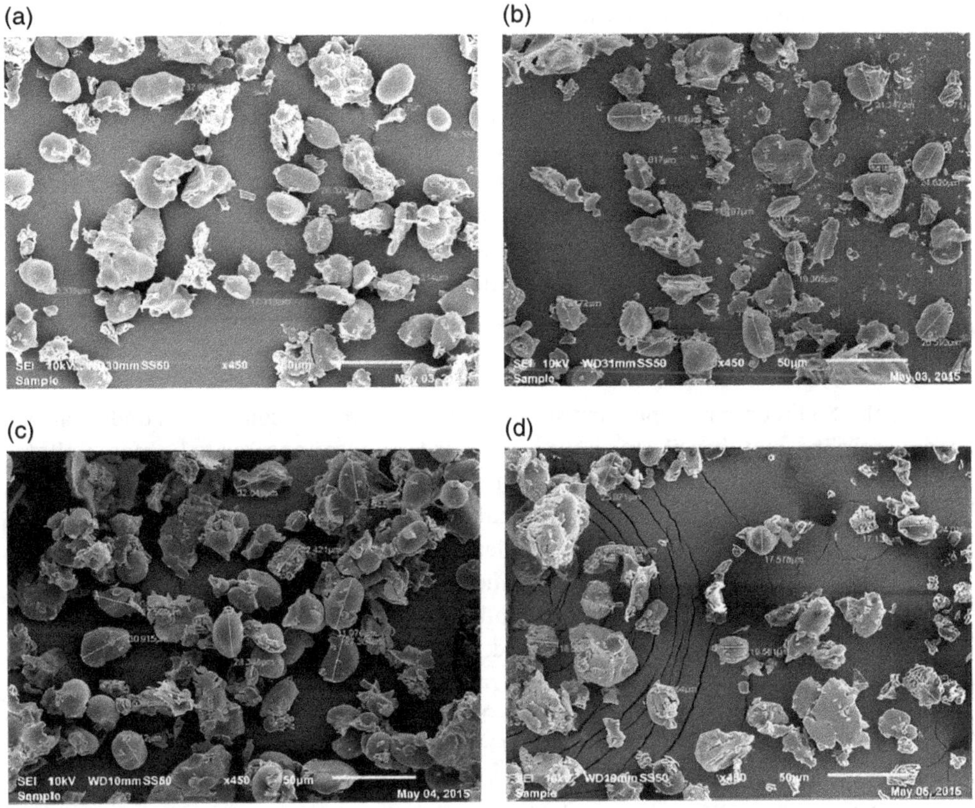

Figure 8.2 SEM micrographs for lentil flour samples. (a) Indian LF 210-μm fraction. (b) Turkish LF 210-μm fraction. (c) Indian LF 74-μm fraction. (d) Turkish LF 74-μm fraction. *Source:* Ahmed et al. (2016), with permission of Elsevier.

The influence of various thermal treatments (e.g., soaking, cooking, roasting, and dehydration) on the microstructural characteristics of lentil flour and isolated starches have been reported by various authors (Aguilera et al. 2009; Ma et al. 2011). Soaked lentil flour resulted in an inflated, rougher, and slightly eroded granular structure. The granules, however, inflated extensively to about 121 μm with a flattened surface when cooked. For dehydrated samples, micrographs were similar to cooked legumes; however, smaller starch granules were appeared as compared to cooked seeds (Aguilera et al. 2009). The influence of roasting and boiling on the microstructural analysis of green lentil flours showed that the roasted pulse flours had similar microstructure to the raw samples (Ma et al. 2011). No starch granule was visualized in the micrograph in the boiled and freeze-dried samples. Furthermore, the flakes converted to an amorphous stage because of the gelatinization of starches and cross-linking between protein and starch.

The micrographs of the isolated lentil starch under various treatments (untreated [ULS], roasted [RLS], germinated [GLS], fermented [FLS], microwave cooked [MLS], conventional cooked [CLS], and autoclaved [ALS]) suggested that the roasting and germination had little effect on the starch granule morphologies when compared with the ULS, which maintained a smooth surface varying in shape from spherical to void (Yin et al. 2018). For FLS, the original shape of the starch granule collapsed and formed densely packed flakes with rough surfaces. Such changes occurred because of the degradation of starch aggregation induced by the endogenous enzyme activity during fermentation as well as the rearrangement of the chain structures. For MLS, CLS, and ALS samples, morphological characterization with expanded and tightly packed gully shapes and rough surfaces was observed. The change in the microstructure could be attributed to the leaching of amylose, the loss of amylopectin crystalline during the autoclaving, microwave, and conventional cooking, as well as the subsequent reassociation of molecular chains within the granules.

8.4.2 Proteins

Joshi et al. (2011) compared the microstructure of lentil protein isolate (LPI) under various modes of drying, and they observed the spray-dried samples appeared to be more uniform in particle size distribution than the freeze-dried and vacuum-dried samples (Figure 8.3a,f). The surface of the spray dried isolates appeared to be folded or wrinkled, and some of the particles were hollow, whereas a smooth indented surface was apparent at higher magnification (Figure 8.3b,d,f). Under high magnification, freeze-dried and vacuum-dried samples appear like thin flakes or a crumbled multi-layered sheet-like structure, respectively.

Recently, protein aggregates have created interest in food applications for improving functionality. Jo et al. (2020) investigated thermal aggregation and gelation in LPI as a function of pH (2–9), protein concentration (1–13%, w/w), and heating time (0.5–16 hours) using various microscopic measurements. The TEM morphologies showed the formation of aggregates at 5% LPI concentration and a heating time of two hours at a selected pH (Figure 8.3a–f). The LPI aggregates had a fibrillar structure at pH 2, featuring flexible curved thin stands with a length of 10–100 nm and a thickness of a few nanometers. At pH 7, particulate aggregates formed with an average diameter of less than 50 nm (Figure 8.4), which resulted from the random aggregation of globulin proteins by hydrophobic and other interactions. The confocal laser scanning microscopy (CLSM) exhibited that the microstructure of the lentil gel was more homogeneous and compact at pH 2 than the gel at pH 7 (Figure 8.4b,g). On the other hand, SEM images of the fibrillar aggregates at pH 2 confirmed a gel network and densely crosslinked microstructures with a uniform distribution of pores (Figure 8.4c,d). Additionally, the wall of the pores in the fibrillar gel (pH 2) is thicker (yellow arrow in Figure 8.4d) than the gel produced at pH 7. The fibrillar gels were translucent (Figure 8.4e), which can be attributed to the narrow width of the strands that form the building blocks of the gel networks. The 3D networks were also visible at pH 7, but the gels were less homogeneous and loosely bound.

8.4.3 Starch–Protein Composite Gels

The microstructure of thermally induced pastes and gels of lentil starch and lentil protein composites at various starch-to-protein ratios suggests that the starch fraction (ϕ_s) has a significant influence on the composite gel morphology (Figure 8.5) (Joshi et al. 2014). The microstructure of single component gels of protein (a and b) and starch (c and d) are typical of globular protein gels and starch gels. Both polymer gel microstructures reveal a very

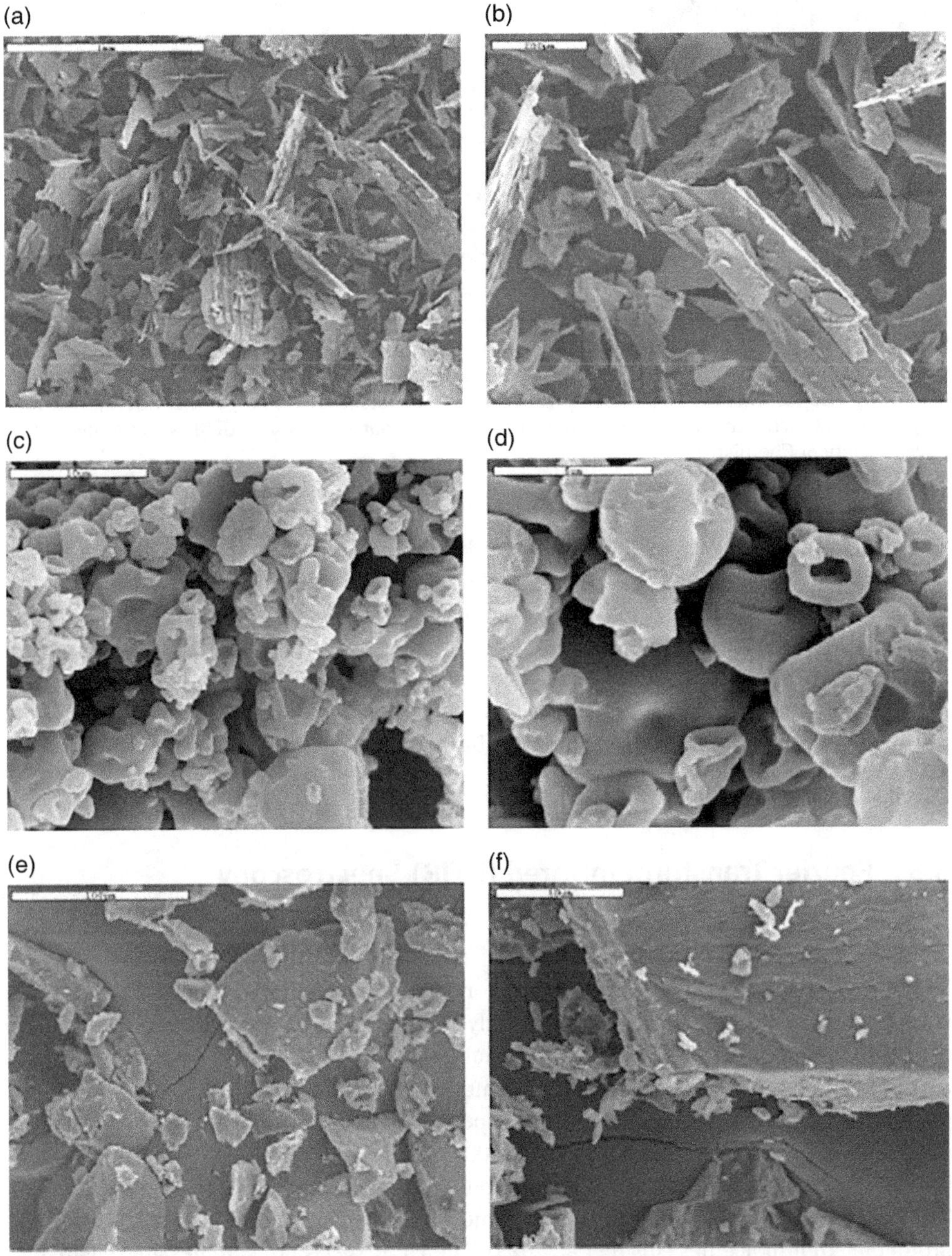

Figure 8.3 SEM pictures of lentil protein isolate powders: (a and b) freeze-dried; (c and d) spray-dried; and (e and f) vacuum-dried. *Source:* Joshi et al. (2011), with permission of Elsevier.

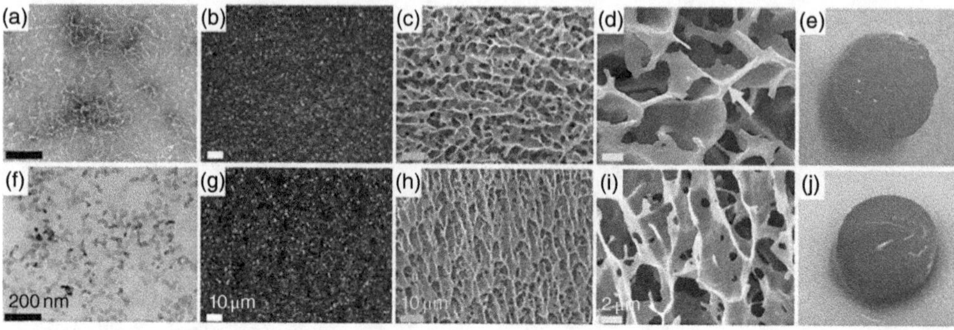

Figure 8.4 Microstructures and appearance for LPI gels at different pH levels: (a and f) TEM images of LPI aggregates; (b and g) CLSM images of LPI gels; (c, d, h, and i) SEM images of LPI gels; (e and j) gel appearance; (a–f) pH 2; (f–j) pH 7. LPI aggregates (5%, w/v LPI) and LPI gels (10%, w/v LPI) were heat incubated for 2 and 16 hours, respectively. *Source:* Jo et al. (2020), with permission of Royal Society of Chemistry.

well-developed three-dimensional gel network, and the honey-comb structure is exhibited as tunnels at higher magnification. The starch gel structure appeared much finer and denser than the protein gel structure. The composite gel at $\phi_s = 0.3$ (e and f) demonstrated the gel network becomes flake-like in appearance and the wall of the gel network becomes thicker compared with that in protein gel structures. Both the starch and protein were able to form more uniform gels quite effectively in formulations having $\phi_s < 0.5$. This might be due to the formation of a strong gel network by starch before the protein could establish its gel network in the composite gels.

8.5 Fourier Transform Infrared (FTIR) Spectroscopy

8.5.1 Starch

FTIR provides information on the structural organization of starch chains near the granule surface since the infrared beam penetrates only to a depth of 2-μm into the granule (Sevenou et al. 2002). The infrared (IR) spectra of starches can be divided into five spectral regions ascribed to the stretching and deformation modes created by the OH, CH/CH_2, and C–O/C–C groups (Al-Attar et al. 2022). These spectral regions are divided into five parts: I (<800 cm^{-1}), II (800–1500 cm^{-1}), III (1500–1200 cm^{-1}), IV, C–H stretching region (2800–3050 cm^{-1}), and V, OH stretching region (3050–3600 cm^{-1}) (Wiercigroch et al. 2017). The band at ~1000 cm^{-1} is related to intramolecular hydrogen bonding of the hydroxyl group of starch molecules (Joshi et al. 2013). The FTIR spectra of starches are sensitive to the molecular level, such as crystallinity, helicity, and confrontation (Lu et al. 2019). The IR bands at 1047 and 1022 cm^{-1} have been associated with ordered and amorphous structures of starch, respectively (Capron et al. 2007; van Soest et al. 1995). The ratio of the heights of bands at 1047 and 1022 cm^{-1} represents the amount of ordered starch to amorphous starch, and the ratio changed significantly with the thermal treatment (Chung et al. 2010). For example, the ratio increased on annealing (ANN) of lentil starch but dropped on

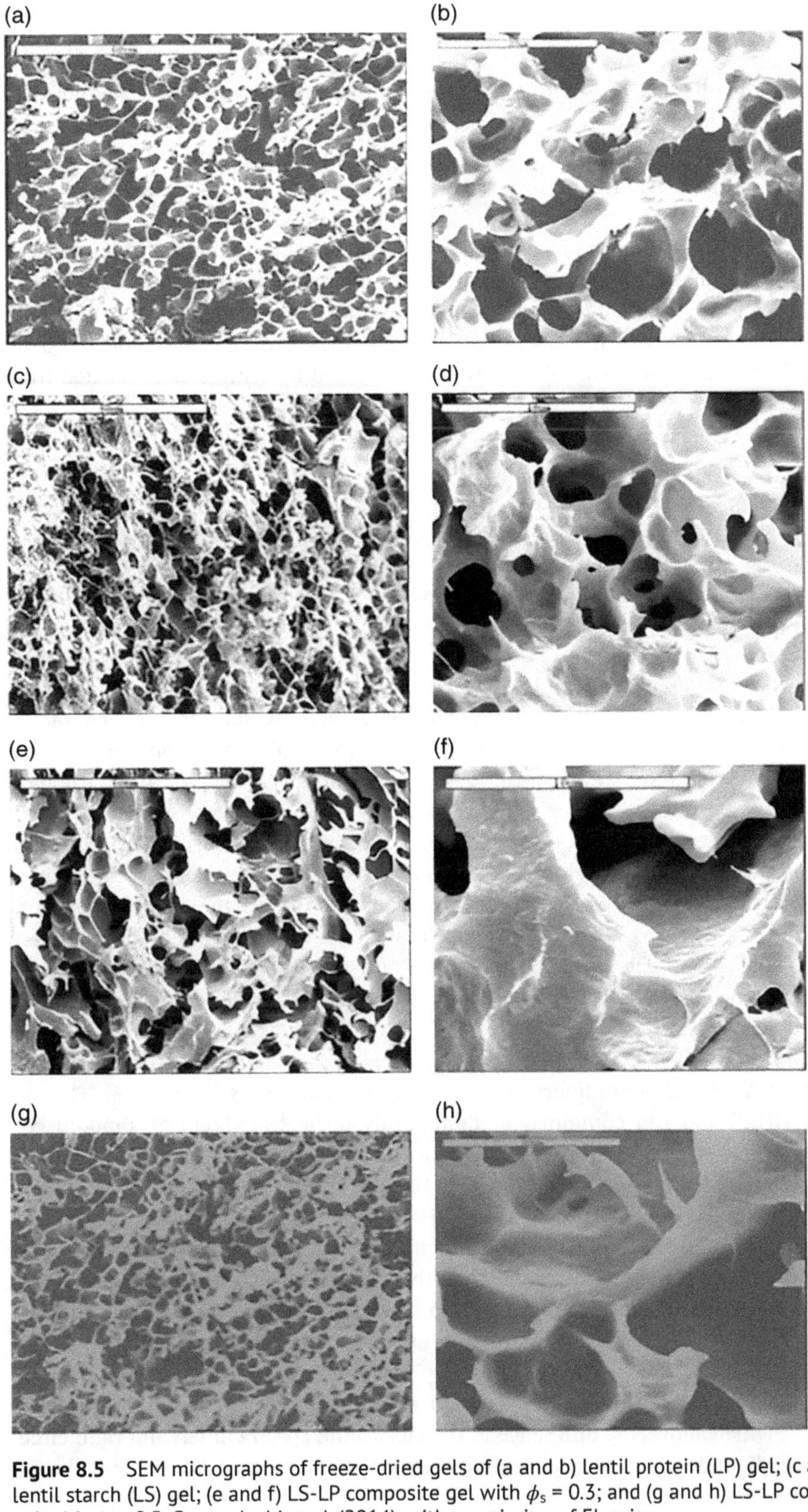

Figure 8.5 SEM micrographs of freeze-dried gels of (a and b) lentil protein (LP) gel; (c and d) lentil starch (LS) gel; (e and f) LS-LP composite gel with ϕ_s = 0.3; and (g and h) LS-LP composite gel with ϕ_s = 0.5. *Source:* Joshi et al. (2014), with permission of Elsevier.

heat-moisture treatment (HMT). On the other hand, the ratio of HMT followed by ANN (ANN–HMT) decreased against ANN starches, which is an indication of disruption of crystallites. ANN followed by HMT (HMT–ANN), however, showed an increase in the ratio as compared to HMT starches. This indicates the perfection of those crystallites, which were disrupted on HMT.

The FTIR spectra of the isolated starch obtained after various treatments of lentil seeds indicated the intensity of the absorptions related to hydrogen bonding (3000–3600 cm^{-1}, and ~1000 cm^{-1}) was markedly decreased when compared to the native starch (Yin et al. 2018). The intensity of the absorption between 1300 and 1200 cm^{-1} and the band at 2930 cm^{-1} of treated seeds was also markedly decreased for lentil starch isolated from processed seeds.

8.5.2 Proteins

FTIR spectroscopy of lentil proteins and peptides exhibited several amide bands (A, B, I, II, III, IV, V, VI, and VII) and measured the vibrational transitions due to N–H, C=O, and C–N stretching and bending modes (Aryee and Boye 2017). The NH stretching vibrations of amide A and B are absorbed near 3310 and 3270 cm^{-1}, whereas amide I, II, and III vibrations are absorbed near 1650 cm^{-1}, 1550 cm^{-1}, and near 1400–1200 cm^{-1}, respectively (Carbonaro et al. 2008). The amide I band is used for elucidating the lentil protein conformation and its secondary structure. Additionally, the amide I regions of LPI is characterized by the presence of anti-parallel β-sheets (1632 and 1684 cm^{-1}), overlapped α-helix and random coil signals (1651 cm^{-1}), and turns (1670 cm^{-1}) (Barbana and Boye 2013; Carbonaro et al. 2008; Jarpa-Parra et al. 2014). A higher intensity band at 1632 cm^{-1} over 1651 cm^{-1} suggests the presence of more β-sheet structures. Reducing peak intensities at 1651, 1670, and 1684 cm^{-1} with increasing pH from 8 to 10 suggests a partial unfolding of the protein structure because of the disruption of hydrogen bonds and dissociation of hydrogen from carbonyl and sulfate groups at alkali conditions that facilitate partial protein denaturation (Jarpa-Parra et al. 2014). The higher content of β-sheet structure retention during processing has been associated with the unaltered amide I spectral profile (Ahmed et al. 2009; Carbonaro et al. 2008).

Aryee and Boye (2017) observed significant changes in the secondary structure of thermally induced (25–95 °C) lentil flours and protein isolates at pH 7.4 (Figure 8.6). The raw lentil flour (RLF) and LPI contained predominantly β-sheet (1632 cm^{-1}), random coil (1643 cm^{-1}), which decreased during heating, accompanied by the formation of β-turns (1677 and 1669 cm^{-1}) and non-native β-sheet (1621 cm^{-1}). The intensities of the β-sheet structures decreased around 70 and 80 °C in RLF and LPI, respectively, while the intensity of the intermolecular β-sheet increased at 1621 cm^{-1} at 75 and 85 °C. The native β-sheet in the RLF and LPI was more heat sensitive than the α-helical structure. The protein in the cooked flour had intense bands at 1621 cm^{-1}, indicative of thermal aggregation, a significant loss of native secondary structure, relative extensive unfolding, and protein aggregation (Saguer et al. 2013).

The secondary structure of the proteins was further analyzed by circular dichroism (CD) spectroscopy, which shows the difference in the adsorption spectra of left and right circularly polarized light derived from the peptide backbone and aromatic side chains. The CD

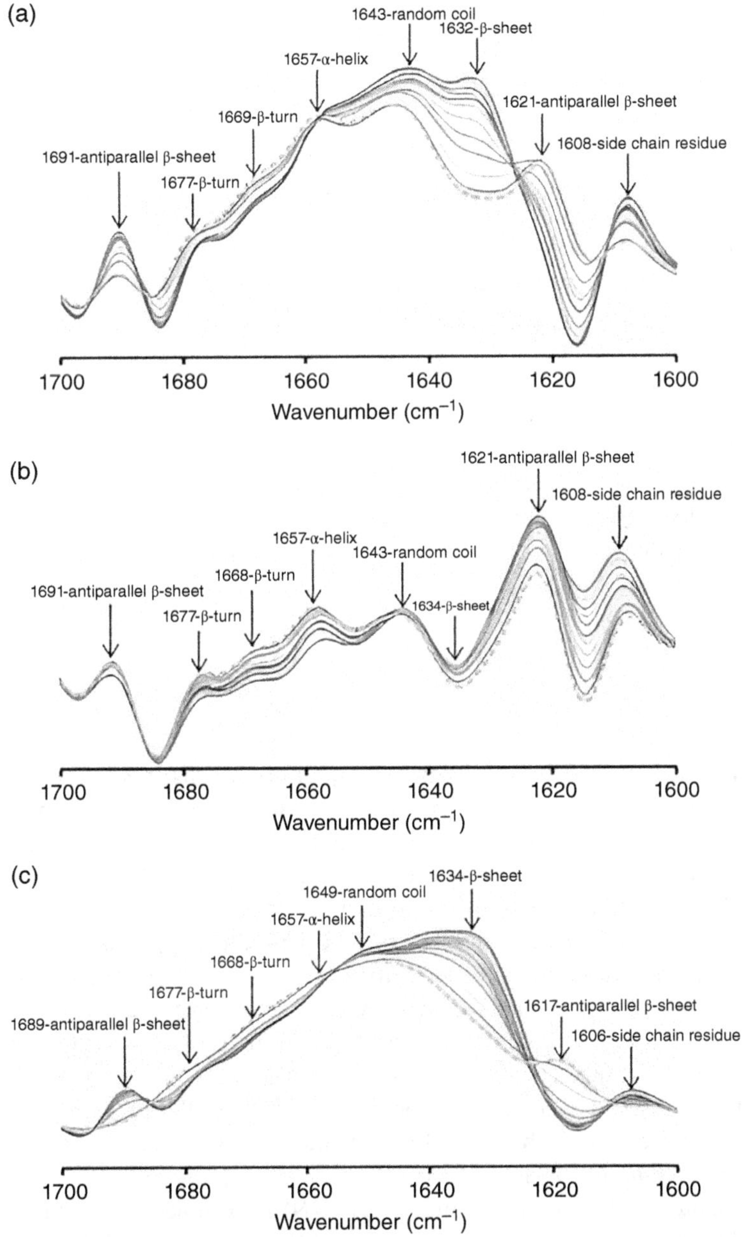

Figure 8.6 Deconvoluted FTIR spectra in the amide 1′ absorption region on heating from 25 to 95 °C of (a) RLF, raw lentil flour; (b) CLF, cooked lentil flour; and (c) LPI, lentil protein isolate in D_2O-phosphate buffer at pH 7.4. *Source:* Aryee and Boye (2017), with permission of John Wiley & Sons.

results suggest that cooked lentil proteins underwent greater secondary structure conformational changes than the fractionated protein isolate and confirm the FTIR results. The weaker CD signal, especially below 204 nm, confirms the low content of α-helical structure in CLF.

8.5.3 Conformational Changes of Protein During Gel Formation

Native lentil protein isolate (LPI) was converted into particulate aggregates and fibrillar aggregates and their spectra were compared with the native proteins (Jo et al. 2020). The native lentil proteins exhibited five major bands at 1608, 1632, 1646, 1659, and 1692 cm^{-1}. The particulate aggregates of LPI at pH 7 did not show any change in the spectra over the native proteins confirming the insignificant loss of the secondary structure; however, at pH 2, the native protein and particulate aggregates differed significantly. The absorption peak at 1692 cm^{-1} disappeared, and at the same time, new absorption bands at 1619 and 1684 cm^{-1} appeared. The peak at 1619 cm^{-1} describes the intermolecular β-sheets formation via aggregation and hydrogen bonding. On the other hand, the peak at 1684 cm^{-1} indicates β-sheets remained anti-parallel.

8.6 X-ray Diffraction (XRD) Patterns

8.6.1 Starch

X-ray diffraction (XRD) is used to describe qualitative and quantitative changes in crystalline domains in starch. The XRD patterns confirmed that the lentil starch has a C-type diffraction pattern, which is a blend of type-A and type-B unit cell crystalline (Hoover and Ratnayake 2002). Many diffraction peaks were detected at 2θ values of 5.6°, 15°, 17°, 17.8°, 23°, and 26.3° for lentil starch (Ahmed et al. 2016; Joshi et al. 2013; Kaur et al. 2010; Lu et al. 2018) (Figure 8.7). Higher peak values suggest a better orientation and/or more compactly packed crystallites within the granules (Hoover and Manuel 1995). The relative crystallinity (RC) obtained from XRD demonstrated that the isolated starch from red lentils had a significantly higher RC than that of green lentil starch (Lu et al. 2018). The degree of crystallinity of the lentil starches ranged from 27% to 33% (Kaur et al. 2010; Joshi et al. 2013), which was influenced by the size of the crystals, the number of crystalline regions, the orientation of the double helices within the crystalline domains, and the extent of interaction between double helices (Ambigaipalan et al. 2011).

The XRD diffraction patterns of natural and annealed (ANN)/heat moisture treatment (HMT) of lentil starches resulted in a modification compared to the native starch (Chung et al. 2010). The relative crystallinity of lentils improved on annealing (ANN) but was reduced on heat moisture treatment (HMT). The enhancement of resistant starch after annealing might happen because of the compact packing of double helices inside the crystalline lamella. The relative crystallinity of lentils after ANN–HMT was lower than ANN, which infers that a combined treatment would have damaged the crystalline structure. The octenyl succinylated lentil starch exhibited a similar crystalline pattern as their respective native starches. However, the modification resulted in a decrease in crystallinity, which

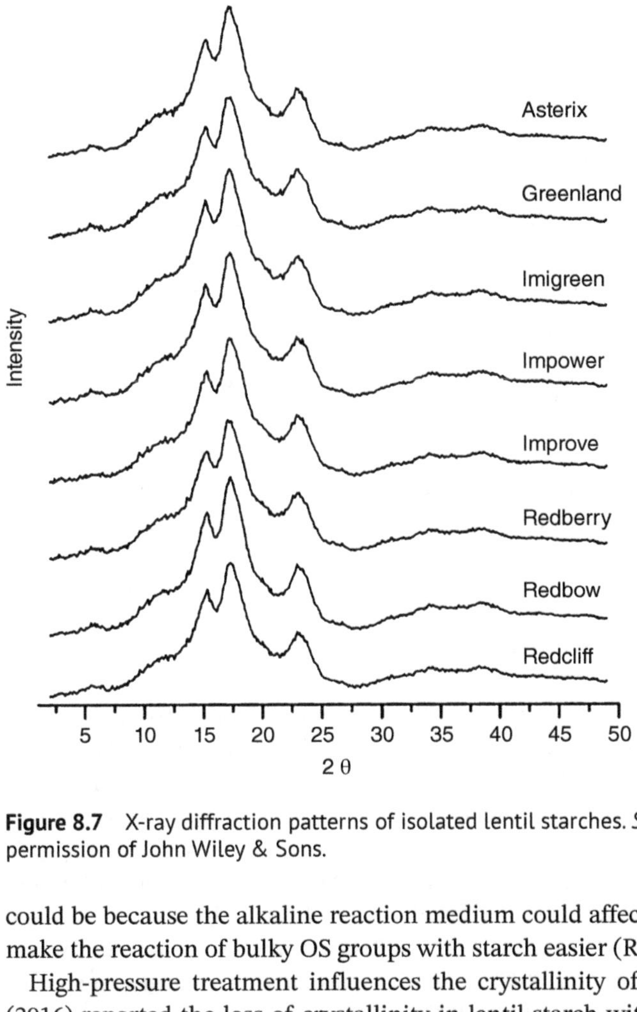

Figure 8.7 X-ray diffraction patterns of isolated lentil starches. *Source:* Lu et al. (2018), with permission of John Wiley & Sons.

could be because the alkaline reaction medium could affect the crystallinity of starch and make the reaction of bulky OS groups with starch easier (Remya et al. 2018).

High-pressure treatment influences the crystallinity of legume starch. Ahmed et al. (2016) reported the loss of crystallinity in lentil starch with pressure application. The 2θ values of the native starches changed gradually after the pressure treatment, and the loss of crystallinity was evident at >400 MPa. A complete loss of crystallinity was detected at 600 MPa with the disappearance of the peak intensities at 14.9° and 22.8°. The third peak at 17.2° reduced its intensity significantly as compared to the untreated sample. Such changes in 2θ values indicate the change in crystallinities, which further translates into structural transformation from the C-type to B-type XRD pattern (Guo et al. 2015).

8.6.2 Proteins

Lentil protein isolate samples showed two characteristic X-ray diffraction peaks around 10° (small intensity) and 24° (high intensity) (Joshi et al. 2011). The intensity of the diffraction peak for the vacuum-dried LPI was the highest, while the intensities of the spray and FD LPIs were similar to each other. This reason could be that vacuum-dried LPI might have partially crystallized due to longer drying time.

8.7 X-ray Tomography

The microstructures, including mean cell size (MCS) and mean wall thickness (MWT) of red lentil extrudates produced by injecting N_2 gas, were studied using X-ray microtomography (Guillermic et al. 2021; Luo et al. 2020). The microstructure of the extrudates was strongly correlated with the gas pressure and, to a lesser extent, to the moisture and screw speed. Different pairs of extrudates produced at various extrusion conditions are shown in Figure 8.8, demonstrating individual effects of screw speed, moisture content, and pressure. The screw speed had the least impact on porosity and radial expansion index in comparison to moisture content and N_2 injection (Luo et al. 2020). The extrudates exhibited many evenly distributed cells with relatively smaller and thicker cell walls at a higher moisture content due to even hydration of proteins and carbohydrates, which in turn created a stronger network that resists expansion at the die exit. Tomography results also exhibited that N_2 injection produced numerous small and more evenly distributed cells. The high number of small cells has also been improved by N_2 injection in extrudates made from red lentil flour. There was no strong correlation between separately calculated MCS and MWT values, so these quantities can be considered independent.

8.8 Nuclear Magnetic Resonance (NMR)

Nuclear magnetic resonance (NMR) has been considered a versatile nondestructive method and useful tool due to its ability to provide a complete view of the structure and composition of food, providing qualitative and quantitative information on major and minor compounds (Mannina et al. 2012). NMR spectroscopy records the interaction of radiofrequency (RF) electromagnetic radiation with the nuclei of molecules placed in a strong magnetic field. Mostly, ^{1}H and ^{13}C NMR are used in food analysis. NMR in combination with multivariate statistical analysis can be used as a powerful tool for determining food quality and geographical origin, especially when used as an untargeted method, where the whole spectra can be used as fingerprints without assigning particular resonances to specific metabolites (Baiano et al. 2012; Ferrara et al. 2013).

Longobardi et al. (2017) used ^{1}H NMR fingerprinting in combination with chemometrics to build models able to classify lentils according to their geographical origin. A series of statistical techniques (soft independent modelling of class analogy [SIMCA], k-nearest neighbor [k-NN], principal component analysis followed by linear discriminant analysis [PCA-LDA], and partial least squares-discriminant analysis [PLS-DA]) were used to establish the authenticity of lentils. The best combination of average recognition (100%) and cross-validation prediction abilities (96.7%) was obtained for the PCA-LDA, with external prediction rates of 100% and 95.3% on the test set and by the MCCV procedure, respectively. Furthermore, k-NN and PLS-DA discriminant models also highlighted that the NMR data contained enough information to build adequate models. Finally, a pattern of metabolites that mostly contributed to the lentil discrimination based on their geographical origin was identified.

Tsopmo and Muir (2010) developed a high-performance liquid chromatography (HPLC) method to obtain fingerprints of secondary metabolites of 12 lentil cultivars grown in Canada. Sixteen compounds were found in the LMI cultivar and characterized by a 100%

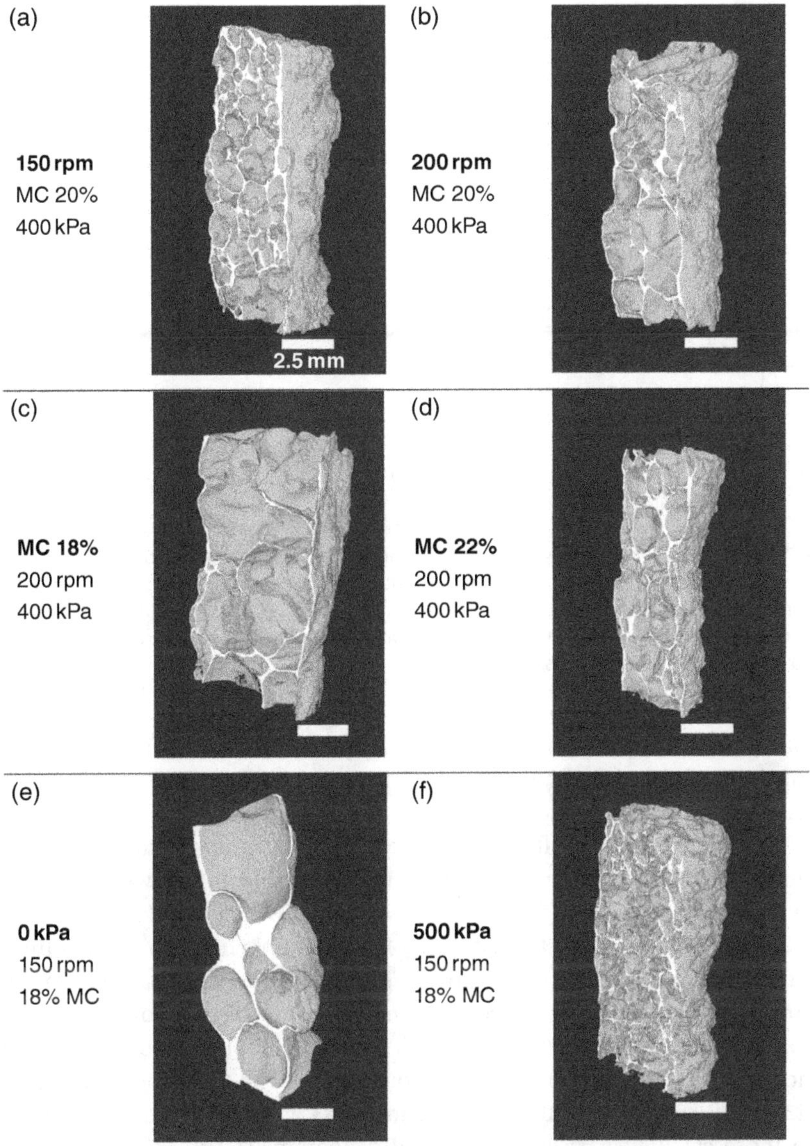

Figure 8.8 3D cross-sectional microstructure of red lentil extrudates at (a) 150 rpm, and (b) 200 rpm, at the constant feed moisture content (20%) and constant N_2 injection pressure (400 kPa); at (c) 18% moisture content, and (d) 22% moisture content, at constant screw speed (200 rpm) and constant N_2 injection pressure (400 kPa); and at (e) 0 kPa and (f) 500 kPa N_2 injection pressure at constant screw speed (150 rpm) and constant moisture content (18%). The scale bar in (a) is 2.5 mm and applies to all images. *Source:* Luo et al. (2020), with permission of Elsevier.

methanol extract based on NMR and MS spectra. The presence of stachyose and pinitol 2-*O*-*R*-D-galactoside, raffinose, sucrose, and D-glucose were identified, which are well-characterized constituents of lentils. The most important compounds, which are reported for the first time from lentils, are the novel 4-chloro-1*H*-indole-3-*N*-methyl acetamide (13), itaconic acid, arbutin, gentisic acid 5-O-[β-D-apiofuranosyl-(1f2)-β-D-xylopyranoside], and

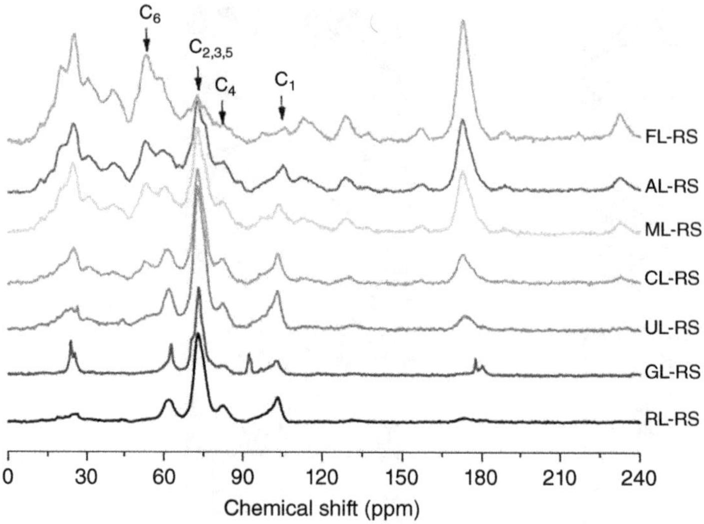

Figure 8.9 ^{13}C CP/MPS NMR spectra of resistant starch from processed and untreated lentil seeds. UL-RS (untreated lentil resistant starch); RL-RS (roasted lentil resistant starch); GL-RS (germinated lentil resistant starch); FL-RS (fermented lentil resistant starch); ML-RS (microwaved lentil resistant starch); AL-RS (autoclaved lentil resistant starch); CL-RS (conventionally cooked lentil re-resistant starch). *Source:* Ma et al. (2018), with permission of Elsevier.

(6*S*,7*Z*,9*R*)-9-hydroxymegastigma-4,7-dien-3-one-9-*O*-β-D-apiofuranosyl-(1f2)-β-D glucopyranoside. Cluster analysis can be successfully employed for metabolic profiling of legumes or food components before separation is attempted.

^{13}C CP/MAS NMR of lentil starches after selected treatments (mentioned earlier) showed the RS from fermented, microwaved, and autoclaved lentils exhibited the lowest values of crystallinity and double helix content (Figure 8.9) (Ma et al. 2018). The peaks observed at 70–79 ppm (C2, C3, and C5) are characterized as the B-type double helices resulting from the residues of free amylose, and the peak appearing at 80–84 ppm was assigned to the partial fraction of C4 carbons with the V-type single helical conformation dominating the amorphous state. The resonance peak at 102 ppm was related to the simple helix organized in the crystalline phase or dispersed in the amorphous region. The relative crystallinities obtained from 13^{C} NMR spectra for different LRS samples (C2: 4.33–11.41%) were generally higher than those calculated based on X-ray diffractograms (C1: 0.483–2.82%). The high-resolution solid-state 13^{C} NMR can detect the short-range ordered structure in RS, including the sub-crystal region, the overlapped crystal, and amorphous region, and the double helix structure arranged in the amorphous region.

8.9 Atomic Force Microscopy (AFM)

AFM has been successfully employed for the nanoimaging of lentil proteins and structure of fibrils (Li et al. 2021). AFM results (Figure 8.10) demonstrate that the vicilin fractions were responsible for the formation of fibrils as well as semiflexible fibrils and rigid fibrils. The lentil proteins exhibited spherical particles with a diameter of about 50 nm, and those spherical particles were transformed into short fibrils after heating for 0.5 hour, and the contour length of the LP fibrils increased when the heating time increased from 6 to 24 hours (Figure 8.10 C1–C5). The long and semiflexible LP fibrils with a maximum contour length of 816 nm obtained after 24 hours heating. Additionally, the LP fibrils assembled into larger and tighter entangled aggregates when the heating time lasted for 12 and 24 hours. Overall, the fibril length increased as the heating time increased.

The surface topography of lentil protein concentrate (LPC) and enzyme-mediated hydrolyzed (LPH) samples demonstrated that significant changes occurred after the hydrolysis of LPC (Rezvankhah et al. 2021). High peaks were visible for the control sample (Figure 8.11a) while the hydrolysates resulted in short peaks that depended on the DH values. Alcalase (Al)-2% and a holding time of 180 minute and Al (120 minute)-Flavourzyme (Fl) (60 minute)-2% produced higher DH values and shorter peaks (Figure 8.11b,d). Fl-2%-180 minute produced a lower DH value and higher peaks than other hydrolysates (Figure 8.11c). Hydrolysis led to the production of bioactive peptides with different MWs with different topographies. In conclusion, it could be possible to visualize hydrolyzed powder with higher or lower MWs peptides using AFM images.

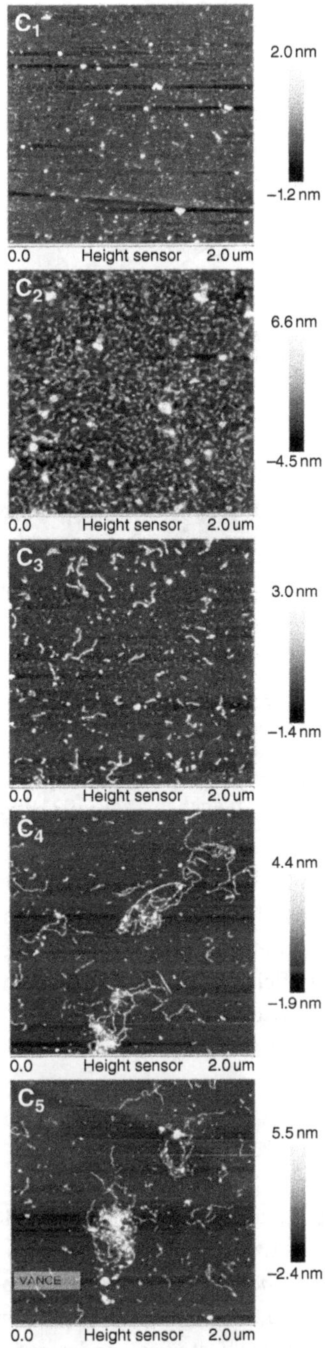

Figure 8.10 The AFM images of lentil proteins during fibril formation (C1: 0 hour, C2: 0.5 hour, C3: 6 hours, C4: 12 hours, and C5: 24 hours). *Source:* Li et al. (2021), with permission of Elsevier.

8.10 Conclusions

Industrial applications of lentils depend upon their pasting, thermal, and structural properties. Lentil starch and protein isolates are the most valuable products from lentils. With increasing consumer demand for a meat substitute or meat-analog, research on the structural and functional properties of lentil ingredients has increased manifold. Various structural analysis techniques and their applications have been elucidated in the chapter. The gelatinization, retrogradation, and setback viscosity provide a guideline on how to use lentil starch in product development. On the other hand, structural analysis of lentil proteins using innovative analytical tools including tomography, AFM, XRD, NMR, and CLM brings new information on the surface properties and microstructure of

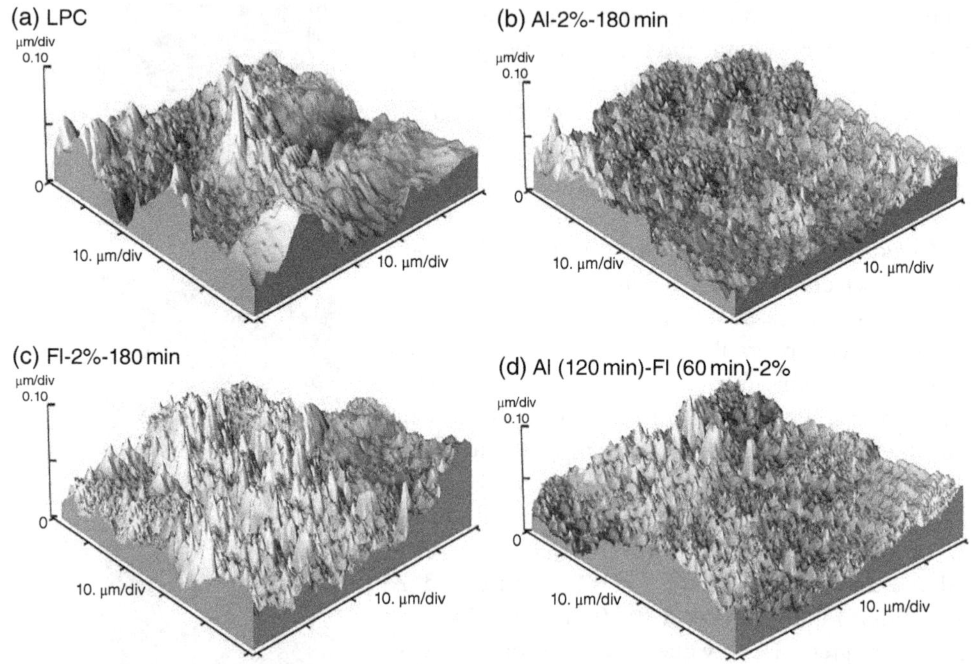

Figure 8.11 AFM images of lentil protein concentrate (LPC) and generated bioactive peptides (powdered by spray dryer). LPC is a non-hydrolyzed protein (a), Al-2%-180 minute (b) and Fl-2%-180 minute (c) indicate lentil protein hydrolysate samples produced through the single hydrolysis, Al (120 minute)-Fl (60 minute)-2% (d) was produced through the sequential hydrolysis. *Source:* Rezvankhah et al. (2021), with permission of John Wiley & Sons.

lentils that could be translated not only into manufacturing food products but also into food and drug delivery. Antinutritional factors and the structures of lentil ingredients could be better established using structural analysis tools. Overall, all of these techniques will be beneficial in the formulation and processing of lentil-based novel foods and could achieve the sustainability and security requirements of the EU Horizon 2021–2026 and the UN Sustainable Development Goals 2030.

References

Aguilera, Y., Esteban, R.M., Benítez, V. et al. (2009). Starch, functional properties, and microstructural characteristics in chickpea and lentil as affected by thermal processing. *Journal of Agricultural and Food Chemistry* 57 (22): 10682–10688.

Ahmed, J. (2012). Rheometric non-isothermal gelatinization kinetics of mung bean starch slurry: effect of salt and sugar – part 1. *Journal of Food Engineering* 109 (2): 321–328.

Ahmed, J. and Auras, R. (2011). Effect of acid hydrolysis on rheological and thermal characteristics of lentil starch slurry. *LWT Food Science and Technology* 44 (4): 976–983.

Ahmed, J. and Rahman, M.S. (2014). Glass transition in foods. In: *Engineering Properties of Foods* (ed. M.A. Rao, S.S.H. Rizvi, A.K. Datta, and J. Ahmed), 94–116. CRC Press.

Ahmed, J., Ramaswamy, H.S., Ayad, A., and Alli, I. (2008). Thermal and dynamic rheology of insoluble starch from basmati rice. *Food Hydrocolloids* 22 (2): 278–287.

Ahmed, J., Varshney, S.K., and Ramaswamy, H.S. (2009). Effect of high pressure treatment on thermal and rheological properties of lentil flour slurry. *LWT Food Science and Technology* 42 (9): 1538–1544.

Ahmed, J., Al-Jassar, S., and Thomas, L. (2015). A comparison in rheological, thermal, and structural properties between Indian Basmati and Egyptian Giza rice flour dispersions as influenced by particle size. *Food Hydrocolloids* 48: 72–83.

Ahmed, J., Thomas, L., Taher, A., and Joseph, A. (2016a). Impact of high pressure treatment on functional, rheological, pasting, and structural properties of lentil starch dispersions. *Carbohydrate Polymers* 152: 639–647.

Ahmed, J., Taher, A., Mulla, M.Z. et al. (2016b). Effect of sieve particle size on functional, thermal, rheological and pasting properties of Indian and Turkish lentil flour. *Journal of Food Engineering* 186: 34–41.

Al-Attar, H., Ahmed, J., and Thomas, L. (2022). Rheological, pasting and textural properties of corn flour as influenced by the addition of rice and lentil flour. *LWT Food Science and Technology* 160: 113231.

Ambigaipalan, P., Hoover, R., Donner, E. et al. (2011). Structure of faba bean, black bean and pinto bean starches at different levels of granule organization and their physicochemical properties. *Food Research International* 44 (9): 2962–2974.

Aryee, A.N.A. and Boye, J.I. (2017). Comparative study of the effects of processing on the nutritional, physicochemical and functional properties of lentil. *Journal of Food Processing and Preservation* 41 (1): e12824.

Asif, M., Rooney, L.W., Ali, R., and Riaz, M.N. (2013). Application and opportunities of pulses in food system: a review. *Critical Reviews in Food Science and Nutrition* 53 (11): 1168–1179.

Baiano, A., Terracone, C., Longobardi, F. et al. (2012). Effects of different vinification technologies on physical and chemical characteristics of Sauvignon blanc wines. *Food Chemistry* 135 (4): 2694–2701.

Barbana, C. and Boye, J.I. (2011). Angiotensin I-converting enzyme inhibitory properties of lentil protein hydrolysates: determination of the kinetics of inhibition. *Food Chemistry* 127 (1): 94–101.

Barbana, C. and Boye, J.I. (2013). In vitro protein digestibility and physico-chemical properties of flours and protein concentrates from two varieties of lentil (*Lens culinaris*). *Food & Function* 4 (2): 310–321.

Bhatty, R.S. (1984). Relationship between physical and chemical characters and cooking quality in lentil. *Journal of Agricultural and Food Chemistry* 32: 1166–1172.

Biliaderis, C.G., Page, C.M., Maurice, T.J., and Juliano, B.O. (1986). Thermal characterization of rice starches: a polymeric approach to phase transitions of granular starch. *Journal of Agricultural and Food Chemistry* 34 (1): 6–14.

Bresciani, A., Giuberti, G., Cervini, M., and Marti, A. (2021). Pasta from yellow lentils: how process affects starch features and pasta quality. *Food Chemistry* 364: 130387.

Bresciani, A., Emide, D., Saitta, F. et al. (2022). Impact of thermal treatment on the starch-protein interplay in red lentils: connecting molecular features and rheological properties. *Molecules* 27 (4): 1266.

Cai, R., McCurdy, A., and Baik, B.K. (2002). Textural property of 6 legume curds in relation to their protein constituents. *Journal of Food Science* 67 (5): 1725–1730.

Capron, I., Robert, P., Colonna, P. et al. (2007). Starch in rubbery and glassy states by FTIR spectroscopy. *Carbohydrate Polymers* 68 (2): 249–259.

Carbonaro, M., Maselli, P., Dore, P., and Nucara, A. (2008). Application of Fourier transform infrared spectroscopy to legume seed flour analysis. *Food Chemistry* 108 (1): 361–368.

Chavan, J.K. and Kadam, S.S. (1989). Nutritional improvement of cereals by sprouting. *Critical Reviews in Food Science and Nutrition* 28 (5): 401–437.

Chedid, L.L. and Kokini, J.L. (1992). Influence of protein addition on rheological properties of amylose-and amylopectin-based starches in excess water. *Cereal Chemistry* 69 (5): 551–555.

Chung, H.J., Liu, Q., Hoover, R. et al. (2008). *In vitro* starch digestibility, expected glycemic index, and thermal and pasting properties of flours from pea, lentil and chickpea cultivars. *Food Chemistry* 111 (2): 316–321.

Chung, H., Liu, Q., and Hoover, R. (2010). Effect of single and dual hydrothermal treatments on the crystalline structure, thermal properties, and nutritional fractions of pea, lentil, and navy bean starches. *Food Research International* 43 (2): 501–508.

Cooke, D. and Gidley, M.J. (1992). Loss of crystalline and molecular order during starch gelatinisation: origin of the enthalpic transition. *Carbohydrate Research* 227 (C): 103–112.

Ek, P., Gu, B.J., and Ganjyal, G.M. (2021). Whole seed lentil flours from different varieties (Brewer, Crimson, and Richlea) demonstrated significant variations in their expansion characteristics during extrusion. *Journal of Food Science* 86 (3): 942–951.

Ferrara, G., Mazzeo, A., Matarrese, A.M.S. et al. (2013). Application of abscisic acid (S-ABA) to "Crimson Seedless" grape berries in a mediterranean climate: effects on color, chemical characteristics, metabolic profile, and S-ABA concentration. *Journal of Plant Growth Regulation* 32 (3): 491–505.

Gharibzahedi, S.M.T., Emam-Djomeh, Z., Razavi, S.H., and Jafari, S.M. (2014). Mechanical behavior of lentil seeds in relation to their physicochemical and microstructural characteristics. *International Journal of Food Properties* 17 (3): 545–558.

Guillermic, R.M., Aksoy, E.C., Aritan, S. et al. (2021). X-ray microtomography imaging of red lentil puffed snacks: processing conditions, microstructure and texture. *Food Research International* 140: 109996.

Guo, Z., Zeng, S., Lu, X. et al. (2015). Structural and physicochemical properties of lotus seed starch treated with ultra-high pressure. *Food Chemistry* 186: 223–230.

Hall, A.E. and Moraru, C.I. (2021). Structure and function of pea, lentil and faba bean proteins treated by high pressure processing and heat treatment. *LWT Food Science and Technology* 152 (April): 112349.

Hall, C., Hillen, C., and Robinson, J.G. (2017). Composition, nutritional value, and health benefits of pulses. *Cereal Chemistry* 94 (1): 11–31.

Hoover, R. and Manuel, H. (1995). A comparative study of the physicochemical properties of starches from two lentil cultivars. *Food Chemistry* 53 (3): 275–284.

Hoover, R. and Ratnayake, W.S. (2002). Starch characteristics of black bean, chick pea, lentil, navy bean and pinto bean cultivars grown in Canada. *Food Chemistry* 78 (4): 489–498.

Hoover, R., Hughes, T., Chung, H.J., and Liu, Q. (2010). Composition, molecular structure, properties, and modification of pulse starches: a review. *Food Research International* 43 (2): 399–413.

Jarpa-Parra, M., Bamdad, F., Wang, Y. et al. (2014). Optimization of lentil protein extraction and the influence of process pH on protein structure and functionality. *LWT Food Science and Technology* 57 (2): 461–469.

Jenkins, P.J. and Donald, A.M. (1995). The influence of amylose on starch granule structure. *International Journal of Biological Macromolecules* 17 (6): 315–321.

Jo, Y.J., Huang, W., and Chen, L. (2020). Fabrication and characterization of lentil protein gels from fibrillar aggregates and the gelling mechanism study. *Food & Function* 11 (11): 10114–10125.

Joshi, M., Adhikari, B., Aldred, P. et al. (2011). Physicochemical and functional properties of lentil protein isolates prepared by different drying methods. *Food Chemistry* 129 (4): 1513–1522.

Joshi, M., Aldred, P., McKnight, S. et al. (2013). Physicochemical and functional characteristics of lentil starch. *Carbohydrate Polymers* 92 (2): 1484–1496.

Joshi, M., Aldred, P., Panozzo, J.F. et al. (2014). Rheological and microstructural characteristics of lentil starch-lentil protein composite pastes and gels. *Food Hydrocolloids* 35: 226–237.

Kaur, M., Sandhu, K.S., and Lim, S.T. (2010). Microstructure, physicochemical properties and in vitro digestibility of starches from different Indian lentil (*Lens culinaris*) cultivars. *Carbohydrate Polymers* 79 (2): 349–355. https://doi.org/10.1016/J.CARBPOL.2009.08.017.

Lee, H. (2007). Physio-chemistry and rheology of Australian lentil flour and starch, and their implications for extrusion. PhD dissertations. UNSW Sydney.

Li, C. and Ganjyal, G.M. (2017). Chemical composition, pasting, and thermal properties of 22 different varieties of peas and lentils. *Cereal Chemistry* 94 (3): 392–399.

Li, L., Yuan, T.Z., Setia, R. et al. (2019). Characteristics of pea, lentil and faba bean starches isolated from air-classified flours in comparison with commercial starches. *Food Chemistry* 276 (April 2018): 599–607.

Li, T., Wang, L., Zhang, X. et al. (2021). Assembly behavior, structural characterization and rheological properties of legume proteins based amyloid fibrils. *Food Hydrocolloids* 111: 106396.

Longobardi, F., Innamorato, V., Di Gioia, A. et al. (2017). Geographical origin discrimination of lentils (*Lens culinaris* Medik.) using ^{1}H NMR fingerprinting and multivariate statistical analyses. *Food Chemistry* 237: 743–748.

Lopez-Rubio, A., Clarke, J.M., Scherer, B. et al. (2009). Structural modifications of granular starch upon acylation with short-chain fatty acids. *Food Hydrocolloids* 23 (7): 1940–1946.

Lu, Z.H., Donner, E., Tsao, R. et al. (2018). Physicochemical and digestion characteristics of flour and starch from eight Canadian red and green lentils. *International Journal of Food Science and Technology* 53 (3): 735–746.

Lu, H., Ji, N., Li, M. et al. (2019). Preparation of borax cross-linked starch nanoparticles for improvement of mechanical properties of maize starch films. *Journal of Agricultural and Food Chemistry* 67 (10): 2916–2925.

Luo, S., Chan, E., Masatcioglu, M.T. et al. (2020). Effects of extrusion conditions and nitrogen injection on physical, mechanical, and microstructural properties of red lentil puffed snacks. *Food and Bioproducts Processing* 121: 143–153.

Ma, Z., Boye, J.I., Simpson, B.K. et al. (2011). Thermal processing effects on the functional properties and microstructure of lentil, chickpea, and pea flours. *Food Research International* 44 (8): 2534–2544.

Ma, Z., Yin, X., Hu, X. et al. (2018). Structural characterization of resistant starch isolated from Laird lentils (*Lens culinaris*) seeds subjected to different processing treatments. *Food Chemistry* 263: 163–170.

Majeed, T., Wani, I.A., and Hussain, P.R. (2017). Effect of dual modification of sonication and γ-irradiation on physicochemical and functional properties of lentil (*Lens culinaris* L.) starch. *International Journal of Biological Macromolecules* 101: 358–365.

Mannina, L., Sobolev, A.P., and Viel, S. (2012). Liquid state ^{1}H high field NMR in food analysis. *Progress in Nuclear Magnetic Resonance Spectroscopy* 66: 1–39.

Masatcioglu, M.T. and Koksel, F. (2019). Functional and thermal properties of yellow pea and red lentil extrudates produced by nitrogen gas injection assisted extrusion cooking. *Journal of the Science of Food and Agriculture* 99 (15): 6796–6805.

Pathiratne, S.M., Shand, P.J., Pickard, M., and Wanasundara, J.P.D. (2015). Generating functional property variation in lentil (*Lens culinaris*) flour by seed micronization: effects of seed moisture level and surface temperature. *Food Research International* 76 (P1): 122–131.

Ratnayake, W.S., Hoover, R., and Warkentin, T. (2002). Pea starch: composition, structure and properties – a review. *Starch – Staerke* 54 (6): 217–234.

Rebello, C.J., Greenway, F.L., and Finley, J.W. (2014). Whole grains and pulses: a comparison of the nutritional and health benefits. *Journal of Agricultural and Food Chemistry* 62 (29): 7029–7049.

Remya, R., Jyothi, A.N., and Sreekumar, J. (2018). Morphological, structural and digestibility properties of RS4 enriched octenyl succinylated sweet potato, banana and lentil starches. *Food Hydrocolloids* 82: 219–229.

Ren, N., Ma, Z., Li, X., and Hu, X. (2021). Preparation of rutin-loaded microparticles by debranched lentil starch-based wall materials: structure, morphology and *in vitro* release behavior. *International Journal of Biological Macromolecules* 173: 293–306.

Rezvankhah, A., Yarmand, M.S., Ghanbarzadeh, B., and Mirzaee, H. (2021). Characterization of bioactive peptides produced from green lentil (*Lens culinaris*) seed protein concentrate using Alcalase and Flavourzyme in single and sequential hydrolysis. *Journal of Food Processing and Preservation* 45 (11): e15932.

Rodríguez-Damian, A.R., De La Rosa-Millán, J., Agama-Acevedo, E. et al. (2013). Effect of different thermal treatments and storage on starch digestibility and physicochemical characteristics of unripe banana flour. *Journal of Food Processing and Preservation* 37 (5): 987–998.

Saguer, E., Alvarez, P.A., Sedman, J., and Ismail, A.A. (2013). Study of the denaturation/aggregation behaviour of whole porcine plasma and its protein fractions during heating under acidic pH by variable-temperature FTIR spectroscopy. *Food Hydrocolloids* 33 (2): 402–414.

Sevenou, O., Hill, S.E., Farhat, I.A., and Mitchell, J.R. (2002). Organisation of the external region of the starch granule as determined by infrared spectroscopy. *International Journal of Biological Macromolecules* 31 (1–3): 79–85.

Shand, P.J., Ya, H., Pietrasik, Z., and Wanasundara, P.K.J.P.D. (2007). Physicochemical and textural properties of heat-induced pea protein isolate gels. *Food Chemistry* 102 (4): 1119–1130.

van Soest, J.J.G., Tournois, H., de Wit, D., and Vliegenthart, J.F.G. (1995). Short-range structure in (partially) crystalline potato starch determined with attenuated total reflectance Fourier-transform IR spectroscopy. *Carbohydrate Research* 279: 201–214.

Tsopmo, A. and Muir, A.D. (2010). Chemical profiling of lentil (*Lens culinaris* Medik.) cultivars and isolation of compounds. *Journal of Agricultural and Food Chemistry* 58 (15): 8715–8721.

Tsutsui, K., Katsuta, K., Matoba, T. et al. (2005). Effect of annealing temperature on gelatinization of rice starch suspension as studied by rheological and thermal measurements. *Journal of Agricultural and Food Chemistry* 53 (23): 9056–9063.

Wang, N., Warkentin, T.D., Vandenberg, B., and Bing, D.J. (2014). Physicochemical properties of starches from various pea and lentil varieties, and characteristics of their noodles prepared by high temperature extrusion. *Food Research International* 55: 119–127.

Wiercigroch, E., Szafraniec, E., Czamara, K. et al. (2017). Raman and infrared spectroscopy of carbohydrates: a review. *Spectrochimica Acta – Part A: Molecular and Biomolecular Spectroscopy* 185: 317–335.

Xu, M., Jin, Z., Simsek, S. et al. (2019). Effect of germination on the chemical composition, thermal, pasting, and moisture sorption properties of flours from chickpea, lentil, and yellow pea. *Food Chemistry* 295: 579–587.

Yin, X., Ma, Z., Hu, X. et al. (2018). Molecular rearrangement of Laird lentil (*Lens culinaris* Medikus) starch during different processing treatments of the seeds. *Food Hydrocolloids* 79: 399–408.

9

Lentil Protein: A Sustainable and Green Alternative to Animal Meat Protein

Fatema H. Brishti[1], Tareq Mzek[2], Nazamid Saari[3], Elenjikkal J. Rifna[4], Madhuresh Dwivedi[4], Kulwinder Kaur[5], and Preetinder Kaur[5]

[1] *Department of Food Technology, Faculty of Food Science & Technology, Universiti Putra Malaysia, Serdang, Selangor, Malaysia*
[2] *School of Business and Economics, Universiti Putra Malaysia, Serdang, Selangor, Malaysia*
[3] *Department of Food Science, Faculty of Food Science and Technology, Universiti Putra Malaysia, Serdang, Selangor, Malaysia*
[4] *Department of Food Process Engineering, National Institute of Technology, Rourkela, Odisha, India*
[5] *Department of Processing and Food Engineering, Punjab Agricultural University, Ludhiana, India*

9.1 Introduction

Proteins are one of the important nutrients needed for human growth and sustenance. As the world's population and food demand increase, current trends in animal protein consumption cannot be sustained (Godfray et al. 2018). Plant-based diets are produced in environmentally friendly manner, which reduce greenhouse gas emissions, water consumption, and deforestation (Kim et al. 2020; Willett et al. 2019). Legumes are among the more environmentally friendly alternatives to meat, pork, poultry, and dairy products (Bleakley and Hayes 2017). Moreover, plant proteins are highly sustainable and healthier options as meat substitutes compared to other protein sources (Santo et al. 2020). Agri-Food Innovation Council (2019) projected that plant proteins will occupy approximately one third of the protein market by 2024. The plant protein-based meat analogues have similar properties to meat, such as their fibrous appearances when fried or cooked (do Carmo et al. 2021), and are also cheaper and less processed than cell-engineered meat analogues (Santo et al. 2020). Meat analogues based on protein-rich pulse crops like lentils can help to slowdown climate change and contribute toward sustainable agricultural production systems (Clune et al. 2017) in addition to enhancing biodiversity, farmer livelihoods, food security, and human nutrition.

Lentils: Production, Processing Technologies, Products, and Nutritional Profile, First Edition.
Edited by Jasim Ahmed, Muhammad Siddiq, and Mark A. Uebersax.

9.1.1 Why Lentil?

Lentil (*Lens culinaris* L. Medikus), commonly known as "masoor" in Indian subcontinent, has all the features that make it the sustainable protein-rich source of future (Figure 9.1). Many farmers in arid zones, such as the northern Great Plains, rely on lentils as a livelihood strategy, as it is a drought-tolerant pulse that grows well in rain-fed and water-limited conditions (Thornton and Cramer 2012). By fixing nitrogen from the atmosphere by legume-rhizobia symbiosis in soil, lentils, and other pulse crops can reduce the need for inorganic nitrogen fertilizer during crop growth and are suitable in diverse crop rotation systems (Peoples et al. 2015). The lentil is also capable of recycling water and nutrients and controlling weeds and pests (Lupwayi and Kennedy 2007). Incorporating lentils into production systems potentially enhances soil building capacity and promotes nitrogen fixation, which might enhance reduced-tillage practices (Tanaka et al. 2010). As a result of these benefits, nitrogen fertilizer and pesticide use can be reduced significantly (MacWilliam et al. 2014) and the water consumption may also be reduced (Lupwayi and Kennedy 2007). By integrating lentils in low-tillage agricultural systems, labor and time are also reduced, as well as fossil fuel usage (van Kessel and Hartley 2000).

Food security is also supported by cultivating lentils (*L. culinaris* Medik), which are consumed in more than 70 countries. There has been widespread consumption of lentils in its whole form, dehulled form, and split form throughout the world. Even though lentils are relatively more expensive than most other pulses, the high growth rate is attributed to the fast cooking time and more culinary uses than most pulses. Lentils have nutrient-dense profile (Dhull et al. 2022; USDA 2018), which may support healthy pregnancy by supporting micronutrient deficiencies (Didinger and Thompson 2022; Ganesan and Xu 2017; Singh 2018) and can also help prevent stress-related death during pregnancy (sen Gupta et al. 2013). Lentils also possess high antioxidant activity compared to other food legumes. This is mainly due to their high concentration of phenolic compounds.

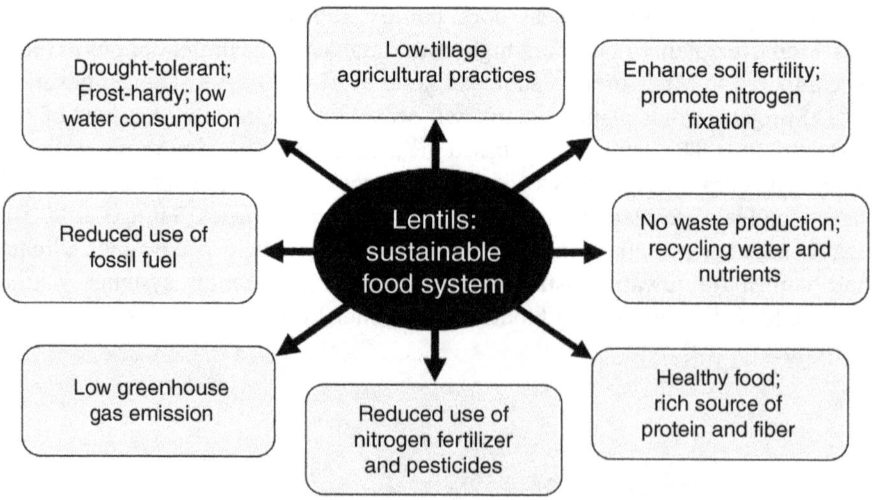

Figure 9.1 Lentil as a sustainable source of protein for developing meat analogues. *Source:* Schematic diagram by authors, adapted from Warne et al. (2019).

Due to cultural, religious, and ethical and dietary restrictions, lentil protein ingredients offer a potential and viable alternative to animal-derived proteins and soybean proteins. People are more likely to completely or partially substitute their meat consumption with meat analogues that mimic both the nutritional and physical properties of meat. In recent research and development on meat analogues, it has been proven that plant-based proteins such as lentil protein can be used to produce meat-like textures because of their excellent emulsifying properties, and water and oil retention capacities. The water holding capacity of lentil proteins was equivalent to that of pea proteins and greater than that of chickpea. In comparison to other pulses, red lentil protein concentrate had a higher capacity to absorb oil (Khazaei et al. 2019). People with high blood pressure and diabetes will benefit from these products since they contain low amounts of saturated fats and high amounts of polyunsaturated fats, fiber, and vitamins. In addition to providing good texture, flavor, appearance, and binding capacity, textured proteins are also an excellent source of protein and insoluble fiber for nutrition. Consumers prefer these characteristics, as they contribute to the versatility of meat analogues without compromising their function as food products when cooked.

Despite the growing production of lentils worldwide, and many benefits lentils offer (both on the production and consumption sides of the food system), a relatively small number of studies have examined this food legume as sustainable sources of protein for the development of meat alternatives. As an alternative to meat consumption, this chapter examines how legumes could become an important, sustainable source of dietary protein. The chapter provides an overview of lentil-based proteins, their functions, novel processing options, and potential applications for designing plant-based meat alternatives or meat analogues. Nutritionists, food security experts, food system professionals, farmers, and climate-change professionals will find this chapter useful.

9.2 Extraction Techniques for Plant Protein

Macronutrient protein is essential to human and animal diets due to its nutritional and health benefits. The proteins in lentils range from 20.6% to 31.4% (Urbano et al. 2007). A majority of these are storage proteins, i.e., albumins (16%), globulins (70%), glutelins (11%), and prolamins (3%) based on their solubility (Jarpa-Parra 2018). Lentil proteins have different physicochemical properties depending on their molecular structure (20 kDa albumin, 17–46 kDa glutelin, and 16–64 kDa prolamin). The hydrophilic surface of the albumin fraction renders the proteins water soluble, while a hydrophobic surface restricts the solubility of globulins in aqueous media. In recent years, increasing interest and research have been focused on the use of lentil protein isolates and concentrates to develop meat analogues for use in burgers, patties, sausages, and nuggets. In order to achieve high-quality protein isolates, it is crucial to use the right protein isolation and purification procedures. Ratnayake and Naguleswaran (2022) reported that during pulse protein extraction processes, starch, fiber, and soluble material are also generated as side streams. These side streams offer additional value-utilization opportunities and economic and sustainability benefits.

Table 9.1 presents an overview of different extraction techniques for protein isolation out of lentil flour. The detailed process to obtain legume protein concentrates and isolates through different extraction technologies is discussed below.

Table 9.1 Commonly used protein extraction techniques for lentil flour.

Method	Process conditions	Salient findings	References
Wet extraction	• pH (6–12), room temp., solid:solvent (5 : 100)	• Protein enrichment (21.5–80.0%)	Ahmad El-Sohaimy and Sitohy (2007)
	• pH (8–9.5), temp. (22–40 °C), solid:solvent (1 : 10)	• Protein yield (56.6–59.3%)	Lee et al. (2007)
	• pH (9), temp. (25 °C), solid:solvent (1 : 10)	• Protein yield (50.3–69.1%)	Boye et al. (2010a)
	• pH (4.6), room temp., solid:solvent ratio (1 : 10)	• Protein enrichment (14.5–82.0%)	Johnston et al. (2015)
	• pH (8–10), room temp., solid:solvent (1 : 10)	• Protein enrichment (12.3–85.7%)	Jarpa-Parra et al. (2014)
Dry fractionation	• Air flow (52 m^3/h), feed flow (1.5–3.5 kg/h) with the internal pressure (30–39 mbar), and air classifier speed (3000, 6000, and 12,000 rpm)	• Protein content in different fractions varied from 11% to 54%	Funke et al. (2022a)
	• Airflow (52 m^3/h), feed rate (~0.5 kg/h), wheel speed 10,000 rpm. Electrode distance (10 cm), electrode voltage (20 kV), dose rate (1.25 kg/h)	• Protein content in flour (~22%), coarse fraction (~18%), fine fraction (~60%)	Xing et al. (2018)
	• Air classifier speed 8036 rpm and air flow rate of 46 m^3/h	• Protein enrichment from flour to fine fraction (24.8–53.0%)	Afam-Mbah et al. (2018)
	• Air flow of 65 m^3/h, feed rate between 0.75 and 1 kg/h, air classifier speed as 11,000, 9000, 7000, 3000 rpm	• Protein enrichment from flour to fine fraction (26.2–82.1%)	Funke et al. (2022a)
Enzymatic hydrolysis	• Enzyme/substrate ratio(E/S) of 0.1 AU/mg of soluble protein at 40 °C and pH 8 with different enzymes (Alcalase, Savinase, Protamex, and Corolase 7089)	• Alcalase showed the highest degree of hydrolysis (23%)	Garcia-Mora et al. (2014)
	• Alcalase (1%, 2%, and 4%) and Flavourzyme (2%), processing time (180 min with an interval of 30 min)	• No change in amino acid profile	Rezvankhah et al. (2021)

High-pressure-assisted enzymatic hydrolysis	• Pressure 300 MPa E/S of 0.1 AU/mg of soluble protein at 40 °C and pH 8 with different enzymes (Alcalase, Savinase, Protamex, and Corolase 7089)	• Alcalase showed the highest content of peptides <3 kDa (65.6 mg/g hydrolysate)	Garcia-Mora et al. (2015)
	• Hydrolysis by Alcalase (0.5–1%) on HP pre-treated (300, 450, and 600 MPa for 15 min)	• Maximum degree of hydrolysis (57.66%) with HP (300 MPa/15 min) assisted proteolysis (0.05% Alcalase)	Ahmed et al. (2019)
Heat and freeze–thaw pre-treatment enzymatic hydrolysis	• Hydrolysis (3-h) by Alcalase (E/S of 90 AU/kg protein, 55 °C), heat pre-treated (65, 75, 85 °C, for 15 min) and freeze–thaw samples (three freezing cycles at −20 °C and thawing at room temp.)	• Highest value of free amino groups at 75 °C pretreatment	Ghasemi et al. (2021)

9.2.1 Wet-Extraction Processes

Wet separation is traditionally used to prepare plant protein isolates since it is highly soluble under alkaline and acidic conditions. It is possible to separate proteins from their sources using wet methods, but it requires large quantities of water, chemicals, and energy (Pelgrom et al. 2015). The most commonly used method for wet extraction is alkaline extraction. In comparison with organic solvents, it is widely used to improve protein's extraction ability. The use of alkali has the capability of breaking disulfide cross-linkages in proteins (e.g., in proteins derived from legumes). The basic pH enhances the solubility of neutral and acidic amino acids (Sari 2015). A wet extraction under alkaline conditions involves three steps: pre-treatment of the seeds for cell wall disruption (alteration of cellulose and hemicellulose compositions), solubilization of proteins in alkaline solutions (at pH >8), and selective precipitation of the proteins by adjusting pH to the isoelectric point (pH of ~4.5). A variety of methods are available for alkaline extraction, such as isoelectric precipitation, ultrafiltration, salt extraction, and micellization. Subsequently, the insoluble material is centrifuged and the proteins are extracted by adjusting the supernatant pH to ~4.5, where they precipitate isoelectrically. An additional process for separating the insoluble fractions (centrifugation or filtration) and preparing the protein concentration (spray-drying or freeze-drying) is also necessary (Figure 9.2).

During the wet-extraction method, it is essential to control the solid-to-liquid ratio, extraction time, pH, temperature, and concentration parameters, since these parameters influence protein yield (Bou et al. 2022). Generally, flour and water are mixed in a ratio of 1 : 5–1 : 20 during protein extraction. In order to maximize the soluble and extracted proteins, time and temperature are typically increased, between 30 and 120 minutes and 25 and 50 °C (Boye et al. 2010b). Temperature plays a critical role in the stability of folded states, and covalent bonds can be damaged in proteins at high temperatures. Hydrophobic interactions also can lose their effectiveness at 100 °C due to thermal damage. Increasing the temperature above 140 °C may cause the protein to denature and may also cause their hydrolyzation (Celiktas et al. 2014). A second way to improve recovery is to add NaCl, but the resulting effect is dependent on pH (Fetzer et al. 2018).

There have been several studies on lentil protein extracted with diluted NaOH. However, the pH conditions vary from one study to another (Table 9.1). Joshi et al. (2012), for example, isolated lentil protein by alkaline extraction at pH 8.0, room temperature, and 1 : 10 solid-to-solvent ratio, and compared the physicochemical properties of the isolated proteins using freeze, spray, and vacuum drying methods. In a similar study, Boye et al. (2010a) extracted proteins from red and green lentils using 1 : 10 solid-to-solvent ratio by means of isoelectric precipitation or ultrafiltration. Taking into account the protein content of dehulled flour as starting point, their concentrates had 78.2–88.6% protein content and yields of 50.3–62.8%. In another study by Ahmad El-Sohaimy and Sitohy (2007), lentil proteins' recovery was achieved using three different methods (isoelectric precipitation, ammonium sulfate precipitation, and alcohol precipitation) at room temperature using a solid/solvent ratio of 5 : 100.

A combination of ammonium sulfate extraction and alcohol precipitation produced the best protein recovery (93% and 100%, respectively). According to Lee et al. (2007), different pH and temperature conditions (deionized water, pH 8.0, 8.5, 9.0, and 9.5) with constant

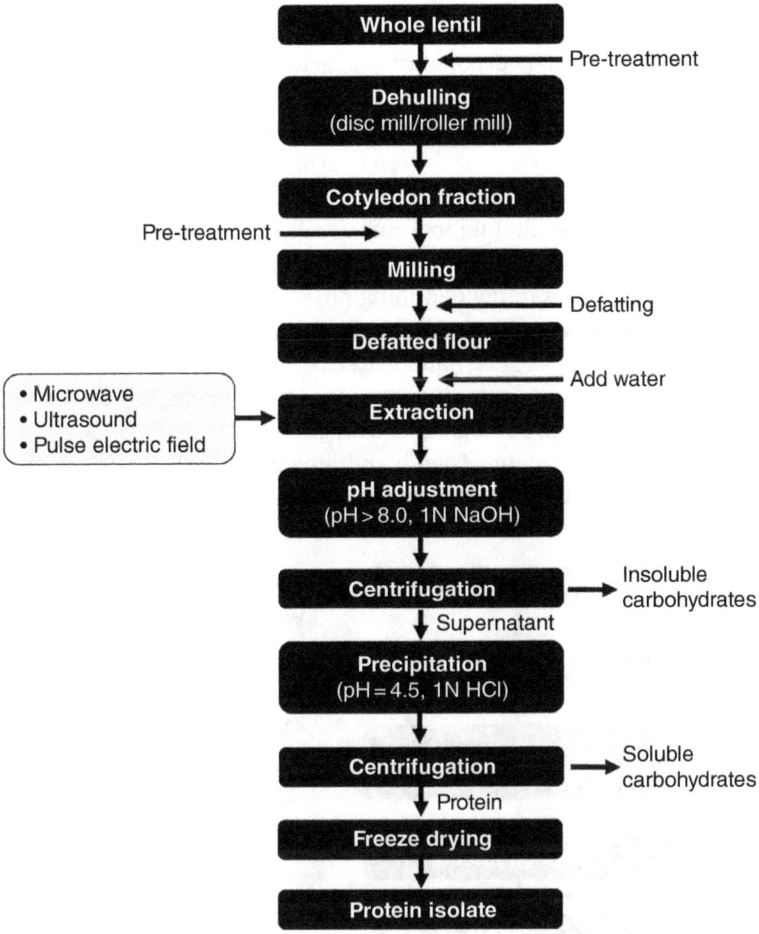

Figure 9.2 Schematic diagram for wet protein extraction from food legumes. *Source:* Adapted from Ferreira et al. (2022).

solid-to-solvent ratios (1 : 10) were utilized to extract protein from green and red lentils. A 56.6% recovery of green lentil protein was successfully extracted at pH 9.0 at 30 °C while 59.3% red lentil of protein were extracted at pH 8.5 at 35 °C.

Generally, it is more convenient to use wet-extraction procedures because the protein isolates obtained are more pure, digestible, and of higher quality (Khazaei et al. 2019). This can be attributed to the efficient removal of antinutritional factors during wet extraction in comparison to the dry fractionation process (Vogelsang-O'Dwyer et al. 2020). Furthermore, drying assists in safe and convenient transport, storage, as well as application of high purity protein as major ingredient in developing functional foods (Alavi et al. 2021; Pelgrom et al. 2013). However, the pH fluctuation and drying process may result in extreme alkaline conditions, which may reduce the native functionality of proteins, primarily the solubility and degradation of amino acids, in addition to the depletion of essential amino acids (Fetzer et al. 2018); however, seeds can be pre-treated to overcome these problems.

9.2.2 Dry Separation

In recent years, dry separation has gained popularity due to its more environmentally friendly nature. The seed coat of the whole legume contains antinutritional factors, such as soluble phenolics, which may affect the recovery of protein. Dehulling by hammer mill, roller mill, or disc mill is an essential process for the removal of seed coat from the legumes (Wang 2005). Before dehulling, legumes are pre-treated in order to loosen the seed coat from cotyledons. It is common to soak and temper pulse-milled grains with water or oil before dehulling. In addition to ultrasound and microwave, chemical and enzyme treatments have also been studied as an aid in the dehulling process. As a result of these pre-treatments, the adherence of seed coat to cotyledon become weak, allowing the seed coat to be readily removed by mechanical force (Fernando 2021).

Dehulled and pre-treated lentils are then subjected to dry fractionation, which is comprised of two-unit operations mainly, i.e., pin milling and separation. Separation is employed for classification of different legume fractions (starch and protein) according to their density, size, and electrostatic properties (Figure 9.3). Separation using dry processes can involve

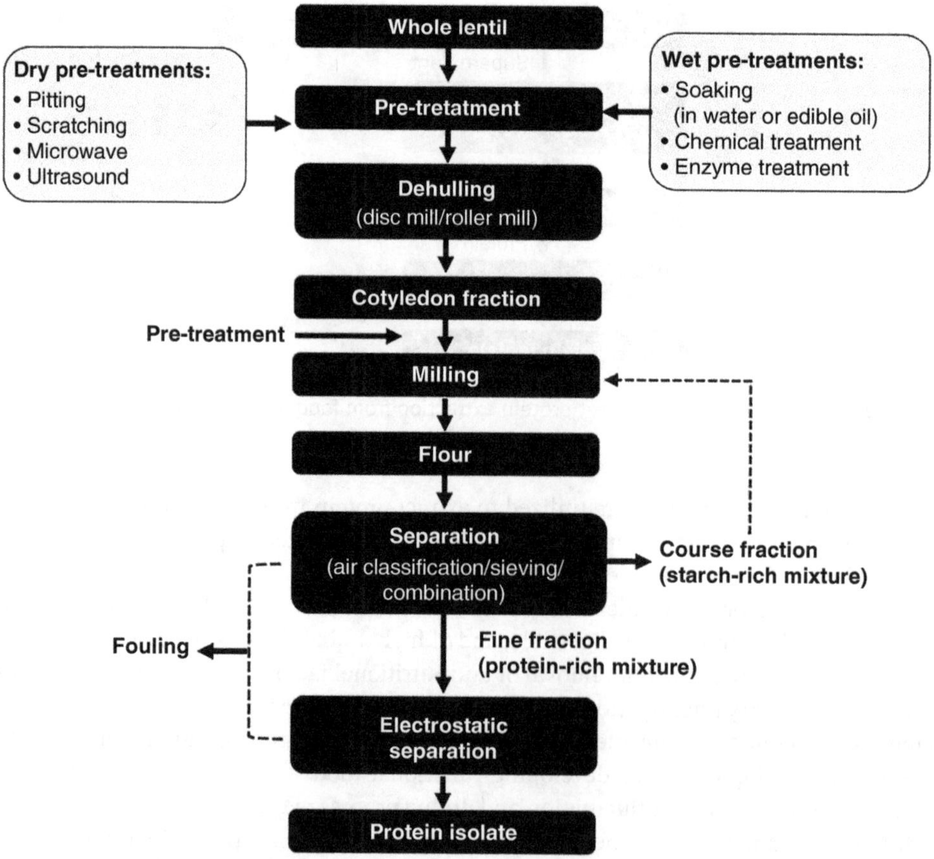

Figure 9.3 Schematic diagram for dry protein fractionation from food legumes. *Source:* Adapted from Fernando (2021).

purely physical mechanical methods, such as ultrafine milling, or chemical methods, such as sieving, air classification, or electrostatic separation (Elalami et al. 2018). There are two types of lentil concentrates: a light to fine protein fraction and a heavy or coarse starch fraction. It has been reported that air-classified protein concentrates have a protein content of 38–68%. Air classification processes have historically been well suited for separating lentil isolates due to their large diameter and uniform distribution (Afam-Mbah et al. 2018; Funke et al. 2022a,b; Xing et al. 2018). The process does not require the use of chemicals, preserves protein native structure and functionality, reduces antinutritional factors, removes bitter/astringent compounds, and improves color (do Carmo et al. 2021; Vogelsang-O'Dwyer et al. 2020). However, despite the relatively simple and easy process, the separation does not yield high protein concentrations when used without the pretreatments (Fernando 2021). Sozer et al. (2017) reported that pretreatments, such as drying, soaking, freezing, thawing, or defatting/de-oiling, can also enhance yield and contribute to higher levels of protein. The reason is that carbohydrates and fibers are fragmented during grinding resulting in their accumulation in the fine fraction (Wang et al. 2016).

Besides air classification, tribo-electrostatic separation has lately appeared as a method for producing protein isolates in the food industry that is solvent-free (Pelgrom et al. 2015). In this method, flour particles are analyzed differently based on their dielectric properties instead of their size and density. This technique involves electrostatic charging proteins more than carbohydrates, because their amino acid residues have ionizable N-terminal and C-terminal groups. The magnitude of the charge and type of the proteins in a protein-rich fraction can therefore be separated by the application of an electric field. This process is more energy efficient than air classification because it effectively separates particles with similar sizes and densities, but different charges. It is important to note that this approach has some major limitations. For example, the possibility of similar charges exhibiting under certain conditions, which can reduce concentrate's purity, as well as the possibility of particles becoming airborne under certain conditions. In order to obtain pulse protein concentrates with higher purity, it has been suggested to combine air classification with electrostatic separation.

9.2.3 Enzyme-Assisted Extraction

Although legume protein is better extracted under alkaline conditions, excessive alkaline extraction can result in several negative effects, including degradation of amino acids, lowered protein bioavailability, and destruction of some amino acids (e.g., lysine and cysteine). Furthermore, the concentrated alkaline condition leads to the formation of salt, which should be avoided whenever possible. Enzyme-assisted extraction (EAE) can be a suitable substitute for alkaline extraction because it minimizes side effects and environmental problems. The EAE method is intended to be a milder process and is also environmentally friendly since it is a low-energy extraction method (Shen et al. 2008). In this method, protein recovery from plant sources is facilitated by enzymes while extraction is carried out with water. Optimal conditions of temperature and pH allow a specific enzyme activity that hydrolyzes carbohydrates to degrade the cell walls of plants, isolating intracellular compounds of interest from within the cell (Nadar et al. 2018). Specific enzymes, e.g., proteases, enhance the recovery of valuable by-products, such as protein hydrolysates, by disintegrating polysaccharides and lipoprotein bodies within the cell wall (Baker and Charlton 2020). The EAE process steps include milling of protein seeds, fatty acid removal, hydrolysis (mixing ground seeds at a defined temperature for a specified period of time with water and enzyme), precipitation and separation of the oil phase, the insoluble solids, and the aqueous phase from the water-soluble phase, as shown in Figure 9.4. Drying techniques, such as freeze drying and spray drying, can be used to recover hydrolyzed proteins, which are soluble fragments of peptides. Protein extraction from cereal crops usually includes defatting, but pulse seeds have relatively low oil contents (Sathe 2002), thus defatting can be eliminated in most cases.

There have been a number of studies on lentil protein extracted with enzymatic hydrolysis. It is possible to improve extraction yields by choosing enzymes with synergistic effects and by applying the right conditions (e.g., pH, solid to liquid ratio, extraction time, temperature, NaCl concentration, and number of extraction cycles). Ghasemi et al. (2021) subjected lentils to heat pretreatments at varying temperatures (65, 75, and 85 °C, for about 15 minutes) followed by freezing (three freezing cycles at −20 °C) and thawing at room temperature. Pre-treated samples were then exposed to enzymatic hydrolysis at 55 °C for three hours by Alcalase, with an enzyme to substrate (E/S) of 90 AU/kg protein. A pre-treatment of flour at 75 °C prior to extraction increased the amount of free amino acids due to the

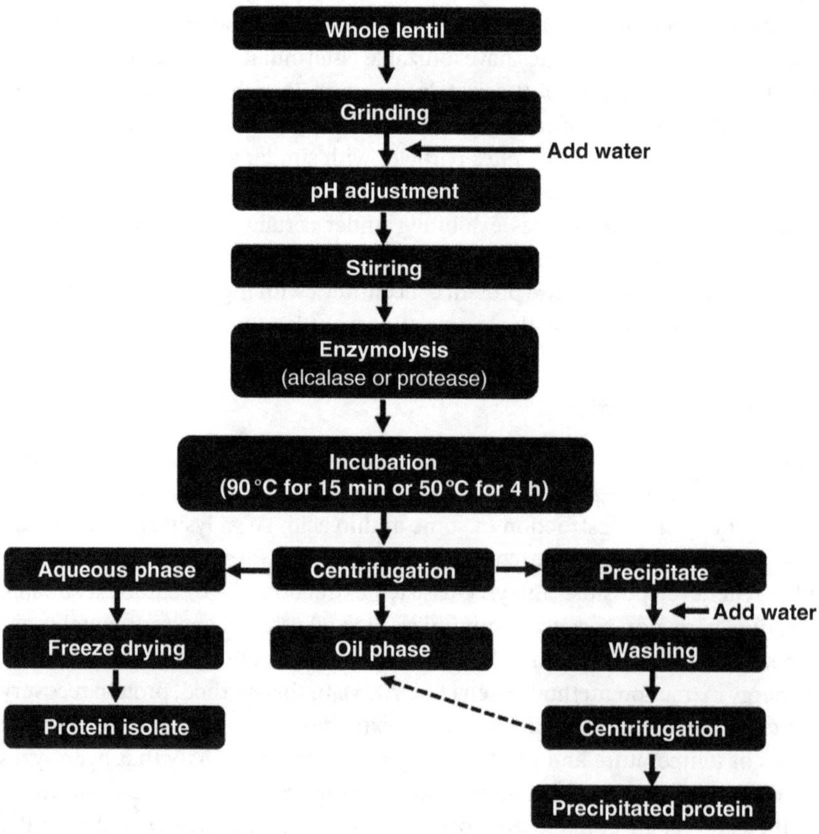

Figure 9.4 Schematic diagram of the enzyme-assisted aqueous extraction of protein hydrolysates from food legumes. *Source:* Adapted from Liu et al. (2020).

creation of peptide fragments, which enhance the protein solubility. Furthermore, it exhibits a rapid extraction rate and reduces the need for solvents compared to traditional methods. For the proteolysis of lentil protein, Garcia-Mora et al. (2015) used high pressure (HP) and Savinase, but that study was limited to allergens.

Ahmed et al. (2019) used HP processing as a pre-treatment step to induce Alcalase digestion of lentil protein. As compared to the control hydrolysates, the protein hydrolysates with HP-assisted (300 MPa, 15 minutes) proteolysis (with 0.5% Alcalase) exhibited improved foaming properties (≈1.5 times) as well as antioxidant activity (≈2 times) in addition to maximum DH (degree of hydrolysis) of 57.66%. The improvement in protein solubility is useful for the development of pulse protein ingredients-based food products (Dent and Maleky 2022). The concentration of free amino acids during enzymatic hydrolysis is highly dependent upon enzyme/substrate ratio and reaction time. Avramenko et al. (2013) reported an increase in the creation of peptide bonds at highest E/S ratio and shorter reaction time during trypsin enzyme-assisted hydrolysis of lentil protein. Even though enzyme-assisted extraction is less efficient than chemical extraction (e.g., alkaline extraction), pre-treatment (cell wall disruption) followed by chemical extraction would be interesting. Therefore, alkaline extraction combined with it could be an extremely valuable tool rather than one-step extraction process. However, there are drawbacks to EAE technique, e.g., the high cost or scaling-up required to optimize them (Baker and Charlton 2020). Thus, EAE could be potentially used in conjunction with other novel and emerging techniques in order to mitigate these disadvantages.

Overall, lentil protein and peptide production are viable value-addition strategies for food processing, because lentil protein possess excellent nutritional and techno-functional properties. In recent years, protein extraction methods have been integrated with sustainable and emerging technologies, including pre-treatment (removing unwanted compounds or disrupting cell walls). In addition to their multiple advantages, such treatments reduce solvent usage, processing time, waste generation, and energy consumption. These approaches were developed at a laboratory scale or using conceptual designs, so they do not have large-scale feasibility that requires further study. Generally, extraction processes should be cost-efficient, and the interactions between proteins during extraction (hydrolysis, denaturation, and formation of undesirable species) should be elucidated, particularly for food applications.

9.3 Techno-Functional Properties of Lentil Protein

The functional properties of proteins are their physicochemical properties that govern the protein's behavior during the production of food products and aid in providing the final structure, desired color, flavor, and texture to the end product (Ahmed et al. 2022; Keskin et al. 2022). These physicochemical properties serve as final quality indicators for protein-based food products developed via novel technologies and are termed as techno-functional properties (Ustunol 2014). Therefore, it is crucial to determine these properties of protein before application in different food systems. For instance, the successful production of meat analogue or TVP (texturized vegetable protein) requires a protein-based raw material in the form of either protein concentrate or isolate with good techno-functional properties, such as high solubility, high water and oil absorption capacities, good gelling capacity, along with a flexible protein structure to rearrange into fibril structure during extrusion processing. Thus, in following sub-sections, lentil protein's techno-functional properties will be discussed as a sustainable, alternative source to traditional meat proteins.

9.3.1 Solubility

Protein solubility is the main techno-functional property on which other functional properties such as emulsification, foaming capacity, and gelling ability are dependent. The solubility of protein is an indicator of the equilibrium between protein–protein and protein–solvent interactions, which are facilitated by the non-covalent bonds (electrostatic and hydrophobic). As shown in Table 9.2, lentil proteins, despite different varieties from Nepal, United States, and Turkey, showed a typical U-shaped solubility profile meaning high solubility in acidic and alkaline pH solutions and low at isoelectric pH 4–6 (Bora 2002; Lee et al. 2021). However, heat and high-pressure treatment can alter the solubility profile of the lentil protein due to the exposure and subsequent aggregation of the hydrophobic groups which was previously embedded within the core of the protein structure. For example, Hall and Moraru (2021) reported that the solubility of native lentil protein (65%) decreased significantly (to 22%) under heat treatment (95 °C for 15 minutes) and to 25% after high-pressure treatment (600 MPa for 4 minutes at 5 °C).

Drying method that is applied during the production of protein isolates can also alter the solubility significantly. Joshi et al. (2011) determined the effect of freeze-, spray-, and vacuum-drying on lentil protein solubility at mixing times of 5–60 minutes at 20 °C, pH 7. During the first 10 minutes of mixing, solubility of all lentil protein samples (freeze-, spray-, and vacuum-dried samples) increased and reached an equilibrium condition. Vacuum-dried lentil protein showed lower solubility compared to the other drying techniques. Another important characteristic is that the spray-dried lentil protein sample showed a higher solubility profile after 15 minutes of mixing compared to the freeze-dried sample even though the freeze-drying condition is much gentler compared to spray-drying processing. Authors suggested that the spray-dried lentil protein was less denatured as the heating temperature was below that of the denaturation and residence time in the drying chamber was quick, while for freeze and vacuum drying, the drying period was much longer. Moreover, Boye et al. (2010a) reported that protein extraction technique can also influence the solubility profile of lentil protein and showed that isoelectric precipitation yielded lower protein solubility compared to the ultrafiltration technique.

It can be noted from the above discussion that a particular protein extraction and drying technique has a significant effect on the lentil protein solubility. Therefore, it is necessary to evaluate the solubility profile of lentil protein and verify that the solubility is at the optimum level for the production of meat analogue products.

9.3.2 Water and Oil Absorption Capacity

The water absorption capacity (WAC) is one of the useful techniques to assess the technological suitability of lentil proteins. For instance, WAC can provide insight into the texture and juiciness of protein ingredients in a meat-based food system. High water absorption

Table 9.2 Techno-functional properties of lentil protein.

Lentil cultivar	Extraction/ (drying) method	Protein content (%)	Techno-functional properties						References
			Solubility	WAC	OAC	EAI	ESI	LGC	
Red lentils, USA, Nepal, and Turkey variety	AE-IEP (Freeze)	—	U-shaped solubility profile = high solubility at acidic pH (2, 3) and alkaline pH (6, 7, 8, 9, 10) and low solubility at isoelectric point- pH 5.2	3.1–3.5 g/g	5.8–7.3 g/g	USA = 17.1–23.8 m^2/g Nepal = 9.6–24.5 m^2/g Turkey lentils = 14.0–25.1 m^2/g	USA = 9.2–21.9 min Nepal = 9.8–25.9 min Turkey = 9.1–23.7 min	—	Lee et al. (2021)
Lentils, Brewer variety	Protein (globulin) extraction by K_3PO_4 buffer and IEP (Freeze)	—	U-shaped solubility profile for native protein = high solubility at acidic (2, 3) and alkaline pH (6, 7, 8) and low solubility at isoelectric point- pH 4.5	1.1 ml/g	2.6 ml/g	54.1%	52.2%	—	Bora (2002)
Lentil protein, AGT food and ingredients, Canada	Air classification(−)	53.3 g/100 g DM	Native protein = 65% at pH 6.42	32%	—	Native protein = 8 m^2/g	Native protein = 2303 min	—	Hall and Moraru (2021)
			High-pressure-treated protein = 25% at pH 6.47	96%	—	High-pressure-treated protein = 7 m^2/g	High pressure treated protein = 2894 min		
			Heat-treated protein = 22% at pH 6.42	100%	—	Heat-treated protein = 6 m^2/g	Heat-treated protein = minimal visible separation only		

(*Continued*)

Table 9.2 (Continued)

Lentil cultivar	Extraction/ (drying) method	Protein content (%)	Techno-functional properties						References
			Solubility	WAC	OAC	EAI	ESI	LGC	
Australian lentil cultivar (Aldinga)	AE-IEP (Freeze)	90.2%	78% at pH 7	0.48 (g/g)	—	—	—	—	Joshi et al. (2011)
	AE-IEP (Spray)	91.4%	81% at pH 7	0.43 (g/g)	—	—	—		
	AE-IEP (Vacuum)	91.9%	50% at pH 7	0.47 (g/g)	—	—	—		
Red and green lentils (pulse growers, Saskatchewan)	Ultrafiltration and IEP (spray)	Red lentil UF = 82.7%; IEP = 78.2 Green lentil UF = 88.6%; IEP = 79.1%	U-shaped solubility profile = UF treated protein showed high solubility at pH 1–3, 7–10 and low solubility at pH 4–6 compared to IEP protein concentrate	Red lentil IEP = 3.7 ml/g and UF = 4.1 ml/g; Green lentil IEP = 3.9 ml/g and UF = 3.5 ml/g	Red lentil IEP = 110% and UF = 226%; Green lentil IEP = 130% and UF = 140%	Red lentil IEP = 5 m^2/g and UF = 5.3 m^2/g; Green lentil IEP = 4.9 m^2/g and UF = 5.1 m^2/g	Red lentil IEP = 18 min and UF = 19 min; Green lentil IEP = 17 min and UF = 19 min	Red lentil IEP = 12% LGC and UF = 10% LGC; Green lentil IEP = 12% LGC and UF = 8% LGC	Boye et al. (2010a)
Red lentil cultivars: Ali dayı (L-1) = LPE1, Çiftçi (L-2) = LPE2, Fırat (L-3) = LPE3, & Kafkas (L-4) = LPE4; Green lentil cultivars: Meyveci (L5) = LPE-5 & Pul II (L6) = LPE-6	AE-IEP (Freeze)	Red lentil cultivars: LPE1 = 0.90 g/g LPE2 = 0.95 g/g LPE3 = 0.88 g/g LPE4 = 0.93 g/g; Green lentil cultivars: LPE5 = 0.90 g/g LPE6 = 0.87 g/g	—	—	Red lentil cultivars: LPE1 = 6.90 g/g LPE2 = 7.57 g/g LPE3 = 9.67 g/g LPE4 = 8.55 g/g; Green lentil cultivars: LPE5 = 8.57 g/g LPE6 = 10.47 g/g	Red lentil cultivars: LPE1 = 12% LGC LPE2 = NHGF* LPE3 = 12% LGC LPE4 = NHGF* Green lentil cultivars: LPE5 = 14% LGC; LPE6 = 14% LGC		—	Aydemir and Yemenicioğlu (2013)

– = not recorded; AE-IEP = alkaline extraction with isoelectric precipitation; WAC = water absorption capacity; OAC = oil absorption capacity; EAI = emulsifying activity index; ESI = emulsifying stability index; LGC = least gelation concentration; NHGF* = no hard gel formation.

capacity is needed for the production of meat analogue so that the lentil protein can hold the water during high-moisture texturization processing and form a 3D network supporting gel structure. The WAC of different proteins is also determined to facilitate adjustments in food formulations when substituting protein sources. For instance, proteins with high water absorption capacities can absorb the majority of the water from the food system and partially dehydrate other components in it. The most widely used method to measure water absorption capacity is mixing protein samples with water and centrifuging the protein dispersion. The amount of water retained by the protein sample after centrifugation is termed WAC. A wide range of WAC of lentil protein was found due to the differences in protein extraction and drying techniques, heat treatment, and pH alternation. It is also important to note that the method for the determination of WAC varies among different studies, i.e., mixing and holding time, temperature, and protein concentration, thus yielding variable results (Ma et al. 2022).

Boye et al. (2010a) reported that protein extraction and concentration method can substantially affect the WAC of lentil protein, as shown in Table 9.2. Authors reported that red lentil protein prepared via IEP (isoelectric precipitation) had a lower WAC capacity (3.7 ml/g) compared to the UF (ultrafiltration) technique (4.1 ml/g), this pattern was contrary to green lentil protein wherein IEP had higher WAC (3.9 ml/g) compared to UF (3.5 ml/g). On the other hand, Bora (2002) reported significantly lower WAC of 1.1 ml/g when protein was extracted using potassium phosphate buffer. Lee et al. (2021) reported that cultivar and growth conditions of lentils can significantly influence protein's WAC. Authors found high WAC for Turkish varieties compared to those from USA and Nepal. Joshi et al. (2011) reported that the Australian lentil variety demonstrated the least WAC. In another study, Hall and Moraru (2021) measured WAC of concentrated native, high-pressure, and heat-treated lentil gelled protein samples and found WAC of 32%, 96%, and 100%, respectively. Both heat and high-pressure treatments improved WAC due to the formation of a strong gel or network-like structure. These results showed that lentil protein can be an excellent raw material for the production of juicy meat analogue products.

Oil absorption capacity (OAC) is the amount of fat/oil absorbed per weight of protein isolates or protein concentrates (PI or PC). It is measured by thoroughly mixing PI or PC in excess oil, followed by centrifugation and measuring the amount of oil absorbed. The OAC finds numerous applications in meat emulsion and meat analogue products as it reveals the fat and flavor retention in food products (Zayas 2012). The OAC depends on the number of hydrophobic amino acids, protein's source, size, processing methods, and protein–lipid interactions. Lee et al. (2021) reported that OAC (7.3 g/g protein) of Turkish red lentil was significantly higher compared to that of United States (5.8 g/g protein) and Nepal (7.1 g/g protein) red lentils due to the differences in protein composition (Table 9.2). The presence of a high number of hydrophilic groups in the protein structure can be associated with low OAC (Jarpa-Parra 2018). In another study, Aryee and Boye (2017) reported that small- and low-density lentil proteins can absorb and retain higher amount of fat than large and high-density proteins. For instance, lentil protein with a volume-weighted mean (VWM) of 44.03 μm retained fat higher than the cooked (VWM = 183.20 μm) and raw lentil flour (VWM = 243.08 μm). Boye et al. (2010a) reported that OAC was greatly influenced by the protein extraction method and variety. Their results that red lentil proteins extracted via UF had higher OAC (226%) compared to the IEP (110%) method. In contrast, OAC for UF green lentil protein was similar to IEP protein (Table 9.2). Bora (2002) reported OAC for lentil Brewer variety was 2.6 ml/g protein. While Aydemir and Yemenicioğlu (2013) reported that OAC varied among the cultivars whereby green lentil cultivar LPE6 recorded the highest OAC. These variation in OAC among different studies could be due to the differences in OAC assay or protein compositions of different cultivars.

Conclusively, lentil protein exhibited high WAC and OAC values and thus can be an excellent raw material for the production of meat analogue and improve end product qualities in terms of juiciness and texture.

9.3.3 Emulsifying Properties

Emulsion is a mixture of immiscible liquids, which is thermodynamically unstable and tends to separate over time during storage. Hence, emulsifiers are generally used to stabilize the emulsion. Protein act as an excellent emulsifier due to its amphiphilic nature, reducing the interfacial tension between the oil–water interfaces (Damodaran and Parkin 2017). The emulsifying property is usually investigated by measuring the emulsifying activity index (EAI) and emulsifying stability index (ESI). EAI is a measure of the unit of oil that can be emulsified per unit of protein, and the ESI examines the resistance of the emulsion over a definite time period. Protein acts as an excellent emulsifier by adsorbing at the interface, coating the oil droplets, and developing stable films. When protein adsorbs at the interface, they relocate their structure and expose hydrophobic amino acid residues to the oil phase and the hydrophilic amino acids to the aqueous phase, which reduces the interfacial tension. Hence, the ratio of hydrophobicity–hydrophilicity and ease of protein unfolding have considerable influence on their emulsifying property (Zayas 2012).

Several studies reported that the EAI and ESI for lentil protein are significantly influenced by the variety or the experimental methods used for their preparation. Lee et al. (2021) recorded EAI and ESI of red lentil protein in pH 2–10 using the turbidimetric method and found that EAI and ESI were high in acidic and alkaline regions due to their high protein solubility in this pH region. Authors also reported that EAI and ESI were cultivar dependent for red lentil proteins (Table 9.2). Canadian lentil protein variety showed comparatively lower EAI of 8, 7, 6 m^2/g and higher ESI of 2303, 2984 minutes and minimal visible separation for native, high pressure- and heat-treated samples, respectively (Hall and Moraru 2021). Heating and high-pressure treatment reduced EAI due to the protein denaturation and their subsequent aggregation. According to Boye et al. (2010a), EAI and ESI of red and green lentil cultivars were not significantly affected by the protein extraction methods, i.e., IEP and UF. In another study, Bora (2002) recorded EAI and ESI values for lentil protein of 54.1% and 52.2%. Author measured EAI by comparing the height of the emulsion formed to the height of the total contents of the container where ESI was calculated similarly but heated at 80 °C for 30 minutes and expressed in percentage.

In brief, EAI and ESI of lentil protein are governed by a number of factors such as protein composition and cultivar, protein's structure, shape and flexibility, protein's denaturation, and aggregation state, experimental methods for the test, processing treatments (heat or pressure), and environmental factors (pH, protein concentration, and temperature). Therefore, it is recommended that prior to choosing the lentil protein sample for meat analogue production, a thorough assay needs to be conducted using a standard test procedure.

9.3.4 Gelation

Gelation is the crucial functional property for meat-based products revealing the viscoelastic properties of foods. A strong gelled meat analogue can retain a high amount of water and mimic the mouthfeel of lipids (Kew et al. 2020). Typically, a gel is a 3-D network-like structure, made up of cross-linked macromolecules via covalent and non-covalent bonds and capable of trapping water and other small molecular weight compounds. The gelation mechanism involves linking protein macromolecules and forming a branched polymer structure containing water, which is called a "sol." As the structure dimensions increases, the solubility of the polymer decreases, which is called "gel." This transformation from the "sol" state to "gel" state is called gelation. Protein can form two types of gel, that is coagulum and translucent gel. The coagulum type of gel is formed in the presence of high amount of nonpolar amino acids via hydrophobic aggregation upon denaturation while translucent gels are formed in the presence of small amount of nonpolar amino acids via hydrogen bonding. Translucent gels require heating followed by cooling wherein soluble protein complexes associate to from and set into an ordered gel network (Damodaran and Parkin 2017).

The most widely used method to determine protein gelation capacity is recording the least gelation concentration (LGC) wherein minimum protein concentration is determined, which can form a strong gel and set in a test tube instead of sliding along the wall when inverted. Boye et al. (2010a) recorded that green lentil protein extracted using the UF technique showed the best gelling capacity and formed a strong gel at a protein concentration of 8% compared to IEP (12%), whereas IEP- and UF-extracted red lentil protein formed gel at a protein concentration of 12% and 10%, respectively. Similarly, Aydemir and Yemenicioğlu (2013) reported that red and green lentil protein showed LGC in the range of 12–14% for varieties LPE1, LPE13, LPE15, and LPE16, while LPE2 and LPE4 were unable to form a hard gel within protein concentration of 1–14% (Table 9.2). Apart from lentil cultivar and protein extraction approach, different drying techniques that are applied to prepare protein isolates or concentrates can alter protein structure, eventually affecting protein gelation capacity. Joshi et al. (2011) reported that the LGC for freeze- and spray-dried lentil protein isolate was 11%, while that for vacuum-dried lentil protein isolate was 14%. Authors reported that vacuum-dried lentil protein isolate had low protein solubility compared to freeze- and spray-dried lentil protein, thus requiring a higher concentration to form a gel.

Conclusively, techno-functional properties of lentil protein are prominently influenced by the different compositions shown by different lentil cultivars. Moreover, different protein extraction and drying techniques, as well as other conditions (pH, protein concentration, temperature), further influenced lentil proteins' techno-functional properties. Therefore, an optimization process can assist in getting the best techno-functional properties from lentil protein by analyzing different factors concurrently.

9.4 Lentil Protein-Based Meat Analogues and Extenders

The restructuring of plant protein materials into a three-dimensional, cross-linked mass with a meat-like texture is termed as the protein texturization process and the end product is known as texturized vegetable protein (TVP). It has the potent potential to serve as a healthy alternative to animal meat as it is cholesterol-free and environment friendly. TVP is categorized into two groups, viz, meat extender (serves as a meat extender, low-moisture extruded product) and meat analogue (serves as complete meat substitute, high-moisture extruded product) (Riaz 2001). Plant protein has been reported to play a significant role in anti-aging, and reducing blood pressure as they contain functional compounds such as amino acids, isoflavones, and saponins (Zhang et al. 2019). Hence, regular consumption of TVP as meat extenders or analogues can prevent noncommunicable diseases such as obesity, hypertension, diabetes, and cardiovascular diseases (Singh et al. 2007; Zhang et al. 2019). For protein texturization, extrusion and fiber spinning methods are commonly used. Fiber formation by electrospinning is expensive and requires a highly refined protein of 90%, whereas extrusion processing requires a protein content of 50–70% protein (Berk 2018). Additionally, texturized protein produced by the extrusion technique has biological digestion potency that can reach 93–97% in the human body (Zhang et al. 2019). Extrusion technology can be used to produce meat analogues on an industrial scale. Currently, shear cell technology is introduced to develop fibrous products that can mimic animal muscle meat structure and texture, i.e., steak-like meat or smaller fillers like chops (Krintiras et al. 2016) (Figure 9.5). This technology is practiced up to pilot scale only. 3D printing is another novel technology that is used to develop plant protein-based meat analogue. Through this technology desirable fibrous structure can be created; however, 3D-printed plant meat product is limited due to the lack of suitable raw materials. Extensive research is required to identify potential protein ingredients for meat analogue production using these novel techniques (Ramachandraiah 2021).

In recent years, few studies have been conducted on the lentil protein-based meat analogues and extenders (Table 9.3). Kim (2018) developed texturized vegetable protein from lentil, faba, pea, and soy protein as an alternative to animal meat. The texturized lentil protein showed a lower water holding capacity compared to all the other samples, i.e., texturized faba, pea, and soy protein. Author stated that texturized lentil protein had a mushy structure and was unable to retain its shape when cut after hydration. Arshad et al. (2022) prepared nuggets using pea and lentil protein and reported that lentil protein incorporation improves

(a) (b) (c)

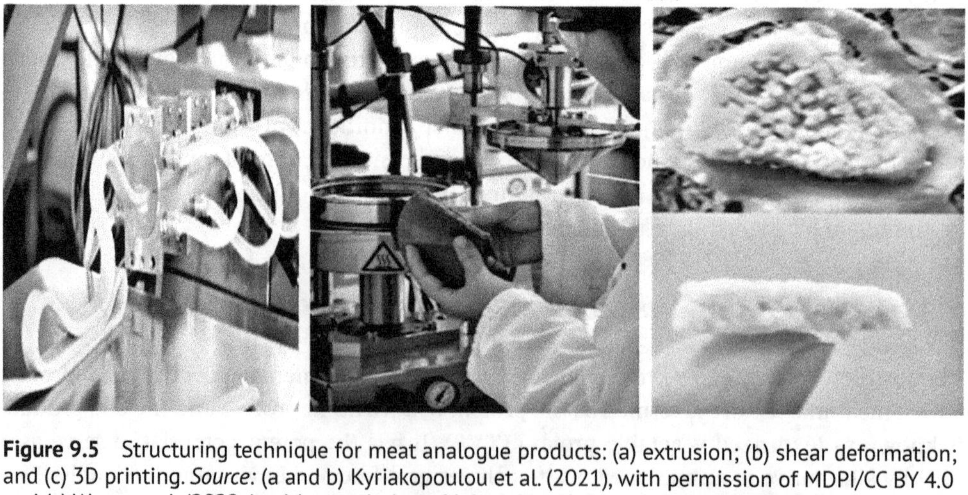

Figure 9.5 Structuring technique for meat analogue products: (a) extrusion; (b) shear deformation; and (c) 3D printing. *Source:* (a and b) Kyriakopoulou et al. (2021), with permission of MDPI/CC BY 4.0 and (c) Wang et al. (2022a), with permission of John Wiley & Sons, Inc./CC BY-NC-ND 4.0.

Table 9.3 Lentil protein-based meat alternative products.

Protein source	Additives	Equipment and processing conditions	Reference
Texturized lentil protein: Lentil protein isolate (Vitessence™ Pulse 2550)	Sodium bisulfite, xanthan gum, calcium hydroxide, and water	Twin screw extruder Screw speed: 495 rpm Feed moisture: 3 kg/h Exit temperature: 100 °C	Kim (2018)
Nugget: Pea and lentil protein used in the ratio of 60 : 40, 40 : 60, and 50 : 50	Onion powder, white pepper, cumin, milk powder, garlic powder. Egg white and bread flour used for coating before frying	Ingredients mixed, held for 30 min, and then cooked for 3 min	Arshad et al. (2022)
Meatball extender: Minced beef and lentil flour (10%)	Beef fat, onion powder, black pepper, red pepper, white pepper, and salt	Ingredients blended in a food processor and shaped into meatballs (1-cm thick, 80-mm diameter)	Serdaroğlu et al. (2005)

all the sensory properties of the nugget. Serdaroğlu et al. (2005) used 10% lentil flour as a beef meatball extender and reported that lentil flour increased the cooking yield, moisture, and fat retention capacity of the meatball products compared to blackeye bean, chickpea, and rusk flour, thus demonstrating the potential of using lentils as animal meat substitute.

Conclusively, the behavior of lentil protein solely depends on its techno-functional properties. Lentil protein extraction and drying method can significantly shift its techno-functional properties; therefore, it is crucial to investigate its techno-functional properties prior to producing meat analogues.

9.5 Consumer Preferences and Willingness to Pay for Lentil Protein Food

The main concept of consumers' choice behavior is based on the assumption of rational choice of products and services that maximize the utility of consumers (McFadden 2001). In this regard, goods and services have various attributes and consumers have different preferences for these attributes. Therefore, explaining consumer behavior when choosing from a set of alternatives containing different attributes can be based on random utility theory (McFadden 1974). Moreover, if goods and services contribute positively to human well-being, they have economic value (Bateman et al. 2002). Therefore, the attributes of goods and services need to be defined to understand their values and to do so the most commonly used approach is based on the concept of total economic value (TEV) as explained by ten Brink et al. (2011). In the TEV approach, the utilization of environmental resources is divided into two main categories: the use value (UV) and nonuse value (NUV). The UV could be direct (e.g., consuming the goods and engaging in a recreational activity) or indirect (e.g., water purification and carbon storage). While the NUV could be driven by cultural values and the satisfaction of knowing that the goods and services are available for future generations (Bateman et al. 2002). Therefore, considering only the market prices of goods and services could lead to incorrect choices, especially when the goods and services have several attributes. The traditional approaches to measuring economic value are by using revealed preferences (RP) and stated preferences (SP) methods as shown in Figure 9.6.

Accordingly, the production of plant-based meat is considered an emerging industry, and there is a scarcity of research to identify consumers' preferences, willingness to pay (WTP) and factors that can drive consumer acceptance to purchase and consume such products (Ho et al. 2022). This gap led to the investigation of these factors from a socio-economic perspective. From this perspective, a systematic review of 91 research articles has been done by Onwezen et al. (2021) to identify the drivers of consumer acceptance of alternative plant proteins, i.e., pulses, algae, insects, plant-based meat alternatives, and cultured meat. Authors defined the product-related factors, psychological factors, and external attributes (e.g., social environment, trust, and culture). The results showed that pulses and plant-based alternative proteins were accepted by consumers more than algae and insects whereby product- and health-related factors, i.e., healthiness and taste, were the two major drivers of product acceptance. In another study, Ishaq et al. (2022) categorized the factors into traditional, e.g., convenience, cost, and taste, and emerging drivers, e.g., safety, health, environmental impact, and animal welfare. Notably, Apostolidis and McLeay (2016) stated that social, health, and environmental benefits could have a significant impact on choosing plant-based alternatives and reducing meat consumption. For example, Ho et al. (2022) found that health and environmental consciousness were crucial precursors to gaining consumer acceptance for willingness to pay more for plant-based meat in Singapore.

In addition, De Oliveira Padilha et al. (2022) compared consumers' attitudes and behavior toward meat alternatives of lab-grown meat, conventionally raised meat, and plant-based

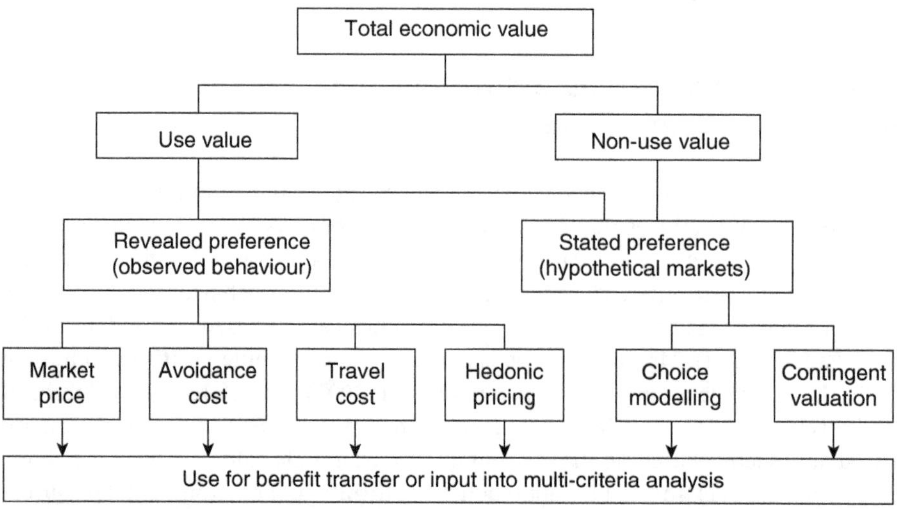

Figure 9.6 The economic valuation techniques for goods and services. *Source:* Tapsuwan et al. (2009), with permission of CSIRO.

protein alternatives through several attributes. These attributes were health, safety, affordability, eating enjoyment, animal welfare, and environmental friendliness. The results showed that plant-based meat alternatives were positively relative to all products and have the greatest potential to replace meat in consumers' diets. Moreover, Lea et al. (2005) highlighted the undesirable barriers to eating plant foods were skill and length of preparation time, which could be considered attributes that affect choosing plant proteins over meat proteins. The authors suggested focusing on promoting "easy-to-prepare" foods to increase the consumption of plant-based meals. This attribute can be applied to extruded meat analogue products, produced from pulses and legume protein to increase their acceptance in the market as these proteins are hard to cook and require a long cooking time if cooked traditionally. Furthermore, Attwood and Hajat (2020) raised food safety concerns, especially after zoonotic outbreaks, such as COVID-19, SARS, and the swine flu pandemic. This led to changes in animal-based meat consumption patterns and searching for potential health and environmental benefits in the longer term, which can be attained through the development of plant-based meat alternatives.

Marette (2021) conducted a study to record the reactions of 127 randomly selected respondents on their willingness to pay for lentils having different production labels. The result showed that organic and local production labelling has a weak yet significant influence on consumers' willingness to pay. It was concluded that the consumers' acceptance of lentil food is driven not only by the product's value but also by consumers' preferences and WTP of the products and their attributes which do not have an explicit price (e.g., safety, environmental impact, and animal welfare). Thus, future research needs to consider the socioeconomic factors that maximize consumers' well-being as well as environmental wellness when consuming lentil-protein meat alternates.

9.6 Quality of Lentil Protein-Based Meat Analogue

Customers are currently establishing augmenting importance in meat analogues produced from plant-based sources, including lentils. The aspects that vary the quality of protein involve amino acid composition, protein content, and protein digestibility. Similar to other pulse crops, lentils also have some antinutritional elements, e.g., tannins, trypsin inhibitors, raffinose family oligosaccharides, and lectins (Nosworthy et al. 2018). Some of these antinutrients have a negative impact on the protein bioavailability by deactivating significant gastric enzymes or complexing with the dietary proteins (Wang et al. 2009). Therefore, scientists have pursued to identify fast, effortless, precise, and accurate procedures for evaluating lentil protein meat analogue quality and value in consumed foods for numerous reasons (Zhou et al. 2021). These techniques should quantify the basic parameters that assess the quality of protein such as protein digestibility, along with the bioavailability of essential and nonessential amino acids (Tan et al. 2021).

The assessment of protein superiority and ensuing valuation of content claim legitimacy are dissimilar relying on the authority being debated. North America necessitates the usage of protein efficiency ratio (El-Niely 2007). In European countries, the source for assessment of the quality of protein is the quantity of protein that adds to the whole energy extant in a food product (El-Niely 2007). Nevertheless, a more reliable method for evaluation of protein quality was proposed by the United States Food and Drug Administration, which mandates the application of modified protein requirements as % daily value using protein digestibility-corrected amino acid score (PDCAAS) (Mefleh et al. 2022).

9.6.1 Protein Digestibility-Corrected Amino Acid Score (PDCAAS) Analysis

The PDCAAS of food is dependent on the amino acid score of a dietary protein and the exact fecal nitrogen digestibility value. The true fecal digestibility value of nitrogen is the amount of nitrogen expelled through feces with respect to the amount consumed (modified for interior losses). The value of amino acid score is calculated as the bottom-most proportion value when a distinct amino acid percentage in food is divided using a reference requisite profile of amino acids (Schaafsma 2005). Diets with reduced amounts of essential amino acids and/or lower coefficients of true fecal digestibility value of nitrogen will show decreased PDCAAS values and, henceforth, be considered as an inferior protein with respect to quality. When the exact fecal digestibility value and the profiling of amino acids of food or constituents are voluntarily accessible, the PDCAAS value can be measured.

The PDCAAS value had been accepted in the United States as a procedure for assessing the suitability of foods directed at kids (≥3 years) to grown-ups intended for claims of protein content (Mathai et al. 2017). Additionally, US guidelines postulate that the necessities of amino acid quantity for playschool kids aged between three and six years – as delineated in the Food and Agricultural Organization report, *Quality of Protein Assessment: Statement of Mutual WHO/FAO Proficient Consultation* (Schaafsma 2000), be applied as the orientation pattern to regulate the PDCAAS of foods. Subsequently, measuring the revised amount of protein among a controlled portion of food otherwise reference quantity normally consumed, the percentage daily value is calculated by means of 50 g to be the day-to-day reference amount for protein. In case food subsidizes 11–19.8% or ≥20% of the daily value for protein, the food succeeds to fall into the category that title food to be a "decent source" and a "tremendous basis" of protein, correspondingly. However, compared to the protein efficiency ratio (PER), there exist few benefits and drawbacks to PDCAAS both being a measure to assess the quality of protein (Schaafsma 2012).

Most remarkably, PDCAAS delivers comprehensive evidence concerning the profiling of amino acid and protein digestibility, with regard to the growth calculation of PER. The values of PDCAAS are condensed/compared to 1.00, signifying that no protein test could have a greater value compared to the reference protein, dissimilar to PER in which the ultimate value could be greater than that for casein (which is a value of 2.5) (Schaafsma 2000). Table 9.4 shows the PDCAAS and true protein digestibility of lentils with some other food sources. The aforementioned two parameters stand to be an important standard in evaluating the lentil protein quality with regards to human amino acid requirement and their ability to digest it. Generally, proteins are considered good if their PDCAAS value falls near 1.0.

Few studies have reported on the PDCASS values of extruded lentil proteins. For instance, *in vitro* PDCAAS values varied from 47.14% for baked green lentils to 63.01% for extruded red lentils. Extruded green and red lentils possessed maximum protein digestibility values of 57.09% and 63.01% followed by boiled green and red lentils 52.92% and 57.40% and baked green and red lentils 47.14% and 53.84%, respectively (Nosworthy et al. 2018). Extruded red

Table 9.4 Protein digestibility-corrected amino acid score (PDCAAS) for selected legume and cereal seeds.

Lentils	PDCASS	True protein digestibility (%)
Green lentils (whole)	0.63	87.9
Red lentils (split)	0.54	90.6
Chickpea	0.52	85.0
Black bean	0.53	70.0
Kidney bean	0.55	78.6
Navy bean	0.67	80.0
Pinto bean	0.59	76.2
Rice flour[a]	0.50	92.0
Wheat flour[a]	0.43	92.3
Casein[b]	1.00	96.6

[a] Given for comparison.
[b] Reference protein.
Source: Adapted from Pulse Canada (2016).

lentils (the same process used to produce plant meat analogue) recorded maximum PDCAAS values due to having a greater amino acid score and higher protein digestibility value. In another study, Nosworthy et al. (2017) found that green lentils scored a PDCASS value of 0.63, which was higher compared to the value of 0.51 recorded for average different lentil types.

9.6.2 Blood Serum Analysis

Blood serum analysis is a prominent basis of biomarkers with regards to disease identification and inequities in endogenous serum ingredients, which is of significant clinical prominence. With respect to protein molecules with high molecular weight fraction (HMWF), albumin amount, the utmost ample plasma protein, are classically higher in cases of critical hydration (Busher 1990) and decline significantly in patients prone to hypoalbuminemia (Fleck et al. 1985), major surgical stress, and prior to sepsis (Sun et al. 1993). Various clinical reports have related prominent stages of fibrinogen in patients having history of stroke (Chitsaz et al. 2012) along with circulatory disease (Van Der Bom et al. 1998) and pulmonary embolism (Klovaite et al. 2013). Dissimilarities of stages of minor molecular weight fraction elements, namely glucose, need to be examined at a consistent rate in situations where a new form of food is taken into the body, e.g., meat analogues made from pulse, cereal, lentil, or any other plant sources (Orsonneau et al. 1992).

Based on the past studies, it is well recognized that lentil proteins are well known to decrease the influence of triacylglycerols and cholesterol concentrations when compared to animal proteins. A comparison between casein protein diet, lentil protein, and chickpea protein was carried out by Boualga et al. (2009). It was observed that the consumption of lentil protein lowered the triacylglycerol value by 2.5-folds. Furthermore, Gowder (2015) reported that the consumption of lentil proteins or use of lentil proteins as a substitute to animal protein results in numerous positive effects, i.e., mitigation and protection against diabetes, cardiovascular, and metabolic diseases.

Conclusively, assessment of protein quality is imperative since it provides valuable information regarding the bioavailability of essential amino acids at a safe level for body growth as well as alternation in the protein nutritive values during food processing. More research is needed to predict the quality of texturized lentil protein to use as a safe and nutritious alternative to animal protein.

9.7 Future Trends and Suggestions for Lentil Protein-Based Meat Analogue

In spite of the significant nutritional value and continued progress and application of lentil-based meat analogues, the palatability of lentil-based meat analogues presents some hindrance to consumer acceptance (Dekkers et al. 2018). To enhance the flavor and texture of lentil-based meat analogues, various food additives are needed to be incorporated at the time of the manufacturing process. Concerning texture, various methods, namely thermoplastic extrusion, spinning, and steam texturization, have been used for the fundamental arrangement of lentil protein, as lentils are chiefly comprised of amorphous structure (Tabe and Higgins 1998). Among the aforementioned methods, extrusion processing is the primary method used, since it is a cost-effective process and could produce various sizes and shapes of meat analogues. The method is centered on using a screw system inside a barrel with the aid of which plant proteins are heated to be denatured and rearranged into a cross-linked, layered, and striated structure, eventually resulting in the creation of texturized lentil protein (Kim 2005).

The application of lentil protein as texturized lentil protein components might convey a texture, appearance, nutritional value and taste comparable to that of original meat has been recommended by previous research (Kim 2018). In the year 2001, an US patent has been filed for the invention of food products containing lentil and pea flour to prepare and process various food products including meat analogue. The proposed method proved that the high coagulation and gelatinizing properties of lentil flour make it widely suitable for the preparation of meat analogues (Thresher 2004). Baugreet et al. (2016) assessed the influence of functional protein constituents: rice protein, pea protein isolate, and lentil flour at 3% and 7% levels on the shelf life and textural properties of beef patties over 12-day storage. Authors observed that beef patties developed using lentil flour reduced the food hardness, gumminess, chewiness, and cohesiveness, compared with 100% beef patties. Further in-depth studies are required to explore the effect of lentil protein isolates in mimicking muscle meat texture under different processing conditions.

Meaty flavor is another challenge yet to be harnessed for meat analogue products prepared from plant proteins. The flavor of meat analogues from lentil protein is an important parameter of their reception and steady ingestion (Wang et al. 2022b). The disagreeable sensory characteristics of lentil proteins are accredited to grassy smell and astringent, or bitter flavor. The ingredients leading to such off-flavors are physically and chemically different and display a fluctuating binding affinity toward lentil proteins. Some of the techniques that can be used to efficiently remove these off flavors include chiefly solvent extraction, salting out, and removal by β-cyclodextrin (Damodaran and Arora 2013).

Guldiken et al. (2021) utilized five diverse adsorbents (XAD7HP, XAD4, SP207, XAD16N, and HP20) to minimize off-odors and flavors of lentil protein isolates. It was shown that XAD16N, HP20, and SP207 decreased the main volatile aldehyde groups present in lentils by 32%, 41%, and 32%, respectively.

9.8 Market Status and Prospect of Lentil Protein-Based Meat Analogues

The market call for lentil-based meat analogue products has significantly enlarged throughout the world owing to its nutritive and ecological benefits. Until 1960, the alternatives used for meat namely soy proteins as well as tofu were limited to Western nations; however, with time and variations in human food behaviors, the consumption of legume-based meat analogues has developed in various regions of the globe. In 2005, Burger King (USA) announced burger patties made with veggies at more than 7500 stores around the world. Currently, the worldwide meat auxiliary bazaar is controlled by European nations, e.g., Germany, the United Kingdom, France, Italy, Sweden, Belgium, and Norway being prominent in pioneering works on lentil based meat substitute proteins (Machovina et al. 2015). Since 2015, about 4500 plant-related meat alternative food products have been introduced in the market that resemble the characteristics of meat.

The USA and European countries are leading meat alternatives markets worth around US$1000 million followed by China (Fresán et al. 2019). Australia is also in lentil-based meat alternate, with a market value of US$200 million. In 2019, the global plant-based meat analogue market was projected at US$10 billion and projected to reach US$30 billion by the year 2025. Increasing environmental concerns and mounting governing stress have forced food producers to discover the prospective of plant-based proteins. Furthermore, the exponential increase in the vegan consumers across the globe (predominantly, countries in North America, Latin America, the Middle East, and Europe) is another important element which is projected to deliver significant enhancement to the lentil-protein market. However, in Asian nations, poorer recognition of meat substitutes has been perceived owing to the ethnic preferences and food habits, e.g., food constituents, food foundations, and accessibility to various foods. The situation could alter in the coming years as individuals get more accustomed to novel food products. A study on the consumers' recognition of meat substitutes in India and China, revealed that about 60–85% of consumers have shown their interest in meat alternative products (Bryant et al. 2019). It is anticipated that by 2050, the demand for lentil-based meat analogue would increase by about 70%. In this regard, numerous companies or brands had initiated the development of lentil-based meat analogues at the global level and numerous start-ups are focusing on the lentil-based meat analogues in recent years.

9.9 Conclusion

The lentil proteins are gaining increasing attention with respect to value-added utilization in diverse food applications. The worldwide lentil-based protein market is projected to grow at a net annual growth rate of 7.2% by the year 2026 (Boukid et al. 2021). The shift to a new plant-based diets is helping to decrease climate variations, as such diets require decreased energy, land resources, and water when compared to animal-based foods. Furthermore, the intake of lentils has also elucidated a numeral of health benefits, namely decreased chance of evolving type 2 diabetes, lesser low-density lipoprotein and total cholesterol, and a positive effect on blood pressure and weight management. The ingestion of protein-rich lentil-based meat alternatives food swings the responsiveness from the unsustainable meat production and processing industry. Both of these sectors produce enormous waste, which results in pollution, making animals vulnerable to illnesses, and negatively impacting the environment. The appropriate enhancements in lentil-based meat alternatives have already placed great emphasis on nutritious eminence, palatability, flavor, and shelf-life. For successful commercialization and sustainable demand, future research must focus on enhancing the quality, ensuring consumer acceptance, addressing technical obstructions (including regulatory matters), and making reasonable cost estimation of lentil-based meat analogues.

References

Afam-Mbah, C., Ketabi, L., Emami, S. et al. (2018). Air classification of by-product of lentil dehulling into protein-rich and starch-rich fractions. *ASABE 2018 Annual International Meeting* (29 July–1 August 2018), 1–14. St. Joseph, MI: American Society of Agricultural and Biological Engineers (Paper Number: 1801275).

Agri-Food Innovation Council (2019). Plant based protein market global and Canadian market analysis. https://nrc.canada.ca/en/research-development/research-collaboration/programs/plant-based-protein-market-global-canadian-market-analysis (accessed 18 September 2022).

Ahmad El-Sohaimy, S. and Sitohy, M. (2007). Isolation and partial characterization of chickpea, lupine and lentil seed proteins lipolytic lactic acid bacteria view project probiotic properties of lactic acid bacteria view project. *World Journal of Agricultural Sciences* 3: 123–129.

Ahmed, J., Mulla, M., Al-Ruwaih, N., and Arfat, Y.A. (2019). Effect of high-pressure treatment prior to enzymatic hydrolysis on rheological, thermal, and antioxidant properties of lentil protein isolate. *Legume Science* 1: e10.

Ahmed, J., Mulla, M.Z., Siddiq, M., and Dolan, K.D. (2022). Micromeritic, thermal, dielectric, and microstructural properties of legume ingredients: a review. *Legume Science* 4: e123.

Alavi, F., Chen, L., and Emam-Djomeh, Z. (2021). Effect of ultrasound-assisted alkaline treatment on functional property modifications of faba bean protein. *Food Chemistry* 354: 129494.

Apostolidis, C. and McLeay, F. (2016). Should we stop meating like this? Reducing meat consumption through substitution. *Food Policy* 65: 74–89.

Arshad, M., Anwar, S., Pasha, I. et al. (2022). Development of imitated meat product by utilizing pea and lentil protein isolates. *International Journal of Food Science & Technology* 57 (5): 3031–3037.

Aryee, A.N. and Boye, J.I. (2017). Comparative study of the effects of processing on the nutritional, physicochemical and functional properties of lentil. *Journal of Food Processing and Preservation* 41: e12824.

Attwood, S. and Hajat, C. (2020). How will the COVID-19 pandemic shape the future of meat consumption? *Public Health Nutrition* 23: 3116–3120.

Avramenko, N.A., Low, N.H., and Nickerson, M.T. (2013). The effects of limited enzymatic hydrolysis on the physicochemical and emulsifying properties of a lentil protein isolate. *Food Research International* 51: 162–169.

Aydemir, L.Y. and Yemenicioğlu, A. (2013). Potential of Turkish Kabuli type chickpea and green and red lentil cultivars as source of soy and animal origin functional protein alternatives. *LWT Food Science and Technology* 50: 686–694.

Baker, P.W. and Charlton, A. (2020). A comparison in protein extraction from four major crop residues in Europe using chemical and enzymatic processes – a review. *Innovative Food Science and Emerging Technologies* 59: 102239.

Bateman, I., Carson, R.T., Day, B. et al. (2002). *Economic Valuation with Stated Preference Techniques: A Manual.* Cheltenham, UK: Edward Elgar.

Baugreet, S., Kerry, J.P., Botineştean, C. et al. (2016). Development of novel fortified beef patties with added functional protein ingredients for the elderly. *Meat Science 122*: 40–47.

Berk, Z. (2018). *Food Process Engineering and Technology*. Academic Press.

Bleakley, S. and Hayes, M. (2017). Algal proteins: extraction, application, and challenges concerning production. *Food* 6: 1–34.

Bora, P.S. (2002). Functional properties of native and succinylated lentil (*Lens culinaris*) globulins. *Food Chemistry* 77: 171–176.

Bou, R., Navarro-Vozmediano, P., Domínguez, R. et al. (2022). Application of emerging technologies to obtain legume protein isolates with improved techno-functional properties and health effects. *Comprehensive Reviews in Food Science and Food Safety* 21: 2200–2232.

Boualga, A., Prost, J., Taleb-Senouci, D. et al. (2009). Purified chickpea or lentil proteins impair VLDL metabolism and lipoprotein lipase activity in epididymal fat, but not in muscle, compared to casein, in growing rats. *European Journal of Nutrition* 48: 162–169.

Boukid, F., Rosell, C.M., Rosene, S. et al. (2021). Non-animal proteins as cutting-edge ingredients to reformulate animal-free foodstuffs: present status and future perspectives. *Critical Reviews in Food Science and Nutrition* 62 (23): 6390–6420.

Boye, J.I., Aksay, S., Roufik, S. et al. (2010a). Comparison of the functional properties of pea, chickpea and lentil protein concentrates processed using ultrafiltration and isoelectric precipitation techniques. *Food Research International* 43: 537–546.

Boye, J., Zare, F., and Pletch, A. (2010b). Pulse proteins: processing, characterization, functional properties and applications in food and feed. *Food Research International* 43 (2): 414–431.

ten Brink, P., Badura, T., Bassi, S. et al. (2011). Estimating the overall economic value of the benefits provided by the Natura 2000 network. Institute for European Environmental Policy/GHK/Ecologic. Report no. Final Report to the European Commission, DG Environment on Contract ENV. B, 2.

Bryant, C., Szejda, K., Parekh, N. et al. (2019). A survey of consumer perceptions of plant-based and clean meat in the USA, India, and China. *Frontiers in Sustainable Food Systems* 3: 11.

Busher, J.T. (1990). Serum albumin and globulin. *Clinical Methods: The History, Physical, and Laboratory Examinations* 3: 497–499.

do Carmo, C.S., Knutsen, S.H., Malizia, G. et al. (2021). Meat analogues from a faba bean concentrate can be generated by high moisture extrusion. *Future Foods* 3: 100014.

Celiktas, M., Kirsch, C., and Smirnova, I. (2014). Cascade processing of wheat bran through a biorefinery approach. *Energy Conversion and Management* 84: 633–639.

Chitsaz, A., Mousavi, S.A., Yousef, Y., and Mostafa, V. (2012). Comparison of changes in serum fibrinogen level in primary intracranial hemorrhage (ICH) and ischemic stroke. *ARYA Atherosclerosis* 7: 142.

Clune, S., Crossin, E., and Verghese, K. (2017). Systematic review of greenhouse gas emissions for different fresh food categories. *Journal of Cleaner Production* 140: 766–783.

Damodaran, S. and Arora, A. (2013). Off-flavor precursors in soy protein isolate and novel strategies for their removal. *Annual Review of Food Science and Technology* 4: 327–346.

Damodaran, S. and Parkin, K.L. (ed.) (2017). Amino acids, peptides and proteins. In: *Fennema's Food Chemistry*, 235–356. Boca Raton: CRC Press.

De Oliveira Padilha, L.G., Malek, L., and Umberger, W.J. (2022). Consumers' attitudes towards lab-grown meat, conventionally raised meat and plant-based protein alternatives. *Food Quality and Preference* 99: 104573.

Dekkers, B.L., Boom, R.M., and van der Goot, A.J. (2018). Structuring processes for meat analogues. *Trends in Food Science & Technology* 81: 25–36.

Dent, T. and Maleky, F. (2022). Pulse protein processing: the effect of processing choices and enzymatic hydrolysis on ingredient functionality. *Critical Reviews in Food Science and Nutrition* 1–12.

Dhull, S.B., Kinabo, J., and Uebersax, M.A. (2022). Nutrient profile and effect of processing methods on the composition and functional properties of lentils (*Lens culinaris* Medik): a review. *Legume Science* 2022: e156.

Didinger, C. and Thompson, H.J. (2022). The role of pulses in improving human health: a review. *Legume Science* 4: e147.

Elalami, D., Carrère, H., Abdelouahdi, K. et al. (2018). Combination of dry milling and separation processes with anaerobic digestion of olive mill solid waste: methane production and energy efficiency. *Molecules* 23: 3295.

El-Niely, H.F. (2007). Effect of radiation processing on antinutrients, in-vitro protein digestibility and protein efficiency ratio bioassay of legume seeds. *Radiation Physics and Chemistry* 76: 1050–1057.

Fernando, S. (2021). Production of protein-rich pulse ingredients through dry fractionation: a review. *LWT Food Science and Technology* 141: 110961.

Ferreira, K.C., Bento, J.A.C., Caliari, M. et al. (2022). Dry bean proteins: extraction methods, functionality, and application in products for human consumption. *Cereal Chemistry* 99: 67–77.

Fetzer, A., Herfellner, T., Stäbler, A. et al. (2018). Influence of process conditions during aqueous protein extraction upon yield from pre-pressed and cold-pressed rapeseed press cake. *Industrial Crops and Products* 112: 236–246.

Fleck, A., Hawker, F., Wallace, P. et al. (1985). Increased vascular permeability: a major cause of hypoalbuminaemia in disease and injury. *The Lancet* 325: 781–784.

Fresán, U., Mejia, M.A., Craig, W.J. et al. (2019). Meat analogs from different protein sources: a comparison of their sustainability and nutritional content. *Sustainability* 11: 3231.

Funke, M., Boom, R., and Weiss, J. (2022a). Dry fractionation of lentils by air classification – composition, interfacial properties and behavior in concentrated O/W emulsions. *LWT Food Science and Technology* 154: 112718.

Funke, M., Loeffler, M., Winkelmeyer, C. et al. (2022b). Emulsifying properties of lentil protein preparations obtained by dry fractionation. *European Food Research and Technology* 248: 381–391.

Ganesan, K. and Xu, B. (2017). Polyphenol-rich lentils and their health promoting effects. *International Journal of Molecular Sciences* 18: 2390.

Garcia-Mora, P., Peñas, E., Frias, J., and Martínez-Villaluenga, C. (2014). Savinase the most suitable enzyme for releasing peptides from lentil (*Lens culinaris* var. Castellana) protein concentrates with multifunctional properties. *Journal of Agricultural and Food Chemistry* 62: 4166–4174.

Garcia-Mora, P., Peñas, E., Frias, J. et al. (2015). High-pressure improves enzymatic proteolysis and the release of peptides with angiotensin I converting enzyme inhibitory and antioxidant activities from lentil proteins. *Food Chemistry* 171: 224–232.

Ghasemi, P., Mirzaei, M., and Mirdamadi, S. (2021). The effect of heat and freeze-thaw pretreatment on the alcalase enzymatic hydrolysis of lentil protein and production of antioxidant peptides. *Journal of Food Technology and Nutrition* 18 (Summer 2021): 17–32.

Godfray, H.C.J., Aveyard, P., Garnett, T. et al. (2018). Meat consumption, health and the environment. *Science* 361 (6399): eaam5324.

Gowder, S.J.T. (2015). *Basic Principles and Clinical Significance of Oxidative Stress*. BoD–Books on Demand.

Guldiken, B., Green, R., and Nickerson, M.T. (2021). The impact of different adsorbents on flavour characteristics of a lentil protein isolate. *European Food Research and Technology* 247: 593–604.

Hall, A.E. and Moraru, C.I. (2021). Structure and function of pea, lentil and faba bean proteins treated by high pressure processing and heat treatment. *LWT Food Science and Technology* 152: 112349.

Ho, S.S., Chuah, A.S., Koh, E.L. et al. (2022). Understanding public willingness to pay more for plant-based meat: environmental and health consciousness as precursors to the influence of presumed media influence model. *Environmental Communication* 16 (4): 520–534.

Ishaq, A., Irfan, S., Sameen, A., and Khalid, N. (2022). Plant-based meat analogs: a review with reference to formulation and gastrointestinal fate. *Current Research in Food Science* 5: 973–983.

Jarpa-Parra, M. (2018). Lentil protein: a review of functional properties and food application. An overview of lentil protein functionality. *International Journal of Food Science and Technology* 53: 892–893.

Jarpa-Parra, M., Bamdad, F., Wang, Y. et al. (2014). Optimization of lentil protein extraction and the influence of process pH on protein structure and functionality. *LWT Food Science and Technology* 57: 461–469.

Johnston, S.P., Nickerson, M.T., and Low, N.H. (2015). The physicochemical properties of legume protein isolates and their ability to stabilize oil-in-water emulsions with and without genipin. *Journal of Food Science and Technology* 52: 4135–4145.

Joshi, M., Adhikari, B., Aldred, P. et al. (2011). Physicochemical and functional properties of lentil protein isolates prepared by different drying methods. *Food Chemistry* 129: 1513–1522.

Joshi, M., Adhikari, B., Aldred, P. et al. (2012). Interfacial and emulsifying properties of lentil protein isolate. *Food Chemistry* 134: 1343–1353.

Keskin, S.O., Ali, T.M., Ahmed, J. et al. (2022). Physico-chemical and functional properties of legume protein, starch, and dietary fiber – a review. *Legume Science* 4: e117.

van Kessel, C. and Hartley, C. (2000). Agricultural management of grain legumes: has it led to an increase in nitrogen fixation? *Field Crops Research* 65: 165–181.

Kew, B., Holmes, M., Stieger, M., and Sarkar, A. (2020). Review on fat replacement using protein-based microparticulated powders or microgels: a textural perspective. *Trends in Food Science & Technology* 106: 457–468.

Khazaei, H., Subedi, M., Nickerson, M. et al. (2019). Seed protein of lentils: current status, progress, and food applications. *Food* 8: 1–23.

Kim, C. (2005). Development of meat substitute food using vegetable protein. *EASDL Conference*. Seoul, Korea: The East Asian Society of Dietary Life.

Kim, T. (2018). Texturization of pulse proteins: peas, lentils, and faba beans. Doctoral dissertation. Texas A&M University. https://hdl.handle.net/1969.1/173522 (accessed 14 August 2022).

Kim, B.F., Santo, R.E., Scatterday, A.P. et al. (2020). Country-specific dietary shifts to mitigate climate and water crises. *Global Environmental Change* 62: 101926.

Klovaite, J., Nordestgaard, B.G., Tybjærg-Hansen, A., and Benn, M. (2013). Elevated fibrinogen levels are associated with risk of pulmonary embolism, but not with deep venous thrombosis. *American Journal of Respiratory and Critical Care Medicine* 187: 286–293.

Krintiras, G.A., Diaz, J.G., Van Der Goot, A.J. et al. (2016). On the use of the Couette cell technology for large scale production of textured soy-based meat replacers. *Journal of Food Engineering* 169: 205–213.

Kyriakopoulou, K., Keppler, J.K., and van der Goot, A.J. (2021). Functionality of ingredients and additives in plant-based meat analogues. *Food* 10: 600.

Lea, E., Worsley, A., and Crawford, D. (2005). Australian adult consumers' beliefs about plant foods: a qualitative study. *Health Education & Behavior* 32: 795–808.

Lee, H.C., Htoon, A.K., Uthayakumaran, S., and Paterson, J.L. (2007). Chemical and functional quality of protein isolated from alkaline extraction of Australian lentil cultivars: Matilda and Digger. *Food Chemistry* 102: 1199–1207.

Lee, H.W., Lu, Y., Zhang, Y. et al. (2021). Physicochemical and functional properties of red lentil protein isolates from three origins at different pH. *Food Chemistry* 358: 129749.

Liu, C., Hao, L., Chen, F., and Yang, C. (2020). Study on extraction of peanut protein and oil bodies by aqueous enzymatic extraction and characterization of protein. *Journal of Chemistry* 2020: 1–11.

Lupwayi, N.Z. and Kennedy, A.C. (2007). Grain legumes in Northern Great Plains: impacts on selected biological soil processes. *Agronomy Journal* 99: 1700–1709.

Ma, K.K., Greis, M., Lu, J. et al. (2022). Functional performance of plant proteins. *Food* 11: 594.

Machovina, B., Feeley, K.J., and Ripple, W.J. (2015). Biodiversity conservation: the key is reducing meat consumption. *Science of the Total Environment* 536: 419–431.

MacWilliam, S., Wismer, M., and Kulshreshtha, S. (2014). Life cycle and economic assessment of Western Canadian pulse systems: the inclusion of pulses in crop rotations. *Agricultural Systems* 123: 43–53.

Marette, S. (2021). Sustainability and consumer willingness to pay for legumes: a laboratory study with lentils. *Sustainability* 13: 3408.

Mathai, J.K., Liu, Y., and Stein, H.H. (2017). Values for digestible indispensable amino acid scores (DIAAS) for some dairy and plant proteins may better describe protein quality than values calculated using the concept for protein digestibility-corrected amino acid scores (PDCAAS). *British Journal of Nutrition* 117: 490–499.

McFadden, D. (1974). Conditional logit analysis of qualitative choice behavior. In: *Frontiers in Econometrics* (ed. P. Zarembka), 105–142. New York: Academic Press.

McFadden, D. (2001). Economic choices. *American Economic Review* 91 (3): 351–378.

Mefleh, M., Pasqualone, A., Caponio, F., and Faccia, M. (2022). Legumes as basic ingredients in the production of dairy-free cheese alternatives: a review. *Journal of the Science of Food and Agriculture* 102: 8–18.

Nadar, S.S., Rao, P., and Rathod, V.K. (2018). Enzyme assisted extraction of biomolecules as an approach to novel extraction technology: a review. *Food Research International* 108: 309–330.

Nosworthy, M.G., Neufeld, J., Frohlich, P. et al. (2017). Determination of the protein quality of cooked Canadian pulses. *Food Science & Nutrition* 5: 896–903.

Nosworthy, M.G., Medina, G., Franczyk, A.J. et al. (2018). Effect of processing on the in vitro and in vivo protein quality of red and green lentils (*Lens culinaris*). *Food Chemistry* 240: 588–593.

Onwezen, M., Bouwman, E., Reinders, M., and Dagevos, H. (2021). A systematic review on consumer acceptance of alternative proteins: pulses, algae, insects, plant-based meat alternatives, and cultured meat. *Appetite* 159: 105058.

Orsonneau, J.-L., Massoubre, C., Cabanes, M., and Lustenberger, P. (1992). Simple and sensitive determination of urea in serum and urine. *Clinical Chemistry* 38: 619–623.

Pelgrom, P.J.M., Vissers, A.M., Boom, R.M., and Schutyser, M.A.I. (2013). Dry fractionation for production of functional pea protein concentrates. *Food Research International* 53: 232–239.

Pelgrom, P.J.M., Boom, R.M., and Schutyser, M.A.I. (2015). Functional analysis of mildly refined fractions from yellow pea. *Food Hydrocolloids* 44: 12–22.

Peoples, M., Swan, T., Goward, L., et al. (2015). Legume effects on available soil nitrogen and comparisons of estimates of the ap-parent mineralisation of legume nitrogen. *Proceedings of the 17th ASA Conference on "Building Productive, Diverse and Sustainable Landscapes"* (20–24 September 2015), 971–974. Victoria, Australia: Australian Society of Agronomy Inc.

Pulse Canada (2016). Protein quality of cooked pulses. https://pdf4pro.com/amp/cdn/protein-quality-of-cooked-pulses-pulse-canada-5a3c38.pdf (accessed 18 September 2022).

Ramachandraiah, K. (2021). Potential development of sustainable 3d-printed meat analogues: a review. *Sustainability* 13: 938.

Ratnayake, W.S. and Naguleswaran, S. (2022). Utilizing side streams of pulse protein processing: a review. *Legume Science* 4: e120.

Rezvankhah, A., Yarmand, M., Ghanbarzadeh, B., and Mirzaee, H. (2021). Characterization of bioactive peptides produced from green lentil (*Lens culinaris*) seed protein concentrate using Alcalase and Flavourzyme in single and sequential hydrolysis. *Journal of Food Processing and Preservation* 45.

Riaz, M.N. (2001). Selecting the right extruder. In: *Extrusion Cooking-Technologies & Applications* (ed. R. Guy), 29–50. Woodhead Publishing.

Santo, R.E., Kim, B.F., Goldman, S.E. et al. (2020). Considering plant-based meat substitutes and cell-based meats: a public health and food systems perspective. *Frontiers in Sustainable Food Systems* 4: 134.

Sari, W.Y. (2015). *Biomass and Its Potential for Protein and Amino Acids; Valorizing Agricultural by-Products*. Wageningen University.

Sathe, S.K. (2002). Dry Bean Protein Functionality. *Critical Reviews in Biotechnology* 22: 175–223.

Schaafsma, G. (2000). The protein digestibility–corrected amino acid score. *The Journal of Nutrition* 130: 1865S–1867S.

Schaafsma, G. (2005). The protein digestibility-corrected amino acid score (PDCAAS) – a concept for describing protein quality in foods and food ingredients: a critical review. *Journal of AOAC International* 88: 988–994.

Schaafsma, G. (2012). Advantages and limitations of the protein digestibility-corrected amino acid score (PDCAAS) as a method for evaluating protein quality in human diets. *British Journal of Nutrition* 108: S333–S336.

sen Gupta, D., Thavarajah, D., Knutson, P. et al. (2013). Lentils (*Lens Culinaris* L.), a rich source of folates. *Journal of Agricultural and Food Chemistry* 61: 7794–7799.

Serdaroğlu, M., Yıldız-Turp, G., and Abrodímov, K. (2005). Quality of low-fat meatballs containing legume flours as extenders. *Meat Science* 70: 99–105.

Shen, L., Wang, X., Wang, Z. et al. (2008). Studies on tea protein extraction using alkaline and enzyme methods. *Food Chemistry* 107: 929–938.

Singh, J. (2018). Folate content in legumes. *Biomedical Journal of Scientific & Technical Research* 3 (4): 3475–3480.

Singh, S., Gamlath, S., and Wakeling, L. (2007). Nutritional aspects of food extrusion: a review. *International Journal of Food Science & Technology* 42: 916–929.

Sozer, N., Holopainen-Mantila, U., and Poutanen, K. (2017). Traditional and new food uses of pulses. *Cereal Chemistry* 94: 66–73.

Sun, X., Iles, M., and Weissman, C. (1993). Physiologic variables and fluid resuscitation in the postoperative intensive care unit patient. *Critical Care Medicine* 21 (4): 555–561.

Tabe, L. and Higgins, T. (1998). Engineering plant protein composition for improved nutrition. *Trends in Plant Science* 3: 282–286.

Tan, M., Nawaz, M.A., and Buckow, R. (2021). Functional and food application of plant proteins – a review. *Food Reviews International* 1–29.

Tanaka, D., Lyon, D., Miller, P. et al. (2010). Soil and water conservation advances in the semiarid Northern Great Plains. *Soil and Water Conservation Advances in the United States* 60: 81–102.

Tapsuwan, S., Ranjan, R., McFarlane, D. et al. (2009). Economic and social values of land and water uses on the Gnangara groundwater system. Report prepared for the Gnangara Taskforce, Department of Water, Western Australia. Gnangara Sustainability Strategy and CSIRO Water for a Healthy Country National Research Flagship Report.

Thornton, P., and Cramer, L. (2012). Impacts of climate change on the agricultural and aquatic systems and natural resources within the CGIAR's mandate. In *CCAFS Working Paper 23.* www.ccafs.cgiar.org

Thresher, W. C. (2004). Food products comprising pea or lentil flours and the process of making the same. Google Patents.

United States Department of Agriculture (USDA) National Agricultural Statistics Service (NASS) (2018). 2018 state agriculture overview for Montana [Online]. https://www.nass.usda.gov/Quick_Stats/Ag_Overview/stateOverview.php?state=MONTANA (accessed 27 July 2022).

Urbano, G., Porres, J.M., Frias, J., and Vidal-Valverde, C. (2007). Nutritional value. In: *Lentil: An Ancient Crop for Modern Times* (ed. S.S. Yadav, D.L. McNeil, and P.C. Stevenson), 47–94. Springer.

Ustunol, Z. (ed.) (2014). *Applied Food Protein Chemistry*. Wiley.

Van Der Bom, J.G., De Maat, M.P., Bots, M.L. et al. (1998). Elevated plasma fibrinogen: cause or consequence of cardiovascular disease? *Arteriosclerosis, Thrombosis, and Vascular Biology* 18: 621–625.

Vogelsang-O'Dwyer, M., Petersen, I.L., Joehnke, M.S. et al. (2020). Comparison of Faba bean protein ingredients produced using dry fractionation and isoelectric precipitation: techno-functional, nutritional and environmental performance. *Food* 9.

Wang, N. (2005). Optimization of a laboratory dehulling process for lentil (*Lens culinaris*). *Cereal Chemistry* 82: 671–676.

Wang, N., Hatcher, D., Toews, R., and Gawalko, E. (2009). Influence of cooking and dehulling on nutritional composition of several varieties of lentils (*Lens culinaris*). *LWT Food Science and Technology* 42: 842–848.

Wang, J., Zhao, J., de Wit, M. et al. (2016). Lupine protein enrichment by milling and electrostatic separation. *Innovative Food Science and Emerging Technologies* 33: 596–602.

Wang, T., Kaur, L., Furuhata, Y. et al. (2022a). 3D printing of textured soft hybrid meat analogues. *Food* 11: 478.

Wang, Y., Tuccillo, F., Lampi, A.M. et al. (2022b). Flavor challenges in extruded plant-based meat alternatives: a review. *Comprehensive Reviews in Food Science and Food Safety* 21: 2898–2929.

Warne, T., Ahmed, S., Byker Shanks, C., and Miller, P. (2019). Sustainability dimensions of a North American lentil system in a changing world. *Frontiers in Sustainable Food Systems* 3: 88.

Willett, W., Rockström, J., Loken, B. et al. (2019). Food in the Anthropocene: the EAT–Lancet Commission on healthy diets from sustainable food systems. *The Lancet* 393: 447–492.

Xing, Q., de Wit, M., Kyriakopoulou, K. et al. (2018). Protein enrichment of defatted soybean flour by fine milling and electrostatic separation. *Innovative Food Science and Emerging Technologies* 50: 42–49.

Zayas, J.F. (2012). *Functionality of Proteins in Food*. Springer Science & Business Media.

Zhang, J., Liu, L., Liu, H. et al. (2019). Changes in conformation and quality of vegetable protein during texturization process by extrusion. *Critical Reviews in Food Science and Nutrition* 59 (20): 3267–3280.

Zhou, H., Hu, Y., Tan, Y. et al. (2021). Digestibility and gastrointestinal fate of meat versus plant-based meat analogs: an in vitro comparison. *Food Chemistry* 364: 130439.

10

Utilization of Lentils in Different Food Products

Ayca A. Emir[1], Eda Yildiz[2], and Gulum Sumnu[2]

[1] *Department of Food Technology, Faculty of Canakkale Applied Sciences, Çanakkale Onsekiz Mart University, Canakkale, Turkey*
[2] *Department of Food Engineering, Middle East Technical University, Ankara, Turkey*

10.1 Introduction

Pasta, bread, biscuit, and other bakery products are regarded as an indispensable part of our diets because they provide a substantial source of macronutrients such as carbohydrate, protein, micronutrients, dietary fiber, and antioxidants. However, there has been a noticeable change in consumers' interest looking for healthier bakery products (Cappelli and Cini 2021). Furthermore, number of people suffering from celiac disease, a small intestine inflammation from eating wheat, barley, and rye proteins, have been on the increase. Celiac disease causes a serious malabsorption problem of vitamins, iron, folate, and calcium, and the only way to avoid these undesirable problems is to exclude these proteins. Therefore, there is a growing interest in gluten-free products that are nutritious and healthy (Yildiz et al. 2018a; Didinger and Thompson 2022; Dhull et al. 2022). In this regard, legumes, such as nutrient-rich lentils, offer a great potential to develop gluten-free products as well as for incorporation into a variety of food products. The consumption of this favorite legume has an increasing trend around the world with the increasing interest to functional and high nutrition food (Hernandez-Aguilar et al. 2020; Keskin et al. 2022).

Besides plant foods, animal-based products (meat and dairy products) are a major part of human diet. Animal production has serious negative impact on the environment besides ethical concerns for slaughtering animals. Moreover, the animal-based products cause health problems such as cardiovascular diseases and colon cancer (Substitutes 2020). Therefore, reducing animal-based product consumption as much as possible can decrease their negative impact on the environment and human health. To replace animal-based products, legumes are an excellent alternative. Such shift toward legume-based foods can reduce greenhouse gas emissions and water usage for animal agriculture.

According to World Health Organization (WHO), legumes are categorized as a critical component of the diet offering an economical, sustainable source of proteins, and other

Lentils: Production, Processing Technologies, Products, and Nutritional Profile, First Edition.
Edited by Jasim Ahmed, Muhammad Siddiq, and Mark A. Uebersax.

essential nutrients. Lentils have higher protein, carbohydrate, and energy compared to most other legumes. Further, they are a well-known source of essential amino acids including lysine and arginine. Lentils also have high content of dietary fiber, mineral, and are very low in saturated fat and sugar (Perri et al. 2021). Therefore, lentil can become the most commonly utilized legume to produce healthier bakery and other products, and a potential alternative to animal-based protein. In this chapter, the effect of lentil incorporation on product quality, consumer perception, and other possible methods to enhance nutritional and health aspects of the products are summarized.

10.2 Bakery Products

Baking is a process that includes both heat and mass transfer. While heat is transferred through batter, it caused vaporization of moisture through the surface, which leads to weight loss of cakes. Because moisture loss increased hardness of cakes, high starch, fiber containing flour, and formulations with hydrocolloids decreased this property (Berk et al. 2017b). Lentil flour and fractions can be successfully used for baked products development. The processing methods (heat treatment, germination, and fermentation) to produce lentil ingredients can significantly reduce or eliminate the content of undesirable antinutrients, thereby improving functionality and sensorial properties. The effect of fermentation conditions on functional properties on wheat-fermented lentil blends bread was investigated. Both water hydration and oil binding capacities of fermented lentil flour were higher than non-fermented ones because longer soaking time during fermentation process led to swelling of starch and protein molecules. Although degradation of starch reduced swelling ability the same time, all in all, overall water hydration capacity increased (Sadowska et al. 1999).

10.2.1 Bread

Bread, as a global food, dates back to Mesopotamia or Ancient Egypt approximately 5000 years ago (De Boni et al. 2019). It is produced and consumed in different varieties and huge quantities in worldwide and considered as a source of main food constituents like proteins, carbohydrates, fibers, and vitamins. Therefore, it has a direct influence on human health (Mollakhalili-Meybodi et al. 2021). Bread making involves three main steps: dough kneading, dough fermentation, and baking (Gancarz et al. 2021). The main aim of the kneading process is to form viscoelastic dough matrix, proper development of gluten network, and incorporate gas in the dough (Cappelli et al. 2020). Microorganisms have a key role in the formation of volatile compounds in bread making in fermentation process. The diversity of microbial species causes formation of different metabolic substances, which leads to unique flavors, taste, and aroma (Suo et al. 2020). Baking transforms bread dough into a digestible, flavorful product (Ureta et al. 2019).

The viscoelastic structure of bread dough is an indicator of the dough machinability and textural characteristic of final product. Water absorption capacity (WAC) is a parameter giving an idea about water absorbing ability of component to reach optimal consistency

(Bojňanská et al. 2021). To alter WAC of bread dough, lentil flour use was explored as an alternative option, e.g., blending wheat flour with carob bean flour (CBF) and lentil flour (LF), which affected WAC, viscoelastic, and mixing characteristics of the dough. Although addition of CBF increased WAC, it was nearly the same in sample with LF. This effect was probably due to different hydrophilic polymer constituents since CBF with more soluble fiber have more hydroxyl side than LF, which provided more possibility to interact with water than gluten and starch (Turfani et al. 2017). In another study on the enrichment in Sangak bread with LF, WAC of the flour increased in parallel with the level of germinated LF due to the production of hydrophilic parts in germination and fibers in LF. Moreover, WAC also was a function of protein content and pentosane amount in flour (Razavi et al. 2020). Bojňanská et al. (2021) analyzed the effect of chickpea, broad bean, common bean, and red lentil on bread dough and observed that WAC of composite flours was dependent on the ratio of protein-to-starch and other constituents in the legume flour. Flour containing common bean and broad bean increased WAC, whereas lentil and chickpea flour slightly reduced WAC (Bojňanská et al. 2021).

Dough expansion is correlated with protein levels in flour that has been hydrated. Flour having higher particle size and higher WAC causes an increase in peak time or development time of dough (Razavi et al. 2020). Dough development time (DDT) which is defined as required time for the dough to reach maximum consistency (Huang et al. 2016) increased from 1.45 to 5.15 minute with addition of LF. Dough with LF received lower tenacity, extensibility and strength (Turfani et al. 2017). Softness of blend dough increased by approximately 15% which was associated with dilution of gluten and excessive amount of protease enzyme. Protease enzyme weakened the gluten network, which resulted in decreased elasticity in the gluten network and its gas-holding capacity (Razavi et al. 2020). Additionally, lentil proteins likely competed with gluten for water thereby retarding gluten hydration. Furthermore, high amylose content in lentil flour (35%) was possibly another reason for the reduction of peak viscosity of the wheat–LF blend. In contrast to amylose, amylopectin has an ability to enhance viscosity and strength of starch–protein mixtures (Turfani et al. 2017). Previtali et al. (2014) reported similar results, where incorporation of LF not only caused decreasing tenacity but also lowered the force required for rupture, and corresponding deformation especially at high LF concentrations. To improve structural characteristics, different structuring agents, i.e., carboxymethyl cellulose, guar seed flour, pectin, and tapioca starch, were added in the bread formulations at 1%, 2%, or 3%. Deformation at maximum force, deformation corresponding rupture, and tenacity parameters were significantly affected by the hydrocolloid type and their concentrations.

The addition of fermented LF is known to cause deterioration of dough rheology (development and stability times) as evidenced by changes in microstructure. Wheat dough has a fiber-like structure that traps globular particles. Although large starch particles with smooth surface are packed closely among fibrils without any interaction with the proteins, small starch particles are embedded directly in the protein network. This structure of dough was mainly responsible from higher mixing resistance and low dough weakening. However, dough with fermented LF had no fibril-like structure. Starch granules both small and larger ones entrapped into the gluten matrix appeared to be underdeveloped or unmixed. Moreover, gluten network structure was in the form of flat and elongated rather than fibrillary (Sadowska et al. 1999). All these changes in dough rheology significantly affected final product quality since weakening gluten network mostly resulted in increasing in crumb firmness and bubble size, which limits the acceptability of the bread. The weakened interaction between LF proteins and gluten can restrict the formation of elastic structure, which is crucial for the retaining gas cell integrity, gas retention during fermentation, and finally acceptable crumb porosity (Previtali et al. 2014). Increasing fiber-like constituents and dilution of gluten were the most important reasons for coarser crumb-like structure (Paladugula et al. 2021). Gas cell characteristic was also influenced from particle size of flour. Although the number of cells per slice area was the same for the bread with wheat flour and the bread with commercial LF with defined particle size, with that property increasing with decreasing particle size of LF. Contrary to findings by Bourré et al. (2019), cell wall thickness of both regular size and finer LF breads were not different from wheat flour bread. However, cell diameter of regular particle size lentil bread (1.97 mm) was greater than that of finer size (1.87 mm) one. Furthermore, breads made with finer LF had higher firmness than regular LF bread (Bourré et al. 2019).

The firmness of bread is mostly a function of the substitution level of LF, as reported in various studies. Increasing substitution level of LF caused steady increase in bread's firmness, e.g., bread with 20% LF showed almost five times higher firmness than the wheat bread. This effect was primarily correlated with the lower WAC from higher level of LF addition. However, incorporation of gluten to LF bread formulation kept firmness at the same level with wheat bread containing 20% LF addition. These results showed the restorative effect of gluten on bread quality. Loaf volumes decreased with addition of LF, independent from presence of gluten and level of LF. The most drastic change was observed for the bread formulated without gluten. A reduction in specific volume was due to the lower hydration of gluten. Both dilution gluten and competition between gluten–gliadin, lentil proteins, albumin, and globulin toward water were responsible for these effects since lentil protein had numerous number of hydroxyl groups with high water binding affinity (Portman et al. 2018).

Color is another important sensory attribute that influences consumer acceptability of bread. In general, LF substituted breads have more brown color due to brown color of LF (Portman et al. 2018). Phenolic compounds in the seed coat are colored and mostly responsible from color of seed, e.g., lentils (490 mg catechin equivalent, CE/100 g) had more polyphenols than pinto beans (264.7 mg CE/100 g) and pita bread (Borsuk et al. 2012). Furthermore, acceleration of Maillard reaction during baking affects the color, which was due to the increasing lysine content in LF (Portman et al. 2018).

Incorporation of LF was shown to enhance bread's nutritional quality by increasing the protein, dietary fiber, bioactive substances, vitamins, and minerals content. Although LF contains less amount of sulfur-containing amino acid, it is richer in lysine. Added LF enhanced polyphenols providing higher total phenolic content. Insoluble dietary fiber percentages increased with increasing LF level in bread. Their health benefits include lowering cholesterol, sugar level, blood pressure, and reducing intestinal disorders. Antioxidant activity of bread also increased from 6.4 to 23.9 μg/100 g (Turfani et al. 2017). LF obtained from germinated lentils greatly enhanced the fiber, fat, and ash content compared to raw LF substitution. Since starch, protein, and fat break down into smaller fragments during germination, their digestibility and absorbability increased significantly. Furthermore, vitamin B_1, B_3, and B_6 increased in germinated LF breads compared to non-germinated ones. Since the metabolism of plant seeds increases during germination, concentration of B-group vitamins also increases, which are part of the oxidation reduction vitamins (Razavi et al. 2020).

10.2.2 Cake

Lentil ingredients or flour can be incorporated into cake formulation to enhance cake's nutritional quality. Thus, cake with only lentil flour (LF) might be a good alternative for the celiac patients who experiences genetically auto-immune disorder related to intolerance to indigestion of gluten (Marston et al. 2016). In addition, LF can be substituted for wheat flours at variable levels. The batter quality depends on the density, specific gravity, and rheological properties, which determine the final quality of cake and can be impacted with LF addition. For example, specific gravity gives information about how much air is incorporated into batter during mixing. Thus, lower specific gravity indicates more air aeration, which is always desirable for cake batter (Berk et al. 2017a).

The specific gravity of angel cakes increased from 0.33 to 0.47 g/l with added lentil protein to the mix. Like all other proteins, lentil proteins also have hydrophilic and hydrophobic groups that are responsible from amphiphilic character of protein. However, solubility, emulsifying ability, foaming capacity, and gelling property are some other important parameters for proteins to evaluate performance in batter matrix (Berk et al. 2017b). Contrary to previous example, in the case of muffin batter comparison, lentil protein incorporation did not cause any changes in specific gravity. Therefore, similar results of muffins of with/out lentil protein indicated comparable performance of lentil protein compared to milk and egg (Jarpa-Parra et al. 2017). Based on the particle size of flours, batter density of LF cakes was also affected significantly, e.g., 0.94 g/cm^3 versus 0.94 g/cm^3 with fines and coarse LF, respectively. However, the characteristics of air incorporation was as important as the quantity of air. Batters with fine flour had higher number of gas bubbles (smaller in size) than batters with coarse flour. Furthermore, both number and size of bubbles of fine LF batter were more similar to wheat flour sponge batter than batters made from coarse flour (De la Hera et al. 2012).

Lentil protein-based batters also exhibited shear thinning characteristic, similar to most cake batters, which is the indication of alignment of batter microstructure in the direction of flow. As a characteristic of shear thinning behavior in which flow behavior index is less than unity, apparent viscosity of sample decreases with increasing shear rate. Flow behavior index of control angel batter, which did not contain any lentil protein, was reported as 0.26, compared to 0.17 of that with lentil protein. The incorporation of lentil protein caused increment in consistency index of batter, which was interpreted as more entangled structure. This effect was also correlated with higher water holding capacity of lentil protein (~4.0 ml/g) than egg white (1.68 ml/g) (Jarpa-Parra et al. 2017).

In a study of gluten-free cakes, density of lentil (0.9 g/cm^3) and bean batters (0.9 g/cm^3) were significantly lower than those with chickpea (1.0 g/cm^3), pea (1.0 g/cm^3), and control (1.0 g/cm^3), which was an indication of more air incorporation into structure. Furthermore, bean and lentil batters exhibited higher number of gas cells with uneven structure. Generally, low batter density was interpreted as high cake volume (Gularte et al. 2012). On the other hand, viscosity of lentil/wheat composite batters increased after incorporation of

LF. The possible reason for this increase was the higher protein-to-starch ratio in batter after incorporation of lentil, which has higher protein content than wheat flour. Moreover, particle size of LF was another influencing factor for batter viscosity. Batter made with coarse fraction had higher protein content than fine fraction, which resulted in increasing batter viscosity of layer cake (De la Hera et al. 2012).

Aeration of gas bubbles into the batter mainly relies on the mixer speed, viscosity, and surface tension of batter. Moreover, efficiency of air incorporation depended on both film-forming capacity and rising speed of gas cells. In addition, emulsifier type and amount affected air bubble structure and distribution. Emulsifiers that are critical for the interfacial properties are mainly divided into two main classes: lipid and protein based (Yildiz et al. 2018b). Like other proteins, lentil protein has amphiphilic structure providing stabilizing an emulsification ability.

Jarpa-Parra et al. (2017) highlighted that lentil protein addition led to more success in cake production than bread making. This might be due to different dough formation requirement of bread compared to cake. Although the mean area of air cells had a tendency to reduce in the presence of lentil protein, the number of that cells per unit area increased, which resulted in similar volume characteristic of lentil-added and control samples. This effect was an indication of accomplished preservation of air cells and formation of strong cell walls around them, which prevented collapsing during baking process. Furthermore, due to similar volume characteristics, it has been also noted that lentil protein had a comparable stabilization function to egg white, whey protein, and other hydrocolloids. Specific volume, reported as cm^3/g, of lentil flour gluten-free cakes (3.2) was higher than chickpea (2.7), pea (2.9), common bean (2.9), and control (wheat) cakes (2.7). This was the indication of more air incorporation in cake and lower batter density (Gularte et al. 2012). In layer cake samples, increasing of lentil flour proportion in formulation resulted in a decrease in volume due to an increase in cake density. Therefore, there might be an optimum protein content to obtain the most desirable cake property. Flour with finer particle size achieved better volume and density values compared to coarse size flour (De la Hera et al. 2012).

The starch granule size is another influencing factor for the moisture loss of the samples. Smaller size of starch granules creates higher specific surface area by the way that of granules absorbed more water, disrupted which resulted in swelling of starch granules and hindered the moisture loss (Yildiz et al. 2018a). Incorporation of lentil flour in cake formulation also affected the macrostructure including number of pores and pore area fraction. Although the number of pores in lentil cakes decreased in microwave-infrared (MW-IR) oven compared to wheat cake, the pore area fractions increased. Higher pressure gradient occurring inside the cakes in microwave oven was mainly responsible from this result. Similarly, in conventional heating, total number of pores of lentil cake was lower than wheat cake. However, pore is fraction of wheat cake was higher than lentil cake (Ozkahraman et al. 2016).

Although texture analysis (compressive stress, N/m^2) of fresh control and lentil protein replaced angel cakes showed similar trend, their compressive stress values drastically increased during storage, which was mainly due to airy and light structure of the product that caused rapid loss of moisture from crumb to crust and finally to surrounding.

Furthermore, at higher substitution level of lentil protein, chewiness of angel cakes were higher than control because of reinforcement of cell walls due to higher protein entanglement network (Jarpa-Parra et al. 2017). However, in the case of muffin cakes, lentil-protein-added samples received higher compressive stress than control throughout storage time. These results suggested that lentil protein and gluten interaction caused stronger network and increase that value (Jarpa-Parra et al. 2017). Hardness of gluten-free cakes (chickpea, pea, and bean) was higher than lentil and control cakes. Hardness of cakes were negatively correlated to volume. Cohesiveness of legume cakes were higher due to their higher protein content (Gularte et al. 2012). Springiness is a term that defines shape recovery of the food between first and second bite. Cohesiveness is defined as a rate at which food will disintegrate under mechanical stress. The springiness of cakes decreased with incorporation lentil flour in layer cakes independent from flour particle size and kind of lentil (orange or yellow). Similar trend was also valid for the cohesiveness of the cakes (De la Hera et al. 2012).

The volume of cakes has been shown to be inversely correlated to hardness values. Ozkahraman et al. (2016) cakes made with only wheat flour had relatively lower hardness value compared to those containing lentil flour. The specific volume of wheat flour cake was relatively higher than lentil flour cake baked in both conventional and MW ovens. These results were attributed to lower gelatinization degrees of the cakes. Higher fiber and protein content of lentil flour were responsible for the higher water binding capacity. Higher water binding capacity of legume flour, however, decreased available water in the structure due to higher fiber and protein content, which is necessary for starch gelatinization during baking. Therefore, gelatinization temperature of starch shifted higher values, so cakes with legume flour might gelatinize less compared to wheat flour cake. Therefore, cakes with legume flour might result in lower volume due to deficiency structure formation in cake (Ozkahraman et al. 2016). Although gelatinization degrees of wheat cakes were higher than lentil cake, more uniform sheet of deformed starch as seen in SEM images (Figure 10.1). The possible reason of this effect was that lentil proteins covered the surface of starch granules and caused formation of veil-like structure, restricted the available water amount, and so starch swelling and gelatinization.

The color of food products, such as cakes, is an important quality attribute for acceptance of a product by consumers. When lentil protein was included in angel food cake and muffin, ΔE^* value changed in parallel with the concentration of lentil protein, which was possibly due to the phenolic compounds bound to lentil protein and Maillard reactions during baking (Jarpa-Parra et al. 2017). Lightness of gluten-free cakes was listed from the brightest to darkest as following order: chickpea, pea, bean, lentil. Since protein composition of these legume flours almost same, the difference between the crust color was attributed to different sugar content. Flour with higher sugar had tendency to Maillard browning and caramelization reactions on the cake surface (Gularte et al. 2012). In the comparison of MW-IR and conventional cake study, higher crust color change was observed in MW-IR than conventional oven because halogen lamp in MW-IR increased surface temperature significantly, which resulted in higher browning. Regardless of the type, cakes with any legume flour received higher crust color change than wheat flour, which was due to effect of original color of legume flour (Ozkahraman et al. 2016).

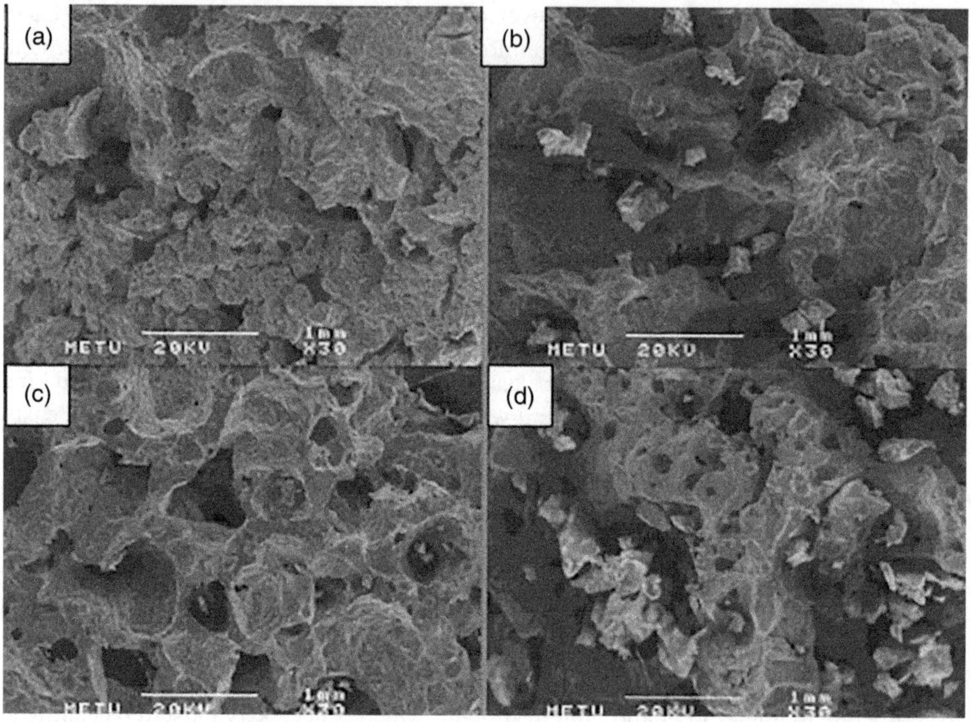

Figure 10.1 SEM micrographs (x30) of cake samples baked in different ovens. (a) Wheat cake baked in conventional oven, (b) wheat cake baked in MW-IR oven, (c) lentil cake baked in conventional oven, and (d) lentil cake baked in MW-IR oven. *Source:* Ozkahraman et al. (2016), with permission of Springer Nature.

10.2.3 Other Bakery Products (Cookie, Cracker, Wafer, Biscuit)

Since wafer is one of the widely consumed product due to its unique sensory attributes, flat structure, light weight, and crunchy structure, to improve its nutritional value and product quality, some modifications have been adopted (Nasabi et al. 2021). Tufan et al. (2020) partially replaced wheat flour in wafer sheet formulations by bean, chickpea, and lentil flour. In addition to rheological characterization, textural attributes of the wafer were assessed. Consistency index of batters increased with increasing fiber (especially insoluble fiber) concentration. Wafer batter with bean flour reached the highest consistency index (4.953 $Pa{\cdot}s^n$), whereas batter with lentil flour (1.484 $Pa{\cdot}s^n$) had the lowest texture profile analysis of wafer showed that samples with high protein concentration had higher hardness values. Wafer with lentil flour had the second highest value after bean-containing samples and control achieved the lowest hardness (Tufan et al. 2020).

Since wafer is a carbohydrate-based snack, it does not meet the consumer demand due to high glycemic index (GI) loading and high fat level. Therefore, to modify this drawback, wafers composed of different proportion of red lentil and corn grits were investigated. Hardacre et al. (2006) showed that the amount of glucose released from sample depended on the amount of starch present in wafer. *In vitro* analysis showed that after 120-minute digestion, wafers containing corn had higher GI than wafers with added lentil.

Biscuit as a versatile product has a remarkable position due to its longer shelf life, wide range of taste, and texture. Gluten-free biscuits composed from different proportions of buckwheat, millet, chickpea, and lentil flour were investigated in terms of physical properties, starch digestibility, and GI. Since resistant starch is not digested in small intestine, it does not contribute the glucose release. Di Cairano et al. (2021) showed that resistant starch content changed depending on legume flour type, with lentil flour having relatively lower resistant starch than chickpea samples.

To decrease glycemic response of biscuits, Chávez-Murillo et al. (2021) applied the heat moist treatment to black bean, chickpea, lentil, and faba bean flour and analyzed *in vitro* digestion. The highest protein digestion rate was observed for biscuits with lentil flour (80.11%) and heat treatment improved this rate to 85.16%. However, among the heat-treated samples, biscuits with faba bean flour had the highest digestibility of 87.43%. Furthermore, heat treatment decreased the predicted GI of all samples, including a reduction from 77.01 to 71.27 in biscuits with added lentil flour (Chávez-Murillo et al. 2021). Di Cairano et al. (2022) analyzed the effect of maltitol and inulin as a replacer of sucrose (sugar). Furthermore, the substitution of resistant starch on characteristic of biscuits made up of three different flours (buckwheat, lentil, and sorghum) was also investigated. Since inulin has higher water retention capacity than sucrose, samples with inulin had higher water activity and moisture content. The color of the samples was also affected with the replacement and samples with polyols resulted in a paler crust due to less tendency to Maillard reaction. Among the samples, the highest glucose release was recorded for samples containing sucrose (control), whereas the lowest release was observed with maltitol (Di Cairano et al. 2022).

Oskaybaş-Emlek et al. (2021) studied the effects of flour from germinated lentil seeds on properties of cookies. Protein content of raw lentil flour cookies was lower than germinated ones and the fat content of germinated samples also decreased from 1.10% to 0.75%, due to activation of intrinsic enzymes such as lipase. Germinated lentil seed flour cookies had darker color than raw lentil flour cookies, which was due to higher protein and phenolic compounds. For example, total phenolic content increased significantly from 149.5 to 183.3 mg gallic acid equiv. (GAE)/100 mg with germination. The peak viscosity of samples after germination decreased due to swelling of starch (Oskaybaş-Emlek et al. 2021).

The effects of lentil seed quality (premium and frost-damaged) on cookies was investigated by Portman et al. (2020). Wheat flour was replaced with 25%, 50%, 75%, or 100% lentil flour. Increasing substitution level in formulation caused reduction in cookie height and increased hardness and darker color formation. Regardless of the lentil type, total protein content and non-dietary fiber increased significantly compared to 100% wheat cookie. Raffinose family oligosaccharide concentration significantly increased with addition of lentil flour independent of seed quality. Antioxidant activity of cookies increased with corresponding increase in lentil flour addition level. Overall, there was no significant difference between cookies made from premium and frost-damaged lentil.

To develop functional legume-based cookies, navy bean, pinto bean, green lentil, and yellow pea flours were partially or totally substituted for wheat flour, and influence of particle size of flour (coarse and fine) on cookie quality characteristic was investigated. Cookie width was affected from both particle size and type of flour. Cookies made from green lentil had the higher width compared to the cookies with navy bean and pinto bean, and had harder texture than pinto and navy bean samples – both of these effects were independent of the flour's particle size. Protein content of cookies containing green lentil was higher than navy bean and pinto bean (Zucco et al. 2011).

Crackers are another bakery product popular among consumers due to low cost and long shelf-life. The manufacturing of processed foods including crackers needs some improvements according to consumer demands such as addition of bioactive compounds (Morales-Polanco et al. 2017). To improve nutritional quality, germinated lentil flour (GLF) was incorporated into formulation of crackers. Total phenolic content results showed that germination improved this property due to biosynthesis during germination and release of cell-wall-bound phenolics. Furthermore, GLF addition also increased the protein content. However, increasing GLF level resulted in increasing hardness of crackers (Polat et al. 2020). The performance of nine commercially available pulse fraction, including lentil, was evaluated in a model cracker formulation. Although sensory panelists indicated good crispiness of the crackers with chickpea, red lentil, pinto bean, and pea protein isolate, they also stated that other textural characteristics, e.g., hardness and chewiness, should be improved. Crackers with red lentil and pinto bean flour were significantly darker in color than other samples (Han et al. 2010).

It is noteworthy that despite all nutritional benefits of the lentil flour, a study focused on the formation of acrylamide in breads, cookies, and extruded products made with lentil-wheat composite flour. Results indicated that increasing lentil flour amount in formulations led to significantly higher acrylamide formation for all products due to the presence of higher amount of asparagine and reducing sugars such as glucose, fructose, and maltose (Portman et al. 2021).

10.3 Extruded Products

Extrusion, a high-shear and high-temperature and short-time (HTST) process, has been commonly used to produce food products with different shape, size, texture products (e.g., pasta, noodles, snacks, breakfast cereals). The quality of extruded food products depends on feed composition and extrusion system parameters (barrel temperature, screw speed etc.). In general, consumers' preferences for extruded products are high expansion, and crispy and crunchy texture. High amount of starch composed of cereal flours has been used to extrusion due to their superior expansion properties. Although legumes provide better nutritional value, high dietary fiber and protein content could suppress expansion characteristics, which makes it somewhat challenging to produce legume-based extruded foods (Alam et al. 2016).

Ek et al. (2021) extruded three lentil varieties (Brewer, Crimson, and Richlea) and evaluated extruded flour properties and extrusion parameters (barrel temperature (110, 125, and 140 °C) and screw speed (150, 200, and 250 rpm) of a corotating twin screw extruder) for optimum quality. All lentil flours showed the highest expansion ratio at the lowest temperature (110 °C) and the highest screw speed (250 rpm). Richlea variety flour exhibited the highest expansion ratio (3.6) and the highest water solubility index (45.4%). Lazou and Krokida (2011) investigated the effect of extrusion process parameters – feed rate

(2.52–6.84 kg/h) and extrusion temperature (170–230 °C) – feed composition (lentil flour–corn flour blends, at ratios ranging from 10% to 50%), and 13–19% feed moisture content on the glass transition temperature and melting temperatures of extruded snacks. It was found that the glass transition temperature of extrudates increased with lentil-to-corn ratio, feed rate, and extrusion temperature. Moreover, due to plasticizing effect of water, high moisture content depressed the glass transition temperature.

Although cereals are the most common ingredients for extruded products due to puffing effect, they have high glycemic impact due to higher starch content. To enrich the nutritional value of extruded products, Patil et al. (2016) mixed four different legume flours (lentil, chickpea, green pea, and yellow pea) at different levels (5%, 10%, and 15%) with wheat-based extrudates. Legume flour addition increased protein content of extrudates and extrusion process increased the protein digestibility. In a similar study, Lazou and Krokida (2010) mixed lentil flour with corn flour at 10–50% levels and the effects of process conditions, i.e., extrusion temperature (170–230 °C), feed rate (2.52–6.84 kg/h), and feed moisture content (13–19% wb), on the structural and textural properties of extrudates were examined. While extrusion temperature improved the crispness of extrudates, as the temperature increased, diameter and hardness of extrudates decreased. Higher moisture content and feed rate resulted in harder and denser extrudates. Lentil flour addition negatively influenced expansion ratio of extrudates. It was suggested that lentil flour could be used to produce acceptable snacks by optimizing extrusion parameters and feed composition.

In a recent study, Kesre and Masatcioglu (2022) produced extruded snacks from corn flour supplemented with red lentil bran (RLB) and evaluated the effects of exit-die temperature (110, 130 or 150 °C), feed moisture content (19% or 21%), and bran levels (5% or 10%) on the physical and textural properties of extrudates. It was found that 5% RLB-fortified snacks showed higher radial expansion index and crunchiness values than corn flour extrudates (control), showing that the utilization of lentil by-product as fiber source for extruded products with acceptable properties. In another study on corn-legume extrudates, the mechanical properties of corn-legume extrudates were examined at different extrusion parameters. Corn flour was blended with white bean and lentil flour at a ratio of 10 : 90–90 : 10 (corn-legume). The increase in extrusion temperature resulted in decreasing breaking stress and increasing the crispness of extrudates. The use of lentil in mixtures for the production of snacks showed higher breaking stress than the white bean (Lazou et al. 2010). Dogan et al. (2013) also extruded lentil flour–corn starch–corn oil mixtures. The effects of protein, lipid, and moisture contents of the feed mixture, cooking temperature, and screw speed on expansion ratio, cell size distribution, and cutting strength were investigated. Optimum feed composition was found to be 67% lentil flour, 30% corn starch, and 3% corn oil (db).

Morales et al. (2015) investigated the effects of extrusion conditions on some properties (soluble and insoluble fiber, total available carbohydrates, total phenolics, tocopherols, and antioxidant capacity) of different fiber-enriched lentil flour. Lentil flours were blended with specific food ingredients (starch, soluble fiber – Nutriose®, and/or insoluble fiber) from wheat bran, corn, and apple, and flavoring agents. Extrusion resulted in an increase in soluble fiber and total phenolic amount of fiber-enriched flour, and due to increase in phenolic bioavailability, higher antioxidant activity was observed. Legume-based flours formulated with fiber ingredients were suggested for producing gluten-free extruded snack-type products. Similarly, Ciudad-Mulero et al. (2018) formulated lentil flours mixed with bioactive compounds (dietary fiber, arabinoxylans, individual phenolic compounds and tocopherols) and fortified with nutritional yeast. Depending on extrusion process temperature, individual phenolic compounds decreased, while the total dietary fiber and arabinoxylan content significantly increased. Thus, extruded lentil flours fortified with nutritional yeast could be anew snack-type product.

Nasereddine et al. (2017) manufactured precooked rice pasta fortified with legume flours (yellow pea, chickpea, and lentil). Single-screw extruder was used to produce gluten-free spaghetti-type pasta and physical properties, sensory attributes, and texture properties were evaluated. While addition of legume flours to rice pasta decreased hardness, lightness, and the expansion ratio, the adhesiveness, the firmness, and the yellowness increased. All legume-flour-enriched rice pasta showed acceptable scores for overall acceptability and homogenous microstructure. Similarly, Wojtowicz and Moscicki (2014) mixed white bean, split yellow pea, and lentil flours with wheat flour to produce extruded pasta and studied the effect texture, microstructure, and sensory of extrudates. The addition of legume flours decreased the hardness of wheat-flour-based pasta, with the legume-flour-enriched pasta showing a dense and stable structure. Pasta containing up to 30% of legume flour exhibited high nutritional value and acceptable sensory scores.

Rathod and Annapure (2017) produced nutrition-rich extruded noodle with acceptable physical and sensory properties. Lentil flour was mixed with rice flour at different ratios and response surface methodology was used to establish the correlation between process variables (moisture content, die temperature, and screw speed) and physical properties (bulk density, water soluble index, water absorption index, cooking time, cooking gain, cooking loss, and hardness) of noodle. Lentil–rice blend (at 40 : 60) was found to be the most acceptable noodle sample. Wang et al. (2014) isolated starches from pea and lentil using a laboratory wet milling procedure, and those starches were used to prepare noodles by high-temperature extrusion process. Noodles with pea starch showed lower expansion ratio, lower cooking loss, and higher cooked weight than noodles with lentil starch. Cooked noodles from pea and lentil starches exhibited superior texture in comparison to commercial mung bean starch noodles.

10.4 Dairy Products

The demand by consumers toward plant-based milk products has increased over the past years due to health, environmental, and ethical concerns. Almond, soy, coconut, rice, and oat have been commonly used as raw materials to produce commercially plant-based milk. However, some consumers do not consider plant-based milk products not having the same taste, stability, and foaming ability as cow milk when especially added to hot beverages or boiled. Therefore, the main concern of plant-based milk production is to simulate cow milk properties (McClements et al. 2019). Generally, two methods are used to produce plant-based milk: (i) breakdown of plant material to release oil bodies and colloidal materials, and (ii) simulation of fat globules from plants materials (Sethi et al. 2016; Do et al. 2018). Among legumes, lentils can be distinguished with 20–30% protein content, so they could be an excellent raw material for milk substitutes since proteins exhibit valuable emulsifying ability. Stable emulsions could be produced from lentils by applying suitable homogenization technique.

Jeske et al. (2019) produced emulsions for lentil-based milk substitutes, which showed similar properties to that of cow's milk. High-pressure homogenization and heat treatments were used to lentil-based emulsions containing sunflower oil. Sucrose (2.5%) and salt (0.08%) were added to prepare emulsion containing 3.3% protein and 1.5% oil that was similar to cow's milk. The sensory analysis was conducted to compare prepared lentil-based milk substitutes with commercial plant-based milk substitutes (based on soy, almond, oat, rice, and hemp). The sensory rating of lentil-based milk substitute was not different from the commercial plant-based samples. Thus, showing that lentil-based emulsions are a good choice for plant-based milk substitutes as an alternative to dairy products.

Plant-based yogurts (made from soy, almond, and coconut) are an important part of plant-based products and they are produced by the fermentation of plant-based milk. As the demand for plant-based yogurts increases, the alternatives for different raw material are being explored. Since plant-based yogurts have low protein content, lentils are a promising alternative source due to high protein content. Boeck et al. (2021) prepared lentil protein isolate emulsion containing sunflower oil, sucrose, and salt using high-pressure homogenizer. Lentil protein emulsion was then fermented with mixed culture of *Streptococcus thermophilus* and *Lactobacillus bulgaricus* like traditional yogurt process. Since rheological and textural properties are the most significant yogurt criteria, plant-based yogurts must possess similar rheological and textural properties to that of dairy yogurt. In that study, textural and rheological analyses showed that the structure of lentil-based yogurt was found comparable to that of dairy yogurt. Besides lentil-based yogurt, lentil flour could be added to yogurt to improve nutritional values and positive health effects.

Benmeziane et al. (2021) examine the effect of roasted and unroasted lentil flour on dairy-based yogurt quality. Partly skimmed milk and roasted/unroasted lentil flour were mixed and inoculated with thermophilic yogurt culture. Syneresis, i.e., whey separation due to disruption of gel structure's integrity, is a common quality defect during the storage of yogurt. Lentils are rich in starch, which as an absorbing polymer plays a role of stabilizer in yogurt. The incorporation of roasted or unroasted lentil flour reduced the syneresis of dairy yogurt. According to sensory evaluation results, lentil-flour-fortified yogurt showed similar acceptability with the control yogurt, which showed that lentil flour could be added as a functional ingredient for nutritionally fortified yogurt. In another study, Zare et al. (2011) mixed skimmed milk powder and lentil flour (1–3%, w/v), followed by fermentation using a yogurt culture. Although 1% and 2% lentil flour incorporation did not decrease syneresis of yogurt during storage, yogurt supplemented with 3% lentil flour showed the lowest syneresis. Moreover, lentil-flour-supplemented yogurt had similar sensory results as the dairy yogurt. Haq et al. (2019) also studied adding 1–4% (w/v) lentil flour to yogurt. Lentil flour addition improved nutritional profile (i.e., mineral and protein content). Since syneresis of yogurt is an indicator of low quality, water holding capacity (WHC, a crucial parameter) was improved due to higher WHC of lentil flour owing to high protein and starch content in lentil. Moreover, yogurt containing 2% lentil flour exhibited the highest overall sensory acceptability score.

Zare et al. (2012) investigated the effect of adding pulses (pea, chickpea, soy, and lentil) on the acidifying activity of yogurt starter and probiotic cultures. The fermentation efficiency depends on the acidification rate since it directly affects processing time and the process cost. Analysis of acidification levels showed that the most remarkable enhancement in acidification rates was achieved by lentil and soy flour addition. Pulses being rich in carbohydrates, proteins, minerals, and vitamins provide vital nutrients for probiotic and yogurt starter cultures. When lentil flour was added to milk, it improved the growth of *Lactobacilli*. Similarly, Agil et al. (2013) also reported that lentil flour addition was helpful in promoting the growth of probiotic bacteria. Moreover, lentil-fortified yogurt also showed higher antioxidant activity due to higher polyphenolic content in green, red, and dehulled red lentils Thus, lentil addition not only provided usability as a prebiotic but it also enhanced antioxidant potential, making it a valuable source for yogurt fortification.

10.5 Meat Products

The main sources of proteins and amino acid are animal origin but due to health concerns (e.g., high cholesterol) there has been increasing trend to consume vegetable proteins. Therefore, in recent years, meat analogs have gained popularity among consumers. Meat analogs, which are also called as meat substitutes or imitation meat, are protein-based products that are typically produced by extrusion. Generally, soy protein, mushrooms, wheat gluten, egg albumin, and gums have been used to formulate high-moisture meat analogs. Legumes having high protein content are an excellent source for developing meat analogs. A summary of the studies that represent the use of lentil in meat products is presented in Table 10.1. Kim et al. (2021) produced meat analogs by mixing pea protein (55.4% protein), lentil protein (55.4% protein), or faba bean protein (61.5% protein) with other ingredients (pea isolates, and wheat gluten and canola oil) using a twin-screw extruder. With the addition of spices and binders to meat analogs, vegetable hamburger patties were prepared for consumer sensory tests by trained panelists. Vegetable hamburger patties containing soy concentrate and soy isolates were served as a control. Results showed that samples containing pulse proteins showed no differences in consumer liking for the cooked appearance and overall flavor.

Dried meat and meat products are served as protein sources, especially for malnourished people in underdeveloped and developing countries. To improve texture and protein content, legumes could be good ingredients to dried processed meat products. Nayar et al. (2015) prepared hot-air oven-dried goat meat cubes extended with different legume-based binder mixes using Bengal gram, green gram, black gram, and lentil. All legume-fortified goat meat cubes were found to show better shelf stability due to having low water activity (a_w, ~0.6). According to texture profile analysis, 80% minced meat with 10% lentil-based binder mix showed the highest yield, hardness, gumminess, and chewiness parameters.

For older adults, higher protein intake is much more important than younger adults to reduce sarcopenia and other age-related diseases risks. Therefore, to increase protein content in meals, there is a tendency to prepare products extended with legumes having high protein content. Baugreet et al. (2016) studied the effects of pea protein isolate, rice protein, and lentil flour on quality and shelf-life parameters of beef patties. In addition to

Table 10.1 The recent studies about meat products supplemented with lentil ingredients.

Lentil ingredient	Other added legume ingredients for comparison	Type of meat product	Aim of lentil addition	Quality and sensory parameters	Salient findings	References
Lentil protein	Pea protein, faba protein, soy protein	Vegetable hamburger patties	As high moisture meat analog	Cooking yield, hardness, cohesiveness, gumminess, consumer liking, color	• No significant difference between overall liking, cooking yield, texture properties • higher redness	Kim et al. (2021)
Lentil granule	Bengal gram, green gram, black gram granules	Dried goat meat cubes	As protein source	Sensory attributes, water holding capacity, rehydration capacity, texture profile, color parameters, water activity	• No significant difference between water activity and sensory parameters • Black gram granules containing ones showed the highest flavor, texture, juiciness, and overall acceptability scores	Nayar et al. (2015)
Lentil flour	Pea protein isolate, rice protein	Restructured beef steaks	As protein source	Release of free amino acid during *in vitro* gastrointestinal digestion	• Only lentil-enriched restructured beef steaks were found to have significant amounts of released, free amino acids, namely isoleucine, lysine, phenylalanine, and valine after GI digestion	Baugreet et al. (2019)
Lentil flour	Chickpea, pea, and bean	Low-fat burgers	Reduce fat content	Color, texture and sensory parameters	• Lentil flour showed highest water absorption capacity • Up to 44%, low-fat-content burgers were prepared with pulse flours	Argel et al. (2020)
Lentil flour	Corn and chickpea flour	Chicken meatball	As coating material	Oxidation parameters	• 1 : 1 lentil:chickpea flours or lentil suggested for coating to reduce oxidation	Kılınççeker et al. (2015)

requirement high protein amount, meat texture is also an important property, especially for older people with dysphagia or diminished muscular strength. Lentil-flour-extended meat products demonstrated softer structure and good textural properties. Meat products enhanced proteins from plant sources could not only improve textural properties but also be served as high protein meal alternative for older people. Similarly, Baugreet et al. (2017) used mixture design to optimize composition of lentil flour and rice protein fortified beef patties and it was expressed that lentil flour (at 6.05% and 6.22%) and rice protein (at 3.63% and 3.95%) produced acceptable textural parameters.

The most common meat replacement in meat analogues is based on the use of soy proteins; however, some people are allergic to soy. Parniakov et al. (2018) added different protein sources (soy, bean, lentil, broad bean, Spirulina, and Chlorella) to chicken rotti and nutritional, physicochemical, and sensory properties were evaluated. Rotti is a product in which minced chicken is wrapped in chicken breasts and then baked in an oven. Sensory acceptability of broad bean and lentil was found similar to soy protein. Based on chemical composition and sensory evaluation, broad beans and lentil proteins were suggested for making chicken rotti. In a similar study, the addition of vegetable protein sources (beans, peas, and lentils) and microalgae (Chlorella and Spirulina) on physicochemical properties and amino acid profiles of pork sausages prepared was evaluated. Beans, peas, and lentils showed similar properties to the soy-protein-added sausages (control) (Marti-Quijal et al. 2019).

Hydrocolloids, starches, and fibers have been commonly used to reduce fat content of meat products and improve WHC and textural properties. Argel et al. (2020) studied the effect of partial pork meat replacement by different pulse flours (pea, chickpea, lentil, and bean) on physical and sensory characteristics of low-fat burgers made with pork meat, water, sunflower oil, and salt. Prepared burger was an emulsion, thus emulsifying activity and emulsion stability were important parameter for the quality of burgers. WHC was the highest for lentil flour, whereas no significance differences were found for oil holding capacity (OHC) with all studied pulses. Pea, bean, and lentil flour showed similar emulsion activity and stability. Burgers having lentil flour indicated the lowest total color difference compared to the commercial type. Pulse flours were suggested as promising ingredients to produce low-fat burgers. Chaudhary and Tremorin (2020) reformulated beef burger patty with lentil (66% raw ground beef and 26.5% whole cooked lentils). The prepared lentil-reformulated burger was found to have significantly higher dietary fiber, total folate, manganese, and selenium than the 100% beef burger.

Zamuz et al. (2019) prepared pork liver pâté, turkey patties, cooked turkey breast, fresh pork sausages, and marinated turkey breast with soybean (control), and other protein sources, such as lentil, bean, Chlorella, Spirulina, and dairy protein. Physicochemical parameters (pH, color, and hardness) and sensorial characteristics were investigated, and bean and lentil protein were suggested as alternatives for the replacement of soy protein in meat products. As an alternative to low-fat meat products, Serdaroğlu et al. (2005) used legume flours (blackeye bean, chickpea, and lentil) to fortify meatballs. The protein content of meatballs was shown to increase significantly with the addition of legume flours.

Due to their linoleic and linolenic acid content, lentil and chickpea flours are susceptible to oxidation. The lipoxygenase enzyme promotes oxidation of lentil and chickpea flours and causes flavors due to volatile organic compounds formation. Shariati-Ievari et al. (2016)

investigated the effect of infrared micronization at 130 and 150 °C on lipoxygenase activity of lentil and chickpea flour and assessed volatile organic compounds formation. Moreover, pulse flours were added to low-fat beef burgers and physicochemical properties and consumer acceptability of burgers were determined. For both flours, micronization at 130 °C significantly decreased lipoxygenase activity. For chickpea flour, volatile organic compounds formation was lowered with micronization at 130 °C but for lentil flour, micronization at 150 °C was found more effective in decreasing volatile organic compounds. No significant change in fatty acid characteristic of both flour was observed after micronization treatment. Acceptability of micronized flour-fortified low-fat beef burgers were found higher than non-micronized flour-fortified burgers. The low-fat beef burgers containing 6% micronized pulse flours were suggested as healthy option than 100% beef burgers.

Kılınççeker et al. (2015) used yellow lentil and chickpea flours as breading materials for chicken meatballs and evaluated the effects on physical, chemical, and sensory properties of meatballs during frozen storage. Meatballs coated with lentil and chickpea flour and raw and fried meatballs were stored at −13 °C for four months. Corn-flour-coated meatballs were chosen as control. It was observed that lentil and chickpea coating increased moisture retention and decreased oxidation stability of meatballs compared to the corn flour. Sensory attributes of meatballs coated with lentil, chickpea, and corn were found similar. Based on color, moisture content, and oxidation parameters, lentil flour and 1 : 1 lentil–chickpea flour mixture was suggested as the suitable coating for the chicken meatballs.

10.6 Salad Dressing

Salad dressing, essentially an emulsion, is widely popular product to enhance the taste of foods. Egg yolk is commonly used as an emulsifier in production of salad dressing by reducing interfacial tension and preventing aggregated droplets. In addition to showing strong emulsifier property, egg yolk enhances color, mouthfeel, and flavor of salad dressing. However, egg yolk usage has some health concerns due to its high cholesterol. Therefore, researchers have explored alternate resources that could stabilize emulsions (Ma et al. 2016). Boye et al. (2010) reported that legume proteins, due to their good emulsifying, gelling, and fat/water binding properties, could be potential emulsifier for food applications.

Ma et al. (2016) used pulse protein (green lentil, pea, and chickpea at 3–8%, w/w) was used to prepare salad dressing. Salad dressing exhibited non-Newtonian, pseudoplastic fluid behavior with yield stress. Power law, Herschel–Bulkley, and Carreau models are used to describe the flow behavior of salad dressing emulsions. The rheological behavior of emulsions is one of the crucial parameters that affects sensory attributes and shelf life of salad dressings. In addition, firmness as a texture indicator must be examined to produce salad dressings with physical properties similar to commercial types. Therefore, to optimize formulation of pulse-protein-added salad dressing compared with egg-yolk-emulsified ones, response surface methodology (RSM) was used by Ma et al. (2016). The suggested RSM models could be used to develop pulse-protein-emulsified salad dressings. In another study, Ma et al. (2015) prepared salad dressings containing different concentration of egg yolk (3%, 5%, and 7%) and oil (20%, 35%, and 50%) and were supplemented with green lentil, yellow pea, and chickpea (3.5%, 7%, and 10.5%). Rheological and sensory properties of thus prepared salad dressings were analyzed using a three-factor face-centered central composite design. Salad dressings supplemented with whole green lentils (3.5% flour, 3% egg yolk, and 50% oil), yellow pea with low flour content (3.5% flour, 5% egg yolk, 35% oil), and chickpea with high oil content (3.5% flour, 5% egg yolk, 35% oil) showed promising physico-chemical properties and consumer acceptability. In another study, Ma et al. (2013) applied different thermal processing treatments (i.e., ground roasted seeds; roasted flours; pre-cooked, ground, and freeze-dried seeds; pre-cooked, ground, and spray-dried seeds) to raw lentil flour and seeds and they were used to prepare salad dressing emulsions. In general, lentil addition increased the yellowness and the total color intensity of the salad dressings and promoted a thickening effect. Salad dressings supplemented with pre-cooked, freeze-, and spray-dried lentil flours had the best rheological properties.

10.7 Conclusion

Lentil seems an indispensable component of the diet due to high protein including lysine, arginine, carbohydrate, dietary fiber, and essential nutrient content. Moreover, significant number of people are suffering from celiac disease, so there is a tendency to consume gluten-free legumes to balance diets. Lentils have been popular ingredients for gluten free bakery and extrudate products. High amount of protein content is the most attractive reason for the consuming animal-based foods. However, animal-based product diets are implicated in coronary diseases. Legumes contain a considerable amount of protein, so they could be a promising alternative to animal-based products. Diet based on legumes has also shown significant health benefits such as lower coronary heart disease and diabetes (Afshin et al. 2014; Boeck et al. 2021). Among legumes, lentils are gaining increasing popularity among consumer, which offers potential to be used in a range of food products, e.g., bakery, dairy, extruded, and animal-meat alternates. Furthermore, reducing animal-based product consumption can potentially decrease their negative impact on the environment. Such shift toward legume-based foods can reduce greenhouse gas emissions and promote agricultural sustainability.

References

Afshin, A., Micha, R., Khatibzadeh, S., and Mozaffarian, D. (2014). Consumption of nuts and legumes and risk of incident ischemic heart disease, stroke, and diabetes: a systematic review and meta-analysis. *The American Journal of Clinical Nutrition* 100: 278–288.

Agil, R., Gaget, A., Gliwa, J. et al. (2013). Lentils enhance probiotic growth in yogurt and provide added benefit of antioxidant protection. *LWT Food Science and Technology* 50: 45–49.

Alam, M.S., Kaur, J., Khaira, H., and Gupta, K. (2016). Extrusion and extruded products: changes in quality attributes as affected by extrusion process parameters: a review. *Critical Reviews in Food Science and Nutrition* 56: 445–473.

Argel, N.S., Ranalli, N., Califano, A.N., and Andrés, S.C. (2020). Influence of partial pork meat replacement bypulse flour on physicochemical and sensorycharacteristics of low-fat burgers. *Journal of the Science of Food and Agriculture* 100: 3932–3941.

Baugreet, S., Kerry, J.P., Botineştean, C. et al. (2016). Development of novel fortified beef patties with added functional protein ingredients for the elderly. *Meat Science* 122: 40–47.

Baugreet, S., Kerry, J.P., Allen, P., and Hamill, R.M. (2017). Optimisation of protein-fortified beef patties targeted to the needs of older adults: a mixture design approach. *Meat Science* 134 (July): 111–118.

Baugreet, S., Gomez, C., Auty, M.A.E. et al. (2019). in vitro digestion of protein-enriched restructured beef steaks with pea protein isolate, rice protein and lentil flour following sous vide processing. *Innovative Food Science and Emerging Technologies* 54 (April): 152–161.

Benmeziane, F., Raigar, R.K., Ayat, N.E.H. et al. (2021). Lentil (*Lens culinaris*) flour addition to yogurt: impact on physicochemical, microbiological and sensory attributes during refrigeration storage and microstructure changes. *LWT Food Science and Technology* 140: 110793.

Berk, E., Sumnu, G., and Sahin, S. (2017a). Usage of carob bean flour in gluten free cakes. *Chemical Engineering Transactions* 57: 1909–1914.

Berk, E., Şumnu, S.G., and Şahin, S. (2017b). Effects of hydrocolloids and carob bean flour on rheological properties of batter and cake quality. *Gida/the Journal of Food* 42: 754–762.

Boeck, T., Zannini, E., Sahin, A.W. et al. (2021). Nutritional and rheological features of lentil protein isolate for yoghurt-like application. *Food* 10: 1692.

Bojňanská, T., Musilová, J., and Vollmannová, A. (2021). Effects of adding legume flours on the rheological and breadmaking properties of dough. *Food* 10: 1087.

Borsuk, Y., Arntfield, S., Lukow, O.M. et al. (2012). Incorporation of pulse flours of different particle size in relation to pita bread quality. *Journal of the Science of Food and Agriculture* 92: 2055–2061.

Bourré, L., Frohlich, P., Young, G. et al. (2019). Influence of particle size on flour and baking properties of yellow pea, navy bean, and red lentil flours. *Cereal Chemistry* 96: 655–667.

Boye, J.I., Aksay, S., Roufik, S. et al. (2010). Comparison of the functional properties of pea, chickpea and lentil protein concentrates processed using ultrafiltration and isoelectric precipitation techniques. *Food Research International* 43: 537–546.

Cappelli, A. and Cini, E. (2021). Challenges and opportunities in wheat flour, pasta, bread, and bakery product production chains: a systematic review of innovations and improvement strategies to increase sustainability, productivity, and product quality. *Sustainability (Switzerland)* 13: 1–16.

Cappelli, A., Bettaccini, L., and Cini, E. (2020). The kneading process: a systematic review of the effects on dough rheology and resulting bread characteristics, including improvement strategies. *Trends in Food Science and Technology* 104 (April): 91–101.

Chaudhary, A. and Tremorin, D. (2020). Nutritional and environmental sustainability of lentil reformulated beef burger. *Sustainability (Switzerland)* 12: 1–18.

Chávez-Murillo, C.E., Aceves-Flores, M.S., Verástegui-Quevedo, M., and de la Rosa-Millán, J. (2021). Influence of starch and protein molecular interactions on the in vitro digestion characteristics of biscuits partially substituted with pulse flours. *International Journal of Food Science and Technology* 56: 3388–3399.

Ciudad-Mulero, M., Barros, L., Cámara, M. et al. (2018). Function extruded snack-type products developed from. *Food & Function* 9: 819–829.

De Boni, A., Pasqualone, A., Roma, R., and Acciani, C. (2019). Traditions, health and environment as bread purchase drivers: a choice experiment on high-quality artisanal Italian bread. *Journal of Cleaner Production* 221: 249–260.

De la Hera, E., Ruiz-París, E., Oliete, B., and Gómez, M. (2012). Studies of the quality of cakes made with wheat-lentil composite flours. *LWT Food Science and Technology* 49: 48–54.

Dhull, S.B., Kinabo, J., and Uebersax, M.A. (2022). Nutrient profile and effect of processing methods on the composition and functional properties of lentils (*Lens culinaris* Medik): a review. *Legume Science* 2022: e156.

Di Cairano, M., Condelli, N., Caruso, M.C. et al. (2021). Use of underexploited flours for the reduction of glycaemic index of gluten-free biscuits: physicochemical and sensory characterization. *Food and Bioprocess Technology* 14: 1490–1502.

Di Cairano, M., Condelli, N., Cela, N. et al. (2022). Formulation of gluten-free biscuits with reduced glycaemic index: focus on in vitro glucose release, physical and sensory properties. *LWT Food Science and Technology* 154: 112654.

Didinger, C. and Thompson, H.J. (2022). The role of pulses in improving human health: a review. *Legume Science* 4: e147.

Do, D.T., Singh, J., Oey, I., and Singh, H. (2018). Biomimetic plant foods: structural design and functionality. *Trends in Food Science and Technology* 82: 46–59.

Dogan, H., Gueven, A., and Hicsasmaz, Z. (2013). Extrusion cooking of lentil flour (*Lens culinaris*–red)–corn starch–corn oil mixtures. *International Journal of Food Properties* 16: 341–358.

Ek, P., Gu, B.-J., and Ganjyal, G.M. (2021). Whole seed lentil flours from different varieties (Brewer, Crimson, and Richlea) demonstrated significant variations in their expansion characteristics during extrusion. *Journal of Food Science* 86: 942–951.

Gancarz, M., Malaga-Toboła, U., Oniszczuk, A. et al. (2021). Detection and measurement of aroma compounds with the electronic nose and a novel method for MOS sensor signal analysis during the wheat bread making process. *Food and Bioproducts Processing* 127: 90–98.

Gularte, M.A., Gómez, M., and Rosell, C.M. (2012). Impact of legume flours on quality and in vitro digestibility of starch and protein from gluten-free cakes. *Food and Bioprocess Technology* 5: 3142–3150.

Han, J.J., Janz, J.A.M., and Gerlat, M. (2010). Development of gluten-free cracker snacks using pulse flours and fractions. *Food Research International* 43: 627–633.

Haq, F.u., Sameen, A., Zaman, Q.u. et al. (2019). Development and evalutation of yogurt supplemented with lentil flour. *Journal of Microbiology, Biotechnology and Food Sciences* 8: 1005–1009.

Hardacre, A.K., Clark, S.M., Riviere, S. et al. (2006). Some textural, sensory and nutritional properties of expanded snack food wafers made from corn, lentil and other ingredients. *Journal of Texture Studies* 37: 94–111.

Hernandez-Aguilar, C., Dominguez-Pacheco, A., Palma Tenango, M. et al. (2020). Lentil sprouts: a nutraceutical alternative for the elaboration of bread. *Journal of Food Science and Technology* 57: 1817–1829.

Huang, G., Guo, Q., Wang, C. et al. (2016). Fenugreek fibre in bread: effects on dough development and bread quality. *LWT Food Science and Technology* 71: 274–280.

Jarpa-Parra, M., Wong, L., Wismer, W. et al. (2017). Quality characteristics of angel food cake and muffin using lentil protein as egg/milk replacer. *International Journal of Food Science and Technology* 52: 1604–1613.

Jeske, S., Bez, J., Arendt, E.K., and Zannini, E. (2019). Formation, stability, and sensory characteristics of a lentil-based milk substitute as affected by homogenisation and pasteurisation. *European Food Research and Technology* 245: 1519–1531.

Keskin, S.O., Ali, T.M., Ahmed, J. et al. (2022). Physico-chemical and functional properties of legume protein, starch, and dietary fiber: a review. *Legume Science* 2022: e117.

Kesre, C. and Masatcioglu, M.T. (2022). Physical characteristics of corn extrudates supplemented with red lentil bran. *LWT Food Science and Technology* 153: 112530.

Kılınççeker, O., Hepsağ, F., and Kurt, Ş. (2015). The effects of lentil and chickpea flours as the breading materials on some properties of chicken meatballs during frozen storage. *Journal of Food Science and Technology* 52: 580–585.

Kim, T., Miller, R., Laird, H., and Riaz, M.N. (2021). Beef flavor vegetable hamburger patties with high moisture meat analogs (HMMA) with pulse proteins-peas, lentils, and faba beans. *Food Science & Nutrition* 9: 4048–4056.

Lazou, A. and Krokida, M. (2010). Structural and textural characterization of corn – lentil extruded snacks. *Journal of Food Engineering* 100: 392–408.

Lazou, A. and Krokida, M. (2011). Thermal characterisation of corn – lentil extruded snacks. *Food Chemistry* 127: 1625–1633.

Lazou, A., Krokida, M., Karathanos, V., and Marinos-Kouris, D. (2010). Mechanical properties of corn-legume based extrudates. *International Journal of Food Properties* 13: 847–863.

Ma, Z., Boye, J.I., Fortin, J. et al. (2013). Rheological, physical stability, microstructural and sensory properties of salad dressings supplemented with raw and thermally treated lentil flours. *Journal of Food Engineering* 116: 862–872.

Ma, Z., Boye, J.I., Swallow, K. et al. (2015). Techno-functional characterization of saladdressing emulsions supplemented with pea, lentil and chickpea flours. *Society of Chemical Industry* 96: 837–847.

Ma, Z., Boye, J.I., and Simpson, B.K. (2016). Preparation of salad dressing emulsions using lentil, chickpea and pea protein isolates: a response surface methodology study. *Journal of Food Quality* 39: 274–291.

Marston, K., Khouryieh, H., and Aramouni, F. (2016). Effect of heat treatment of sorghum flour on the functional properties of gluten-free bread and cake. *LWT Food Science and Technology* 65: 637–644.

Marti-Quijal, F.J., Zamuz, S., Tomašević, I. et al. (2019). Influence of different sources of vegetable, whey and microalgae proteins on the physicochemical properties and amino acid profile of fresh pork sausages. *LWT Food Science and Technology* 110: 316–323.

McClements, D.J., Newman, E., and McClements, I.F. (2019). Plant-based milks: a review of the science underpinning their design, fabrication, and performance. *Comprehensive Reviews in Food Science and Food Safety* 18: 2047–2067.

Mollakhalili-Meybodi, N., Khorshidian, N., Nematollahi, A., and Arab, M. (2021). Acrylamide in bread: a review on formation, health risk assessment, and determination by analytical techniques. *Environmental Science and Pollution Research* 28: 15627–15645.

Morales, P., Cebadera-miranda, L., Cámara, R.M. et al. (2015). Lentil flour formulations to develop new snack-type products by extrusion processing: phytochemicals and antioxidant capacity. *Journal of Functional Foods* 19: 537–544.

Morales-Polanco, E., Campos-Vega, R., Gaytán-Martínez, M. et al. (2017). Functional and textural properties of a dehulled oat (*Avena sativa* L) and pea (*Pisum sativum*) protein isolate cracker. *LWT Food Science and Technology* 86: 418–423.

Nasabi, M., Naderi, B., Akbari, M. et al. (2021). Physical, structural and sensory properties of wafer batter and wafer sheets influenced by various sources of grains. *LWT Food Science and Technology* 149: 111826.

Nasereddine, M., Bouasla, A., and Agnieszka, W. (2017). Gluten-free precooked rice pasta enriched with legumes fl ours: physical properties, texture, sensory attributes and microstructure. *LWT Food Science and Technology* 75: 569–577.

Nayar, R., Mendiratta, S.K., Prabhakaran, P.P. et al. (2015). Development and comparison of shelf stable extended and dehydrated goat meat cubes containing four different legume based binder mixes. *Journal of Food Science and Technology* 52: 4515–4522.

Oskaybaş-Emlek, B., Özbey, A., and Kahraman, K. (2021). Effects of germination on the physicochemical and nutritional characteristics of lentil and its utilization potential in cookie-making. *Journal of Food Measurement and Characterization* 15: 4245–4255.

Ozkahraman, B.C., Sumnu, G., and Sahin, S. (2016). Effect of different flours on quality of legume cakes to be baked in microwave-infrared combination oven and conventional oven. *Journal of Food Science and Technology* 53: 1567–1575.

Paladugula, M.P., Smith, B., Morris, C.F., and Kiszonas, A. (2021). Incorporation of yellow pea flour into white pan bread. *Cereal Chemistry* 98: 1020–1026.

Parniakov, O., Toepfl, S., Barba, F.J. et al. (2018). Impact of the soy protein replacement by legumes and algae based proteins on the quality of chicken rotti. *Journal of Food Science and Technology* 55: 2552–2559.

Patil, S.S., Brennan, M.A., Mason, S.L., and Brennan, C.S. (2016). The effects of fortification of legumes and extrusion on the protein digestibility of wheat based snack. *Food* 5: 26.

Perri, G., Coda, R., Rizzello, C.G. et al. (2021). Sourdough fermentation of whole and sprouted lentil flours: *in situ* formation of dextran and effects on the nutritional, texture and sensory characteristics of white bread. *Food Chemistry* 355: 129638.

Polat, H., Dursun Capar, T., Inanir, C. et al. (2020). Formulation of functional crackers enriched with germinated lentil extract: a response surface methodology box-behnken design. *LWT Food Science and Technology* 123: 109065.

Portman, D., Blanchard, C., Maharjan, P. et al. (2018). Blending studies using wheat and lentil cotyledon flour: effects on rheology and bread quality. *Cereal Chemistry* 95: 849–860.

Portman, D., Maharjan, P., McDonald, L. et al. (2020). Nutritional and functional properties of cookies made using down-graded lentil: a candidate for novel food production and crop utilization. *Cereal Chemistry* 97: 95–103.

Portman, D., Maharjan, P., Blanchard, C. et al. (2021). Impact of thermal processing on levels of acrylamide in a wheat-lentil flour matrix. *Legume Science* 3: 1–8.

Previtali, M.A., Mastromatteo, M., De Vita, P. et al. (2014). Effect of the lentil flour and hydrocolloids on baking characteristics of wholemeal durum wheat bread. *International Journal of Food Science and Technology* 49: 2382–2390.

Rathod, R.P. and Annapure, U.S. (2017). Physicochemical properties, protein and starch digestibility of lentil based noodle prepared by using extrusion processing. *LWT Food Science and Technology* 80: 121–130.

Razavi, S.N., Hojjatoleslamy, M., Molavi, H., and Boroujeni, L.S. (2020). The effect of germinated lentil flour on the physicochemical and organoleptic characteristics of Sangak bread. *Journal of Culinary Science and Technology* 20: 253–265.

Sadowska, J., Fornal, J., Vidal-Valverde, C., and Frias, J. (1999). Natural fermentation of lentils. Functional properties and potential in breadmaking of fermented lentil flour. *Nahrung – Food* 43: 396–401.

Serdaroğlu, M., Yildiz-Turp, G., and Abrodímov, K. (2005). Quality of low-fat meatballs containing legume flours as extenders. *Meat Science* 70: 99–105.

Sethi, S., Tyagi, S.K., and Anurag, R.K. (2016). Plant-based milk alternatives an emerging segment of functional beverages: a review. *Journal of Food Science and Technology* 53: 3408–3423.

Shariati-Ievari, S., Ryland, D., Edel, A. et al. (2016). Sensory and physicochemical studies of thermally micronized chickpea (*Cicer arietinum*) and green lentil (*Lens culinaris*) flours as binders in low-fat beef burgers. *Journal of Food Science* 81: S1230–S1242.

Substitutes, F.P.M. (2020). Development of next-generation nutritionally fortified plant-based milk substitutes: structural design principles. *Food* 9: 421.

Suo, B., Nie, W., Wang, Y. et al. (2020). Microbial diversity of fermented dough and volatile compounds in steamed bread prepared with traditional Chinese starters. *LWT Food Science and Technology* 126: 109350.

Tufan, B., Sahin, S., and Sumnu, G. (2020). Utilization of legume flours in wafer sheets. *Legume Science* 2: 1–9.

Turfani, V., Narducci, V., Durazzo, A. et al. (2017). Technological, nutritional and functional properties of wheat bread enriched with lentil or carob flours. *LWT Food Science and Technology* 78: 361–366.

Ureta, M.M., Diascorn, Y., Cambert, M. et al. (2019). Water transport during bread baking: impact of the baking temperature and the baking time. *Food Science and Technology International* 25: 187–197.

Wang, N., Warkentin, T.D., Vandenberg, B., and Bing, D.J. (2014). Physicochemical properties of starches from various pea and lentil varieties, and characteristics of their noodles prepared by high temperature extrusion. *Food Research International* 55: 119–127.

Wojtowicz, A. and Moscicki, L. (2014). Influence of legume type and addition level on quality characteristics, texture and microstructure of enriched precooked pasta. *LWT Food Science and Technology* 59: 1175–1185.

Yildiz, E., Şumnu, S.G., and Şahin, S. (2018a). Effects of buckwheat flour, gums and proteins on rheological properties of gluten-free batters and structure of cakes. *Quality Assurance & Safety of Crops and Food* 10: 245–254.

Yildiz, E., Guner, S., Sumnu, G. et al. (2018b). Monitoring the effects of ingredients and baking methods on quality of gluten-free cakes by time-domain (TD) NMR relaxometry. *Food and Bioprocess Technology* 11: 1923–1933.

Zamuz, S., Purriños, L., Galvez, F. et al. (2019). Influence of the addition of different origin sources of protein on meat products sensory acceptance. *Journal of Food Processing and Preservation* 43 (5): 1–12.

Zare, F., Boye, J.I., Orsat, V. et al. (2011). Microbial, physical and sensory properties of yogurt supplemented with lentil flour. *Food Research International* 44 (8): 2482–2488.

Zare, F., Champagne, C.P., Simpson, B.K. et al. (2012). Effect of the addition of pulse ingredients to milk on acid production by probiotic and yoghurt starter cultures. *LWT Food Science and Technology* 45: 155–160.

Zucco, F., Borsuk, Y., and Arntfield, S.D. (2011). Physical and nutritional evaluation of wheat cookies supplemented with pulse flours of different particle sizes. *LWT Food Science and Technology* 44: 2070–2076.

11

Nonfood Applications of Lentils and Their Processing By-products

Harshani Nadeeshani[1,2], *Thilini Dissanayake*[1,2], *and Nandika Bandara*[1,2]

[1] *Department of Food and Human Nutritional Sciences, Faculty of Agricultural and Food Sciences, University of Manitoba, Winnipeg, MB, Canada*
[2] *Richardson Centre for Food Technology and Research (RCFTR), University of Manitoba, Winnipeg, MB, Canada*

11.1 Introduction

Lentil (*Lens culinaris* Medik.) is one of the most important and the oldest food legume crops, having been domesticated since ancient times alongside other pulses such as fava beans, peas, and chickpeas. Although lentils are acclimated to semi-arid environments, they are generally cultivated, providing abundant production on all continents except Antarctica (Muehlbauer and McPhee 2005). Lentil is grown in nearly 52 counties, the majority of which are Canada, India, United States, Turkey, Australia, Kazakhstan, Nepal, Russian Federation, Bangladesh, China, and Ethiopia. These 11 countries contribute more than 93% of the global production of lentils; however, more than three-quarters of the total supply comes from Canada, Turkey, United States, Australia, and India (Joshi et al. 2017; Choukri et al. 2020). The global production of lentils was reported to be 6.5 million tonnes grown on 5 million hectares in 2022, with a productivity of 1300 kg/ha. Figure 11.1 illustrates the total global lentil production and the harvested area between 1994 and 2020 (FAOSTAT 2022).

Lentil is a nutrient-dense food legume crop with a relatively high protein content (20.6–31.4 g/100 g) (Jarpa-Parra et al. 2014). In addition, lentils comprise a variety of important dietary components such as carbohydrates, fiber, essential minerals, vitamins, and antioxidants (Table 11.1). Among the minerals, Mg, P, K, Ca, Fe, and S are present in significant quantities in lentils (Faris et al. 2013). It has also been reported that lentils are a source of many vitamins including folate, thiamin, riboflavin, niacin, pantothenic acid, pyridoxine, vitamin E, and vitamin K (USDA 2022). However, the nutritional quality of lentils is limited because of the presence of high antinutritional factors, which result in low digestibility of protein and starch and poor mineral bioavailability (Hefnawy 2011; Bessada et al. 2019).

Lentils: Production, Processing Technologies, Products, and Nutritional Profile, First Edition.
Edited by Jasim Ahmed, Muhammad Siddiq, and Mark A. Uebersax.

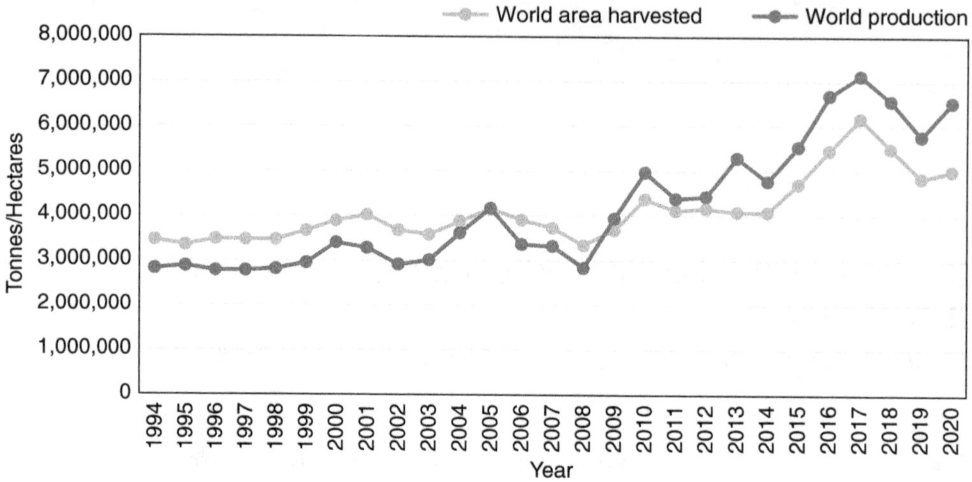

Figure 11.1 Production (hectares) and harvested area (tonnes) of lentils in the world. *Source:* FAOSTAT (Food and Agriculture Organization Corporate Statistical Database) (2022).

Table 11.1 Major nutritional profile of raw lentils (g/100 g dry weight basis).

Nutritional component	Content[a]	Content[b]
Total protein	26.6 ± 0.50	25.63 ± 0.28
Ash	3.4 ± 0.04	2.77 ± 0.16
Fat	1.0 ± 0.08	2.20 ± 0.13
Crude fiber	6.3 ± 0.13	21.70 ± 0.92
Moisture	8.51 ± 0.31	10.70 ± 0.40
Carbohydrate	47.23 ± 0.06	48.70 ± 0.67

[a] Hefnawy (2011).
[b] Fouad and Rehab (2015).

There are five main varieties of lentils, namely red, black, brown, green, and French green lentils. Among them, red and green are the most common varieties. Furthermore, red lentils are mostly cultivated globally with commercial success (Hajas et al. 2022). For postharvest handling, lentils are threshed, dried, and decupled before consumption. During the secondary processing, whole grains are de-hulled using large carborundum emery rollers and split using burr grinders. A peeler assembled with a brusher removes fine bran from the lentil grains in the peeling section. A grader segregates broken split lentils and grinds them to lentil flour (Chelladurai and Erkinbaev 2020). During these unit operations, several by-products are piled up, such as surplus lentils, culled lentils, hull, bran, and lentil screenings, such as whole grains, broken grains, powder, chaff, and dust. These by-products have significant nutritional and functional properties, yet they are discarded or used for animal feed (Roasa et al. 2021).

Lentils have been consumed as the primary plant protein source in many parts of the world due to their low-cost, higher protein content, and complex carbohydrate content (Joshi et al. 2017). The legume has been used in the preparation of different formulations such as traditional curries, dishes, soups, infant formulas, meat analogs, nondairy milk alternatives, bean curd, bakery products, and extruded products (Nadeeshani et al. 2022). In addition, there is an increasing trend for incorporating pulses such as peas, lentils, and beans into diverse food applications due to their excellent nutritional and functional properties (Khazaei et al. 2019).

The nonfood applications of these pulses are not so much because of the availability of the desired materials for industrial applications are readily available (Nadeeshani et al. 2022). In recent times, many emerging potential areas of nonfood applications of lentil grains and by-products have been identified, and some of the examples are industrial starch, biodegradable packaging, protein supplementations, microencapsulation carrier matrices, adhesives, additive ingredients, and cosmetic applications (Figure 11.2) (Aydogdu et al. 2018; Chang and Nickerson 2018). Several studies on these nonfood applications of lentils have been reported; however, in-depth studies are required to improve their industrial acceptability with desired properties by minimizing undesired characteristics. In addition, further research studies are crucial to investigate the possibility of adopting these nonfood applications into large-scale industrial production and the safety of long-term usage by humans. Therefore, this chapter intends to present and critically evaluate the current and potential nonfood applications of lentils with the interest of stimulating further research to optimize the utilization of lentils.

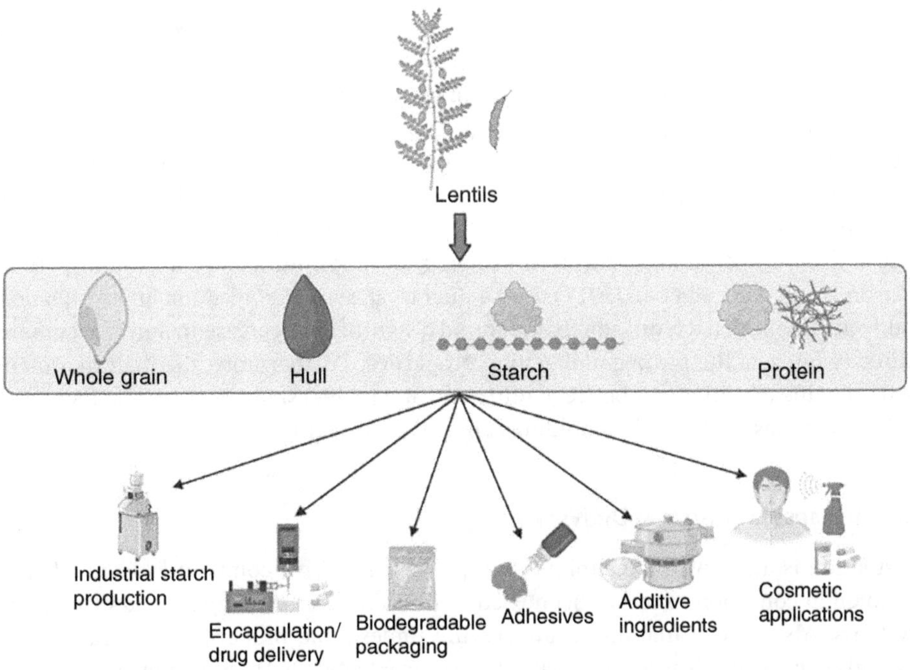

Figure 11.2 Nonfood applications of lentils.

11.2 Nonfood Applications of Lentils and Their Processing By-products

11.2.1 Industrial Starch Production

Lentil is one of the main pulses that acts as a protein source in the food industry. As a co-product, a considerable amount of starch is produced after the protein extraction. The obtained starch has been fractionated by a dry or wet method and used for various purposes (Ren et al. 2021). In general, pulses are consumed as whole seeds, but recent research is moving toward fractionation and using them as ingredients in several food products in the form of lentil proteins, fibers, and starch. Dehulling, soaking, and milling (in a wet state) are the initial steps of wet fractionation. Furthermore, the suspension consists of protein and starch, where protein is extracted by isoelectric precipitation, acid extraction, salt extraction, water extraction, or ultrafiltration. The remaining starch in the suspension is purified by removing protein residues by further washing and removing of fiber residues by filtration (~70 μm). The dry method involves air classification, followed by separation based on the density and particle size differences. The protein fraction becomes more refined and lighter, while starch fraction becomes coarser and denser. The dry fractionation, nonetheless, produces starch with lower purity with extensive starch damage due to intensive milling compared to wet fractionation. Therefore, dry fractionated starch may exhibit poor nutritional and functional properties. In terms of cost, the dry method is inexpensive as it uses less water and lowers operational costs (Boye et al. 2010; Li et al. 2019; Sosulski and Sosulski 1986).

Lentil starch-based ingredients are used as a thickening agent, stabilizer, gelation, texture improver, and ingredient for enhancing water-holding capacity in several food products in the industry. However, the native form of starch has several limitations, such as high viscosity during high-temperature gelatinization, high gel rigidity, swelling and shear thinning resistance, unstable paste, a higher proportion of resistant starch, and fast retrogradation (Singh et al. 2004). Modification of starch can overcome these limitations. Mostly chemical modifications have been practiced for starch; however, food safety remains a key issue for food-related applications. Conversely, physical modifications are more relevant because of the low operational cost, fast processing, and environmental compatibility (Bhat and Karim 2009). Majeed et al. (2017) used a dual treatment of sonication and irradiation to modify alkali-extracted lentil starch and found a significant decrease in amylose content that directly impacts the pasting and gelling properties. Furthermore, the dual treatment showed a significant difference between untreated and treated samples regarding the swelling index, syneresis, water and oil holding capacity, and pasting properties.

11.2.2 Encapsulation/Drug Delivery

Encapsulation is a promising technique that protects sensitive compounds from adverse environmental conditions and has been used in nutrients, nutraceuticals, and drug delivery systems. Also, encapsulation minimizes the undesirable properties of encapsulated compounds such as unpleasant flavor and taste (Afzaal et al. 2021). Proteins have become a promising encapsulation agent due to their biocompatibility, non-toxicity, and high availability. Moreover, functional properties of proteins such as gelation, emulsification, water and oil binding, surface activity, and foaming properties make them excellent encapsulating agents (Dissanayake et al. 2022b). Lentil proteins are ideal materials for this purpose as they are a generally inexpensive source of proteins. The properties of the end products are directly affected by the encapsulating agent/coating material and the encapsulating technique. Legume proteins can be modified by physical, chemical, and enzymatic methods to change their structure to have desirable properties for encapsulation (Afzaal et al. 2021). Lentil shows a high nutritional profile and digestibility rate, increasing its health value in nonfood applications. In addition, lentil proteins possess several excellent properties, such as reducing dynamic interfacial tension and surface hydrophobicity, emulsion stability, and high solubility through several modifications (Yang et al. 2021). Figure 11.3 illustrates the desirable properties of lentils that make them an excellent encapsulating agent.

Encapsulation of ω-fatty acids in legumes could retain desirable properties by masking undesirable odors, protecting from adverse environmental conditions, and providing a platform for easy handling (Chang et al. 2016). Microencapsulation of flaxseed oil in chickpea, and lentil protein isolates with maltodextrin has been reported (Karaca et al. 2013), and those authors assessed the effect of protein type, flaxseed oil concentration (5.3–21%), maltodextrin dextrose equivalent (DE 9 and 18), and concentration (25–40.7%) on the structure and physicochemical properties of the microcapsule. The optimized conditions were maltodextrin-DE 9 (35.5%), 10.5% flaxseed oil, and lentil or chickpea protein, which resulted in an encapsulation efficiency of about 83%, release properties in simulated

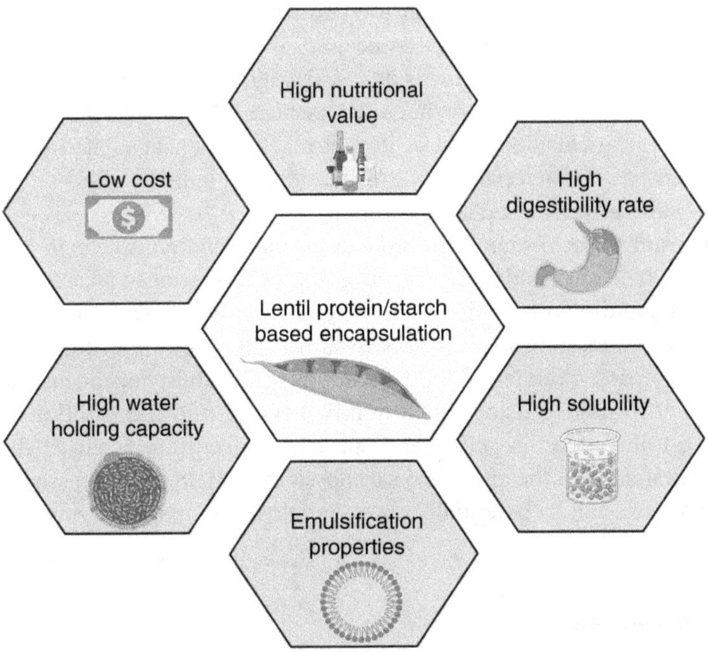

Figure 11.3 Potential of lentils as an encapsulating agent with desirable properties. *Source:* Created with BioRender.com.

gastrointestinal conditions higher than 84%, and surface flaxseed oil content remained below 3%. Furthermore, chickpea and lentil pea protein-based encapsulating systems showed oxidative stability of flaxseed oil over 25 days at room temperature under nitrogen atmosphere.

Chang et al. (2016) used lentil protein isolate and maltodextrin as wall materials to entrap canola oil by spray-drying. They used lecithin and/or sodium alginate as the emulsifier to improve the drying process. The best encapsulation efficiency was achieved (~88%) when 20% oil, 2% lentil protein isolate, and 18% maltodextrin were used. The addition of alginate improved oxidative stability and wall structure in the protein–maltodextrin complex (Chang et al. 2016). Similarly, a range of studies have been conducted on the encapsulation of edible oils in lentil proteins (Avramenko et al. 2016; Gumus et al. 2017a,b; Hadad and Goli 2019).

Since the direct addition of magnesium (Mg) to food products is challenging due to chemical degradation and protein aggregation (DiNicolantonio et al. 2018; Ismail et al. 2018), encapsulation is the appropriate vehicle for the supplementation of Mg. Kabakci et al. (2021) developed a double emulsion system to encapsulate Mg in a lentil flour matrix. Lentil flour acted as the hydrophilic emulsifier at selected concentrations (15%, 20%, 25%, and 30%), and Mg was encapsulated in the inner aqueous phase at a concentration of 1.25 g $MgCl_2$/100 g primary emulsion. The oil phase consisted of poly(glycerol polyricinoleate) (PGPR) and lecithin. Increasing the concentration of lentil flour from 15% to 30% resulted in an increase of 8.4% in the stability of the double emulsion. Additionally, the prepared double emulsion was incorporated into a cake, and there was no significant effect of double emulsion incorporation on the cake quality or taste. Double emulsion resulted in a higher baking stability (79–82%) than the control. *In vitro* bio-accessibility tests demonstrated a similar digestibility for uncoated and encapsulated Mg (Kabakci et al. 2021). Rutin (quercetin-3-rhamnosyl glucoside) is flavanol-rich compound with anticancer, anti-inflammatory, and anti-hemorrhagic properties. The bioavailability of rutin has been improved in an inclusion complex. Ren et al. (2021) encapsulated rutin in debranched lentil starch-based microparticles and debranched lentil starch, which improved targeted delivery and physiological enhancement.

Wang et al. (2015) encapsulated acid-sensitive *Bifidobacterium adolescentis* cells in legume proteins. Mekala et al. (2022) encapsulated a vitamin complex consisting of β-carotene, cholecalciferol, α-tocopherol, and ascorbic acid in a lentil emulsion-filled pectin hydrogel system. High-intensity ultrasound sonication (HIUS) was used as the homogenization technique. Viscoelastic properties, chemical stability, and physical stability of the encapsulated system were further evaluated using HIUS nominal power and the addition of xylooligosaccharides (XOS) and ascorbic acid. Lentil protein concentrate acted as the emulsifier in the delivery system, and the authors have shown that HIUS process-assisted emulsion-filled pectin hydrogels can be potentially used for delivering lipophilic vitamins (Mekala et al. 2022).

11.2.3 Biodegradable Packaging

Despite the excellent properties such as high mechanical, thermal, and barrier properties of fossil fuel-derived packaging, novel research studies are focusing on biodegradable packaging materials due to the negative environmental impact and non-sustainability of synthetic polymer (Dissanayake et al. 2022a; Hadidi et al. 2022). Several studies have used its starch, protein, and fiber fractions to develop packaging (Ochoa-Yepes et al. 2018). Figure 11.4 is a general illustration to show the fabrication process of packaging materials using different fractions of lentils.

Aydogdu et al. (2019) developed an active packaging material using gallic acid encapsulated in lentil flour. The authors have used alkaline conditions to produce lentil flour-based nanofibers where the nanofiber formation was mainly influenced by the viscosity of the solution. The solubility of the lentil protein in flour was adjusted by changing the pH. At the isoelectric point of 4.5 of lentil proteins, it showed the minimum solubility and solubility increases when it is taken away from the isoelectric pH because of the unfolding protein structure. The adjusted pH directly affects the rheological properties and the nanofiber formation. On the other hand, the electrical conductivity of the solution is another crucial parameter that determines the nanofiber formation. Sufficient conductivity should be there to maintain an elongated jet of the solution by electrospinning and is mainly affected by the polymer, solvent type, and amount of ionizable salts. According to the authors, lentil flour proved its potential in active packaging with encapsulated gallic acid (Aydogdu et al. 2019).

Lentil proteins extracted by isoelectric precipitation were used by Bamdad et al. (2006) for the fabrication of packaging films. Fabricated films were semi-transparent and red-brown with a thickness of 0.155 ± 0.04 mm (155 ± 40 μm). The red color is imparted by the pigments leached from the hull and alkaline pH during protein extraction that can extract more pigments. The tensile strength of the films was reported at 4.24 ± 1.26 MPa with 58% elongation at break. The high temperature and alkaline pH caused the protein

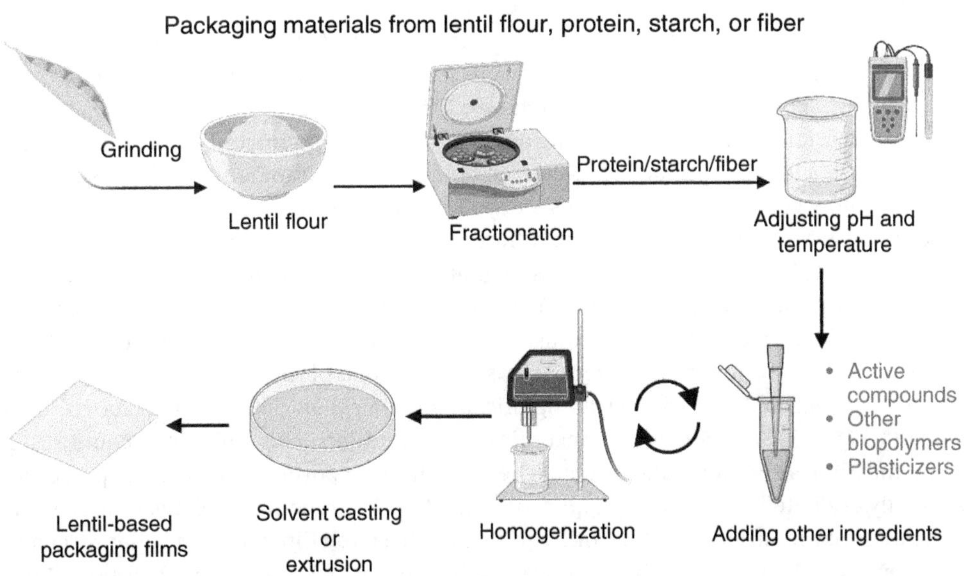

Figure 11.4 Fabrication process of lentil-based packaging materials. *Source:* Created with BioRender.com.

denaturation by forming the covalent crosslinking and, thereby, increasing the tensile strength. The water vapor permeability for the films was reported as $0.3095 \pm 0.002 \times 10^{-9}$ g/m/m^2/s/Pa. The fraction of water-soluble proteins and the presence of hydrophobic amino acids contribute to the permeability of moisture through the films. Compared to the other films (e.g., corn zein) with a higher proportion of hydrophobic amino acids such as leucine, proline, and alanine, lentil protein films show poor moisture barrier properties (Bamdad et al. 2006).

Ochoa-Yepes et al. (2019) produced the starch-lentil protein composite films using extrusion/thermo-compression and solution cast methods. Lentil proteins were added at selected concentrations (0%, 0.75%, and 1.5% wt.) to produce the film, and it has been observed that the protein content influenced the film preparation process. The addition of proteins resulted in the replacement of natural intermolecular H bonds with strong protein crosslinking, ultimately improving water resistance. This is an alternative method to overcome the high moisture sensitivity of starch-based films. Furthermore, the enhanced interactions between protein and polymeric backbone improved the elastic modulus and stress at break properties. On the other hand, the extrusion/thermo-compression method reduced the film's moisture content and water vapor permeability compared to solvent-cast films. The solvent casting method uses a higher amount of water; therefore, there can be more interactions between water molecules and OH groups of the starch that may not be evaporated during drying. Moreover, the extrusion/thermocompression method used higher temperatures than the boiling point of water, whereas the highest temperature used in the casting method was only 80 °C. Overall, blending with lentil proteins and the extrusion/thermo-compression method are potential techniques to overcome the drawbacks of starch-based films.

Compared to potato-based starch, the film properties of the starch-based packaging materials affected by botanical origin were studied by Pająk et al. (2019). Starch extracted from pumpkin, quinoa, and lentil seeds was used to fabricate films, and their mechanical, thermal, and physicochemical properties were compared with potato-starch films. Lentils with a higher starch composition (~67%) have a limited number of research studies to evaluate their feasibility in use in packaging applications. Transparent and clear films are generally accepted for food packaging. However, both lentil and quinoa starch films showed a slightly yellow color. Moreover, lentils showed the highest thickness values compared to other films, which could be due to the differences in amylose and amylopectin molecular orientation. Red-type lentil starch films exhibited the lowest solubility (~11%), which is desirable for packaging applications. Water vapor permeability is another critical parameter that estimates the moisture transfer between the outer environment and food. However, the value reported for lentils was in the range of $1.39 \times 10^{-10} \pm 1.75 \times 10^{-12}$ to $1.58 \times 10^{-10} \pm 4.20 \times 10^{-12}$, which was significantly higher than that of potato starch film's value ($1.27 \times 10^{-10} \pm 5.06 \times 10^{-12}$). According to the authors, this difference could result from the film structure's alterations. The two most important mechanical properties generally evaluated in packaging applications are tensile strength and elongation at break. Regardless of the variety, potato, lentils, and pumpkin starch films showed better mechanical properties. Lentil starch films showed values from 11.31 ± 0.24 to 12.16 ± 0.86 MPa, whereas that was 12.58 ± 1.31 for potato starch films. However, sufficient elongation at break percentage could not be observed for all the films where it was only around 3–5% for all the films. Overall, lentil starch showed satisfactory film properties, but further improvements are required for barrier properties and elongation of the films (Pająk et al. 2019).

11.2.4 Adhesives

Adhesives can be defined as materials (e.g., paste, glue, and cement) that can bind different or the same surfaces together and resist separation when applied to the surfaces. Commercial adhesives contain carcinogenic, nonrenewable petrochemicals; therefore, various protein and starch sources have gained research interest, including protein and starch from legumes (Allen et al. 2010; Cheng et al. 2016). A higher exposure among hydrophobic groups to evade water intrusion, which results from the denaturation and unfolding of protein molecules, leads to water resistance and a performance improvement in protein-based adhesives. It is also facilitated by the reaction of active groups present on protein molecules with the crosslinking agents to construct crosslinking structures. In addition, the enhanced exposure of hydrophilic groups resulting from unfolding the protein structure further enables the formulation of the dense crosslinking structure between the protein and the crosslinking agent to speed up the adhesive properties (Yue et al. 2019).

Native starch is brittle; and it is challenging to convert into bioadhesives because of its large number of hydrophilic hydroxyl groups. The structural and functional properties of native starch are transformed by chemical, physical, or biological modifications to alter its solubility and viscosity, and other relevant properties for preparing starch-based adhesives (Yu et al. 2015). Because of deficiencies in mechanical properties, water resistance, and storage stability, starch and protein-based adhesives are widely used as sustainable solutions after improving their quality and performances with various modifiers (Kumar et al. 2022).

Isolated pulse starch and protein have been added into different formulations of adhesives used for various products, including lignocellulose panels and product labels (Ren et al. 2021). Several researchers reported that the adhesiveness of gels prepared from extruded lentil flour was positively correlated with the starch content and negatively correlated with the protein content (Joshi et al. 2014; Espinosa-Ramírez et al. 2021). Sein (2008) developed a cost-effective and good-quality lentil-flour-based wood adhesive with a formulation of 15.15% of defatted lentil flour, 25.5% of formaldehyde, and 59.4% of phenol, targeting the plywood industry, particleboard making, and parquet flooring. Various particleboards were produced using the developed adhesive, and they observed that the tensile strength and the water resistance values of the particleboards were comparable to those obtained using the standard phenol formaldehyde adhesive. Furthermore, according to Espinosa-Ramírez et al. (2021), the adhesiveness of cold gels prepared from extruded whole grain rice flour was higher than that of cold gels prepared from extruded whole grain lentil flour. Overall, published literature on using lentils and their by-products for the production of bioadhesives are quite scanty; thus, it is an area of research that can be paid more attention.

11.2.5 Additive Ingredients

The food industry progressively tends to replace animal-based food additives, and ingredients derived from milk, egg, or muscle as well as synthetic additives with plant-based ones (Kim et al. 2020). Because of the environmental and health concerns of consumers, plant-based additive ingredients have created a growing interest as sustainable, safe, and natural, which arose from the concept of clean-label food products (Dai et al. 2020).

Protein and starch extracted from pulses have excellent techno-functional properties, making them suitable for use as additive ingredients such as emulsifiers and stabilizers. Emulsions are thermodynamically unstable dispersions of two or more immiscible phases which can separate into discrete aqueous and oil phases owing to a huge magnitude of surface energy between the phases (Jiang et al. 2020). Food emulsions are stabilized by polymers such as protein and starch that reduce steric hindrance, electric repulsion, and interfacial tension, consequently avoiding droplet coalescence via creating a microscopic encapsulating film surrounding the liquid droplets (Shi et al. 2020). Lentil proteins exhibit comparable functional properties, including emulsifying and emulsion stabilizing properties, to soy and pea proteins (Aydemir and Yemenicioğlu 2013). Mekala et al. (2022) reported that producing stable colloidal systems is desired by the food and beverage industry, while lentil protein concentrate is more promising for stabilizing colloidal systems than whey protein concentrate. Foams are common food systems that rely on air entrapment, formed by protein films. Though studies on legumes as the foaming agents are still lacking, it has been suggested that legume flours can produce more foam than cereal flours, which is attributed to a higher content of protein and soluble globulin- and albumin-type proteins allowing greater diffusion to the interface (Stone et al. 2019).

The emulsifying properties of lentil protein reported by Funke et al. (2022) revealed that the dry-fractionated lentil protein could emulsify and stabilize 10 wt% oil-in-water emulsions, while the conventionally solvent-extracted lentil protein could not form a stable emulsion. The authors summarized the development of sustainable additive ingredients with unique compositions from lentils. The lentil proteins act as potential emulsifiers in acidic food formulations, especially with 1.5% added protein and less than 50 mM of NaCl (Ettoumi et al. 2016). In addition, the development of lentil protein fibrillar aggregates provided a new strategy to form stable emulsions without synthetic surfactants, which allowed a wide range of applications in the food industry (Wynnychuk et al. 2021).

Lentil protein isolate exhibited high foaming and emulsifying capacity in a broad pH range (Jarpa-Parra et al. 2015; Lee et al. 2021; Osemwota et al. 2021). The observed foaming capacity was higher than pea and soy proteins at the same pH ranges (Jarpa-Parra et al. 2015). Improvement of the shelf-life of lentil protein foams has been studied at acidic and mild pH in the presence of pectin, xanthan gum, and guar gum, and it was observed the incorporation of polysaccharides did not impact the foaming capacity (Jarpa-Parra et al. 2016). Furthermore, the lentil protein fractions (e.g., albumin, globulin, or glutelin fractions) and pH of the environment exhibited stronger influences on foaming and emulsification properties than the protein concentration (Osemwota et al. 2022). The foaming capacity was superior in globulin, whereas the emulsifying capacity was higher in glutelin fraction.

Starches extracted from plant sources are modified and utilized as food additives, thickeners, and emulsifiers, particularly in the confectionery and bakery industries

(Santana and Meireles 2014). Nomura et al. (2021) reported a high yield of soluble polysaccharide (23.3%) from lentil fiber, and the polysaccharide fractions can disperse and stabilize milk proteins in acidified protein drinks. This study has adequately expanded the use of lentil fibers in the food industry.

Several authors reported the influence of processing on the functionality and processing behavior of the lentil protein and starch and their use as food additive ingredients. Several studies have been reported on the effects of high-pressure processing and heat treatments on the emulsifying and foaming properties of lentil proteins (Primozic et al. 2018; Ahmed et al. 2019; Hall and Moraru 2021), and those results demonstrate that a significant improvement has been observed in those post-process ingredients and their suitability as potential ingredients for the food industry. However, the emulsion capacity and stability of lentil starch produced by nonthermal processing were reduced, in particular, after sonication and irradiation (Majeed et al. 2017). Figure 11.5 illustrates the potential applications of lentils in several food products as an additive ingredient.

11.2.6 Applications in Cosmetics

Cosmetics are categorized based on their protective (moisturizers, lubricants, and sunscreen), corrective (depigmented, epilators, and lighteners), decorative (eye and face makeup and hair dyes), and hygienic (soaps, cleansing forms, and deodorants) functions. Each product comprises an active component, a base substance, and the main ingredient or raw material (Osorio et al. 2021). Several studies focused on using polymers and phytochemicals in legume-based extracts as the ingredients in developing cosmetic products. The emulsification and stabilization properties of natural polymers such as proteins,

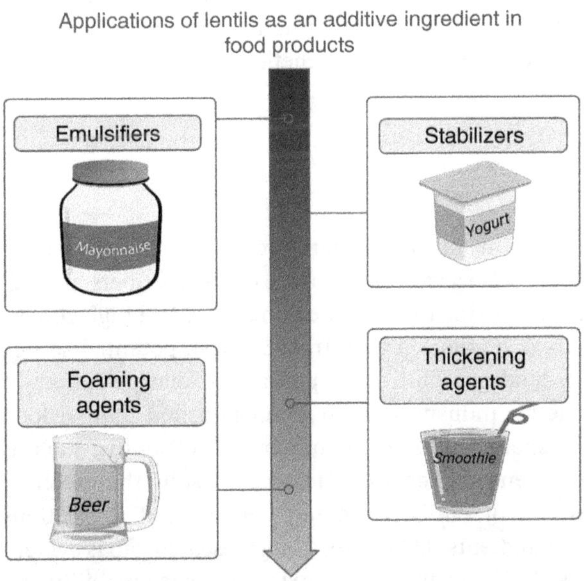

Figure 11.5 Potential applications of lentil-based additive ingredients in different food products. *Source:* Created with BioRender.com.

bioactive peptides, and carbohydrates in pulse proteins are used as emulsion stabilizers in cosmetic products, which can be one of the best alternatives to synthetic nanoemulsions (Tamang et al. 2021). Moreover, there are some established commercial applications as well; nevertheless, none are derived from by-products/wastes (Tassoni et al. 2020).

Many researchers have demonstrated the potential applications of isolated flours, bioactive peptides, phytochemicals, fibers, and enzymes from legumes and their by-products in the cosmetic industry (Aguilar-Toalá et al. 2019; Osorio et al. 2021). Legume proteins have been utilized as an antiperspirant cosmetic agent due to their capacity to significantly reduce axillary sweat (Banowski and Evers 2018). The formation of emulsions by both particulate and fibrillar aggregates from the lentil protein was reported, and it was found that these aggregates could form long-term stable emulsions that could be used in the cosmetic industry (Wynnychuk et al. 2021). On the other hand, the intrusive low molecular weight carbohydrates were successfully removed in inositol-enriched extracts of lentils through a biotechnological approach using *Saccharomyces cerevisiae* (Ruiz-Aceituno et al. 2013). The potential applications of inositol in cosmetics as an ingredient have improved the oxygenation of the cutaneous tissues and enhanced and accelerated the swelling of thickeners in aqueous solutions. Elessawy et al. (2021) investigated the antioxidant potential, iron chelation, and polyphenol profile of some lentil seed coat extracts. As a result, the potential of incorporating the extracted polyphenols as active ingredients in cosmetic formulations and skin care products has been broadened (Elessawy et al. 2021; Piergiovanni 2021).

Primozic et al. (2018) examined the impact of high-pressure homogenization of lentil protein isolate on the formation and stability of nanoemulsions, targeting their use as emulsifiers in cosmetic applications. The safety of lentil-derived proteins and peptides used in hair and/or skin conditioning agents in personal care products has been reviewed by Burnett et al. (2022). Limited research work on the safety and compatibility of lentil extracts in cosmetic applications has been reported. Although the information on techno-functional properties of lentil protein is available, its application to cosmetic industry is still lacking (Khazaei et al. 2019).

11.3 Conclusions

Among the legumes, lentils play an excellent role as a dietary protein source due to their high availability and inexpensiveness. It is also a very good source of dietary fibers, providing several health benefits, including minimizing the risk of colon cancer and high cholesterol. Apart from its dietary applications, lentils can be introduced as a potential source for nonfood applications, including biodegradable food packaging, encapsulation of bioactive compounds, adhesives, cosmetics, industrial starch production, and additive food ingredients. Some studies have been reported on several components of lentils, such as protein, starch, fiber, and flour, in different applications. However, despite the excellent physicochemical and functional properties of lentils, the research on nonfood applications of lentils is limited to starch and food ingredients. Therefore, a huge research gap is evident in the nonfood sector and researchers should focus on the area, so the sustainability and economic feasibility of lentil farmers and industry can be improved. On the other hand, large-scale economic production differs from laboratory-level production, and trials should also be introduced for large-scale production. As for most biopolymer-based products, lentils also face the challenge of the unpredictability of their properties due to climatic impacts, genetics, and batch-to-batch variations.

References

Afzaal, M., Saeed, F., Aamir, M. et al. (2021). Encapsulating properties of legume proteins: recent updates & perspectives. *International Journal of Food Properties* 24 (1): 1603–1614. https://doi.org/10.1080/10942912.2021.1987456.

Aguilar-Toalá, J.E., Hernández-Mendoza, A., González-Córdova, A.F. et al. (2019). Potential role of natural bioactive peptides for development of cosmeceutical skin products. *Peptides* 122: 170170. https://doi.org/10.1016/j.peptides.2019.170170.

Ahmed, J., Mulla, M., Al-Ruwaih, N., and Arfat, Y.A. (2019). Effect of high-pressure treatment prior to enzymatic hydrolysis on rheological, thermal, and antioxidant properties of lentil protein isolate. *Legume Science* 1 (1): 10–23. https://doi.org/10.1002/leg3.10.

Allen, A.J., Marcinko, J.J., Wagler, T.A., and Sosnowick, A.J. (2010). Investigations of the molecular interactions of soy-based adhesives. *Forest Products Journal* 60 (6): 534–540. https://doi.org/10.13073/0015-7473-60.6.534.

Avramenko, N.A., Chang, C., Low, N.H., and Nickerson, M.T. (2016). Encapsulation of flaxseed oil within native and modified lentil protein-based microcapsules. *Food Research International* 81: 17–24. https://doi.org/10.1016/J.FOODRES.2015.12.028.

Aydemir, L.Y. and Yemenicioğlu, A. (2013). Potential of Turkish Kabuli type chickpea and green and red lentil cultivars as source of soy and animal origin functional protein alternatives. *LWT Food Science and Technology* 50 (2): 686–694. https://doi.org/10.1016/j.lwt.2012.07.023.

Aydogdu, A., Kirtil, E., Sumnu, G. et al. (2018). Utilization of lentil flour as a biopolymer source for the development of edible films. *Journal of Applied Polymer Science* 135 (23): 46356–46366. https://doi.org/10.1002/app.46356.

Aydogdu, A., Yildiz, E., Aydogdu, Y. et al. (2019). Enhancing oxidative stability of walnuts by using gallic acid loaded lentil flour based electrospun nanofibers as active packaging material. *Food Hydrocolloids* 95: 245–255. https://doi.org/10.1016/J.FOODHYD.2019.04.020.

Bamdad, F., Goli, A.H., and Kadivar, M. (2006). Preparation and characterization of proteinous film from lentil (*Lens culinaris*): edible film from lentil (*Lens culinaris*). *Food Research International* 39 (1): 106–111. https://doi.org/10.1016/J.FOODRES.2005.06.006.

Banowski, B. and Evers, S. (2018). Antiperspirant cosmetics comprising specific proteins from legumes of the genus *Pisum* and/or *Phaseolus* and/or *Vigna* and/or *Macrotyloma* or from Cruciferous plants of the genus *Brassica* and including no aluminum and/or zirconium halides and/or hydroxy halides. US Patent US 10,123,958 B2. https://patents.google.com/patent/US10123958B2/en.

Bessada, S.M., Barreira, J.C., and Oliveira, M.B.P. (2019). Pulses and food security: dietary protein, digestibility, bioactive and functional properties. *Trends in Food Science & Technology* 93: 53–68. https://doi.org/10.1016/j.tifs.2019.08.022.

Bhat, R. and Karim, A.A. (2009). Impact of radiation processing on starch. *Comprehensive Reviews in Food Science and Food Safety* 8 (2): 44–58. https://doi.org/10.1111/J.1541-4337.2008.00066.X.

Boye, J.I., Aksay, S., Roufik, S. et al. (2010). Comparison of the functional properties of pea, chickpea and lentil protein concentrates processed using ultrafiltration and isoelectric precipitation techniques. *Food Research International* 43 (2): 537–546. https://doi.org/10.1016/J.FOODRES.2009.07.021.

Burnett, C.L., Boyer, I.J., Bergfeld, W.F. et al. (2022). Safety assessment of plant-derived proteins and peptides as used in cosmetics. *International Journal of Toxicology* 41 (2): 5S–20S. https://doi.org/10.1177/10915818221100070.

Chang, C. and Nickerson, M.T. (2018). Stability and *in vitro* release behaviour of encapsulated omega fatty acid-rich oils in lentil protein isolate-based microcapsules. *International Journal of Food Sciences and Nutrition* 69 (1): 12–23. https://doi.org/10.1080/09637486.2017.1336513.

Chang, C., Varankovich, N., and Nickerson, M.T. (2016). Microencapsulation of canola oil by lentil protein isolate-based wall materials. *Food Chemistry* 212: 264–273. https://doi.org/10.1016/J.FOODCHEM.2016.05.136.

Chelladurai, V. and Erkinbaev, C. (2020). Lentils. In: *Pulses* (ed. A. Manickavasagan and P. Thirunathan), 129–143. Cham: Springer. https://doi.org/10.1007/978-3-030-41376-7_8.

Cheng, H.N., Ford, C., Dowd, M.K., and He, Z. (2016). Soy and cottonseed protein blends as wood adhesives. *Industrial Crops and Products* 85: 324–330. https://doi.org/10.1016/j.indcrop.2015.12.024.

Choukri, H., Hejjaoui, K., El-Baouchi, A. et al. (2020). Heat and drought stress impact on phenology, grain yield, and nutritional quality of lentil (*Lens culinaris* Medikus). *Frontiers in Nutrition* 7: 596307. https://doi.org/10.3389/fnut.2020.596307.

Dai, T., Li, T., Li, R. et al. (2020). Utilization of plant-based protein-polyphenol complexes to form and stabilize emulsions: pea proteins and grape seed proanthocyanidins. *Food Chemistry* 329: 127219. https://doi.org/10.1016/j.foodchem.2020.127219.

DiNicolantonio, J.J., O'Keefe, J.H., and Wilson, W. (2018). Subclinical magnesium deficiency: a principal driver of cardiovascular disease and a public health crisis. *Open Heart* 5 (1): e000668. https://doi.org/10.1136/OPENHRT-2017-000668.

Dissanayake, T., Peng Chang, B., Mekonnen, T.H. et al. (2022a). Reinforcing canola protein matrix with chemically tailored nanocrystalline cellulose improves the functionality of canola protein-based packaging materials. *Food Chemistry* 383: 132618. https://doi.org/10.1016/J.FOODCHEM.2022.132618.

Dissanayake, T., Sun, X., Abbey, L., and Bandara, N. (2022b). Recent advances in lipid-protein conjugate-based delivery systems in nutraceutical, drug, and gene delivery. *Food Hydrocolloids for Health* 2: 100054. https://doi.org/10.1016/J.FHFH.2022.100054.

Elessawy, F.M., Vandenberg, A., El-Aneed, A., and Purves, R.W. (2021). An untargeted metabolomics approach for correlating pulse crop seed coat polyphenol profiles with antioxidant capacity and iron chelation ability. *Molecules* 26 (13): 3833–3862. https://doi.org/10.3390/molecules26133833.

Espinosa-Ramírez, J., Rodríguez, A., De la Rosa-Millán, J. et al. (2021). Shear-induced enhancement of techno-functional properties of whole grain flours through extrusion. *Food Hydrocolloids* 111: 106400. https://doi.org/10.1016/j.foodhyd.2020.106400.

Ettoumi, Y.L., Chibane, M., and Romero, A. (2016). Emulsifying properties of legume proteins at acidic conditions: effect of protein concentration and ionic strength. *LWT Food Science and Technology* 66: 260–266. https://doi.org/10.1016/j.lwt.2015.10.051.

FAOSTAT (Food and Agriculture Organization Corporate Statistical Database) (2022). Crops and livestock products. https://www.fao.org/faostat/en/#data/QCL/visualize (accessed 8 September 2022).

Faris, M.E.A.I.E., Takruri, H.R., and Issa, A.Y. (2013). Role of lentils (*Lens culinaris* L.) in human health and nutrition: a review. *Mediterranean Journal of Nutrition and Metabolism* 6 (1): 3–16. https://doi.org/10.3233/s12349-012-0109-8.

Fouad, A.A. and Rehab, F.M. (2015). Effect of germination time on proximate analysis, bioactive compounds and antioxidant activity of lentil (*Lens culinaris* Medik.) sprouts. *Acta Scientiarum Polonorum. Technologia Alimentaria* 14 (3): 233–246. https://doi.org/10.17306/J.AFS.2015.3.25.

Funke, M., Loeffler, M., Winkelmeyer, C. et al. (2022). Emulsifying properties of lentil protein preparations obtained by dry fractionation. *European Food Research and Technology* 248 (2): 381–391. https://doi.org/10.1007/s00217-021-03883-y.

Gumus, C.E., Decker, E.A., and McClements, D.J. (2017a). Gastrointestinal fate of emulsion-based ω-3 oil delivery systems stabilized by plant proteins: lentil, pea, and faba bean proteins. *Journal of Food Engineering* 207: 90–98. https://doi.org/10.1016/J.JFOODENG.2017.03.019.

Gumus, C.E., Decker, E.A., and McClements, D.J. (2017b). Impact of legume protein type and location on lipid oxidation in fish oil-in-water emulsions: lentil, pea, and faba bean proteins. *Food Research International* 100: 175–185. https://doi.org/10.1016/J.FOODRES.2017.08.029.

Hadad, S. and Goli, S.A.H. (2019). Improving oxidative stability of flaxseed oil by encapsulation in electrospun flaxseed mucilage nanofiber. *Food and Bioprocess Technology* 12 (5): 829–838. https://doi.org/10.1007/S11947-019-02259-1/FIGURES/7.

Hadidi, M., Jafarzadeh, S., Forough, M. et al. (2022). Plant protein-based food packaging films; recent advances in fabrication, characterization, and applications. *Trends in Food Science & Technology* 120: 154–173. https://doi.org/10.1016/J.TIFS.2022.01.013.

Hajas, L., Sipos, L., Csobod, É.C. et al. (2022). Lentil (*Lens culinaris* Medik.) flour varieties as promising new ingredients for gluten-free cookies. *Food* 11 (14): 2028–2044. https://doi.org/10.3390/foods11142028.

Hall, A.E. and Moraru, C.I. (2021). Structure and function of pea, lentil and faba bean proteins treated by high pressure processing and heat treatment. *LWT Food Science and Technology* 152: 112349. https://doi.org/10.1016/j.lwt.2021.112349.

Hefnawy, T.H. (2011). Effect of processing methods on nutritional composition and anti-nutritional factors in lentils (*Lens culinaris*). *Annals of Agricultural Science* 56 (2): 57–61. https://doi.org/10.1016/j.aoas.2011.07.001.

Ismail, A.A.A., Ismail, Y., and Ismail, A.A. (2018). Chronic magnesium deficiency and human disease; time for reappraisal? *QJM: An International Journal of Medicine* 111 (11): 759–763. https://doi.org/10.1093/QJMED/HCX186.

Jarpa-Parra, M., Bamdad, F., Wang, Y. et al. (2014). Optimization of lentil protein extraction and the influence of process pH on protein structure and functionality. *LWT Food Science and Technology* 57 (2): 461–469. https://doi.org/10.1016/j.lwt.2014.02.035.

Jarpa-Parra, M., Bamdad, F., Tian, Z. et al. (2015). Impact of pH on molecular structure and surface properties of lentil legumin-like protein and its application as foam stabilizer. *Colloids and Surfaces B: Biointerfaces* 132: 45–53. https://doi.org/10.1016/j.colsurfb.2015.04.065.

Jarpa-Parra, M., Tian, Z., Temelli, F. et al. (2016). Understanding the stability mechanisms of lentil legumin-like protein and polysaccharide foams. *Food Hydrocolloids* 61: 903–913. https://doi.org/10.1016/j.foodhyd.2016.07.017.

Jiang, H., Sheng, Y., and Ngai, T. (2020). Pickering emulsions: versatility of colloidal particles and recent applications. *Current Opinion in Colloid & Interface Science* 49: 1–15. https://doi.org/10.1016/j.cocis.2020.04.010.

Joshi, M., Aldred, P., Panozzo, J.F. et al. (2014). Rheological and microstructural characteristics of lentil starch: lentil protein composite pastes and gels. *Food Hydrocolloids* 35: 226–237. https://doi.org/10.1016/j.foodhyd.2013.05.016.

Joshi, M., Timilsena, Y., and Adhikari, B. (2017). Global production, processing and utilization of lentil: a review. *Journal of Integrative Agriculture* 16 (12): 2898–2913. https://doi.org/10.1016/S2095-3119(17)61793-3.

Kabakci, C., Sumnu, G., Sahin, S., and Oztop, M.H. (2021). Encapsulation of magnesium with lentil flour by using double emulsion to produce magnesium enriched cakes. *Food and Bioprocess Technology* 14 (10): 1773–1790. https://doi.org/10.1007/S11947-021-02672-5/TABLES/6.

Karaca, A.C., Nickerson, M., and Low, N.H. (2013). Microcapsule production employing chickpea or lentil protein isolates and maltodextrin: physicochemical properties and oxidative protection of encapsulated flaxseed oil. *Food Chemistry* 139 (1–4): 448–457. https://doi.org/10.1016/J.FOODCHEM.2013.01.040.

Khazaei, H., Subedi, M., Nickerson, M. et al. (2019). Seed protein of lentils: current status, progress, and food applications. *Food* 8 (9): 391–414. https://doi.org/10.3390/foods8090391.

Kim, W., Wang, Y., and Selomulya, C. (2020). Dairy and plant proteins as natural food emulsifiers. *Trends in Food Science & Technology* 105: 261–272. https://doi.org/10.1016/j.tifs.2020.09.012.

Kumar, M., Tomar, M., Punia, S. et al. (2022). Plant-based proteins and their multifaceted industrial applications. *LWT Food Science and Technology* 154: 112620. https://doi.org/10.1016/j.lwt.2021.112620.

Lee, H.W., Lu, Y., Zhang, Y. et al. (2021). Physicochemical and functional properties of red lentil protein isolates from three origins at different pH. *Food Chemistry* 358: 129749. https://doi.org/10.1016/j.foodchem.2021.129749.

Li, L., Yuan, T.Z., Setia, R. et al. (2019). Characteristics of pea, lentil and faba bean starches isolated from air-classified flours in comparison with commercial starches. *Food Chemistry* 276: 599–607. https://doi.org/10.1016/J.FOODCHEM.2018.10.064.

Majeed, T., Wani, I.A., and Hussain, P.R. (2017). Effect of dual modification of sonication and γ-irradiation on physicochemical and functional properties of lentil (*Lens culinaris* L.) starch. *International Journal of Biological Macromolecules* 101: 358–365. https://doi.org/10.1016/j.ijbiomac.2017.03.110.

Mekala, S., Silva, E.K., and Saldaña, M.D.A. (2022). Ultrasound-assisted production of emulsion-filled pectin hydrogels to encapsulate vitamin complex: impact of the addition of xylooligosaccharides, ascorbic acid and supercritical CO_2 drying. *Innovative Food Science & Emerging Technologies* 76: 102907. https://doi.org/10.1016/J.IFSET.2021.102907.

Muehlbauer, F.J. and McPhee, K.E. (2005). Lentil (*Lens culinaris* Medik.). In: *Genetic Resources and Chromosome Engineering and Crop Improvement* (ed. R.J. Singh and P.P. Jauhar), 219–230. CRC Press. https://doi.org/10.1201/9780203489284.

Nadeeshani, H., Senevirathne, N., Somaratne, G., and Bandara, N. (2022). Recent trends in the utilization of pulse protein in food and industrial applications. *ACS Food Science & Technology* 2 (5): 722–737. https://doi.org/10.1021/acsfoodscitech.1c00448.

Nomura, K., Sakai, M., Ohboshi, H., and Nakamura, A. (2021). Extraction of a water-soluble polysaccharide fraction from lentils and its potential application in acidified protein dispersions. *Food Hydrocolloids* 117: 106740. https://doi.org/10.1016/j.foodhyd.2021.106740.

Ochoa-Yepes, O., Medina-Jaramillo, C., Guz, L., and Famá, L. (2018). Biodegradable and edible starch composites with fiber-rich lentil flour to use as food packaging. *Starch – Stärke* 70 (7, 8): 1700222. https://doi.org/10.1002/STAR.201700222.

Ochoa-Yepes, O., Di Giogio, L., Goyanes, S. et al. (2019). Influence of process (extrusion/thermo-compression, casting) and lentil protein content on physicochemical properties of starch films. *Carbohydrate Polymers* 208: 221–231. https://doi.org/10.1016/J.CARBPOL.2018.12.030.

Osemwota, E.C., Alashi, A.M., and Aluko, R.E. (2021). Comparative study of the structural and functional properties of membrane-isolated and isoelectric pH precipitated green lentil seed protein isolates. *Membranes* 11 (9): 694–716. https://doi.org/10.3390/membranes11090694.

Osemwota, E.C., Alashi, A.M., and Aluko, R.E. (2022). Physicochemical and functional properties of albumin, globulin and glutelin fractions of green lentil seed. *International Journal of Food Science and Technology*. https://doi.org/10.1111/ijfs.15608.

Osorio, L.L.D.R., Flórez-López, E., and Grande-Tovar, C.D. (2021). The potential of selected agri-food loss and waste to contribute to a circular economy: applications in the food, cosmetic and pharmaceutical industries. *Molecules* 26 (2): 515–557. https://doi.org/10.3390/molecules26020515.

Pająk, P., Przetaczek-Rożnowska, I., and Juszczak, L. (2019). Development and physicochemical, thermal and mechanical properties of edible films based on pumpkin, lentil and quinoa starches. *International Journal of Biological Macromolecules* 138: 441–449. https://doi.org/10.1016/J.IJBIOMAC.2019.07.074.

Piergiovanni, A.R. (2021). Nutritional characteristics of black lentil from Soleto: a single-flower vetch landrace of Apulia region (Southern Italy). *Food* 10 (11): 2863–2872. https://doi.org/10.3390/foods10112863.

Primozic, M., Duchek, A., Nickerson, M., and Ghosh, S. (2018). Formation, stability and in vitro digestibility of nanoemulsions stabilized by high-pressure homogenized lentil proteins isolate. *Food Hydrocolloids* 77: 126–141. https://doi.org/10.1016/j.foodhyd.2017.09.028.

Ren, Y., Yuan, T.Z., Chigwedere, C.M., and Ai, Y. (2021). A current review of structure, functional properties, and industrial applications of pulse starches for value-added utilization. *Comprehensive Reviews in Food Science and Food Safety* 20 (3): 3061–3092. https://doi.org/10.1111/1541-4337.12735.

Roasa, J., De Villa, R., Mine, Y., and Tsao, R. (2021). Phenolics of cereal, pulse and oilseed processing byproducts and potential effects of solid-state fermentation on their bioaccessibility, bioavailability and health benefits: a review. *Trends in Food Science & Technology* 116: 954–974. https://doi.org/10.1016/j.tifs.2021.08.027.

Ruiz-Aceituno, L., Rodríguez-Sánchez, S., Ruiz-Matute, A.I. et al. (2013). Optimisation of a biotechnological procedure for selective fractionation of bioactive inositol in edible legume extracts. *Journal of the Science of Food and Agriculture* 93 (11): 2797–2803. https://doi.org/10.1016/j.peptides.2019.170170.

Santana, Á.L. and Meireles, M.A.A. (2014). New starches are the trend for industry applications: a review. *Food and Public Health* 4 (5): 229–241. https://doi.org/10.5923/j.fph.20140405.04.

Sein, T. (2008). Process development of lentil flour-based adhesive for woodworking industries. Doctoral dissertation. University of Yangon. Myanmar Education Research and Learning Portal. https://meral.edu.mm/record/2333/preview/Process%20development%20of%20Lentil%20flour-based%20Adhesive%20for%20Woodworking%20Industries.pdf.

Shi, A., Feng, X., Wang, Q., and Adhikari, B. (2020). Pickering and high internal phase Pickering emulsions stabilized by protein-based particles: a review of synthesis, application and prospective. *Food Hydrocolloids* 109: 106117. https://doi.org/10.1016/j.foodhyd.2020.106117.

Singh, N., Sandhu, K.S., and Kaur, M. (2004). Characterization of starches separated from Indian chickpea (*Cicer arietinum* L.) cultivars. *Journal of Food Engineering 63* (4): 441–449. https://doi.org/10.1016/J.JFOODENG.2003.09.003.

Sosulski, F.W. and Sosulski, K. (1986). Composition and functionality of protein, starch, and fiber from wet and dry processing of grain legumes. In: *Plant Proteins: Applications, Biological Effects, and Chemistry* (ed. R.L. Ory), 176–189. ACS Publications. https://doi.org/10.1021/BK-1986-0312.CH015.

Stone, A.K., Nosworthy, M.G., Chiremba, C. et al. (2019). A comparative study of the functionality and protein quality of a variety of legume and cereal flours. *Cereal Chemistry* 96 (6): 1159–1169. https://doi.org/10.1002/cche.10226.

Tamang, N., Shrestha, P., Khadka, B. et al. (2021). A review of biopolymers' utility as emulsion stabilizers. *Polymers* 14 (1): 127–147. https://doi.org/10.3390/polym14010127.

Tassoni, A., Tedeschi, T., Zurlini, C. et al. (2020). State-of-the-art production chains for peas, beans and chickpeas-valorization of agro-industrial residues and applications of derived extracts. *Molecules* 25 (6): 1383–1404. https://doi.org/10.3390/molecules25061383.

United States Department of Agriculture (USDA) (2022). USDA National Nutrient Database for Standard Reference, Release 23. https://data.nal.usda.gov/dataset/usda-national-nutrient-database-standard-reference-legacy-release (accessed 10 September 2022).

Wang, J., Korber, D.R., Low, N.H., and Nickerson, M.T. (2015). Encapsulation of Bifidobacterium adolescentis cells with legume proteins and survival under stimulated gastric conditions and during storage in commercial fruit juices. *Food Science and Biotechnology* 24 (2): 383–391. https://doi.org/10.1007/S10068-015-0051-X.

Wynnychuk, C., Jo, Y.J., Chu, Y., and Chen, L. (2021). Long-term stable emulsions prepared from lentil protein fibrillar aggregates. *Food Structure* 29: 100204. https://doi.org/10.1016/j.foostr.2021.100204.

Yang, R., Zhu, L., Meng, D. et al. (2021). Proteins from leguminous plants: from structure, property to the function in encapsulation/binding and delivery of bioactive compounds. *Critical Reviews in Food Science and Nutrition* 62 (19): 5203–5223. https://doi.org/10.1080/10408398.2021.1883545.

Yu, H., Cao, Y., Fang, Q., and Liu, Z. (2015). Effects of treatment temperature on properties of starch-based adhesives. *BioResources* 10 (2): 3520–3530.

Yue, L., Meng, Z., Yi, Z. et al. (2019). Effects of different denaturants on properties and performance of soy protein-based adhesive. *Polymers* 11 (8): 1262–1275. https://doi.org/10.3390/polym11081262.

12

Innovative Processing Technologies for Lentil Flour, Protein, and Starch

Jasim Ahmed[1], *Sanju Bala Dhull*[2], *and Madhuresh Dwivedi*[3]

[1] *Food and Nutrition Program, Environment and Life Sciences Research Center, Kuwait Institute for Scientific Research, Safat, Kuwait*
[2] *Department of Food Science and Technology, Chaudhary Devi Lal University, Sirsa, Haryana, India*
[3] *Department of Food Process Engineering, National Institute of Technology, Rourkela, Orissa, India*

12.1 Introduction

Food processing technologies have been significantly improved with time and consumer demands toward safety and the retention of nutrition at the highest level. Thermal processing, undoubtedly, produces safe food with a longer shelf life. Nutritional and sensory evaluations of thermally processed foods (e.g., canned foods) have demonstrated their significant detrimental effects. Scientists and technologists are continuously working together to bring new inventions to market so that food products are safe with the minimum loss of nutrients and retain sensory attributes. In the last few decades, several nonthermal processing technologies (e.g., high-pressure, dielectric heating, ultraviolet light, isoelectric precipitation, and ultrasound) have been experimented for this purpose (Ahmed et al. 2009, 2016, 2019; Aryee and Boye 2017; Heydari et al. 2022a,b; Li et al. 2021). Interestingly, some of those technologies have emerged as front-runners with the potential to replace the conventional technologies. At the beginning, emerging technologies were mostly applied to premier foods to recover the cost of the equipment. Some of the equipment is versatile (e.g., high-pressure processing), and the cost is recovering and even reducing with extensive research and competition among manufacturers. Therefore, innovative technologies have been extended to many other foods.

Legumes have recently received tremendous attention from the food and allied industries as they are excellent sources of protein and could have the potential to replace animal-based proteins. Among legumes, lentils (*Lens culinaris*) are one of the most consumed protein sources (>30% protein content) in many parts of the world. The important storage proteins reported in lentils are oligomeric globulins (49–72%) and albumins (11–16.8%), which belong to the two groups of 7S (vicilin, 175–180 kDa) and 11S (legumin, 340–360 kDa),

Lentils: Production, Processing Technologies, Products, and Nutritional Profile, First Edition.
Edited by Jasim Ahmed, Muhammad Siddiq, and Mark A. Uebersax.

respectively (Scippa et al. 2008; Swanson 1990). The growing interest in plant-based products has led to an increase in the production of legume proteins (e.g., protein concentrates and protein isolates) due to their excellent functional and nutritional properties (Khazaei et al. 2019). On the other hand, starch remains the largest component of legume seeds and is extensively used as one of the most important ingredients in the food industry. The presence of antinutritional factors (e.g., protease inhibitors, lectins, phytates, and alkaloids) remains the bottleneck of lentils or other legumes, and these antinutritional factors produce adverse nutritional effects on their use as human food (Bou et al. 2022). Heat processing, sometimes, can trigger the antinutritional factors that make it worse. Furthermore, the presence of some undesired materials (e.g., fiber, tannins, and lipids) can influence the purity of either protein, starch, or their extraction yield. Emerging technologies are currently applied to food products to alter protein or starch structure at desirable levels for tailoring functional foods with potential health benefits. Therefore, selection of adequate emerging and sustainable facilitating technologies, including high-pressure (HP) treatment, dielectric heating (e.g., microwave and radio-frequency), ultrasound, irradiation, and ultrafiltration and isoelectric precipitation, can be applied to lentils to remove undesired compounds, reduce allergenicity, and improve digestibility and sensory acceptability among consumers. This chapter elucidates the potential of emerging technologies for the processing of lentil as a whole flour or in isolated forms as protein or starch.

12.2 High-Pressure Treatment

High-pressure (HP) processing is a nonthermal treatment in which packaged food materials are subjected to pressurization for a short time in a vessel containing a suitable pressure-transmitting fluid (e.g., pure water). The process has received a "cleaner" ingredient label. According to the U.S. Dept. of Agriculture, Food Safety and Inspection Service (USDA-FSIS), HP processing is an acceptable food safety intervention for eliminating pathogens (e.g., *Listeria monocytogenes*, *Salmonella*, *Escherichia coli*) in processed food products. Typically, a HP unit comprises a vessel, which varies from 2 to 420 l with an enclosure, HP intensifier pump(s), a temperature sensor, a material handling system, and a pressure-transmitting medium (e.g., water). Generally, perforated baskets are employed to load and unload pre-packaged food products from the pressure vessels. Systems also have provisions for filtering and reusing the compression fluid (usually water or a food-grade solution). Recently, HP machines have been available in an expandable mode so that the machine can be tailored based on food materials and production volume. The pressure is transmitted uniformly throughout the product, so unlike the cold point of thermal processing, the HP ensures food safety at the highest level throughout the product. Pressurization assists deprotonation of charged groups and disruption of salt bridges and hydrophobic bonds, resulting in conformational changes and denaturation of proteins (Barbosa-Canovas et al. 1998). HP vessels are available in both vertical and horizontal configurations. A typical HP unit with a vertical configuration is shown in Figure 12.1. HP treatment can successfully modify the functional and sensory characteristics of many food components, in particular, by denaturing proteins and gelatinizing starch. More details of the HP technology are available elsewhere.

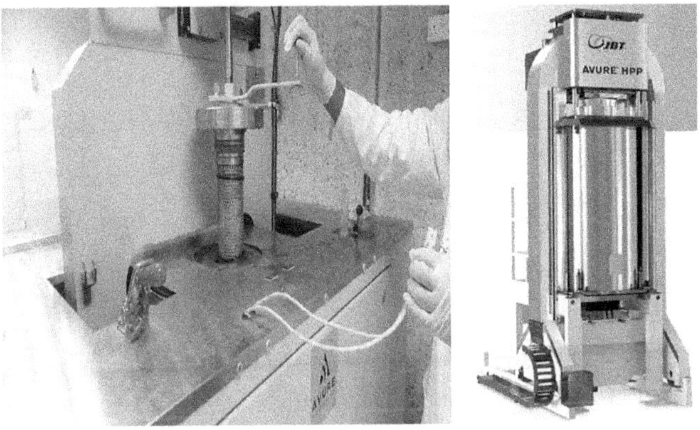

Figure 12.1 Laboratory scale and commercial vertical designed high-pressure unit. *Source:* Avure Technology, with permission.

12.3 HP-Treatment to Lentil Ingredients

The HP treatment of lentil flour, isolated starch, and proteins is described in this section. At the end of the section, HP homogenization is also elucidated. A brief review of published data on the HP treatment of lentil products is presented in Table 12.1.

12.3.1 Lentil Flour

Lentil flour contains a high concentration of starch and proteins, both of which are susceptible to alteration during the high-pressure treatment. Ahmed et al. (2009) studied the influence of HP treatment on the thermal, rheological, and structural properties of lentil flour dispersions as a function of moisture content (14–58 g/100 g of flour) and pressure level (350–650 MPa/10 minutes). Differential scanning calorimetric (DSC) measurement of HP-treated lentil dispersions demonstrated incomplete denaturation of lentil proteins, and the protein-denaturation temperature (T_d) shifted non-systematically with the applied pressure and moisture content. The starch gelatinization peak was not detected during the thermal scanning of lentil slurries. HP treatment of lentil dispersion significantly reduced the retrogradation behavior when compared with a thermally gelatinized sample.

Dynamic rheological measurement of HP-treated lentil slurries demonstrated that the resultant slurries exhibited a true viscoelastic fluid (Ahmed et al. 2009). Slurries gradually transformed from solid-like behavior to liquid-like behavior as a function of moisture content and pressure level. The mechanical spectra of lentil dispersions (37 g water/100 g sample) displayed the predominance of the elastic modulus (G') over the viscous modulus (G'') in the studied frequency range for both pressure-treated and pressure-untreated samples (Figure 12.2). Pressurized dispersion developed a firm gel, and both moduli exhibited frequency independence because of the extensive molecular entanglement. The viscoelasticity of the lentil dispersion can be calculated from the slopes of the linear regression of the

Table 12.1 Application of high-pressure treatment on lentil ingredients and major outcomes.

Lentil component	Process condition	Major observations	References
Whole flour in aqueous dispersion (14.9–57.6 g water/100 g flour)	350–650 MPa/15 min; 22–26 °C	Incomplete denaturation of lentil proteins; the denaturation temperature shifted with pressure and moisture content Pressure treated lentil slurries exhibited a true viscoelastic fluid Insignificant change in the amide band of HP-treated slurry	Ahmed et al. (2009)
Starch dispersions (flour to water 1:4 w/w)	400–600 MPa/10 min; 26–38 °C	Particle size and water holding capacity increased after HP-treatment Resistant starch content increased at 600 MPa Starch was completely gelatinized at 600 MPa Pasting properties were significantly influenced by the HP-treatment	Ahmed et al. (2016)
Protein concentrate (5, 10, and 15 g/100 g water)	HP processing (600 MPa/4 min at 5 °C) and thermal treatment (95 °C/15 min) on the structure and functionality of lentil protein concentrates	Both HP and thermal treatments increased protein surface hydrophobicity Conformational changes resulted in prominent increases in foaming and emulsifying properties The particle size increased along with some particle aggregation	Hall and Moraru (2021)
Protein isolate	Pressurized at 300–600 MPa/15 min followed by hydrolysis with Alcalase (0.5% and 1%, w/w).	HP treatment improved the % degree of hydrolysis of the LPI HP-assisted Alcalase hydrolysis imparted significant changes on the secondary structure of protein with a shift of amide I and amide II bands A significant drop in the DSC enthalpy (ΔH) and apparent viscosity was observed for hydrolyzed proteins	Ahmed et al. (2019)
Protein isolate	Pressurization (100–300 MPa) followed by enzymatic hydrolysis using Protamex, Alcalase, Savinase, and Corolase 7089	Enzymatic proteolysis at 300 MPa led to a complete degradation of lentil proteins and increased peptide (<3 kDa) concentration Proteolysis at 300 MPa by Savinase gave rise to lentil hydrolysates (S300) with the highest ACE inhibitory and antioxidant activities, which were retained upon *in vitro* gastrointestinal digestion	Garcia-Mora et al. (2015)
Protein isolate	High-pressure homogenization (HPH) (25–150 MPa) of lentil protein isolate (LPI) suspensions (4% protein concentration; 4.76 g LPI/100 ml of water)	Protein patterns remained unchanged with HPH treatment Particle size significantly decreased up to 100 MPa treatment and size distribution was mono-modal after 50 MPa. Microstructural images revealed that increasing pressure from 50 to 150 MPa caused unfolding of protein particles LPI suspensions had shear-thinning behavior and form weak-gel	Saricaoglu (2020)
Protein isolate	HPH (50–300 MPa) of LPI using 1–3 passes and emulsifier to oil ratios of 1:1, 1:2, 2:1	The PSD of 1:1 lentil nanoemulsion changed from multimodal to bimodal at >100 MPa Two homogenization passes and ratios of 1:1 and 2:1 improved reduction of droplet size (587–149 nm) and PdI (0.821–0.168) A ζ-potential value of −30 mV indicated a stable emulsion formation	Tabilo-Munizaga et al. (2019)

LPI, Lentil protein isolate; DSC, differential scanning calorimetry; ACE, angiotensin converting enzyme; PSD, particle size distribution.

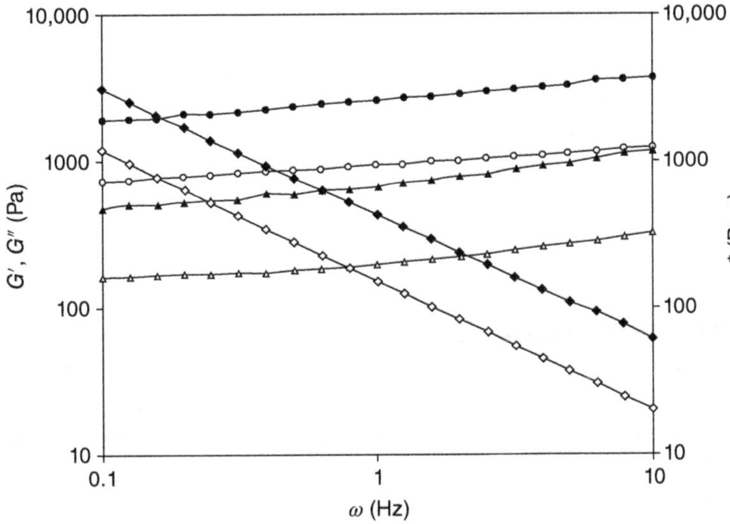

Figure 12.2 Changes in the viscoelastic properties of lentil flour dispersion (37 g water/100 g flour) at 450 MPa (open circle: control G'; closed circle: HP G'; open triangle: control G''; closed triangle HP G''; open diamond: control η^*; closed diamond: HP η^*). *Source:* Ahmed et al. (2009), with permission of Elsevier.

power-type relationship of ln ω versus ln G' and ln G'' (Eq. (12.1)), which is generally used to predict the solid-like behavior of polymers.

$$G' \text{ or } G'' = A\omega^n \tag{12.1}$$

$$\text{After linearization: } \ln G' \text{ or } \ln G'' = \ln A + n \ln \omega \tag{12.2}$$

where n is the slope (dimensionless) and A is the intercept (Pa·s^n).

The magnitude of the slope supported the transformation of the solid-like character of lentil slurry into a liquid-like character with increasing moisture content due to the plasticization effect attributed to water. Additionally, pressurized samples behaved as true viscoelastic fluids with almost similar magnitudes of slopes between ln ω versus ln G' and ln G''. High pressure converted lentil flour dispersion into a viscous gel by gelatinizing starch and/or denatured protein. The slopes increased with the increase in pressure level. For example, the slope of samples having moisture content of 58 g water/100 g flour and processed at 650 MPa for 15 minutes increased from 0.11 to 0.15 and from 0.15 to 0.18 for ln ω versus ln G' (solid-like) and ln ω versus ln G'' (liquid-like), respectively. This observation indicated a shift of viscoelastic solid to viscoelastic fluid in lentil flour dispersion after HP treatment.

Structural changes in the HP-treated lentil flour were assessed by Fourier transform infrared (FTIR) spectroscopy (Ahmed et al. 2009). The untreated samples showed four intense bands located around 1163, 1408, 1550, and 1658 cm^{-1}, which were not significantly changed after the pressurization. The spectral shape at 1652 cm^{-1} (amide I) remained unchanged after pressure treatment, while an insignificant change was observed at 1545 cm^{-1}. This study has provided complementary information on pressure-induced structural changes on both the molecular and the sub-molecular levels of lentil protein.

12.3.2 Lentil Starch

The functional properties of lentil starch can be improved by different processing techniques. The influence of HP treatment (400–600 MPa/10 minutes) on the particle size, rheological, pasting, and structural properties of isolated lentil starch dispersions (flour to water 1 : 4 w/w) was reported by Ahmed et al. (2016). The native starch granules were large and oval to spherical in shape. The starch surface was smooth without any cracks. The particle size distribution (PSD) of the lentil starch was bimodal. Pressurization up to 500 MPa did not affect the particle size; however, changes were distinct at 600 MPa. The peak particle size of the untreated sample increased significantly from 22 to 38 μm following the pressure treatment. At 600 MPa, the PSD curve exhibited a plateau because of the complete gelatinization of starch, and many granules were disintegrated into a gel-like structure. Oscillatory rheology of lentil starch dispersions indicated that the extent of starch gelatinization increased with increasing pressure levels, and thereafter, the dispersion (sol) transformed into gel with increasing rigidity (G'). Pressurized samples showed typical weak gel structure with predominating solid-like behavior ($G' > G''$) in the frequency range of 0.1–10 Hz. The gels were frequency independent in the lower pressure range (0.1–400 MPa) and, however, became frequency dependent above 500 MPa. The frequency dependency is associated with the gelatinization of starch and strong network formation.

The HP treatment significantly influenced the microstructure of lentil starch. Microstructures of HP-treated lentil starches demonstrated that the majority of the starch granules retained their granular shapes with smooth surfaces at 500 MPa, but the surfaces changed and cracked at 600 MPa (Figure 12.3) (Ahmed et al. 2016). The FTIR spectra of HP-treated lentil starch showed only two bands (993 and 1020 cm^{-1}) were significantly influenced by the treatment (Figure 12.4a). The band at 1020 cm^{-1} seems to increase at 600 MPa related to the predominant amorphous structure of starch granules. Changes in some spectral bands at 600 MPa are associated with the destruction of the crystalline structure in starch granules. X-ray diffraction (XRD) patterns of the native starches demonstrated a C-type crystalline pattern (mixture of A- and B-types) with 2θ values of 15.4°, 17.2°, and 23.1°. The disappearance of the peak intensities at 2θ values of 14.9° and 22.8° at 600 MPa confirmed a complete destruction of crystalline structure (Figure 12.4b). Additionally, the peak observed at 17.2° for the untreated sample reduced significantly its intensity after pressurization.

12.3.2.1 Relationship Between Nonisothermal Rheology and DSC of HP-Treated Starch

To elucidate the extent of lentil starch gelatinization during pressurization, both rheometry and DSC were employed, and the obtained values were compared. The rheometric nonisothermal heating of HP-treated samples exhibited two thermal peaks at 69 and 75 °C, respectively. Conversely, the onset and peak DSC gelatinization temperatures (T_p) of the untreated lentil starch were 55.71 and 63.72 °C, respectively, with an enthalpy value of 8.8 J/g. These thermal events are associated with the breaking down of amylopectin crystallites or gelatinization of LS. The crystalline peaks of lentil starch samples processed at 400 MPa remained unaffected during the nonthermal heating on a rheometer as well as in the DSC thermogram. At 500 MPa, the gelatinization progressed

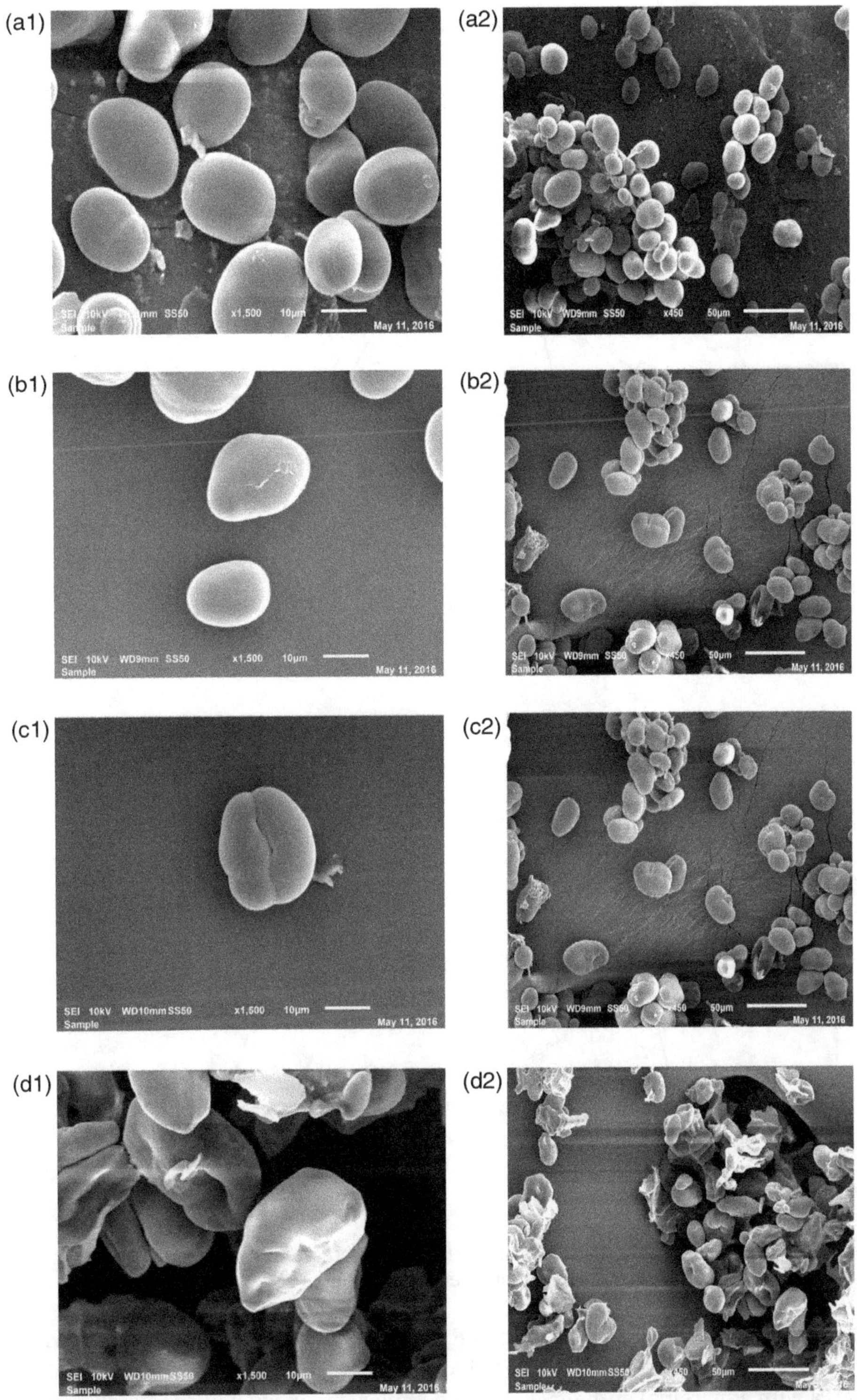

Figure 12.3 Scanning electron micrographs of pressure-treated lentil starch samples (a1) Control-1500×; (a2) control-450×; (b1) 400 MPa-1500×; (b2) 400 MPa-450×; (c1); 500 MPa-1500×; (c2) 500 MPa-450×; (d1); 600 MPa-1500×; (d2) 600 MPa-450×. *Source:* Ahmed et al. (2016), with permission of Elsevier.

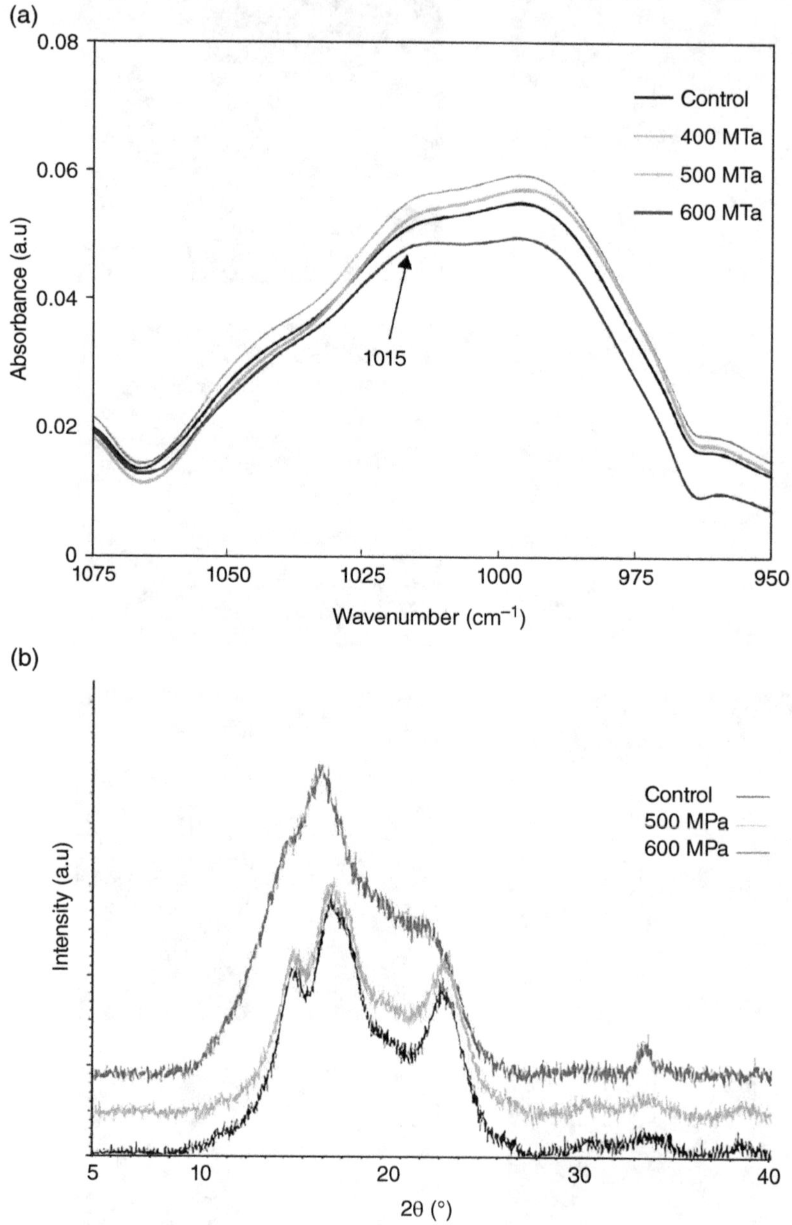

Figure 12.4 FTIR spectra (a) and X-ray diffraction (b) of high-pressure-treated lentil starch. *Source:* Ahmed et al. (2016), with permission of Elsevier.

with the melting of crystallites, resulting in the disappearance of the second peak, while the first crystalline peak appeared with a reducing height. The DSC T_p lowered to 56.65 °C at 500 MPa with a marginal enthalpy value. A complete breakdown of the starch crystallites occurred at 600 MPa, resulting in complete starch gelatinization and the disappearance of both melting peaks, which were captured in both a rheogram and a thermogram.

Both results are complementary to each other, and the difference in magnitudes is attributed to differences in measurement techniques, sample preparation, starch-to-water ratio, and the precision of the instrument.

12.3.3 Pasting Properties

Pasting is an important phenomenon for starch characterization and industrial applications in developing products. It measures starch behavior when heated in the presence of excess water by recording changes in viscosity. Ahmed et al. (2016) provided a comparison of the pasting properties of lentil starch between pre- and post-HP treatment (Table 12.2). HP treatment significantly influenced the pasting parameters of lentil starch ($P < 0.05$). The peak viscosity (PV) of the lentil starch increased at 400 MPa, but dropped significantly at 600 MPa. Such a significant drop in PV has been attributed to the complete gelatinization of starch granules. The pasting temperature decreased gradually from 64.1 to 56.2 °C with increasing pressure. The hot paste viscosity (HPV) decreased from 586 BU to 517 BU when the pressure increased from 0.101 to 600 MPa. The cold paste viscosity (CPV), conversely, decreased significantly from 1462 to 635 BU. The final viscosity (FV) of the lentil starch was reduced to 58% after HP treatment at 600 MPa. Both the breakdown viscosity (BDV) and the setback viscosity (SBV) decreased with the intensity of the pressure. The decrease in BDV indicates that the pressure-modified gel is more heat-stable than the control sample (Karim et al. 2000). The lower BD and SB values of pressurized samples indicate the aggression of starch molecules and a lower retrogradation tendency compared to untreated flour.

12.3.4 Lentil Proteins

The effects of HP processing (600 MPa/4 minutes at 5 °C) and thermal treatment (95 °C/15 minutes) on the structure and functionality of lentil protein concentrates were evaluated by Hall and Moraru (2021) at selected protein concentrations (5, 10, and 15 g/100 g water). Both high pressure and heat treatments resulted in a significant drop in the enthalpy of protein denaturation compared to untreated samples. The degree of denaturation was marginally higher for heat-treated than for HP-treated samples, but these differences were not

Table 12.2 Influence of high-pressure treatment on pasting properties of lentil starch dispersions.

Pressure (MPa)	Paste temp. (°C)	Peak temp. (°C)	Peak time (min)	Peak viscosity (BU)	Hot paste viscosity (BU)	Cold paste viscosity (BU)	Final viscosity (BU)	Break down viscosity (BU)	Setback viscosity (BU)
0.101	64.1	77.7	9.0	958	586	1462	1666	372	1080
400	62.6	80.2	10.2	981	651	1548	1588	330	937
500	65.2	91.5	13.9	860	670	1419	1487	190	817
600	56.2	95.0	44.4	520	517	635	688	3	171

Source: Ahmed et al. (2016), with permission of Elsevier.

statistically significant. Both HP and thermal treatments increased the hydrophobicity of the protein surface of lentil proteins due to denaturation, which exposed hydrophobic residues suppressed within the globular core of the proteins. HP- and heat-induced conformational changes resulted in prominent increases in foaming and emulsifying properties. The particle size of the lentil proteins increased after pressurization and thermal treatment, indicating some particle aggregation took place following the treatment. Frequency sweeps of the selected concentration protein samples indicated that at low concentration (5 g protein/100 g), both HP-treated and untreated samples exhibited a liquid-like behavior ($G'' > G'$), while the HP- and thermally treated samples formed gels at medium and high concentrations ($G' > G''$). Thermally treated samples formed stronger gels than HP-treated samples, and gel rigidity was higher for the high concentration samples than the medium concentration samples. These findings can be used to develop new food products using lentil proteins as ingredients.

Ahmed et al. (2019) studied the effectiveness of high-pressure (HP) treatment (300, 450, and 600 MPa/15 minutes) on the enzymatic hydrolysis (Alcalase [0.5% and 1%, w/w]) of lentil protein isolate (LPI) in order to improve the structural, functional, and antioxidant activities of the hydrolysates. HP treatment improved the %DHH of the LPI ($P \leq 0.05$). The produced hydrolysates improved the foaming properties (≈1.5 times) and antioxidant activities (≈2 times) than the untreated LPI. However, a detrimental effect was pronounced on the emulsifying activity and stability index, foam stability, and water holding capacity of the hydrolysates.

The SDS-PAGE of LPI under various conditions demonstrated that LPI has a complex protein profile between 250 and 10 kDa, which includes the subunits of 7S globulins (e.g., vicilin with 48 kDa) and 11S globulin (71 and 63 kDa) (Ahmed et al. 2019). The HP treatment did not influence the electrophoretic pattern of LPI even at the highest level (600 MPa for 15 minute; Figure 12.5a,b) (Ahmed et al. 2019). While working on HP homogenization (HPH) of LPI suspensions, Saricaoglu (2020) reported that no major differences between the protein patterns of untreated and HPH-treated LPI suspensions were observed, which means that HPH treatment up to 150 MPa was not sufficient for the unfolding of LPI. Generally, the covalent bonds remained unaffected by the pressurization resulting in the retention of the primary protein structure.

The FTIR spectra of lentil protein showed some intense bands in the range from 1235 to 1632 cm^{-1} attributed to the β-sheet and β-turn structures. The bands at 1632 cm^{-1} assigned to the amide-I region representing C=O stretching (Carbonaro et al. 2012), 1526 cm^{-1} for the amide-II region due to CN stretching and NH bending, and the band at 1235 cm^{-1} relates to the amide-III region, also associated with CN stretching and NH bending. HP treatment of LPI demonstrated that the secondary structure remained intact at 300 MPa with no shift of bands (Ahmed et al. 2019). However, the amide I, II, and III bands shifted to 1634, 1532, and 1238 cm^{-1}, respectively, when Alcalase digestion was performed, suggesting a change in the structure (Figure 12.5c).

The steady flow rheology of the LPI dispersions showed shear-thinning behavior. HP-treatment and Alcalase digestion resulted in lowering the *shear stress–shear rate* rheograms in the order of LPI + 0.5% Alcalase < LPI at 300 MPa + 0.5% Alcalase < LPI at 300 MPa (Ahmed et al. 2019). Furthermore, the rheograms exhibited structural breakdown during forward and backward runs and exhibited thixotropy. The structural breakdown was pronounced for both enzyme and combined HP/enzyme hydrolysis. It is postulated that the Alcalase digestion of lentil proteins after HP treatment produced the highest number of

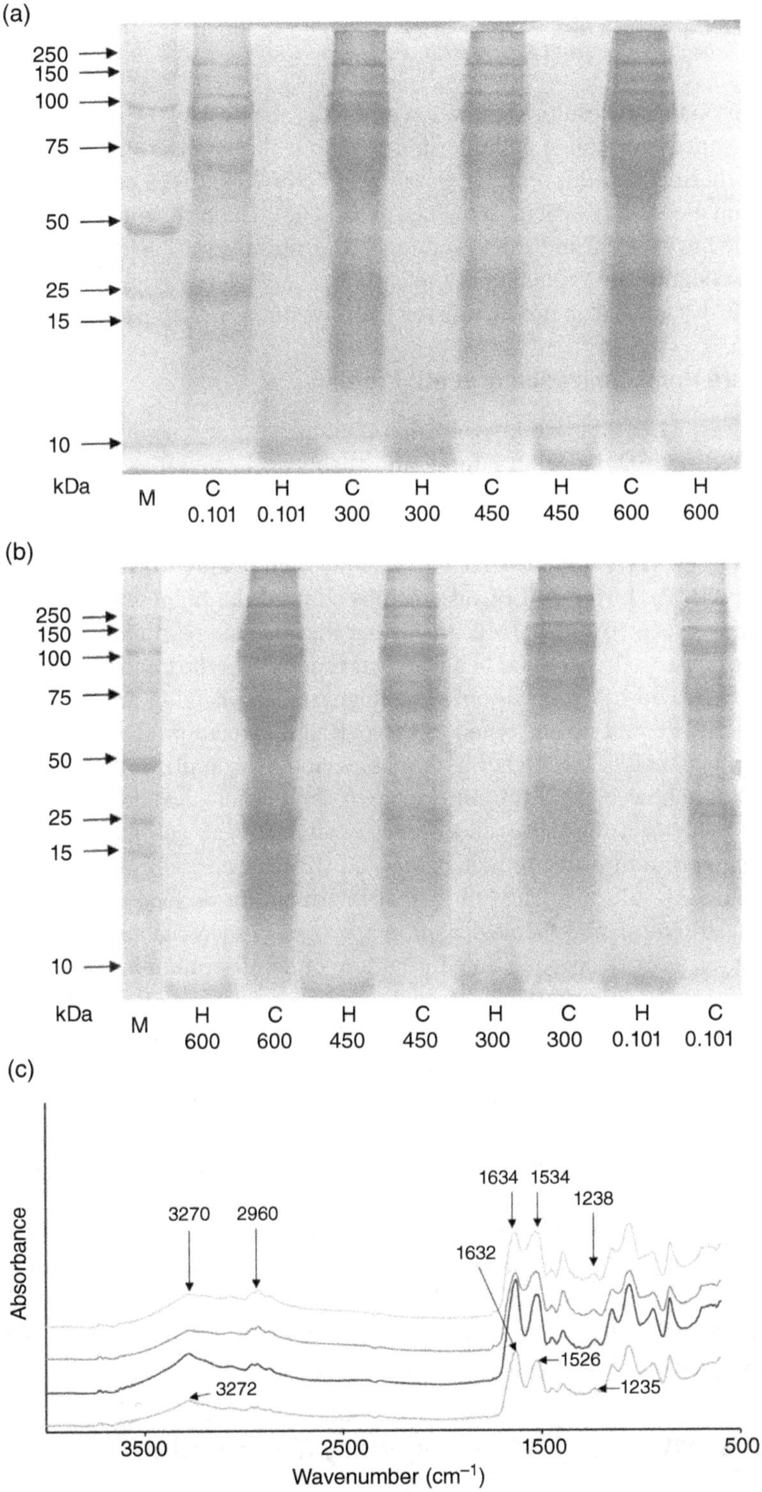

Figure 12.5 SDS-PAGE of lentil protein isolates (LPI) and hydrolysates (LPH) at selected pressures and Alcalase concentrations (a) 0.5% Alcalase and (b) 1% Alcalase concentrations. Lane M: molecular weight marker; Lane C: LPI at 0.101 MPa; Lane H: LPH at 0.101 MPa; Lane C300: LPI treated at 300 MPa; Lane H300: LPH treated at 300 MPa; Lane C450: LPI treated at 450 MPa; Lane H450, LPH treated at 450 MPa; Lane C600: LPI treated at 600 MPa; Lane H600: LPH treated at 600 MPa; (c) FTIR spectra.

low MW peptides (<3 kDa) that resulted in the lowest value of viscosity. DSC measurements showed two distinct irreversible endothermic peaks for the lentil proteins, which represented the peak thermal denaturation temperatures (T_d) of 7S and 11S of lentil proteins, namely lectin and globulin (Cai et al. 2002; Lee et al. 2007). The peak temperatures I and II ranged from 94.82 to 98.46 °C and 108.97 to 111.49 °C, respectively. The hydrolysis of LPI was confirmed by a significant reduction of the enthalpy (ΔH). These results could be useful for the food industry to develop products using hydrolysates as protein supplements.

12.3.5 High-Pressure Homogenization of Lentil Proteins

Saricaoglu (2020) investigated the ability of high-pressure homogenization (HPH) (25–150 MPa) to modify the functional, structural, and rheological properties of LPI suspensions (4% protein concentration; 4.76 g LPI/100 ml of water). Particle size significantly decreased up to 100 MPa treatment, and size distribution was mono-modal after 50 MPa. The emulsifying activity (EAI) of LPI suspensions was not significantly affected by the HPH treatment up to 25 MPa; however, LPI suspensions showed the highest EAI values when pressure ranged between 50 and 100 MPa. A further increase in pressure to 150 MPa caused a significant decrease in EAI. Similar behavior was reported for the emulsion stability index (ESI) of HPH-treated LPI suspensions. The increased emulsifying properties of LPI suspensions with HPH could be attributed to the partial dissociation and unfolding of protein structure, which results in a higher hydrophobic and hydrophilic interaction of proteins. LPI suspensions showed shear-thinning behavior, and rheological data were fitted to the Power law model. Furthermore, all samples exhibited weak gel-like structure ($G' > G''$), and a homogenization pressure higher than 50 MPa caused more pronounced gelation at 51.78 °C. These results suggested that HPH treatment has a good potential to modify the functional, structural, and rheological properties of LPI suspensions.

Tabilo-Munizaga et al. (2019) reported the formation of lentil-protein-based stable nanoemulsions using HPH (50–300 MPa) in a selected number of passes (1–3) and emulsifier to oil ratios (1 : 1, 1 : 2, and 2 : 1). The particle size distribution (PSD) of 1 : 1 lentil nanoemulsion changed from multimodal to bimodal at a pressure higher than 100 MPa, regardless of homogenization passes. Two homogenization passes and ratios of 1 : 1 and 2 : 1 improved the reduction of droplet size (587–149 nm) and PdI (0.821–0.168), with no difference between those ratios. Low ζ-potential values (<−30 mV) indicated formation of a stable emulsion. As emulsifiers, lentil proteins decreased the interfacial tension of the water–oil interphase from 16.5 to <7.0 mN/m irrespective of pressure levels. Nanoemulsions exhibited pseudoplastic characteristics, and a higher apparent viscosity was observed with 2 : 1 ratio. Furthermore, it has been concluded that a stable lentil nanoemulsion can be made by using 1 : 1 emulsifier to oil ratio and pressurization above 200 MPa with two homogenization passes that could be used as novel food systems.

12.4 Microwave (MW) and Radio-Frequency (RF) Heating

Microwave (MW) and radio-frequency (RF) are attractive dielectric heating methods because of their volumetric origin, rapid heating capability, low energy consumption, controllable heat deposition, and ease of cleanup (Yu et al. 2017; Ahmed and Ramaswamy 2020).

The difference between MW and RF heating is in their penetration depths. Microwaves are a type of electromagnetic radio wave, which range between 300 MHz and 300 GHz. Conversely, RF wavelengths cover the range of the electromagnetic spectrum from 3 kHz to 300 MHz. However, in practical applications, narrow frequency bands have been permitted for industrial, scientific, and medical uses so that the frequency range does not interfere with the telecommunication radio frequencies. The most common MW frequencies used for industrial purposes are 915 and 2450 MHz, whereas 2450 MHz used in a domestic microwave oven. The permitted RF frequencies used in the United States are 13.56, 27.12, and 40.68 MHz. Because of its high-penetration depth, RF heating has advantages over MW heating for the industrial processing of materials. However, before dielectric heating (MW or RF) can be applied, the dielectric properties of the food materials should be known so that the appropriate heating can be achieved during the heating process.

Microwave energy is given at the molecular level by molecular interactions with the electromagnetic field, particularly molecular friction caused by polar solvent dipole rotation and the conductive movement of dissolved ions (Oliveira and Franca 2002). During RF heating, dielectric materials are placed in an alternating electric field, and RF energy is generated by a triode valve and applied to the food via a pair of electrodes. Some thermophysical and electrical properties control the absorption of MW energy and the simultaneous heating behavior of food materials during microwave processing (Ahmed and Ramaswamy 2020). Among the controlling factors, the dominant ones are frequency, moisture content, dielectric property, temperature, chemical composition, and geometry of foods. Based on its novelty, MW heating has been employed in various food and agricultural applications, in particular for inactivating enzymes and allergens in foods, germination of seeds, drying, pasteurization, and sterilization of food by inactivating microbes (Ekezie et al. 2017; Elnjikkal Jerome and Dwivedi 2022; Gómez-López and Bhat 2021; Rifna and Dwivedi 2021; Rifna et al. 2019). Conversely, RF is mostly used in the drying, thawing, and tempering of foods. In this section, the influence of MW and RF heating and their applications to lentils are highlighted.

Dielectric property data are important in developing thermal treatments using radiofrequency (RF) and microwave (MW) energy and are essential in estimating heating uniformity in electromagnetic fields. Dielectric properties of lentil flour samples at selected moisture contents (8–21 g/100 g), frequencies (10–1800 MHz), and temperatures (20–90 °C) were measured with an open-ended coaxial probe and impedance analyzer (Guo et al. 2010). The dielectric constant (ε') and loss factor (ε'') of the lentils decreased with increasing frequency, but increased with increasing temperature and moisture content. At low frequencies and high temperatures and moisture contents, negative linear correlations were observed between the ε'' and frequency on a log–log plot, which was mainly influenced by the ionic conductance. The penetration depth (D_p) decreased with increasing frequency, temperature, and moisture content, which was large enough at 27 MHz to develop large-scale industrial RF treatments. The maximum D_p at 27 MHz could help in developing large-scale industrial RF treatments for postharvest insect control or other applications that require bulk heating of lentils with acceptable heating uniformity and throughputs.

The drying kinetics of red lentil seeds were studied in a microwave-assisted fluidized bed dryer (MWFBD) at four selected microwave powers (0, 300, 400, and 500 W) and at 50 and 60 °C (Figure 12.6) and simultaneously the possibility of eradication of seed-borne Botrytis

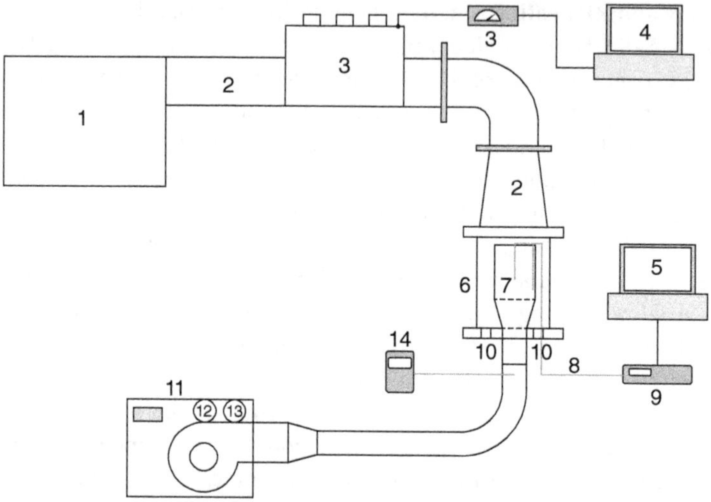

Figure 12.6 Microwave fluidized-bed system for drying and disinfection of particles. 1 – magnetron, 2 – waveguide, 3 –three-stub tuner, 4 – absorbed power monitoring, 5 – data acquisition system and microwave power control, 6 – microwave cavity, 7 – sample holder (fluidized bed), 8 – fiber-optic probes, 9 – temperature monitor, 10 – vents, 11 – air blower and heater, 12 – air speed potentiometer, 13 – air temperature potentiometer, 14 – inlet air temperature monitor. *Source:* Taheri et al. (2020), with permission of Elsevier.

gray mold (BGM) from the seeds during the drying process was evaluated (Taheri et al. 2020). The moisture diffusivities of lentil seeds were described by Fick's second law of diffusion in one direction in a sphere with convective boundary conditions as well as negligible external resistance to mass transfer. The values varied from 0.44×10^{-10} for hot air alone at 50 °C to 3.06×10^{-10} m^2/s for microwave power and air temperature combinations of 500 W and 60 °C by considering convective boundary conditions for the seeds. Drying times obtained from a well-fitted Page's model were applied for the drying of the seeds contaminated with BGM. It was observed that the infected seed percentage (IS%) could be reduced by 30% by applying combinations of microwave power and hot air at 300 W and 60 °C or 400 W and 50 °C, without a significant loss of seed viability. Overall, the MWFBD has good potential to reduce seed-borne BGM in lentil crops.

Oke and Baik (2022) reported the dielectric properties of lentil grain and flour over four moisture contents (MC), seven temperatures, and seven frequencies using a computer-controlled precision LCR meter built for measurements at low frequencies. Both the dielectric constant (ε') and loss factor (ε'') were increased with MC and temperature and, however, decreased with increasing frequency. Table 12.3 reports the dielectric properties of lentil flour at selected frequencies, moisture contents, and temperatures. The decrease was prominent at high MC and temperature as frequency increased from 5 to 30 MHz. At 13.56 MHz, the penetration depth of lentil grain and flour ranged from 1.77 to 34.25 m and 1.78 to 61.07 m, respectively. The resulting penetration depth indicated that thick layers of samples can be processed with adequate heating uniformity using radio-frequency heating as an innovative option in the industrial processing of grains.

Table 12.3 Dielectric properties of lentil flour at selected frequencies, moisture contents (MC), and temperatures.

		Dielectric constant (ε′)				Loss factor (ε″)			
		Frequency (MHz)				Frequency (MHz)			
MC (g/100 g [w.b.])	Temp. (°C)	5	13.56	23	27.12	5	13.56	23	27.12
7.7	30	3.51	3.42	3.34	3.31	0.22	0.11	0.03	0.01
	50	3.96	3.84	3.73	3.70	0.34	0.13	0.08	0.05
	70	4.63	4.39	4.24	4.19	0.65	0.24	0.21	0.17
	90	5.43	5.01	4.78	4.70	1.26	0.50	0.45	0.38
10.6	30	3.82	3.64	3.54	3.50	0.44	0.23	0.12	0.09
	50	4.51	4.20	4.03	3.97	0.87	0.47	0.30	0.25
	70	5.60	4.98	4.69	4.60	1.88	1.03	0.69	0.60
	90	7.05	5.93	5.45	5.31	3.92	2.07	1.41	1.23
12.2	30	4.04	3.79	3.64	3.59	0.77	0.40	0.23	0.19
	50	4.93	4.41	4.17	4.09	1.63	0.86	0.56	0.48
	70	6.35	5.36	4.94	4.82	3.39	1.80	1.21	1.06
	90	8.05	6.49	5.83	5.65	6.55	3.32	2.24	1.97
14.0	30	5.29	4.74	4.47	4.39	1.76	0.94	0.61	0.52
	50	6.63	5.63	5.19	5.06	3.45	1.83	1.24	1.09
	70	8.65	6.94	6.23	6.03	6.60	3.42	2.33	2.06
	90	10.73	8.38	7.38	7.10	12.52	6.07	4.06	3.57

Source: Adapted from Oke and Baik (2022).

Microwave-assisted infrared (MW-IR), an emerging green technology, has been used for the modification of the functional and nutritional properties of lentils to produce high-value plant-based protein ingredients (Heydari et al. 2022a,b). The microwave infrared-assisted Advantium™ was employed to heat lentil seeds tempered to three higher moisture contents to produce modified flours. For MW power, two levels, 0.37 and 0.7 kW, and for infrared power, three levels, 0, 0.375, and 0.75 kW, were selected to perform the thermal processing of tempered lentil seeds. The effect of thermal treatment on the starch structure of lentil samples was observed by varying the absorption intensity at specific bands, in particular the IR peaks at 1047 and 995 cm^{-1}, which are responsive to the molecular order and crystalline region in starch. Results revealed that the degree of order (DO) values of the lentil flours obtained from the processed grains decreased significantly compared to the raw lentil flour. The sharp reductions were observed for the thermally processed samples with the highest initial moisture content of 50%, followed by the tempered samples at 35% and 20%, proving the importance of the initial moisture content during starch gelatinization and showing the higher starch structure alteration for these samples.

The impact of MW and MW-IR thermal processing of lentil seeds pretreated under different conditions on their protein secondary structure can also be observed through the Gaussian spectral deconvolution of the FTIR spectrum (Heydari et al. 2022a,b). The apparent modification demonstrated the reduction and disappearance of the β-band in thermally treated samples. Particle size distributions and scanning electron microscopy analyses illustrated that thermal treatment eased the milling process by breaking down coarse flour particles. The surface of starch granules in raw samples is smooth; they are surrounded by well-defined protein bodies or sections of protein matrix that were cut off during milling. However, after the modification process, it can be observed that the surface of starch granules became rough, and samples tempered to 50% moisture content experienced the most significant impact (Figure 12.7). The modification process was effective in improving the nutritional properties by increasing *in vitro* starch and protein digestibility.

A series of cooking processes (e.g., boiling water, autoclaving, and MW) were conducted for lentils, and furthermore, their quality was assessed (Hefnawy 2011). It was observed that the mineral losses in lentils cooked by MW were lower than those in lentils cooked by boiling or autoclaving. All cooking treatments enhanced the protein efficiency ratio and *in vitro* protein digestibility of cooked lentils in comparison to raw seeds. Autoclaving and MW cooking further reduced the concentrations and activities of trypsin inhibitors, phytic acid, and tannins. In a separate study, Hernandez-Infante et al. (1998) evaluated the effect of MW heating on toxic compounds and protein quality of dry and soaked lentils and found that MW destroyed trypsin inhibitors and hemagglutinins without affecting the protein quality of the legume seeds. Nongmaithem and Meda (2017) studied changes in *in vitro* starch digestibility, phenolic content, and antioxidant activity of germinated red and green lentils by optimizing the processing parameters of pulse-mode microwave-vacuum drying (MWVD). Increased MW power increased *in vitro* starch digestibility significantly. Phenolic content and antioxidant activity levels seemed to differ significantly between the two lentil varieties. There was no discernible effect of vacuum pressure on any of the replies. MWVD showed great potential for the drying of germinated lentils. Divekar et al. (2017) studied the effect of MW treatment on the cooking and macronutrient qualities of red lentils. MW treatment significantly reduced the cooking time, and the lowest cooking time was observed at 600 W, 56 seconds, and 14% moisture content for red lentil.

Ma et al. (2018) focused on the comparative structural characterization of resistant starch after different processing treatments (germination, fermentation, microwave, conventional cooking, and autoclaving) of lentil seeds. The microwaved samples showed lower values for weight-average molecular weight (M_w), average hydrodynamic radius (R_h), double helix content, and crystallinity compared to the untreated and germinated samples. The study suggested that intense degradation and rupture of starch during the above treatments do not favor the formation of ordered crystals, despite the narrow range of molecular size distribution and the greater amount of smaller molecules produced. Also, the degree of double helix increased while the degree of order decreased after microwave processing, suggesting that the destruction of the hydrogen bond linking adjacent helices occurred without causing the disassociation of double helix structure (Yin et al. 2018).

Recently, Najib et al. (2022) performed the drying study of soaked and germinated lentil seeds using the novel MWIR oven and compared the effects of some independent parameters such as variety, physical properties of seeds, germination time, and microwave-infrared

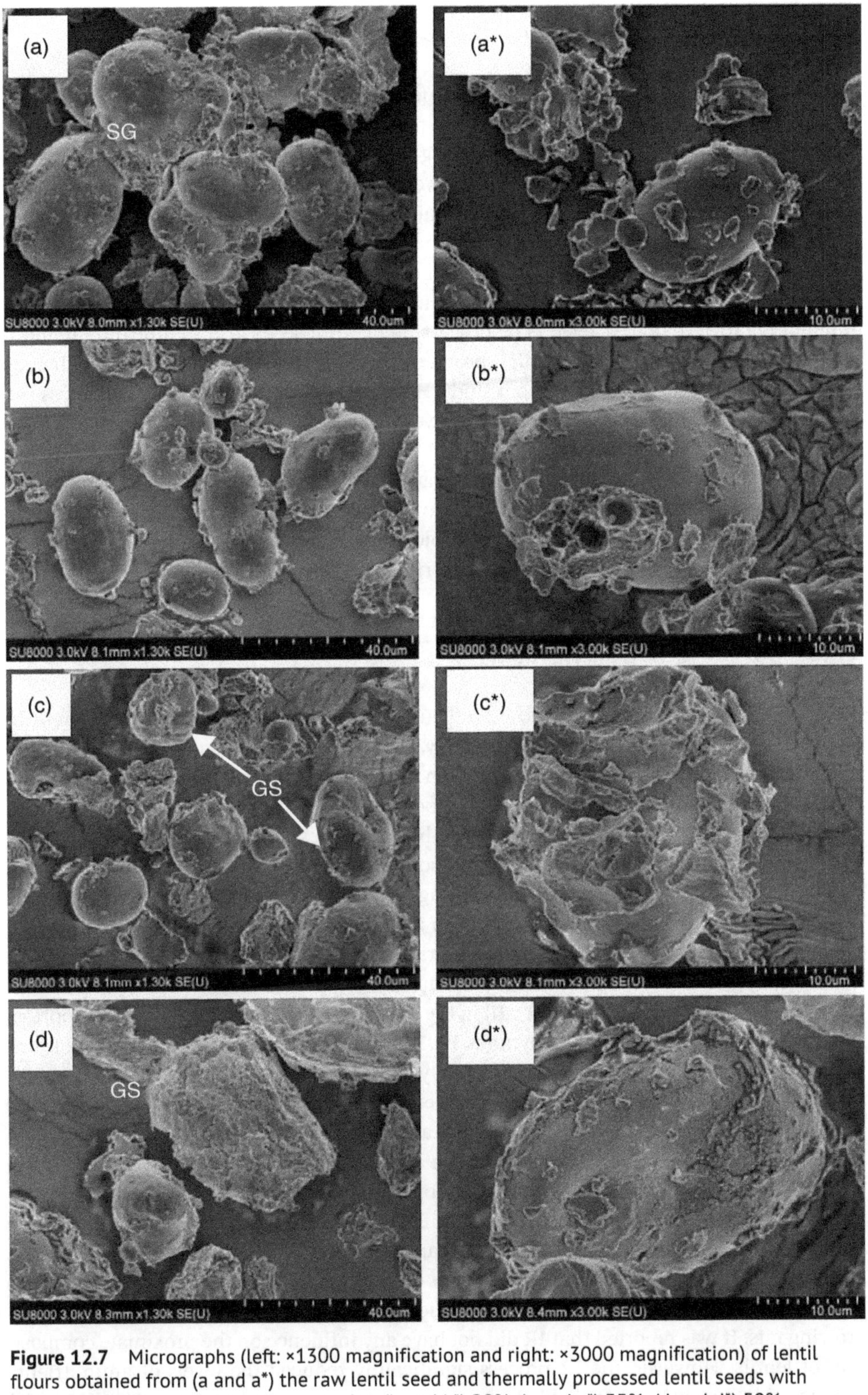

Figure 12.7 Micrographs (left: ×1300 magnification and right: ×3000 magnification) of lentil flours obtained from (a and a*) the raw lentil seed and thermally processed lentil seeds with various tempering conditions: tempered to (b and b*) 20%; (c and c*) 35%; (d and d*) 50%, illustrating SG: starch group, and GS: gelatinized starch. *Source:* Heydari et al. (2022b), with permission of Elsevier/CC BY-NC-ND 4.0.

power on kinetic constants and coefficients and specific energy consumption. The kinetic results showed that the microwave-infrared power significantly affected the drying rate constant (k) of the diffusion approach, while germination time showed no effect on it. The specific energy consumption calculation showed that the experimental condition of microwave power of 0.42 kW and infrared power of 0.375 kW resulted in the least energy consumption, because the transferred water to the surface was in a balanced proportion with infrared radiation. Earlier work by Heydari et al. (2020) concluded that the drying time of lentils can be significantly reduced by a combination of low infrared power (0.375 kW) and different microwave power levels. The processing of lentil seeds also significantly reduced the bioyield force of raw seeds, which provided less firmness to the product and thereby shortened the cooking time. In another study, Heydari et al. (2022a,b) estimated the loss factor and moisture diffusivity during microwave processing of lentils, and the results suggested that the moisture diffusivity increased arithmetically and the loss factor generally rose after processing. It was also observed that the samples tempered up to 50% moisture content showed these transition points in the trend, which can be due to the starch gelatinization. The study confirmed that different bio-chemical reactions can have a noticeable effect on transition points and, therefore, determine microwave energy absorbance.

12.5 Ionizing Irradiation (IR)

Ionizing radiation (IR) has been successfully used as a nonthermal decontamination technology for food products for a while. Generally, gamma rays, X-rays, and electron beams are used for ionizing radiation in food. Irradiation is reported as a safe and effective method to extend the shelf-life of fresh and processed food products. During the process, a predefined dose of IR is passed through food products, which damages the genetic material of the cell by causing a lesion in the DNA or breaking both strands of DNA, resulting in the prevention of multiplication and randomly inhibiting cell functions, resulting in the death of the cell (Lacroix and Follett 2015). According to a Joint Expert Committee convened by the World Health Organization, the Food and Agriculture Organization, and the International Atomic Energy Agency, IR of any food commodity of up to 10 kGy (kilo-Grays) poses no toxicological risk (Anonymous 1981). IR in the range of 0.2–1.0 kGy is efficient in suppressing pest infestation in grains (Anonymous 1991). Many fungi and bacteria spores are killed when the irradiation dose is increased to 5 kGy (Murray 1990).

Most of the research work on lentils has been conducted on the applications of IR doses and their influences on cooking, nutritional, and functional properties. Celik et al. (2007) employed doses of 1, 5, and 10 kGy on green and red lentils and evaluated their cooking quality, water absorption capacity, and electrophoretic patterns of insoluble proteins. The water absorption capacity of lentils was significantly increased at 5 kGy. The wet and dry cooking times of both lentils were significantly reduced with increasing the IR dose. Similarly, El-Niely (2007) examined the effects of IR doses (5–10 kGy) on the nutritive characteristics of lentils, the protein efficiency ratio, and *in vitro* protein digestibility (IVPD) in growing rats. It was reported that IR did not have any influence on the proximate composition of lentils. However, the IR process significantly reduced the levels of antinutrients (phytic acid, tannins) and available lysine compared to untreated samples. The IVPD and protein efficiency ratio of lentils were significantly enhanced in a dose-dependent manner, relative to control samples.

A relatively low-dose gamma irradiation treatment (0.25, 0.50, and 1.0 kGy) and storage time (0–12 months) on the total carotenoids, oligosaccharides, vitamin B content, and sensory properties of green lentils were investigated by Aylangan et al. (2017). The results showed that the IR treatment did not have any adverse effect on the levels of riboflavin and thiamine. The storage period, however, showed effects on the raffinose and stachyose content. All treated lentils were accepted by a sensory panel, and they were not able to differentiate between the treated and the unirradiated samples. Majeed et al. (2017) employed sonication and a combined treatment of sonication and irradiation (5 kGy) for lentil starch. The amylose content of native starch was decreased by individual and dual treatments. The results indicated that the pasting and functional properties of lentil starch were not influenced by sonication; however, they decreased significantly with combined treatments. FT-IR analysis revealed a decrease in the intensities of O–H, C–H, and O=C stretches and CH_2 bending upon sonication and dual treatments.

12.6 Ultrasound (US) Processing

Ultrasound (US) refers to sound waves whose frequency exceeds the human ear's hearing limit of 20 kHz. High-intensity (10–1000 W/cm^2) or low-frequency US (20–100 kHz) is an emerging area in food processing and is considered "green processing," which modifies the physical, biochemical, and mechanical properties of food products by acoustic cavitation while increasing temperature and pressure (Li et al. 2021). The propagation of US in a liquid produces bubble cavitation because of the pressure changes. The collapse of those micro-bubbles is responsible for an increase in temperature and pressure. As a result, the high pressure and concentrated local energy produce a localized pasteurization effect without a considerable rise in macro-temperature (Jimenez-Sanchez et al. 2017). Based on the frequency range, ultrasonic applications in food processing and quality control can be separated into low energy (low power, low intensity) and high energy (high power, high intensity). Low-energy US (frequency > 100 kHz, intensity 11 W/cm^2 per square) has a disruptive impact on the biochemical, physicochemical, and mechanical properties of foods (Awad et al. 2012). High-energy systems are quite attractive for food industries for controlling microstructure and modifying textural characteristics, defoaming, pasteurization, and inactivation or acceleration of enzymatic activity to enhance the quality and shelf-life of products (Awad et al. 2012; Bermúdez-Aguirre et al. 2011; Rifna and Dwivedi 2022). However, due to a lack of effective ultrasonic power system design, the technology remains on a laboratory or pilot scale. US technology has been applied to lentils to increase the germination rate.

Yildirima et al. (2007) reported that the US treatment reduced soaking and cooking time, as well as the amount of energy consumed in the cooking process, thus boosting the consumption of dried legumes. Although US treatment imparts many changes in the processed products, it fails to remove antinutrients like phytic acid in legumes (Kaya et al. 2017). Kaya et al. (2017) evaluated the effect of various US conditions on phenolic content, dietary fiber, oil absorption ability, phytic acid, antioxidant activity, water absorption capacity, and the composition of phenolic acid in the hulls of red and green lentils. A higher dietary fiber content was achieved in red lentils after the US treatment. In comparison to the control, ultrasonically treated lentil hulls had reduced phenolic content and Trolox equivalent antioxidant capacity values, whereas phenolic acids did not change after the treatment. Han and Baik (2006) reported a significant reduction in some antinutrients in lentils, such as oligosaccharides (raffinose, ciceritol, and stachyose), which can be due to increased leaching of these oligosaccharides during the process. Major oligosaccharides were decreased after ultrasonic soaking (47 MHz) of uncooked soaked and cooked soaked lentils. Del Hierro et al. (2018) separated bioactive compounds, including saponins and sapogenins, from lentils using a probe-type (1/2″ diameter) ultrasound-assisted extraction system with sonication output amplitude of 60% and temperature of 75 °C. Thus, ultrasonication can be used as an effective method to treat as well as extract antinutrients and bioactive compounds, thereby modifying the structural and functional properties of lentils in an intensity-dependent manner.

Ultrasonic waves can also be used to soften legumes, which can be used in the preparation of ready-to-eat food products, soft texture, new food products, food materials with high supplementary value, and new healthy foods without antinutrients. Han et al. (2007) treated lentils with ultrasound or high hydrostatic pressure during soaking and then heated them for 30 minutes at 98 °C and demonstrated their *in vitro* protein digestibility (IVPD) using the three selected enzymes (peptidase, trypsin, and chymotrypsin). The IVPD after soaking lentils was increased. The protein molecules of 52 kDa remaining after enzyme hydrolysis of lentil proteins might be responsible for the lower digestibility of insoluble proteins in comparison to soluble proteins.

12.7 Ozone Treatment

The application of ozone treatment to food, either in liquid or gaseous form, has created interest as a novel green processing technology. Ozone is an oxidizing colorless gas and a triatomic allotrope of oxygen that decomposes readily to oxygen atoms and hydroxyl radicals. It has a high oxidation potential of 2.07 V in alkaline solution over chlorine (1.36 V), and, therefore, it can be used as an effective antimicrobial agent. Furthermore, the decomposition of ozone into oxygen without producing toxic residues makes it an environment friendly sanitizer. A trace amount of ozone with a short contact time produces the desired antimicrobial effect. Excess ozone auto-decomposes rapidly to produce oxygen, and thus it leaves no residues in food. Such advantages make ozone attractive to the food industry, and consequently, it has been declared generally recognized as safe for use in food processing by the U.S. Food and Drug Administration in 1997 (Graham 1997). The oxidation effect of ozone has antimicrobial and antiviral activity, and it can modify starch, eliminate pests, and degrade pesticides to improve food safety and quality. An ozone generator is illustrated in Figure 12.8.

De Souza et al. (2021) conducted a study to examine the influence of ozone exposure on water uptake and germination of lentil seeds using a commercial ozone generator that provides a concentration of 1 g/m^3 of ozone. Ten lentil seeds were used per treatment performed at selected times of exposure (2, 3, 5, 10, and 15 minutes) to ozone. It was

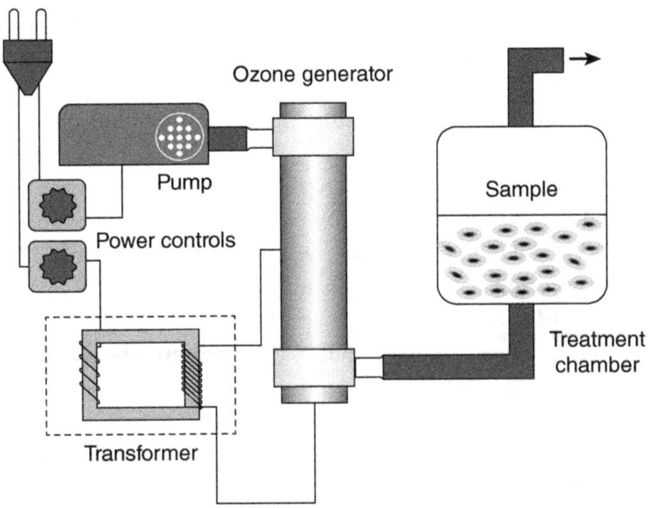

Figure 12.8 Experimental set-up ozone generation unit. *Source:* Anbarasan et al. (2022), with permission of Elsevier.

observed that the samples treated for three and five minutes showed the maximum water uptake in the seeds as a function of the imbibition time. Furthermore, the growth rates of the stem, root, and mass were significantly higher than the control sample after seven days of germination. The FT-IR analysis demonstrated that the formation of bands at 1345 cm^{-1}/(NO_3^-) related to ethylene ozonolysis and also a reduction of the N–H band (amide II) at 1551–1550 cm^{-1} and increase of the C—H bond at 1543 cm^{-1}, evidencing a possible action of ozone on lentil proteins. This fact is probably related to the enhancement of the seed germination process, which allows the germination rates to be 90% for samples treated with ozone for 3 and 10 minutes. The effect of exposure of ozone on the development of lentil seedlings demonstrated that the ozone treatment resulted in controlling the activity of an enzyme that inhibited the seedling growth (Maccarrone et al. 1997). In addition, the reduction of activity of amine oxidase correlated with the rise in putrescine levels in the ozone-treated plantlets, which aids in the regulation of physiological processes such as embryogenesis, which aids in seedling development and maturation.

12.8 Ultrafiltration (UF) and Isoelectric Precipitation (IEP)

Among the wet processing techniques, ultrafiltration (UF) and isoelectric precipitation (IEP) are leading ones. Basically, both are used for either purification or isolating proteins in their purest form so that the isolated products can be used in the processing and formulation of food products, providing texture, taste, and nutrition for a desirable and pleasurable product. IEP involves the fractionation of proteins by altering their solubility by changing the pH of the solution. At the isoelectric point of lentil (pH 4.3), protein–protein interactions that result in precipitation are more favored, allowing separation of proteins from the flour (Aryee and Boye 2017). However, both UF and IEP have significant effects on their structural characteristics, functional properties, and nutritional quality, in particular when inactivating or reducing antinutritional factors (Barbana and Boye 2013).

While comparing between UF and IEP, Boye et al. (2010) reported that UF-processed red and green lentil concentrates generally had higher solubility than the IEP-processed concentrate, with the exception of pH 5 and 6, where solubility was low for both UF and IEP concentrates. The LPI showed high solubility at both acidic and alkaline pH values; however, the maximum solubility was recorded at pH 9.0 (Aryee and Boye 2017). The solubility was very low, between pH 4 and 6. These results correspond to the isoelectric point of the proteins present in lentil. Alonso-Miravalles et al. (2019) produced protein products from lentils using both UF and IEP. It was observed that the UF yielded higher protein concentrations (>90%), whereas the IEP resulted in higher contents of dietary fiber and some minerals (i.e., sodium and phosphorus). There were significant differences in the functional properties between the two ingredients obtained through two different techniques. The isoelectric precipitated ingredient showed lower protein solubility over the studied pH range (3–9), which can be linked to the slightly higher hydrophobicity values and total sulfhydryl groups. Conversely, the protein ingredient obtained by UF was superior in terms of its solubility, fat-binding capacity, water holding capacity, gelling properties, and foam-forming capacity.

In a separate study, Joehnke et al. (2021) reported reductions of about 91% and 58% in total galacto-oligosaccharides when lentil proteins were isolated using UF and IEP, respectively. Additionally, trypsin inhibitor activity decreased while IVPD improved in lentil protein isolated by UF and IEP. IEP involves the fractionation of protein by the solubility modification with a change in pH in the solution. Lentil protein is fractionated from the flour at its isoelectric point due to precipitation caused by protein–protein interactions (Osemwota et al. 2021; Khazaei et al. 2019; Barbana and Boye 2013). Because of the stronger protein–phenol complexion during isoelectric precipitation, it was observed that the protein isolate produced more than double total polyphenolic compounds. On the contrary, a decrease in the antinutritional factors, namely phytic acid-phosphorus (46%) and trypsin inhibitors (69%) were observed in the LPI compared to raw lentil flour. Trypsin inhibitor activity was also reduced in protein isolated by isoelectric precipitation from red and green lentil flours (Barbana and Boye 2013). Therefore, the isoelectric precipitation technique can be used to obtain LPI and improve its digestibility.

12.9 Ultraviolet (UV) and Visible Light Treatment

Ultraviolet (UV) light technology is an inexpensive, nonthermal, simple, and nonchemical approach applied in the food sector for the inactivation of bacteria. UV is an electromagnetic radiation with four wavelength ranges: UV-A (315–400 nm), UV-B (280–315 nm), UV-C (200–280 nm), and vacuum UV (100–200 nm) (Krishnamurthy et al. 2008). The UV-C light treatment uses radiation from the electromagnetic spectrum (200–280 nm) and a powerful surface germicidal method. It is safe to use; however, certain simple precautions should be followed to avoid exposure to light and evacuate the generated ozone. UV-B (280–315 nm) and UV-C (100–280 nm) treatments have also been shown to enhance specific nutrients and nutraceutical compounds. UV has extensive antimicrobial properties with the potential to kill viruses (Eischeid et al. 2009), spores (Silva-Espinoza et al. 2020), and numerous strains of bacteria and yeast (Gómez-López et al. 2005). The UV radiant exposure must be at least 400 J/m^2 in all parts of the product to achieve the desired microbial inactivation. The power, the geometric configuration of the reactor, the physical arrangement, the transmissivity of the product, the wavelength of the ultraviolet source(s), the profile of product flow, and the length of the radiation route are the primary governing parameters in the process. Other alternative process methods, such as powerful oxidizing agents like hydrogen peroxide and ozone, may be employed in conjunction with ultraviolet radiation (FDA 2000).

A computer vision system was successfully employed to quantify the influence of UV-A and visible light on the color of lentils with red, green, and yellow cotyledons (Jackson et al. 2021). The authors selected 20 samples of cotyledons from each of the three-color classes and subjected to six light treatments (ultraviolet, full-spectrum visible, red, green, blue, and dark control) for seven days at room temperature. Results of the analysis indicated that light exposure had a significant influence on all three cotyledon color classes. It was observed that color changes in green lentil cotyledons exposed to all light treatments were significantly different from color changes in the control and had large effect sizes. For red cotyledon lentils, the effect sizes were small, and not all the light treatments resulted in significant changes in all color coordinates. The effect sizes on yellow cotyledon lentil were also small, albeit with significant UV-A, full-visible, and green light effects on the L*-value; blue, full-visible, and green light on the a*-value; blue, full-visible, and green light on b*-value; and all treatments except red light on overall color change. Furthermore, it was observed that light-induced color loss in the cotyledons was significantly influenced by seed-coat color class. The order of protective effect of lentil seed coat from least to highest was found to be as follows: gray-zero tannin, green, normal gray, and black. Results infer that breeding for seed coat protection may improve the overall quality of green lentils.

12.10 Pulsed Light Treatment

Pulsed light food processing was developed as a nonthermal food processing technique that involves the discharge of high-voltage electric pulses (up to 70 kV/cm) into the food product placed between the two electrodes for a few seconds. It is one of the most promising nonthermal approaches for replacing traditional thermal pasteurization. The method has successfully reduced pathogens, pests, and spoilage microorganisms in food while maintaining its quality (Abida et al. 2014). The effect of light and pulsed light treatments on the germination process was quantified by Vasilean et al. (2018). It was reported that the percentage of germinated seeds in lentils increased after the pulsed light treatment. The most significant increase in the antioxidant capacity of the germinated lentils was observed at a high flux value of 43.2 J/cm^2. After the pulse light treatment, lentils have the same amount of protein as in the case of the control sample.

12.11 Conclusions

Nonthermal technologies are being used in food processing along with conventional processing or as a complement. Several industries have adopted these technologies for their product development and to improve the functionality of foods. Among available technologies, high-pressure (HP) processing and irradiation (IR) have had commercial successes for lentil grains and isolates. For the decontamination of lentils, the employment of IR proves its superiority over others. HP processing, on the other hand, is quite successful for the development of functional foods by modifying lentil proteins and starch into a desired texture. The application of HP homogenization is another area that has been extensively used for the development of micro- and nanoemulsions leading to food and drug delivery. For LPIs, isoelectric precipitation is a simple technique that could be used as a nonthermal processing technology for small and medium industries. The limitations of the employment of nonthermal processing in lentil or related industries are restricted because of the high cost of the equipment. The cost of the equipment is expected to be reduced in the coming days with more adaptation of the technology in the food industry and further technological innovations.

References

Abida, J., Rayees, B., and Masoodi, F.A. (2014). Pulsed light technology: a novel method for food preservation. *International Food Research Journal* 21 (3): 839–848.

Ahmed, J. and Ramaswamy, H.S. (2020). Microwave Pasteurization and Sterilization of Foods. In: *Handbook of Food Preservation*, 3e (ed. M.S. In Rahman). Boca Raton: CRC Press.

Ahmed, J., Varshney, S.K., and Ramaswamy, H.S. (2009). Effect of high pressure treatment on thermal and rheological properties of lentil flour slurry. *LWT Food Science and Technology* 42 (9): 1538–1544. https://doi.org/10.1016/j.lwt.2009.05.002.

Ahmed, J., Thomas, L., Taher, A., and Joseph, A. (2016). Impact of high pressure treatment on functional, rheological, pasting, and structural properties of lentil starch dispersions. *Carbohydrate Polymers* 152: 639–647. https://doi.org/10.1016/j.carbpol.2016.07.008.

Ahmed, J., Mulla, M., Al-Ruwaih, N., and Arfat, Y.A. (2019). Effect of high-pressure treatment prior to enzymatic hydrolysis on rheological, thermal, and antioxidant properties of lentil protein isolate. *Legume Science* 1: e10. https://doi.org/10.1002/leg3.10.

Alonso-Miravalles, L., Jeske, S., Bez, J. et al. (2019). Membrane filtration and isoelectric precipitation technological approaches for the preparation of novel, functional and sustainable protein isolate from lentils. *European Food Research and Technology* 245 (9): 1855–1869. https://doi.org/10.1007/s00217-019-03296-y.

Anbarasan, R., Jaspin, S., Bhavadharini, B. et al. (2022). Chlorpyrifos pesticide reduction in soybean using cold plasma and ozone treatments. *LWT Food Science and Technology* 159: 113193. https://doi.org/10.1016/j.lwt.2022.113193.

Anonymous (1981). Wholesomeness of irradiated food. Report of A Joint FAO/IAEA/WHO Expert Committee. Technical Report Series 659. Geneva: World Health Organization. Wholesomeness of irradiated food: report of a Joint FAO/IAEA/WHO Expert Committee [meeting held in Geneva from 27 October to 3 November 1980].

Anonymous (1991). Analytical detection methods for irradiated foods. A review of the current literature. http://iaea.org.

Aryee, A.N.A. and Boye, J.I. (2017). Comparative study of the effects of processing on the nutritional, physicochemical and functional properties of lentil. *Journal of Food Processing and Preservation* 41 (1): e12824. https://doi.org/10.1111/jfpp.12824.

Awad, T.S., Moharram, H.A., Shaltout, O.E. et al. (2012). Applications of ultrasound in analysis, processing and quality control of food: a review. *Food Research International* 48: 410–427. https://doi.org/10.1016/j.foodres.2012.05.004.

Aylangan, A., Ic, E., and Ozyardimci, B. (2017). Investigation of gamma irradiation and storage period effects on the nutritional and sensory quality of chickpeas, kidney beans and green lentils. *Food Control* 80: 428–434. https://doi.org/10.1016/j.foodcont.2017.04.005.

Barbana, C. and Boye, J.I. (2013). *In vitro* protein digestibility and physico-chemical properties of flours and protein concentrates from two varieties of lentil (*Lens culinaris*). *Food and Function* 4 (2): 310–321. https://doi.org/10.1039/c2fo30204g.

Barbosa-Cánovas, G.V., Pothakamury, U.R., Palou, E., and Swanson, B.G. (1998). *Nonthermal preservation of foods*. New York: Marcel Dekker, Inc.

Bermúdez-Aguirre, D., Mobbs, T., and Barbosa-Cánovas, G.V. (2011). Ultrasound applications in food processing. In: *Ultrasound Technologies for Food and Bioprocessing*, 65–105. New York, NY: Springer. https://doi.org/10.1007/978-1-4419-7472-3.

Bou, R., Navarro-Vozmediano, P., Domínguez, R. et al. (2022). Application of emerging technologies to obtain legume protein isolates with improved techno-functional properties and health effects. *Comprehensive Reviews in Food Science and Food Safety* 21 (3): 2200–2232. https://doi-org.lib-ezproxy.concordia.ca/10.1111/1541-4337.12936.

Boye, J., Zare, F., and Pletch, A. (2010). Pulse proteins: processing, characterization, functional properties and applications in food and feed. *Food Research International* 43: 414–431.

Cai, R., McCurdy, A., and Baik, B.K. (2002). Textural property of 6 legume curds in relation to their protein constituents. *Journal of Food Science* 67: 1725–1730.

Carbonaro, M., Maselli, P., and Nucara, A. (2012). Relationship between digestibility and secondary structure of raw and thermally treated legume proteins: a Fourier transform infrared (FT-IR) spectroscopic study. *Amino Acids* 43 (2): 911–921.

Celik, S., Basman, A., Yalcin, E., and Koksel, H. (2007). Effects of irradiation on protein electrophoretic properties, water absorption and cooking quality of dry bean and chickpea. *Food Science and Technology Research* 10 (4): 410–415. https://doi.org/10.1080/10428190500085701.

De Souza, P., Souza, G., Pinto, J. et al. (2021). Effect of ozone exposure on water uptake and germination of lentil (*Lens culinaris*) seeds. *Ozone: Science & Engineering* 43 (1): 48–59. https://doi.org/10.1080/01919512.2020.1754163.

Del Hierro, J.N., Herrera, T., García-Risco, M.R. et al. (2018). Ultrasound-assisted extraction and bioaccessibility of saponins from edible seeds: quinoa, lentil, fenugreek, soybean and lupin. *Food Research International* 109: 440–447. https://doi.org/10.1016/j.foodres.2018.04.058.

Divekar, M.T., Karunakaran, C., Lahlali, R. et al. (2017). Effect of microwave treatment on the cooking and macronutrient qualities of pulses. *International Journal of Food Properties* 20 (2): 409–422. https://doi.org/10.1080/10942912.2016.1163578.

Eischeid, A.C., Meyer, J.N., and Linden, K.G. (2009). UV disinfection of adenoviruses: molecular indications of DNA damage efficiency. *Applied and Environmental Microbiology* 75 (1): 23–28. https://doi.org/10.1128/AEM.02199-08.

Ekezie, F.-G.C., Sun, D.-W., Han, Z., and Cheng, J.-H. (2017). Microwave-assisted food processing technologies for enhancing product quality and process efficiency: a review of recent developments. *Trends in Food Science & Technology* 67: 58–69. https://doi.org/10.1016/j.tifs.2017.05.014.

El-Niely, H.F. (2007). Effect of radiation processing on antinutrients, *in-vitro* protein digestibility and protein efficiency ratio bioassay of legume seeds. *Radiation Physics and Chemistry* 76 (6): 1050–1057. https://doi.org/10.1016/j.radphyschem.2006.10.006.

Elnjikkal Jerome, R. and Dwivedi, M. (2022). Microwave vacuum drying of pomegranate peel: evaluation of specific energy consumption and quality attributes by response surface methodology and artificial neural network. *Journal of Food Processing and Preservation* 46 (3): e16325. https://doi.org/10.1111/jfpp.16325.

FDA (2000). A report of the Institute of Food Technologists for the Food and Drug Administration of the U.S. Department of Health and Human Services. https://www.fda.gov/files/food/published/Evaluation-and-Definition-of-Potentially-Hazardous-Foods.pdf.

Garcia-Mora, P., Peñas, E., Frias, J. et al. (2015). High-pressure improves enzymatic proteolysis and the release of peptides with angiotensin I converting enzyme inhibitory and antioxidant activities from lentil proteins. *Food Chemistry* 171: 224–232. https://doi.org/10.1016/j.foodchem.2014.08.116.

Gómez-López, V.M. and Bhat, R. (Eds.). (2021). Electromagnetic Technologies in Food Science. Wiley. https://doi.org/10.1002/9781119759522.

Gómez-López, V., Devlieghere, F., Bonduelle, V., and Debevere, J. (2005). Factors affecting the inactivation of micro-organisms by intense light pulses. *Journal of Applied Microbiology* 99 (3): 460–470. https://doi.org/10.1111/j.1365-2672.2005.02641.x.

Graham, D.M. (1997). Use of ozone for food processing. *Food Technology* 51: 121. https://doi.org/10.1002/9781118307472.oth1.

Guo, W., Wang, S., Tiwari, G., Johnson, J.A., and Tang, J. (2010). Temperature and moisture dependent dielectric properties of legume flour associated with dielectric heating. *LWT Food Science and Technology* 43 (2010) 193–201. https://doi.org/10.1016/j.lwt.2009.07.008.

Hall, A.E. and Moraru, C.I. (2021). Structure and function of pea, lentil and faba bean proteins treated by high-pressure processing and heat treatment. *LWT Food Science and Technology* 152: 112349. https://doi.org/10.1016/j.lwt.2021.112349.

Han, I.H. and Baik, B.K. (2006). Oligosaccharide content and composition of legumes and their reduction by soaking, cooking, ultrasound, and high hydrostatic pressure. *Cereal Chemistry* 83 (4): 428–433. https://doi.org/10.1094/CC-83-0428.

Han, I.H., Swanson, B.G., and Baik, B.K. (2007). Protein digestibility of selected legumes treated with ultrasound and high hydrostatic pressure during soaking. *Cereal Chemistry* 84 (5): 518–521. https://doi.org/10.1094/CCHEM-84-5-0518.

Hefnawy, T.H. (2011). Effect of processing methods on nutritional composition and anti-nutritional factors in lentils (*Lens culinaris*). *Annals of Agricultural Sciences* 56 (2): 57–61. https://doi.org/10.1016/j.aoas.2011.07.001.

Hernandez-Infante, M., Sousa, V., Montalvo, I., and Tena, E. (1998). Impact of microwave heating on hemagglutinins, trypsin inhibitors and protein quality of selected legume seeds. *Plant Foods for Human Nutrition* 52 (3): 199–208. https://doi.org/10.1023/A:1008033610737.

Heydari, M.M., Kauldhar, B.S., and Meda, V. (2020). Kinetics of a thin-layer microwave-assisted infrared drying of lentil seeds. *Legume Science* 2 (2): e31. https://doi.org/10.1002/leg3.31.

Heydari, M.M., Najib, T., Baik, O.D. et al. (2022a). Loss factor and moisture diffusivity property estimation of lentil crop during microwave processing. *Current Research in Food Science* 5: 73–83. https://doi.org/10.1016/j.crfs.2021.12.008.

Heydari, M.M., Najib, T., and Meda, V. (2022b). Investigating starch and protein structure alterations of the processed lentil by microwave assisted infrared thermal treatment and their correlation with the modified properties. *Food Chemistry Advances* 1: 100091. https://doi.org/10.1016/j.focha.2022.100091.

Jackson, N., Noble, S.D., and Vandenberg, A. (2021). Evaluating the effect of light exposure and seed coat on lentil cotyledon color by computer vision. *Legume Science* 3 (4): e98. https://doi.org/10.1002/leg3.98.

Jiménez-Sánchez, C., Lozano-Sánchez, J., Segura-Carretero, A., and Fernández-Gutiérrez, A. (2017). Alternatives to conventional thermal treatments in fruit-juice processing. Part 1: techniques and applications, *Critical Reviews in Food Science and Nutrition* 57 (3): 501–523. https://doi.org/10.1080/10408398.2013.867828.

Joehnke, M.S., Jeske, S., Ispiryan, L. et al. (2021). Nutritional and anti-nutritional properties of lentil (*Lens culinaris*) protein isolates prepared by pilot-scale processing. *Food Chemistry: X* 9: 100112. https://doi.org/10.1016/j.fochx.2020.100112.

Karim, A.A., Norziah, M.H., and Seow, C.C. (2000). Method for the study of starch retrogradation. *Food Chemistry* 71: 9–31.

Kaya, E., Tuncel, N.B., and Tuncel, Y.N. (2017). The effect of ultrasound on some properties of pulse hulls. *Journal of Food Science and Technology* 54 (9): 2779–2788. https://doi.org/10.1007/s13197-017-2714-5.

Khazaei, H., Subedi, M., Nickerson, M. et al. (2019). Seed protein of lentils: current status, progress, and food applications. *Foods* 8 (9): 391. https://doi.org/10.3390/foods8090391.

Krishnamurthy, K., Irudayaraj, J., Demirci, A., and Yang, W. (2008). UV pasteurization of food materials. In: *Food Processing Operations Modeling: Design and Analysis*, 2e (ed. S. Jun and J.M. Irudayaraj), 281–299. Boca Raton: CRC Press/Taylor and Francis Group. https://doi.org/10.1201/9781420055542.

Lacroix, M. and Follett, P. (2015). Combination irradiation treatments for food safety and phytosanitary uses. *Stewart Postharvest Review* 11 (3).

Lee, H.C., Htoon, A.K., Uthayakumaran, S., and Paterson, J. (2007). Chemical and functional quality of protein isolated from alkaline extraction of Australian lentil cultivars: Maltida and Digger. *Food Chemistry* 102: 1199–1207.

Li, W., Gamlath, C.J., Pathak, R. et al. (2021). Ultrasound – the physical and chemical effects integral to food processing. *Innovative Food Processing Technologies* 329–358. https://doi.org/10.1016/b978-0-08-100596-5.22679-6.

Ma, Z., Yin, X., Hu, X. et al. (2018). Structural characterization of resistant starch isolated from *Laird lentils* (*Lens culinaris*) seeds subjected to different processing treatments. *Food Chemistry* 263: 163–170. https://doi.org/10.1016/j.foodchem.2018.04.122.

Maccarrone, M., Veldink, G.A., Vliegenthart, J.F., and Finazzi Agrò, A. (1997). Ozone stress modulates amine oxidase and lipoxygenase expression in lentil (*Lens culinaris*) seedlings. *FEBS Letters* 408 (2): 241–244. https://doi.org/10.1016/S0014-5793(97)00433-X.

Majeed, T., Wani, I.A., and Hussain, P.R. (2017). Effect of dual modification of sonication and γ-irradiation on physicochemical and functional properties of lentil (*Lens culinaris* L.) starch. *International Journal of Biological Macromolecules* 101: 358–365. https://doi.org/10.1016/j.ijbiomac.2017.03.110.

Murray, D.R. (1990). *Biology of Food Irradiation*. New York: Wiley.

Najib, T., Heydari, M.M., and Meda, V. (2022). Combination of germination and innovative microwave-assisted infrared drying of lentils: effect of physicochemical properties of different varieties on water uptake, germination, and drying kinetics. *Applied Food Research* 2 (1): 100040. https://doi.org/10.1016/j.afres.2021.100040.

Nongmaithem, R. and Meda, V. (2017). Optimization of microwave vacuum drying parameters for germinated lentils based on starch digestibility, antioxidant activity and total phenolic content. *International Journal of Food Studies* 6 (1): 44–55. https://doi.org/10.7455/ijfs/6.1.2017.a5.

Oke, A.B. and Baik, O.-D. (2022). Role of moisture content, temperature, and frequency on dielectric behaviour of red lentil and Kabuli chickpea in relation to radio frequency heating. *Applied Food Research* 2: 100046. https://doi.org/10.1016/j.afres.2022.100046.

Oliveira, M.E.C. and Franca, A.S. (2002). Microwave heating of foodstuff. *Journal of Food & Engineering* 53: 347–359. https://doi.org/10.1016/S0260-8774(01)00176-5.

Osemwota, E.C., Alashi, A.M., and Aluko, R.E. (2021). Comparative study of the structural and functional properties of membrane-isolated and isoelectric pH precipitated green lentil seed protein isolates. *Membranes* 11 (9): 694. https://doi.org/10.3390/membranes11090694.

Rifna, E. and Dwivedi, M. (2021). Optimization and validation of microwave–vacuum drying process variables for recovery of quality attribute and phytochemical properties in pomegranate peels (*Punica granatum* L. cv. Kabul). *Journal of Food Measurement and Characterization* 15 (5): 4446–4464. https://doi.org/10.1007/s11694-021-01016-3.

Rifna, E. and Dwivedi, M. (2022). Effect of pulsed ultrasound assisted extraction and aqueous acetone mixture on total hydrolysable tannins from pomegranate peel. *Food Bioscience*, 45, 101496. https://doi.org/10.1016/j.fbio.2021.101496.

Rifna, E., Ramanan, K.R., and Mahendran, R. (2019). Emerging technology applications for improving seed germination. *Trends in Food Science & Technology* 86: 95–108. https://doi.org/10.1016/j.tifs.2019.02.029.

Saricaoglu, F.T. (2020). Application of high-pressure homogenization (HPH) to modify functional, structural and rheological properties of lentil (*Lens culinaris*) proteins. *International Journal of Biological Macromolecules* 144: 760–769. https://doi.org/10.1016/j.ijbiomac.2019.11.034.

Scippa, G.S., Trupiano, D., Rocco, M. et al. (2008). An integrated approach to the characterization of two autochthonous lentil (*Lens culinaris*) landraces of Molise (south-central Italy). *Heredity* 101: 136–144. https://doi.org/10.1038/hdy.2008.39.

Silva-Espinoza, B.A., Palomares-Navarro, J.J., Tapia-Rodriguez, M.R. et al. (2020). Combination of ultraviolet light-C and clove essential oil to inactivate Salmonella Typhimurium biofilms on stainless steel. *Journal of Food Safety* 40: e12788. https://doi.org/10.1111/jfs.12788.

Swanson, B.G. (1990). Pea and lentil protein extraction and functionality. *Journal of the American Oil Chemists' Society* 67: 276–280.

Tabilo-Munizaga, G., Villalobos-Carvajal, R., Herrera-Lavados, C. et al. (2019). Physicochemical properties of high-pressure treated lentil protein-based nanoemulsions. *LWT Food Science and Technology* 101: 590–598. https://doi.org/10.1016/j.lwt.2018.11.070.

Taheri, S., Brodie, G., and Gupta, D. (2020). Microwave fluidised bed drying of red lentil seeds: Drying kinetics and reduction of botrytis grey mold pathogen. *Food and Bioproducts Processing* 119: 390–401. https://doi.org/10.1016/j.fbp.2019.11.001.

Vasilean, I., Aprodu, I., Neculau, M., and Patrascu, L. (2018). Effects of pulsed light treatment on germination efficiency of pulses. *Scientific Papers: Series D* 61: 266–274.

Yildirima, A., Bayramb, M., and Durdu Onerb, M. (2007). The application of ultrasonic system to improve legume cooking properties. *Foods* 7 (10): 164. https://doi.org/10.3390/foods7100164.

Yin, X., Ma, Z., Hu, X., Li, X., and Boye, J.I. (2018). Molecular rearrangement of *Laird lentil* (*Lens culinaris* Medikus) starch during different processing treatments of the seeds. *Food Hydrocolloids*, 79, 399–408. https://doi.org/10.1016/j.foodhyd.2018.01.012.

Yu, J., Yu, X.L., Han, Z.Y. et al. (2017). Percutaneous cooled-probe microwave versus radiofrequency ablation in early-stage hepatocellular carcinoma: a phase III randomised controlled trial. *Gut* 66 (6): 1172–1173. https://doi.org/10.1136/gutjnl-2016-312629.

Part III

Nutrition, Antinutrients, Sensory Properties, and Global Consumption Trends

13

Nutritional Profile, Bioactive Compounds, and Health Benefits of Lentils

Sanju Bala Dhull[1], Mark A. Uebersax[2], Joyce Kinabo[3], and Muhammad Siddiq[4]

[1] *Department of Food Science and Technology, Chaudhary Devi Lal University, Sirsa, Haryana, India*
[2] *Department of Food Science and Human Nutrition, Michigan State University, East Lansing, MI, USA*
[3] *Department of Human Nutrition and Consumer Sciences, Sokoine University of Agriculture, Morogoro, Tanzania*
[4] *Food Science Consultant, Windsor, ON, Canada*

13.1 Introduction

Legumes are globally important, nutrient-dense food source for human consumption. Among legumes, lentil (*Lens culinaris* Medik.) is important pulse crop with respect to global production, trade, and consumption. It is an annual legume plant with its lens-shaped seeds. The total world production of lentils was 5.61 million metric tons in 2021, which has increased by about 110% since 1991. Lentil production showed a more rapid growth from 2011 to 2021 (i.e., an increase of 1.16 million metric tons) as compared to the preceding two decades. Traditionally, India had been the leading lentil producer in the world, Canada has emerged as a leader in lentil production in recent years, while India ranks second. Although there have been wide variations in cultivated area of lentils in recent years, overall, it has also increased significantly from 3.27 to 5.59 million hectares since 1991 (FAO 2022). Lentil yield, similar to any other field crop's yield, is affected by climatic conditions (i.e., drought, high temperature, rainfall, flooding) and biotic stresses, i.e., mainly diseases. It is noteworthy that lentil crop's quality management, both at the preharvest and postharvest stages, is critical to maximize lentil's economic value and consumption. In particular, the poor postharvest storage and handling practices can result in quantitative and qualitative losses. A careful monitoring of storage temperature and relative humidity is important during postharvest storage of lentils for optimum quality preservation. A separate chapter in this book, entitled *Preharvest quality management, postharvest handling and consumption trends of lentils* provides more details with respect to postharvest handling, storage, and quality management.

Lentils are an excellent source of proteins, low digestible carbohydrates, dietary fibers, and selected minerals and vitamins. Lentils have significant contents of resistant starch and a number of bioactive phytochemicals (Dhull et al. 2022b; Joshi et al. 2017). However,

Lentils: Production, Processing Technologies, Products, and Nutritional Profile, First Edition.
Edited by Jasim Ahmed, Muhammad Siddiq, and Mark A. Uebersax.

there are wide variations in the composition and nutritional profile of lentils with respect to varietal differences. Furthermore, the wide variations in total fiber content could be attributed partially to differences in analytical methods. Grusak (2009) reported following compositional ranges, compiled from different studies: protein (15.9–31.4%), carbohydrates (43.4–74.9%), fat (0.3–3.5%), total fiber (5.1–26.6%), and ash (2.2–6.4%). According to the United States Department of Agriculture's nutrient database, raw lentils' protein, carbohydrates, and ash content is 24.6%, 63.4%, and 2.7%, respectively (USDA 2022). These USDA data on lentil nutrition are for lentils grown in the United States.

Lentils offer superior nutritional profile when compared to commonly consumed cereal grains globally (wheat, corn, and rice), mainly being high in protein and low in carbohydrate contents (Figure 13.1). Specifically, lentils offer a good balance between protein and carbohydrates, and they contain 1.8-fold, 2.6-fold, and 3.5-fold more protein content compared to wheat, corn, and rice, respectively. Additionally, the ash content is also significantly higher than the major cereal crops. By contrast, the carbohydrate content in wheat, corn, and rice is 12–26% higher than that in raw lentils. These superior nutritional characteristics are attractive to consumers opting for healthy meat alternate proteins.

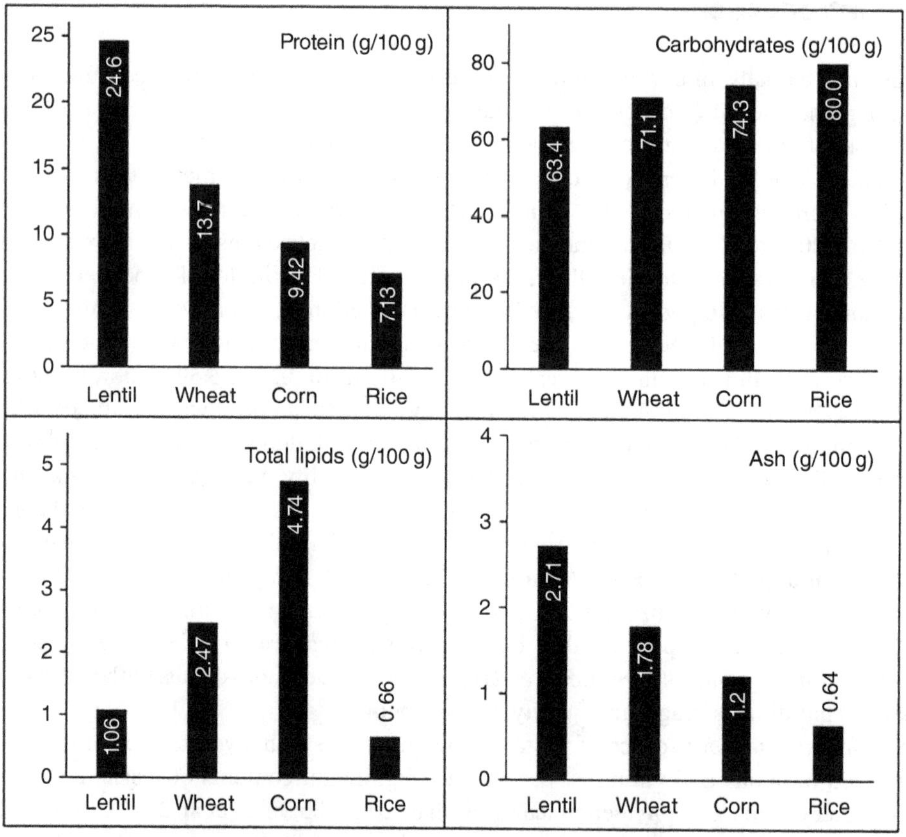

Figure 13.1 A comparison of protein, carbohydrate, total lipids, and ash content of lentils with major cereal grains. *Source:* Adapted from Food-Central data (USDA 2022).

Health-improving benefits of consuming lentils have been reported in a number of biological processes, e.g., inhibiting angiotensin I-converting enzyme (ACE), aiding in reducing blood cholesterol and bioactive role as antioxidant (Khazaei et al. 2017; Patterson et al. 2017). It has also been reported that regular intake of lentils and lentil-based products are effective in mitigating various health problems, e.g., hypertension, cardiovascular diseases, diabetes mellitus, and cancer. As reported by Ganesan and Xu (2017), fiber in lentils is a good source of prebiotics, which can improve gut health and help prevent digestive diseases. There is an increasing consumer awareness, especially among the educated ones, for the role of pre- and probiotics in human health. Although lentils, in different forms, are a common part of regular diets in many countries, in recent years, allergic reactions from consuming lentil have been reported in the literature. However, it is noteworthy that such allergic reactions differ with race/age and level of consumption (Halima et al. 2022; Sackesen et al. 2020).

Lentil consumption has been growing in recent years beyond its traditional regions of consumption since lentils are a healthy food choice for diverse and environmentally conscious consumers, including vegetarians and vegans. Hill (2022) reported that the use of pulses, like lentils, is increasingly gaining recognition as direct use or as ingredients in diverse food product applications. These emerging trends for healthy and sustainable food choices align well with the increased concerns about the environmental impact of producing animal-based foods (Dhull et al. 2022a). In this regard, food legumes, including lentils, have a great potential to become an important part of ever-increasing plant-based food options. This chapter provides a comprehensive review of the composition and nutrition of raw lentils and the impact of different processing methods on lentil composition and quality attributes. Moreover, the health benefits of lentils are also discussed, which could potentially expand direct consumption of lentils and lentil-based ingredients or products.

13.2 Composition and Nutrient Profile

13.2.1 Nutritional Profile

The composition of lentil types and varieties is dependent on its genetics and agro-ecological conditions where a particular variety of lentil is grown (soil characteristics, agronomic practices, temperature, and rainfall). Moreover, the lentil plant has to survive in diverse environmental constraints, that is biotic and abiotic stresses. Thus, the compositional phenotype of harvested lentil seeds is as dependent on genotype-by-environment (GxE) interactions as are the other phenotypic traits, i.e., yield, disease resistance, time to flowering, etc. (Grusak 2009).

Table 13.1 presents raw and cooked lentil composition and nutrient profile. Due to a large number of cultivated varieties globally, a considerable variation in nutrient composition is reported (Joshi et al. 2017; Zia-ul-Haq et al. 2011). The protein content of lentil cotyledons is about twofold higher than that in the seed coat (Rathod and Annapure 2016), therefore, for producing high protein fractions (concentrates or isolates), lentils are typically dehulled, which also reduces some antinutritional factors.

Table 13.1 Composition of raw, sprouted, and cooked (boiled, no salt) lentils and lentil soup (per 100 g).

Composition	Unit	Raw/mature lentils		Sprouted lentil	Cooked lentil	Lentil soup
		Green	Red/pink			
Proximate						
Water	g	8.26	7.82	67.3	69.6	84.8
Energy	Kcal/kJ	352/1470	358/1500	106/444	116/487	64/268
Protein	g	24.6	23.9	8.96	9.02	1.71
Total lipid/fat	g	1.06	2.17	0.55	0.38	
Ash	g	2.71	3	1	0.83	
Carbohydrate, by difference	g	63.4	63.1	22.1	20.1	8.82
Fiber, total dietary	g	10.7	10.8	—	7.9	3.2
Minerals						
Calcium	mg	35	48	25	19	12
Iron	mg	6.51	7.39	3.21	3.33	1.17
Magnesium	mg	47	59	37	36	13
Phosphorus	mg	281	294	173	180	67
Potassium	mg	677	668	322	369	138
Sodium	mg	6	7	11	2	187
Zinc	mg	3.27	3.6	1.51	1.27	0.5
Copper	mg	0.754	1.3	0.352	0.251	0.08
Vitamins[a]						
Vitamin C, total ascorbic acid	mg	4.5	1.7	16.5	1.5	1.3
Niacin	mg	2.6	1.5	1.13	1.06	0.426
Pantothenic acid	mg	2.14	0.35	0.58	0.64	—
Folate, DFE[b]	μg	479	204	100	181	37
Choline, total	mg	96.4	—	-	32.7	13.8
Vitamin A, RAE[c]	μg	2	3	2	0	20
Beta-carotene	μg	23	35	—	5	204
Vitamin A, IU	IU	39	58	45	8	—

[a] Thiamin, riboflavin, B_6, B_{12}, E, and K are <1.0 mg each.
[b] Dietary folate equivalents.
[c] Retinol activity equivalents.
Source: Food-central data (USDA 2022).

The amino acid content of raw lentils is shown in Figure 13.2. Glutamic acid is the most abundant amino acid in lentils, followed by aspartic acid, arginine, leucine, and lysine (making the top-5 list). On the other hand, cystine, tryptophan, and methionine are at the bottom having the lowest concentration (USDA 2022). Thus, sulfur-containing amino

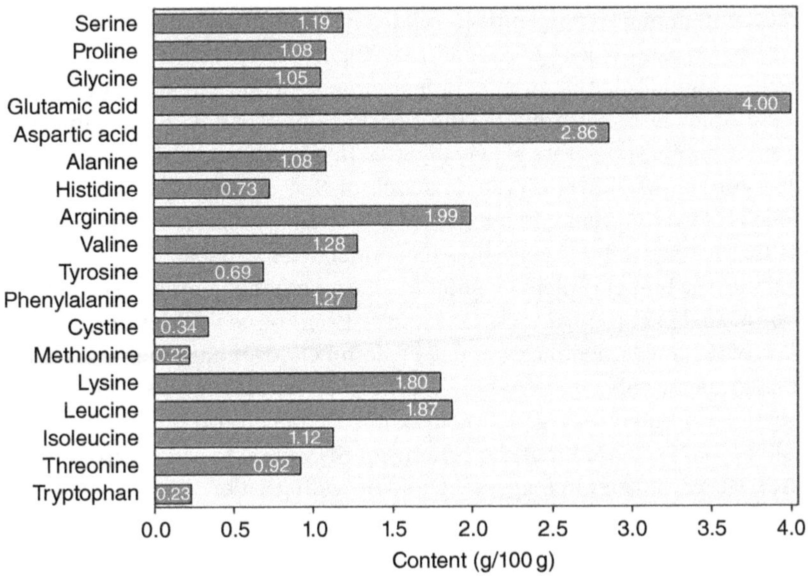

Figure 13.2 Content of amino acids in raw whole lentils. *Source:* Adapted from Food-Central data (USDA 2022).

acids, methionine and tryptophan, are the most limiting amino acids. As summarized in Table 13.2, average essential amino acid (EAA) content of lentil proteins is generally comparable to that in whey protein isolates and casein proteins (Alrosan et al. 2022), and these EAAs levels exceed FAO/WHO recommended daily intake for healthy adults (FAO/WHO 1991). It is noteworthy that lentils contain adequate levels of lysine, the main limiting amino acid in cereals (e.g., wheat, corn, rice). Shewry and Halford (2002) indicated that is why lentils as well as other food legumes are recommended as the most ideal complementary foods to cereals, which are low in lysine, but are relatively good sources of the sulfur-containing EAAs.

Table 13.2 The essential amino acids content (g/100 g) of lentil proteins compared with whey protein isolates, and casein.

Amino acids	Lentil proteins	Whey protein isolates	Casein proteins
Histidine	2.8	1.4	2.8
Isoleucine	4.8	7.6	5.5
Leucine	8.4	11.1	8.3
Lysine	7.1	7.4	7.4
Methionine	0.9	2.6	2.5
Phenylalanine	5.4	2.8	4.5
Threonine	3.5	3.9	4.4
Tryptophan	0.9	–	1.3
Valine	5.1	6.2	6.5

Source: Adapted from Alrosan et al. (2022).

The carbohydrate content of 63.4% in lentils is relatively lower than that found in major cereal grains (71.1–80.0%), e.g., wheat, corn, and rice (Figure 13.1). Lentil carbohydrates contribute a significant amount of energy (calories/serving), while non-starch carbohydrates, proteins, and lipids also contribute to the caloric value (Grusak 2009). However, lentils' carbohydrates (mainly starch) are slowly digested in human gut, which is primarily due to their higher content of slowly digestible starch or SDS. Owing to this beneficial attribute, lentils elicit a lower glycemic index (GI) of 29, when compared to white bread's GI of 100, making them "low GI food" and, thus, recommended for diabetic individuals. It has been suggested that the higher protein and dietary fiber content of lentils aid in a rather slower digestion of available carbohydrates. One cup (~197 g) of cooked lentils contains about 15.6 g of fiber, which meets about 62% of daily recommended requirement. It is noteworthy that the higher dietary fiber content of lentils helps in regulating bowel movement and growth stimulation of probiotics, e.g., *Lactobacilli* and *Bifidobacteria* (Zare et al. 2012). Lehmann and Robin (2007) reported that the benefits of SDS are associated with a stable glucose metabolism, better diabetes management, and providing satiety.

13.2.2 Bioactive Compounds

Lentils also contain a number of bioactive compounds, e.g., flavonoids, extractable, and insoluble-bound phenolics, carotenoids, tocopherols, saponins, phytic acid, and phytosterols. Table 13.3 provides a summary of the main bioactive compounds found in pulses, as well as a description of their positive and negative physiological effects. Lentils, compared to other legumes, are relatively richer in phenolic contents that are known to exhibit significant antioxidant activity. Seed coat from red lentil contains phenolics, with

Table 13.3 Physiological effects of bioactive compounds in legumes.

Compound[a]	Bioactive effect	Antinutrient effect
Oligosaccharides	Prebiotic	Gastrointestinal discomfort
Polyphenols	Antioxidant/prebiotic	Interferes with nutrient absorption
Phytic acid	Antioxidant	Interferes with mineral bioavailability
Lectins/ phytohemagglutinins	Satiety	Gastrointestinal discomfort/irritant
Protease and amylase inhibitors	Anti-diabetic	Interferes with protein bioavailability
Bioactive peptides	Anti-inflammatory	Interferes with protein digestibility
Saponins	Hypocholesterolemic	Gastrointestinal discomfort/irritant

[a] These compounds have a positive (bioactive) as well as negative (antinutrient) health effects.
Source: Wiesinger et al. (2022).

demonstrated antioxidant activity (Bautista-Expósito et al. 2018b; Zhang et al. 2018). The content and biochemical stability of phytochemicals present in lentils may be affected during processing, especially by thermal processing (Zhang et al. 2018).

Lentils are a significant source of extractable phenolics as well as insoluble-bound phenolics. It contains an array of extractable phenolic compounds, such as flavan-3-ols, flavonols, flavones, flavanones, proanthocyanidins, sub-classes of phenolic acids, anthocyanidins, and stilbenes. Among eight different pulses, lentils were found with the highest total phenolic content (TPC) of 7.53 mg gallic acid equivalents (GAE)/g dry weight and total flavonoids content (TFC) of 2.21 mg catechin equivalents (CE)/g (Xu and Chang 2007). Lentils also contain 1.5–2.6 mg/g of total phenolic acids (Xu and Chang 2010). The extractable fractions of red and green lentils were identified to contain several (24 and 20, respectively) phenolic compounds including phenolic acids and flavonoids (Amarowicz et al. 2009, 2010). In red lentils, major phenolic acids include *trans-p*-coumaric acid, *p*-hydroxybenzoic acid, *trans*-ferulic acid, and sinapic acid, while green lentils had mainly *trans-p*-coumaric acid and *trans*-ferulic acid. Zou et al. (2011) identified predominant flavonoids of lentils, including flavan-3-ols, flavonols, and condensed tannins (proanthocyanidins), such as catechin, epicatechin, kaempferol glycoside, quercetin diglycoside, and digallate procyanidin.

The bound-phytochemicals in lentils have been found to possess more antioxidant activity than extractable phytochemicals. The TPC and TFC of insoluble-bound phenolics in lentils ranged from 0.18 to 17.50 mg GAE/g, and 0.03 to 4.13 mg CE/g, respectively (Alshikh et al. 2015; Zhang et al. 2014). A number of phenolic compounds including hydroxycinnamic acids, hydroxybenzoic acids, flavonoids, and their derivatives (Alshikh et al. 2015), gallic acid, protocatechuic acid, catechin, epicatechin, and 3-hydroxycinnamic acid (Zhang et al. 2014) have been confirmed in the insoluble-bound fractions of lentils. Different processing treatments have been found to reduce the phytochemical content, profile, and antioxidant capacities including decortications, cooking, soaking (Han and Baik 2008), and thermal processing (boiling, steaming, pressure boiling, and pressure steaming) (Xu and Chang 2009). By contrast, germination or sprouting (Cevallos-Casals and Cisneros-Zevallos 2010) and solid-state fermentation (Dhull et al. 2020) have been shown to increase the antioxidant properties of lentils.

13.2.3 Antinutritional Factors

Lentils contain a variety of antinutritional factors (Table 13.3), namely lectins, enzyme inhibitors, phytates, saponins, and oligosaccharides. Some of the antinutritional factors are reported to lower digestive enzyme function or sequester essential nutrients, which makes them inaccessible for digestion (Nosworthy et al. 2018; Wiesinger et al. 2022). Therefore, it is important to use appropriate processing methods to either significantly reduce or completely eliminate lectins, phytate, and enzyme inhibitors since they are shown to reduce the protein digestibility and nutrient absorption. It has also been reported that the incorporation of organic acids (ascorbic acid, citric acid) during cooking or thermal processing reduces antinutritional factors significantly. A simple pretreatment of soaking pulses in hot water prior to processing can lower some antinutrient concentration, e.g., lectin and flatulence causing oligosaccharides (Campos-Vega et al. 2010; Khattab and Arntfield 2009).

Ruiz et al. (1997) reported a total saponin content of lentils in the range of 654–1269 mg/kg dry weight with soyasaponin I and VI as main saponins. In another study, soyasaponin I and soyasaponin βg (also known as VI) levels were found to be 28–407 mg/kg and 110–1242 mg/kg, respectively, in 32 Italian lentils (Sagratini et al. 2009). Lentils have been reported to have 86% (4.91 mg/g) phytic acid (or inositol hexaphosphate), 13% (0.72 mg/g) of inositol penta-phosphate, and traces of inositol tetra-phosphate (0.09 mg/g) (Ayet et al. 1997). Trypsin inhibitors are proteins having low molecular weight, which can bind and inactivate digestive enzyme trypsin (Oomah et al. 2011), while tannins form complexes with proteins resulting in low protein digestibility and amino acid availability (Kumar et al. 2021). Different processing methods, e.g., soaking, dehulling, boiling, autoclaving, and germination, are sometimes used to reduce or eliminate these antinutrients (Hefnawy 2011; Pal et al. 2017). Lectins, also called as phytohemagglutinins, are natural carbohydrate binding proteins that react with the surface epithelium of the digestive tract causing mid allergic, antinutritional, or other negative effects in humans (Pusztai et al. 1997). Cuadrado et al. (2002) reported lectin agglutinin (LCA) content of 4.0 g/kg dry matter in raw lentils, which was significantly reduced by fermentation process.

13.3 Processing Effect on Chemical and Nutritional Composition

The use of lentils has increased in recent years, due primarily to their balanced nutritional profile and potential health benefits linked to a lower risk of chronic diseases such as diabetes, obesity, cardiovascular diseases, and cancer (Papandreou et al. 2019). Lentils are either used as cooked for direct consumption or processed into different fractions and ingredients, such as flour, protein, starch, and fiber fractions that can be used in different food applications. Lentil flour and fractions can be prepared by either dry milling or wet milling processes for use in various food applications. In addition, canned lentils are available in some developed countries, offering ready-to-eat convenience for the consumers. The busy lifestyle of consumers in the developed countries makes it difficult for home cooking of legume due to their longer cooking time.

Value-added processing of lentils and development of new products are helpful in expanding economic significance and consumer acceptance of this important food legume. Typical processing methods used for lentils and end-products obtained are summarized in Table 13.4. An understanding of the functional properties (solubility, gelation, emulsification, foaming) of lentil flour and its ingredients (e.g., starch, protein) is critical for preparation of different high-quality foods such as bakery, extruded, dressings, soups, dairy, and meat products (Dhull et al. 2022b). From a food safety perspective, the use of raw lentil flour/ingredients may result in technological problems or introduce some novel allergens. Furthermore, lentils contain several antinutritional factors (e.g., allergic proteins, saponins), which must be minimized or eliminated before consumption (Mo'ez Al-Islam et al. 2020).

A wide variety of preprocessing (dehulling and milling) and processing (soaking, cooking, fermentation, extrusion, germination) methods have been shown to minimize or inactivate the antinutritional factors in lentils (Dueñas et al. 2016; Hefnawy 2011; Wang et al. 2009), which can potentially help in enhancing the use of lentil-based products

Table 13.4 Value-added processing of lentils and typical end-products.

Method	Specific process	End-products
Dry processing	• Direct milling, or • Dehulling → milling	• Flour • Air-classified fractions
Dry processing	• Dehulling → hulls milling	• Fiber-rich fractions
Wet processing	• Water extraction • Alkaline extraction • Acid extraction • Ultrafiltration	• High-protein fractions • High-protein isolates
Other methods	• Cooking/autoclaving • Extrusion • Microwave cooking • Germination or sprouting • Fermentation • Roasting	• Variety of value-added products • Ingredients for food applications

Source: Adapted from Dhull et al. (2022b).

without any negative effects. Most of the antinutritional factors found in lentils can be reduced by some pre-processing treatments, such as dehulling (physical removal) and milling (size reduction) of whole or decorticated seeds (Khazaei et al. 2019). A number of heat-stable and enzyme-resistant allergenic proteins (i.e., Len c 1, Len c 2) are also present in lentils, which can cause immunoglobulin E (IgE)-mediated hypersensitivity reactions (Shaheen et al. 2019). When adding lentil flour and protein-rich fractions for developing novel foods, the potential health risk for allergic consumers posed by these allergens should be duly considered and addressed. Additionally, lentil flour has a high lipoxygenase activity, which can reduce its shelf-life and results in off-flavors during processing as well as storage. All these aspects have a significant impact on the nutritional and sensory properties of lentil-based products, which can be mitigated by using traditional and innovative processing methods.

Table 13.5 presents a comprehensive summary of the effects of different processing methods on the composition and nutrient profile of lentils. In some instances, results reported conflicting patterns (that is both the increasing and decreasing effect), which could be attributed to varietal differences and/or analytical methodology used. In addition, differences in the processing parameters can also result in conflicting results.

13.3.1 Dehulling, Splitting, and Milling

Dehulling or removal of seed coat has been demonstrated to improve the taste and palatability of pulses since it is bitter in taste. The seed coat of lentil seeds is relatively thinner than that in most other food legumes and constitutes about 6–7% of the total seed weight. Usually a gum, e.g., galactomannan, or lignin layer binds the cotyledons to the seed coat (Erskine et al. 1991; Hughes and Swanson 1986; Kurien 1977). Dehulling and splitting

Table 13.5 Changes in the nutritional and non-nutritional constituents of lentils as affected by different processing methods.

Composition	Dehulling	Cooking/boiling	Autoclaving	Microwave cooking	Extrusion	Germination	Fermentation
Proteins	$\uparrow^{a,b}$	$\uparrow^{a,c,d,e}$	$\updownarrow^{f,e}$	$\updownarrow^{f}$	$\uparrow^{g}\downarrow^{h}$	$\downarrow^{c}\uparrow^{i}$	—
Total carbohydrates	$\downarrow^{b}$	$\downarrow^{a,c}$	—	—	$\uparrow^{h,j}$	$\uparrow^{c}$	—
Total starch	$\uparrow^{a}$	$\uparrow^{a,f,e}$	$\updownarrow^{f}\uparrow^{e}$	$\updownarrow^{f}$	—	$\downarrow^{i}$	—
Resistant starch	$\uparrow^{a}$	$\uparrow^{a}\downarrow^{e}$	$\downarrow^{e}$	—	—	—	$\uparrow^{l}$
Fat/lipids	$\uparrow^{b}$	$\uparrow^{a,c,e}\downarrow^{f}$	$\downarrow^{f}\uparrow^{e}$	$\downarrow^{f}$	$\downarrow^{g,h}$	$\uparrow^{c}\downarrow^{i}$	—
Ash	$\updownarrow^{a,b}$	$\downarrow^{a,b,f,e}$	$\downarrow^{f,e}$	$\downarrow^{f}$	$\updownarrow^{h}$	$\uparrow^{c}\updownarrow^{i}$	—
Energy		$\uparrow^{c,d}$	—	—	—	$\updownarrow^{c}$	—
Total dietary fiber	$\downarrow^{a}$	$\uparrow^{a}\downarrow^{b}$	—	—	$\downarrow^{h,j}$	$\downarrow^{c}$	—
Insoluble dietary fiber	$\downarrow^{a}$	$\uparrow^{a}\downarrow^{b}$	—	—	$\downarrow^{h,j}$	$\downarrow^{c}$	—
Soluble dietary fiber	$\downarrow^{a}$	$\downarrow^{a,b}$	—	—	$\downarrow^{h}\uparrow^{j}$	$\downarrow^{c}$	—
Raffinose	—	$\downarrow^{a,f}$	$\downarrow^{f}$	$\downarrow^{f}$	$\uparrow^{k}$	—	$\downarrow^{l}$
Stachyose	—	$\downarrow^{a,f}$	$\downarrow^{f}$	$\downarrow^{f}$	$\uparrow^{k}$	—	—
Verbascose	—	$\downarrow^{a,f}$	$\downarrow^{f}$	$\downarrow^{f}$	$\uparrow^{k}$	—	—
Phenolics/ total polyphenols	—	$\downarrow^{e}$	$\downarrow^{e}$	—	$\downarrow^{m}\uparrow^{j}$	—	$\uparrow^{l,n}$
Trypsin inhibitors	$\downarrow^{a}$	$\uparrow^{a}\downarrow^{f}$	$\downarrow^{f}$	$\downarrow^{f}$	$\downarrow^{k,m}$	—	$\downarrow^{l}$
Tannins	$\downarrow^{a}$	$\downarrow^{a,f}$	$\downarrow^{f}$	$\downarrow^{f}$	$\downarrow^{m}$	—	—
Phytic acid	$\uparrow^{a}$	$\downarrow^{a,f}$	$\downarrow^{f}$	$\downarrow^{f}$	$\downarrow^{m}$	—	$\downarrow^{l}$
Calcium	$\downarrow^{a}$	$\uparrow^{a}\downarrow^{f}$	$\downarrow^{f}$	$\downarrow^{f}$	—	—	$\uparrow^{n}$
Potassium	$\uparrow^{a}$	$\downarrow^{a,f}$	$\downarrow^{f}$	$\downarrow^{f}$	—	—	$\uparrow^{n}$
Phosphorus	$\uparrow^{a}$	$\downarrow^{a,f}$	$\downarrow^{f}$	$\downarrow^{f}$	—	—	—

Copper	$\downarrow^{a}$	$\uparrow^{a,f}$	$\downarrow^{f}$	$\downarrow^{f}$	—	—	$\uparrow^{n}$
Magnesium	$\downarrow^{a}$	$\downarrow^{a,f}$	$\downarrow^{f}$	$\downarrow^{f}$	—	—	—
Zinc	$\updownarrow^{a}$	$\downarrow^{a,f}$	$\downarrow^{f}$	$\downarrow^{f}$	—	—	$\uparrow^{n}$
Iron	$\downarrow^{a}$	$\downarrow^{a,f}$	$\downarrow^{f}$	$\downarrow^{f}$	—	—	$\uparrow^{n}$
Manganese	$\downarrow^{a}$	$\uparrow^{a}$	$\downarrow^{f}$	$\downarrow^{f}$	—	—	—
Sodium	—	$\downarrow^{f}$	$\downarrow^{f}$	$\downarrow^{f}$	—	—	$\uparrow^{n}$

↑ = increase, ↓ = decrease, ↕ = no significant change, — = no data reported.

[a] Wang et al. (2009).
[b] Ma et al. (2011).
[c] Dueñas et al. (2016).
[d] Aguilera et al. (2010).
[e] Piecyk et al. (2012).
[f] Hefnawy (2011).
[g] Nosworthy et al. (2018).
[h] Pasqualone et al. (2021).
[i] Xu et al. (2019a).
[j] Morales et al. (2015).
[k] Ciudad-Mulero et al. (2020).
[l] De Pasquale et al. (2020).
[m] Rathod and Annapure (2016).
[n] Dhull et al. (2020).

is a standard practice for most lentil market classes to satisfy consumers' preference. Splitting involves the separation of the cotyledons in pulses (Singh and Singh 1992). In Southeast Asia (particularly India, Nepal, and Bangladesh), dehulled red lentil and in Sri Lanka, Pakistan, Turkey and the Middle East, dehulled/split red lentils are a major lentil market class (Thavarajah et al. 2008). By removing the relatively water-impermeable seed coat, cooking time of dehulled lentils is reduced significantly. Both dehulling and splitting can increase the acceptance of lentils by consumers by improving the taste, richer green or red color, and convenience of cooking. In Middle-Eastern countries, dehulled/split red lentils are commonly use to prepare soup, which is also served as a popular appetizer in Mediterranean restaurants.

Wang et al. (2009) evaluated the effect of dehulling on different nutritional and antinutritional factors of eight lentil cultivars and found a significant increase in starch, protein, resistant starch, phytic acid, stachyose, verbascose, potassium, and phosphorus content, while soluble dietary fiber (SDF), insoluble dietary fiber (IDF), and total dietary fiber (TDF), tannins, and minerals (calcium, iron, magnesium, copper, manganese) contents were significantly decreased after dehulling. Since seed coat contains insignificant amount of protein and starch, dehulled seeds are reported to have proportionately more protein and starch, whereas it has minimal or no effect on the ash contents (Wang et al. 2009). However, phytic acid decreased slightly, which suggested that phytate is mainly present in cotyledons while decrease in tannin content indicated that tannins are concentrated mainly in seed coat due to their lower content in dehulled lentil.

Dehulling and splitting have been shown to decrease the resistant starch, sugar alcohols, fructo-oligosaccharides, and raffinose-family oligosaccharides in red lentils but had no such effect in large green lentils except sugar alcohols. It is hypothesized that raffinose is present in higher concentration in lentil seed coat while cotyledon is rich in stachyose and verbacose. However, this hypothesis needs further research for confirmation (Siva et al. 2018). Seed coat or hulls have significantly higher phenolic content and antioxidant activity than that in both whole and dehulled lentils. Yeo and Shahidi (2020) noted that since hulls are excellent source of bioactive compounds, they can be used as an ingredient in production of functional foods and gluten-free food products development.

Milling is the main process for producing lentil flour (from whole or dehulled seeds), which can be further fractionated into protein- and starch-rich components for value-added use in different food applications. The utilization of lentil flour as a functional food ingredient primarily depends on its physical, nutritional, and sensory properties, all of which can be impacted with the type of milling process employed (Marchini et al. 2021), thereby impacting lentil flour use in food applications. In recent years, production of legume protein ingredients using dry fractionation has gained attention as alternatives to solvent extraction and can produce sustainable lentil-based ingredients with unique composition and functional properties (Funke et al. 2022).

Milling is favorably associated with nutritional bioaccessibility, since a thorough milling process entails an extended cell rupture (Joshi et al. 2017), which increases nutrient bioaccessibility (Boukid et al. 2019). Raw flour with large particle size has low digestion rates, which could be due to the delayed action of digestive enzymes resulting from large protein bodies or fragments of cell wall. Therefore, an understanding of different pre-processing treatments, milling process, and particle size distribution on physico-chemical and functional properties of lentil flours can benefit its optimal utilization in new product development, including partial replacement of wheat flour in baked goods and extruded products.

13.3.2 Cooking – Boiling, Autoclaving, and Microwaving

Pulses, including lentils, require cooking/processing before their consumption to enhance the nutritional value and reduce the antinutritional components. The most common method of processing in pulses includes heat treatments such as cooking, which significantly reduces the tannins, phytic acid, minerals, phenolics, oligosaccharides, and insoluble dietary fibers (Wang et al. 2009) while it improves the soluble dietary fiber and protein quality significantly (Martín-Cabrejas et al. 2008). Cooking has been shown to result in significant increases in protein digestibility and energy value, but fat, ash, and total carbohydrates showed reduced values (Aguilera et al. 2010; Dueñas et al. 2016). Some other studies also reported a significant increase in protein content while decrease in fat content after cooking of lentils (de Almeida Costa et al. 2006; Wang et al. 2009), which makes lentils a good source of plant-based proteins.

The protein quality, assessed as digestibility, generally measured as *in vitro* protein digestibility (IVPD), protein digestibility corrected amino acid score (PDCAAS), *in vitro* PDCAAS score, and protein efficiency ratio (PER), was reported to be improved after cooking of lentils (Nosworthy et al. 2018). Hefnawy (2011) indicated that total essential amino acids were increased slightly, whereas lysine, tryptophan, and total sulfur-containing and aromatic amino acids reduced after cooking lentils in boiling water. Furthermore, Wang et al. (2009) reported a significant improvement in starch and resistant starch (RS) but decrease in sucrose and oligosaccharide contents in different lentil varieties after cooking. The increased RS content were most likely from the retrogradation of gelatinized starch in fully cooked lentils (de Almeida Costa et al. 2006).

Piecyk et al. (2012) reported that the amylose content of starch in cooked lentils exhibited a decrease; this effect could be attributed to higher solubility and leaching of this component in boiling/cooking water. The rapidly digestible starch (RDS) content increased, while slowly digestible starch (SDS) and resistant starch (RS) all decreased in the lentil flours after cooking versus raw flours. Further, the RDS content values were used in determining the starch digestion index (SDRI) and rapidly available glucose (RAG) (*in vitro* parameters), which are well correlated with glycemic index (Englyst et al. 1996). Similar to RDS, the values of SDRI and RAG also increased in cooked lentil flours. Apart from several other factors, such as amylase inhibitor inactivation, protein denaturation (that form an envelope around starch granules), and removal of other compounds (antinutritional factors) inhibiting protein digestibility, the reported increase in RDS content in thermally treated flours in the presence of water is primarily determined by gelatinization of starch that is also accompanied by partial leaching of amylose (Piecyk et al. 2012).

Lentil's prebiotic carbohydrates, such as raffinose-family oligosaccharides (RFO) and fructo-oligosaccharides (FOS), are reported to be affected by cooking treatments (Wang et al. 2009), including cooking, cooling, and reheating (Johnson et al. 2015). The cooking parameters, e.g., cooking temperature, time, and pH, are known to affect the thermal degradation of these prebiotic carbohydrates (Courtin et al. 2009). In this regard,

Johnson et al. (2015) showed a reduction in the concentration of RFO after cooking and cooling of lentils, while Siva et al. (2018) observed an increase in RFO and FOS following cooking and cooling of lentils. Different oligosaccharides such as reducing sugars, raffinose, and stachyose also showed a decreased concentration after cooking of lentils, which could be due to the diffusion of these oligosaccharides in cooking water (Hefnawy 2011). Wang et al. (2009) observed an increased level of verbascose in eight cooked lentil varieties, while Hefnawy (2011) reported its complete removal after cooking.

Wang et al. (2009) reported that, after cooking, SDF content was decreased but IDF and TDF content increased in eight lentil varieties. It was suggested that cooking resulted in softening of soluble fibers, which consequently reduces its content (Vidal-Valverde and Frias 1991). The increase in IDF may be attributed to the formation of protein–fiber complexes due to possible chemical modification during cooking of lentils (Bressani 1993). However, Dueñas et al. (2016) reported a decrease in the values of IDF and TDF content along with SDF content. In addition, a general reduction in the total phenolic content in dietary fiber fraction of cooked lentil flours was observed compared to that of the raw lentil flours (Dueñas et al. 2016). Wang et al. (2009) and Hefnawy (2011) reported that cooking was also found to significantly reduce the protease inhibitors (e.g., trypsin inhibitors), and phytate and tannin contents compared to the raw seeds.

Among minerals, calcium, copper, and manganese content were increased, but iron, potassium, magnesium, phosphorus, and zinc content were reduced after cooking lentils (Wang et al. 2009). Some other studies have reported a decrease in total mineral content after traditional or commercial cooking (Hefnawy 2011; Ramírez-Ojeda et al. 2018), which can be due to solubilization of minerals into cooking water at variable rates. Viadel et al. (2006) investigated the effect of cooking on the uptake of calcium, iron, and zinc from lentils and concluded that effects such as decrease in tannin, total phenol, and soluble oxalate contents produced by traditional cooking exerted favorable effects only on total cell uptake of calcium. It was concluded that the solubility and uptake of iron and zinc reduced or improved slightly as a result of cooking. On the contrary, a recent study by Ramírez-Ojeda et al. (2018) reported that a considerable increase was observed in the bioaccessibility of iron and zinc in commercially cooked lentils. A significant reduction or complete elimination of some antinutrients during processing was the most likely reason for this effect.

In pulses, protease inhibitors, e.g., trypsin inhibitors, can be effectively inactivated by thermal processing or cooking (Gatta et al. 1989). Wang et al. (2009) and Hefnawy (2011) also showed a substantial reduction in trypsin inhibitor activity after sufficient cooking of lentil varieties. Furthermore, the content of phytic acid and tannin were reduced in cooked lentils in comparison to raw samples.

A 16-hour soaking of lentils, followed by cooking was shown to reduce the oxygen radical absorbing capacity (ORAC), hydroxycinnamics, catechins and procyanidins, flavonols and dihydroflavonols, flavones and flavanones, while hydroxybenzoics and some other phenolic compounds were increased (Aguilera et al. 2010). Additionally, in an investigation on the effect of boiling pretreatment on lentil flour properties (Ma et al. 2011), the water-holding capacity, fat-binding capacity, and gelling capacity of processed flour were improved significantly while solubility of protein was decreased.

Hefnawy (2011) evaluated the effect of different processing methods (boiling in water, autoclaving, and microwave cooking) on lentil quality. It was reported that mineral losses by microwave cooking were relatively lower than those observed in lentils cooked by boiling or autoclaving. All cooking methods enhanced the PER and IVPD values of cooked lentils. Autoclaving and microwave cooking also reduced the trypsin inhibitor activity and contents of phytic acid and tannins (Hefnawy 2011). A later study showed that autoclaving decreased ash, protein, and total phenolic content but increased fat, starch, and amylose content in processed lentils (Piecyk et al. 2012). The autoclaved lentil flours exhibited a significant increase in RDS, SDRI, and RAG values and a decrease in SDS and RS contents compared to that in raw lentil flours (Piecyk et al. 2012).

In a study focusing on effect of conventional cooking and autoclaving on RS of Laird lentils, it was concluded that the values of average hydrodynamic radius ($\bar{R}_h$), weight-average molecular weight (M_w) by size exclusion chromatography (SEC), crystallinity, and double helix content of processed RS were reduced in comparison to untreated RS (Ma et al. 2018). This effect could be due to high degree of degradation and complete destruction of granular structure induced by cooking and autoclaving. Also, B-type crystalline packing was favored by moist-heat treatments (e.g., cooking and autoclaving) in lentil RS. It has been proposed by some studies that high proportion of type B polymorph in RS resulted in better fermentation ability producing higher level of butyrate and short-chain fatty acids in comparison to A-type polymorphs (Schmiedl et al. 2000). Similar to lentil RS, starches extracted from lentils after different cooking treatments, i.e., microwave cooking, conventional cooking, and autoclaving, also showed similar molecular structural changes (Yin et al. 2018).

13.3.3 Extrusion and Baking

Extrusion processing technology, a high-temperature short time (HTST), is a versatile and efficient process that converts agricultural raw materials (in a granular or powdered form), into fully cooked food products (Berrios et al. 2022). During extrusion, ingredients are forced through a specifically designed die after being exposed to expansion-inducing temperatures that can affect the nutritional quality and antinutrients in lentils (Nosworthy et al. 2018; Pasqualone et al. 2021). Extrusion processing has been reported to enhance the IVPD and *in vitro* starch digestibility (IVSD) without affecting the total protein content (Rathod and Annapure 2016). However, Nosworthy et al. (2018) reported an increase in crude protein content and a decrease in crude fat content of extruded lentils while Pasqualone et al. (2021) observed a decrease in protein, lipids, TDF, IDF, and SDF. In addition, extrusion is also shown to result in higher amino acid score, percent true protein digestibility (%TPD), IVPD, PDCAAS, and *in vitro* PDCAAS in comparison to cooked as well as baked lentil flours (Nosworthy et al. 2018).

Kamau et al. (2020) reported a reduction in lipid content during extrusion was due likely to the formation of starch-lipid/amylose-lipid complexes, which reduces their extraction and quantification. The formation of such complexes requires gelatinization of starch that increases amylose availability for complexation. The amylose double helix expands during heating allowing the lipid molecules to form electrostatic bonds that improve stability. During extrusion, high-shear forces and heat transfer can change the physico-chemical properties of dietary fiber that can result in the reduction of TDF and IDF (Kamau et al. 2020). Espinosa-Ramírez et al. (2021) indicated that pseudo-cereal and legume flours that exhibited a decrease in protein content in some studies are rather controversial, as no effluents resulting in fraction lost were produced during extrusion process. Additionally, protein denaturation resulting from extrusion processing does not affect the quantification of proteins. Pasqualone et al. (2021) further reported that, typically, milder or severe extrusion parameters (especially, temperature, and screw speed) have no or minimal impact on the content of total protein, whereas total carbohydrates and dietary fiber concentration could be variable in flours extruded obtained under different extrusion processing parameters.

Morales et al. (2015) reported that some bioactive compounds, e.g., soluble fiber, total phenolic, hydroxybenzoic, and hydroxycinnamic acids, increased while insoluble fiber, flavonols, tocopherols, and most of the organic acids were decreased in the extruded fiber-enriched lentil flours. Different antinutritional factors have also been shown to decrease significantly in extruded lentils: phytic acid, tannins, and trypsin inhibitors were significantly reduced by 99.30%, 98.83%, and 99.54%, respectively (Rathod and Annapure 2016). In another study, extrusion resulted in a significant increase in the total α-galactosides (raffinose, stachyose, verbascose) content, while inositol phosphates, lectins, and trypsin inhibitor activity were all reduced significantly (Ciudad-Mulero et al. 2020). González and Pérez (2002) evaluated the effects of extrusion cooking (150 °C die temperature and 90 rpm screw speed) and microwave irradiation (85 °C, half power of 650 W for six minutes) on some physico-chemical and functional properties of lentil starch. Both the treatments resulted in significant decrease in the crude fiber content and starch retrogradation, while the reducing sugar, ash contents and the absolute density of the extrudates were increased.

Ghumman et al. (2016b) investigated the effect of extrusion temperature (ET) and feed moisture (FM) and reported that the degradation of different polypeptides increased significantly with decreasing in FM and increasing in ET. In addition, the digestibility of extrudates improved at high FM and low ET. The extrudates exhibited lower viscosity, expansion ratio, IVPD with an increase in the FM, whereas at the higher FM oil absorption capacity was improved. The water absorption capacity was shown to decrease with an increase in both the FM and the ET. It was concluded that pulses cooked at the low FM could be potentially utilized in the preparation of ready-to-eat puddings with high protein, enhanced digestibility, and improved viscosity.

The incorporation of lentil flour and protein- and fiber-rich fractions can be easily done in baked goods. Since the majority of baked products are prepared using 100% wheat flour, a partial substitution or replacement of wheat flour with lentil-based ingredients can increased the nutritional quality of such baked products. Nosworthy et al. (2018) reported that baking has been shown to result in a mild increase in crude protein contents. In addition, baking also results in lower amino acid score, PDCAAS, *in vitro* PDCAAS, and %TPD in comparison to cooked and extruded lentil flours.

13.3.4 Germination

Germination is a traditional and inexpensive process to enhance the nutritional quality and health benefits of food legumes. This process is also known to boost antioxidant capacity and reduce the antinutritional factors in the germinated seeds. Besides nutritional improvement, the physiological and textural properties of legume seeds are modified during germination via the activation of endogenous enzymes and breakdown of large molecules (such as starch and proteins), thereby improving the bioavailability and overall digestibility (Oskaybaş-Emlek et al. 2021; Xu et al. 2019b). Germination of lentil has been shown to increase the protein content and decrease starch content significantly (Fouad and Rehab 2015; Xu et al. 2019b). At the start of the germination process, hydrolytic enzymes (e.g., α-amylase, glucosidase, dextranase) that are generated from the aleurone layer of seed coat and β-amylase generated from the endosperm get activated in the seeds (Olaerts et al. 2016). The reduction in starch content during germination is attributed to its conversion to mono- or oligosaccharides by the action of these hydrolytic enzymes.

The non-starch carbohydrates (TDF, IDF and SDF) and lipid content are also decreased after germination (Xu et al. 2019b). Dueñas et al. (2016) reported that germination produced a decrease in protein, IDF, SDF, and TDF and an increase in ash, fat, and total carbohydrate content in lentils. During germination crude lipids in the seed germ are hydrolyzed by enzymes to free fatty acids (FFAs), which produce energy to support sprouting of the seeds. This process is the main reason for a decrease in the crude fat content after the germination process (Cornejo et al. 2015). Germination has also been found to develop distinctive green (grass like), fishy, and beany flavors in some pulses including lentil, which is attributed to an increase in aromatic compounds that are derived from lipid oxidation (Xu et al. 2019a). Xu et al. (2019b) noted that an increase in the activity of lipoxygenase enzyme, which corresponds to a decrease in crude lipid content in germinated seeds, also supports this hypothesis of aromatic compound generation.

Świeca et al. (2013) studied the effect of different sprouting methods on starch content and expected glycemic index (eGI). It was shown that sprouting conditions can have a significant effect on the starch content, eGI, and bioavailability of starch. However, germination under all conditions studied resulted in decreased total starch (TS), α-amylase inhibitor activity (aAI), and eGI. With respect to sprouting conditions, control sprouts resulted in highest TS content, whereas sprouting induced with 300 mM NaCl showed lowest TS content and the highest eGI; while sprouts with 100 mM NaCl showed the lowest eGI. The highest resistant starch content was found in sprouts with 600 mM mannitol. In another study, Świeca and Gawlik-Dziki (2015) observed an increase in available starch, starch digestibility, eGI, total phenolic compounds, and reducing power; however, the TS content and antiradical activity were reduced in germinated lentils (Świeca and Gawlik-Dziki 2015).

The functional properties are improved in the flours prepared from germinated pulses compared to those from raw seeds. Xu et al. (2019b) showed that flours prepared from germinated lentils exhibited higher water absorption capacity, foaming capacity, and final viscosities in comparison to raw lentil flours. Germinated lentil flour also had an improvement in protein content and the breakdown viscosity; however, pasting viscosity, emulsifying activity, and foaming stability were reduced compared to the control (Ghumman et al. 2016a). Also, high-molecular-weight (HMW) proteins decreased, while low-molecular-weight proteins (LMW) increased due to proteases-induced hydrolysis of HMW to LMW proteins. However, germination had no significant effect on polypeptides corresponding to trypsin inhibitors.

13.3.5 Fermentation

Fermentation has the potential to improve the nutritional qualities of lentils by improving the content and bioavailability of bioactive compounds. During fermentation of lentils, endogenous enzymes are activated, which results in α-amylase-induced hydrolysis and breakdown of starch (Zhang et al. 2016). Fermentation increases the digestibility of legume protein by decreasing the contents of antinutritional factors, such as phenolic compounds. This mechanism promotes cross-linking between proteins and also the synthesis of microbial proteases, which potentially results in the breakdown and release of proteins from the protein–enzyme matrix (Alrosan et al. 2021). Bautista-Expósito et al. (2018a) reported that functional properties of food legumes can be improved with fermentation by selected starters.

Alrosan et al. (2021) investigated the effect of fermentation (carried out using water kefir seed) on the quality, structure, and digestibility of lentil proteins. Protein digestibility was shown to increase significantly from 76.4% (control) to 84.1% after five days of fermentation, with changes in protein pH from 6.8 to 3.4 from start to the end of fermentation. Fermentation had a significant effect on the secondary protein structure, i.e., mainly the percentage of alpha-helices and randomly oriented coils. It was shown that the reduction in alpha-helix to beta-sheet ratio with resulting increase in the protein digestibility was related to the longer fermentation time. After two days of fermentation, an increase of 78.8% was observed in total phenolics (792.6 versus 443.4 mg GAE/100 g). Chlorogenic acid and epicatechin were the major phenolic compounds detected, with trace contents of rutin, ferulic acid, and sinapic acid.

Biotransformation processes such as solid substrate fermentation (SSF) can improve the bioaccessibility of bioactive compounds in food or even result in the generation of some new compounds beneficial for health (Dhull et al. 2020). An increase in antioxidant potential and phenolic compounds of lentils after fermentation with starter culture (e.g., *Bacillus subtilis*, *Aspergillus oryzae*, *Aspergillus niger*, and *Aspergillus awamori*) have been reported (Dhull et al. 2020; Torino et al. 2013). Furthermore, an increase in mineral contents (iron, zinc, calcium, copper, sodium, potassium) and *in vitro* bioavailability of iron and zinc in lentils after solid-state fermentation was also reported (Dhull et al. 2020).

Gan et al. (2016) reported that lactic acid bacteria (LAB), *Lactobacillus plantarum*, used as starter in the fermentation of legumes produces a number of carbohydrolases and esterases, which result in a significant increase in the free phenolic compounds. Torino et al. (2013) observed improved antioxidant, antihypertensive, and hypolipidemic activities after fermentation. Fermentation with *L. plantarum* MRS1 and *Lactobacillus brevis* MRS4 enhanced the nutritional profile of processed legume flours by increasing the content of free amino acids, protein digestibility, and the radical scavenging activity, whereas phytic acid, condensed tannins, and raffinose were reduced along with lower trypsin inhibitory activity and starch hydrolysis (De Pasquale et al. 2020). Chen et al. (2021) studied changes in probiotics and bioactive compounds during single fermentation with LAB and co-fermentation with LAB and *B. subtilis* var. Natto. Total phenolic compounds and genistin content increased significantly by co-fermentation process. Based on the results of this study, it was concluded that anaerobic solid-state co-fermentation of lentils using *B. subtilis* (Natto) may promote the multiplication of LAB, thereby enhancing the antioxidant activity and functionality of fermented lentil products.

Fermentation using *L. plantarum* at pH 6.5–8.5 and 5.5–30 hours was reported to affect the proteins, total phenolic contents together with enhancement of the potential antioxidant, hypoglycemic, hypolipidemic, anti-inflammatory, and antihypertensive activities (Bautista-Expósito et al. 2018b). The pH of fermentation was shown to mainly affect the peptides, soluble phenolics, and antioxidant activity, whereas the lipase inhibitory activity was affected negatively. Furthermore, fermentation time showed a dominant positive effect on protein hydrolysis while negatively affected the angiotensin I converting enzyme (ACE) inhibitory activity of fermented lentil. Rui et al. (2016) also reported pH to be a main factor during fermentation since the uncontrolled fermentation by LAB that leads to acidification can result in the precipitation of protein and enzyme inactivation. This effect can potentially cause interference in the release of bound phenolic compounds and the generation of bioactive peptides.

The protein content has been shown to decrease with increasing time during fermentation, which was suggested due to *L. plantarum* proteolytic enzymes resulting in the proteolysis and production of peptides and free amino acids (De Angelis et al. 2016). Bautista-Expósito et al. (2018a) also observed an increase in bioactive peptides, ACE inhibitory activity, and maltase inhibitory activity; however a significant decrease was reported in the total phenolics, α-glucosidase inhibition, and antioxidant activity (assessed as ORAC) of lentils fermented by *L. plantarum*. Fermentation by *Rhizopus oryzae* showed an efficient release of insoluble-bound phenolic compounds from the cell wall matrix, though these released bound-phenolics were not completely transformed into accessible soluble phenolics due to structural changes.

13.3.6 Roasting

Roasting of lentils is not very common as compared to some of the other processing techniques discussed in this chapter. However, roasted lentils with a variety of flavors are available commercially in North America as healthy snack options. Traditionally, effect of roasting has been studied on lentil flours and protein- and starch-rich fractions (Bahnassey and Khan 1986; Baik and Han 2012). Bahnassey and Khan (1986) developed spaghetti fortified with flours from roasted and raw pulses, including lentils. The sensory panelists preferred spaghetti fortified with roasted legume flour compared to the non-roasted flour since beany flavor was observed in spaghetti prepared with non-roasted legume flours or their concentrates (especially, at 25% fortification level). Baik and Han (2012) reported that roasting treatment significantly improved protein digestibility in lentil flours. Bread prepared with the addition of roasted flour had a more appealing aroma and greater loaf volume.

13.4 Health Benefits of Lentils

Lentils are nutrient-dense and possess a number of bioactive polyphenols and are low in sodium and saturated fat (Mo'ez Al-Islam et al. 2020). Regular consumption of lentils has been linked with lower risk of metabolic syndrome, a combination of factors that increases the risk of developing heart disease and diabetes. The major factors associated with the metabolic syndrome include high blood pressure, high insulin levels, over-weight, high levels of triglycerides, and low levels of low-density lipoproteins (LDL), generally referred to as the "bad cholesterol." The higher intake of dietary fiber offers significant protective benefits from metabolic syndrome (McKeown et al. 2004; Pereira et al. 2004), diverticular disease, and constipation. The insoluble component of lentil fiber has been associated with ~40% lower risk of diverticular disease (Aldoori et al. 1998). Fiber in lentils is also a source of prebiotics that helps prevent digestive diseases (Ganesan and Xu 2017). A study of 90,534 subjects showed that high fiber intake reduces the risk of breast cancer (Farvid et al. 2016). Aslani et al. (2015) observed that consumption of lentils may improve cholesterol levels in people with diabetes mellitus and may protect against breast cancer in women. Didinger and Thompson (2022) reported that although the scientific literature supports the role of legumes in many health-related benefits, yet the overall quality of such evidence is often somewhat low.

Potassium is an essential mineral needed for the maintenance of total body fluid volume, acid and electrolyte balance, and normal functioning of cells. The high amounts of potassium aids in offsetting the negative effects of salt (sodium) resulting in lowering of the blood pressure. In this regard, increased potassium intake has been shown to reduce both the systolic and diastolic blood pressure in adults (WHO 2020).

Lentils are also a good source of iron, which is used in the body in the production of red blood cells, which are critical for the transportation of oxygen from the lungs to the cells for use in the generation of energy and prevention of fatigue. Folate is important for maturation of red blood cells, hence prevents megaloblastic anemia and protects the heart by effectively lowering the levels of homocysteine that is linked to heart diseases. For pregnant women, folate is particularly important for normal development of the fetus and preventing birth defects, such as anencephaly and spina bifida (CDC 2021). Anencephaly is a condition that happens when parts of the fetal brain and skull fail to form and develop correctly. Whereas spina bifida is a condition in which the spine of the fetus does not develop correctly, potentially resulting in some severe physical disabilities (Kancherla et al. 2021).

13.5 Conclusion

In recent years, lentils, like other food legumes, have gained increasing popularity among consumers and food processors as nutrient-dense and plant protein food. This changing trend is being driven by the fact that legumes, including lentils, are high in protein content and relatively low in carbohydrates as compared to major cereals (wheat, rice, corn) and high content of resistant starch and their low glycemic index. This balance between relatively high-quality protein and mineral content and lower carbohydrate contents is an important feature of lentils from a human health perspective. Lentils possess a number of bioactive polyphenols and are low in sodium and saturated fat. In addition to being high in proteins and dietary fiber content, lentils are a rich source of many bioactive compounds with proven health benefits. For improving marketability and consumer acceptance, clearly defined quality standards and grades of lentils have been established in the developed countries. Processing is necessary to produce a variety of products, make lentils palatable, and significantly reduce antinutrients. However, different processing methods have a significant effect on the composition and nutritional profile of lentil-based products. Besides traditional consumption, lentil flour and protein- and starch-rich fractions can be prepared for use in new products development. Lentil consumption has been reported to be effective in reducing various health conditions, such as hypertension, cardiovascular diseases, diabetes milieus, and cancer. The nutritional and functional properties of lentil ingredients must be optimized and assessed for food applications to ensure consumer acceptance and sustained increase in the consumption of this important legume. Lentils or lentil-derived ingredients, combined with cereals-based ingredients, can further find potential new applications by offering nutritious and healthy food choices for consumers. The emerging trends of healthy eating align well with the consumer concerns for a negative impact on the environment in producing animal-based foods. In this regard, nutrient-dense food legumes, including lentils, are positioned well to capture significant portion of meat alternate protein market.

References

Aguilera, Y., Duenas, M., Estrella, I. et al. (2010). Evaluation of phenolic profile and antioxidant properties of Pardina lentil as affected by industrial dehydration. *Journal of Agricultural and Food Chemistry* 58: 10101–10108.

Aldoori, W.H., Giovannucci, E.L., Rockett, H.R. et al. (1998). A prospective study of dietary fiber types and symptomatic diverticular disease in men. *Journal of Nutrition* 128: 714–719.

de Almeida Costa, G.E., da Silva Queiroz-Monici, K., Reis, S.M.P.M., and de Oliveira, A.C. (2006). Chemical composition, dietary fibre and resistant starch contents of raw and cooked pea, common bean, chickpea and lentil legumes. *Food Chemistry* 94: 327–330.

Alrosan, M., Tan, T.C., Easa, A.M. et al. (2021). Effects of fermentation on the quality, structure, and nonnutritive contents of lentil (*Lens culinaris*) proteins. *Journal of Food Quality* 2021: 5556450.

Alrosan, M., Tan, T.C., Easa, A.M. et al. (2022). Recent updates on lentil and quinoa protein-based dairy protein alternatives: nutrition, technologies, and challenges. *Food Chemistry* 383: 132386.

Alshikh, N., de Camargo, A.C., and Shahidi, F. (2015). Phenolics of selected lentil cultivars: antioxidant activities and inhibition of low-density lipoprotein and DNA damage. *Journal of Functional Foods* 18: 1022–1038.

Amarowicz, R., Estrella, I., Hernández, T. et al. (2009). Antioxidant activity of a red lentil extract and its fractions. *International Journal of Molecular Sciences* 10: 5513–5527.

Amarowicz, R., Estrella, I., Hernández, T. et al. (2010). Free radical-scavenging capacity, antioxidant activity, and phenolic composition of green lentil (*Lens culinaris*). *Food Chemistry* 121: 705–711.

Aslani, Z., Mirmiran, P., Alipur, B. et al. (2015). Lentil sprouts effect on serum lipids of overweight and obese patients with type 2 diabetes. *Health Promotion Perspective* 5: 215–224.

Ayet, G., Burbano, C., Cuadrado, C. et al. (1997). Effect of germination, under different environmental conditions, on saponins, phytic acid and tannins in lentils (*Lens culinaris*). *Journal of the Science of Food and Agriculture* 74: 273–279.

Bahnassey, Y. and Khan, K. (1986). Fortification of spaghetti with edible legumes. II. Rheological, processing, and quality evaluation studies. *Cereal Chemistry* 63: 216–219.

Baik, B.K. and Han, I.H. (2012). Cooking, roasting, and fermentation of chickpeas, lentils, peas, and soybeans for fortification of leavened bread. *Cereal Chemistry* 89: 269–275.

Bautista-Expósito, S., Peñas, E., Dueñas, M. et al. (2018a). Individual contributions of Savinase and *Lactobacillus plantarum* to lentil functionalization during alkaline pH-controlled fermentation. *Food Chemistry* 257: 341–349.

Bautista-Expósito, S., Peñas, E., Silván, J.M. et al. (2018b). pH-controlled fermentation in mild alkaline conditions enhances bioactive compounds and functional features of lentil to ameliorate metabolic disturbances. *Food Chemistry* 248: 262–271.

Berrios, J.D.J., Losso, J.N., and Albertos, I. (2022). Extrusion processing of dry beans and pulses. In: *Dry Beans and Pulses: Production, Processing, and Nutrition*, 2e (ed. M. Siddiq and M.A. Uebersax), 225–246. Hoboken, NJ: Wiley.

Boukid, F., Zannini, E., Carini, E., and Vittadini, E. (2019). Pulses for bread fortification: a necessity or a choice? *Trends in Food Science & Technology* 88: 416–428.

Bressani, T. (1993). Grain quality of common beans. *Food Review International* 9: 237–297.

Campos-Vega, R., Loarca-Pina, G., and Oomah, B.D. (2010). Minor components of pulses and their potential impact on human health. *Food Research International* 43: 461–482.

CDC (2021). Folic acid helps prevent serious birth defects of the brain and spine. https://www.cdc.gov/ncbddd/folicacid/features/folic-acid-helps-prevent-some-birth-defects.html (accessed 23 March 2021).

Cevallos-Casals, B.A. and Cisneros-Zevallos, L. (2010). Impact of germination on phenolic content and antioxidant activity of 13 edible seed species. *Food Chemistry* 119: 1485–1490.

Chen, K., Gao, C., Han, X. et al. (2021). Co-fermentation of lentils using lactic acid bacteria and *Bacillus subtilis* (Natto) increases functional and antioxidant components. *Journal of Food Science* 86: 475–483.

Ciudad-Mulero, M., Fernández-Ruiz, V., Cuadrado, C. et al. (2020). Novel gluten-free formulations from lentil flours and nutritional yeast: evaluation of extrusion effect on phytochemicals and non-nutritional factors. *Food Chemistry* 315: 126175.

Cornejo, F., Caceres, P.J., Martínez-Villaluenga, C. et al. (2015). Effects of germination on the nutritive value and bioactive compounds of brown rice breads. *Food Chemistry* 173: 298–304.

Courtin, C.M., Swennen, K., Verjans, P., and Delcour, J.A. (2009). Heat and pH stability of prebiotic arabinoxylooligosaccharides, xylooligosaccharides and fructooligosaccharides. *Food Chemistry* 112: 831–837.

Cuadrado, C., Hajos, G., Burbano, C. et al. (2002). Effect of natural fermentation on the lectin of lentils measured by immunological methods. *Food and Agricultural Immunology* 14: 41–49.

De Angelis, M., Calasso, M., Cavallo, N. et al. (2016). Functional proteomics within the genus *Lactobacillus*. *Proteomics* 16: 946–962.

De Pasquale, I., Pontonio, E., Gobbetti, M., and Rizzello, C.G. (2020). Nutritional and functional effects of the lactic acid bacteria fermentation on gelatinized legume flours. *International Journal of Food Microbiology* 316: 108426.

Dhull, S.B., Punia, S., Kidwai, M.K. et al. (2020). Solid-state fermentation of lentil (*Lens culinaris* L.) with *Aspergillus awamori*: effect on phenolic compounds, mineral content, and their bioavailability. *Legume Science* 2020: e37.

Dhull, S.B., Kidwai, M.K., Siddiq, M., and Sidhu, J.S. (2022a). Faba (broad) bean production, processing, and nutritional profile. In: *Dry Beans and Pulses: Production, Processing, and Nutrition*, 2e (ed. M. Siddiq and M.A. Uebersax), 359–381. Hoboken, NJ: Wiley.

Dhull, S.B., Kinabo, J., and Uebersax, M.A. (2022b). Nutrient profile and effect of processing methods on the composition and functional properties of lentils (*Lens culinaris* Medik): a review. *Legume Science* 2022: e156.

Didinger, C. and Thompson, H.J. (2022). The role of pulses in improving human health: a review. *Legume Science* 4: e147.

Dueñas, M., Sarmento, T., Aguilera, Y. et al. (2016). Impact of cooking and germination on phenolic composition and dietary fibre fractions in dark beans (*Phaseolus vulgaris* L.) and lentils (*Lens culinaris* L.). *LWT Food Science and Technology* 66: 72–78.

Englyst, H.N., Veenstra, J., and Hudson, G.J. (1996). Measurement of rapidly available glucose (RAG) in plant foods: a potential *in vitro* predictor of the glycaemic response. *British Journal of Nutrition* 75: 327–337.

Erskine, W., Williams, P.C., and Nakkoul, H. (1991). Splitting and dehulling lentil (*Lens culinaris*): effects of seed size and different pretreatments. *Journal of the Science of Food and Agriculture* 57: 77–84.

Espinosa-Ramírez, J., Rodríguez, A., De la Rosa-Millán, J. et al. (2021). Shear-induced enhancement of technofunctional properties of whole grain flours through extrusion. *Food Hydrocolloids* 111: 106400.

FAO (Food and Agriculture Organization) (2022). Crop production and trade data. http://www.fao.org/faostat/en/#home (accessed 30 December 2022).

FAO/WHO (Food and Agriculture Organization/World Health Organization) (1991). Protein quality evaluation. *Joint FAO/WHO, FAO Food & Nutrition Paper* 51: 1.

Farvid, M.S., Eliassen, A.H., Cho, E. et al. (2016). Dietary fiber intake in young adults and breast cancer risk. *Pediatrics* 137: e20151226.

Fouad, A.A. and Rehab, F.M.A. (2015). Effect of germination time on proximate analysis, bioactive compounds and antioxidant activity of lentil (*Lens culinaris* Medik.) sprouts. *Acta Scientiarum Polonorum. Technologia Alimentaria* 14: 233–246.

Funke, M., Loeffler, M., Winkelmeyer, C. et al. (2022). Emulsifying properties of lentil protein preparations obtained by dry fractionation. *European Food Research and Technology* 248: 381–391.

Gan, R.Y., Shah, N.P., Wang, M.F. et al. (2016). Fermentation alters antioxidant capacity and polyphenol distribution in selected edible legumes. *International Journal of Food Science and Technology* 51: 875–884.

Ganesan, K. and Xu, B. (2017). Polyphenol-rich lentils and their health promoting effects. *International Journal of Molecular Sciences* 18: 2390.

Gatta, C.D., Piergiovanni, A.R., Ng, N.Q. et al. (1989). Trypsin inhibitor levels in raw and cooked cowpea (*Vigna unguiculata*) seeds. *LWT Food Science and Technology* 22: 78–80.

Ghumman, A., Kaur, A., and Singh, N. (2016a). Impact of germination on flour, protein and starch characteristics of lentil (*Lens culinari*) and horsegram (*Macrotyloma uniflorum* L.) lines. *LWT Food Science and Technology* 65: 137–144.

Ghumman, A., Kaur, A., Singh, N., and Singh, B. (2016b). Effect of feed moisture and extrusion temperature on protein digestibility and extrusion behaviour of lentil and horsegram. *LWT Food Science and Technology* 70: 349–357.

Gonzalez, Z. and Perez, E. (2002). Evaluation of lentil starches modified by microwave irradiation and extrusion cooking. *Food Research International* 35: 415–420.

Grusak, M.A. (2009). Nutritional and health-beneficial quality. In: *The Lentil: Botany, Production and Uses* (ed. W. Erskine, F.J. Muehlbauer, A. Sarker, and B. Sharma), 368–390. Wallingford, UK: CABI.

Halima, O., Najar, F.Z., Wahab, A. et al. (2022). Lentil allergens identification and quantification: an update from omics perspective. *Food Chemistry: Molecular Sciences* 4: 100109.

Han, H. and Baik, B.K. (2008). Antioxidant activity and phenolic content of lentils (*Lens culinaris*), chickpeas (*Cicer arietinum* L.), peas (*Pisum sativum* L.) and soybeans (*Glycine max*), and their quantitative changes during processing. *International Journal of Food Science & Technology* 43: 1971–1978.

Hefnawy, T.H. (2011). Effect of processing methods on nutritional composition and anti-nutritional factors in lentils (*Lens culinaris*). *Annals of Agricultural Science* 56: 57–61.

Hill, H. (2022). Utilization of dry beans and other pulses as ingredients in diverse food products. In: *Dry Beans and Pulses: Production, Processing, and Nutrition*, 2e (ed. M. Siddiq and M.A. Uebersax), 307–329. Hoboken, NJ: Wiley.

Hughes, J.S. and Swanson, B.G. (1986). Microstructure of lentil seeds. *Food Microstructure* 5: 241–246.

Johnson, C.R., Thavarajah, P., Payne, S. et al. (2015). Processing, cooking, and cooling affect prebiotic concentrations in lentil (*Lens culinaris* Medikus). *Journal of Food Composition and Analysis* 38: 106–111.

Joshi, M., Timilsena, Y., and Adhikari, B. (2017). Global production, processing and utilization of lentil: a review. *Journal of Integrative Agriculture* 16: 2898–2913.

Kamau, E.H., Nkhata, S.G., and Ayua, E.O. (2020). Extrusion and nixtamalization conditions influence the magnitude of change in the nutrients and bioactive components of cereals and legumes. *Food Science & Nutrition* 8: 1753–1765.

Kancherla, V., Wagh, K., Pachón, H., and Oakley, G.P. Jr. (2021). A 2019 global update on folic acid-preventable spina bifida and anencephaly. *Birth Defects Research* 113: 77–89.

Khattab, R.Y. and Arntfield, S.D. (2009). Nutritional quality of legume seeds as affected by some physical treatments 2 antinutritional factors. *LWT Food Science and Technology* 42: 1113–1118.

Khazaei, H., Podder, R., Caron, C.T. et al. (2017). Marker–trait association analysis of iron and zinc concentration in lentil (*Lens culinaris* Medik.) seeds. *The Plant Genome* 10: 1–8.

Khazaei, H., Subedi, M., Nickerson, M. et al. (2019). Seed protein of lentils: current status, progress, and food applications. *Food* 8: 391.

Kumar, Y., Basu, S., Goswami, D. et al. (2021). Anti-nutritional compounds in pulses: implications and alleviation methods. *Legume Science* 2021: e111.

Kurien, P.P. (1977). Grain legume milling technology. *Paper Presented at FAO Expert Consultation on Grain Legume Processing* (14 November 1977). Rome, Italy: FAO.

Lehmann, U. and Robin, F. (2007). Slowly digestible starch-its structure and health implications: a review. *Trends in Food Science and Technology* 18: 346–355.

Ma, Z., Boye, J.I., Simpson, B.K. et al. (2011). Thermal processing effects on the functional properties and microstructure of lentil, chickpea, and pea flours. *Food Research International* 44: 2534–2544.

Ma, Z., Yin, X., Hu, X. et al. (2018). Structural characterization of resistant starch isolated from Laird lentils (*Lens culinaris*) seeds subjected to different processing treatments. *Food Chemistry* 263: 163–170.

Marchini, M., Carini, E., Cataldi, N. et al. (2021). The use of red lentil flour in bakery products: how do particle size and substitution level affect rheological properties of wheat bread dough? *LWT Food Science and Technology* 136: 110299.

Martín-Cabrejas, M.A., Díaz, M.F., Aguilera, Y. et al. (2008). Influence of germination on the soluble carbohydrates and dietary fibre fractions in non-conventional legumes. *Food Chemistry* 107: 1045–1052.

McKeown, N.M., Meigs, J.B., Liu, S. et al. (2004). Carbohydrate nutrition, insulin resistance, and the prevalence of the metabolic syndrome in the Framingham Offspring Cohort. *Diabetes Care* 27: 538–546.

Mo'ez Al-Islam, E.F., Mohammad, M.G., and Soliman, S. (2020). Lentils (*Lens culinaris* L.): a candidate chemopreventive and antitumor functional food. In: *Functional Foods in Cancer Prevention and Therapy* (ed. Y. Kabir), 99–120. London, UK: Academic Press.

Morales, P., Cebadera-Miranda, L., Cámara, R.M. et al. (2015). Lentil flour formulations to develop new snack-type products by extrusion processing: phytochemicals and antioxidant capacity. *Journal of Functional Foods* 19: 537–544.

Nosworthy, M.G., Medina, G., Franczyk, A.J. et al. (2018). Effect of processing on the *in vitro* and *in vivo* protein quality of red and green lentils (*Lens culinaris*). *Food Chemistry* 240: 588–593.

Olaerts, H., Roye, C., Derde, L.J. et al. (2016). Impact of preharvest sprouting of wheat (*Triticum aestivum*) in the field on starch, protein, and arabinoxylan properties. *Journal of Agricultural and Food Chemistry* 64: 8324–8332.

Oomah, B.D., Caspar, F., Malcolmson, L.J., and Bellido, A.S. (2011). Phenolics and antioxidant activity of lentil and pea hulls. *Food Research International* 44: 436–441.

Oskaybaş-Emlek, B., Özbey, A., and Kahraman, K. (2021). Effects of germination on the physicochemical and nutritional characteristics of lentil and its utilization potential in cookie-making. *Journal of Food Measurement and Characterization* 15: 4245–4255.

Pal, R.S., Bhartiya, A., Yadav, P. et al. (2017). Effect of dehulling, germination and cooking on nutrients, anti-nutrients, fatty acid composition and antioxidant properties in lentil (*Lens culinaris*). *Journal of Food Science and Technology* 54: 909–920.

Papandreou, C., Becerra-Tomás, N., Bulló, M. et al. (2019). Legume consumption and risk of all-cause, cardiovascular, and cancer mortality in the PREDIMED study. *Clinical Nutrition* 38: 348–356.

Pasqualone, A., Costantini, M., Labarbuta, R., and Summo, C. (2021). Production of extruded-cooked lentil flours at industrial level: effect of processing conditions on starch gelatinization, dough rheological properties and techno-functional parameters. *LWT Food Science and Technology* 147: 111580.

Patterson, C.A., Curran, J., and Der, T. (2017). Effect of processing on antinutrient compounds in pulses. *Cereal Chemistry* 94: 2–10.

Pereira, M.A., O'Reilly, E., Augustsson, K. et al. (2004). Dietary fiber and risk of coronary heart disease: a pooled analysis of cohort studies. *Archives of Internal Medicine* 164: 370–376.

Piecyk, M., Wołosiak, R., Drużynska, B., and Worobiej, E. (2012). Chemical composition and starch digestibility in flours from Polish processed legume seeds. *Food Chemistry* 135: 1057–1064.

Pusztai, A., Gelencser, E., Grant, G., and Bardocz, S. (1997). Nutritional manipulation of immune competence in young non-ruminant animals. In: *Recent Advances in Animal Nutrition* (ed. P.C. Garnsworlhy and G. Wisaman), 29–43. Nottingham, UK: Nottingham University Press.

Ramírez-Ojeda, A.M., Moreno-Rojas, R., and Cámara-Martos, F. (2018). Mineral and trace element content in legumes (lentils, chickpeas and beans): bioaccessibility and probabilistic assessment of the dietary intake. *Journal of Food Composition and Analysis* 73: 17–28.

Rathod, R.P. and Annapure, U.S. (2016). Effect of extrusion process on antinutritional factors and protein and starch digestibility of lentil splits. *LWT Food Science and Technology* 66: 114–123.

Rui, X., Xing, G., Zhang, Q. et al. (2016). Protein bioaccessibility of soymilk and soymilk curd prepared with two *Lactobacillus plantarum* strains as assessed by *in vitro* gastrointestinal digestion. *Innovative Food Science and Emerging Technologies* 38: 155–159.

Ruiz, R.G., Price, K.R., Rose, M.E., and Fenwick, G.R. (1997). Effect of seed size and testa colour on saponin content of Spanish lentil seed. *Food Chemistry* 58: 223–226.

Sackesen, C., Erman, B., Gimenez, G. et al. (2020). IgE and IgG4 binding to lentil epitopes in children with red and green lentil allergy. *Pediatric Allergy and Immunology* 31: 158–166.

Sagratini, G., Zuo, Y., Caprioli, G. et al. (2009). Quantification of soyasaponins I and βg in Italian lentil seeds by solid-phase extraction (SPE) and high-performance liquid chromatography – mass spectrometry (HPLC-MS). *Journal of Agricultural and Food Chemistry* 57: 11226–11233.

Schmiedl, D., Bauerlein, M., Bengs, H., and Jacobasch, G. (2000). Production of heat-stable, butyrogenic resistant starch. *Carbohydrate Polymers* 43: 183–193.

Shaheen, N., Halima, O., Akhter, K.T. et al. (2019). Proteomic characterization of low molecular weight allergens and putative allergen proteins in lentil (*Lens culinaris*) cultivars of Bangladesh. *Food Chemistry* 297: 124936.

Shewry, P.R. and Halford, N.G. (2002). Cereal seed storage proteins: structures, properties and role in grain utilization. *Journal of Experimental Botany* 53: 947–958.

Singh, U. and Singh, B. (1992). Tropical grain legumes as important human foods. *Economic Botany* 46: 310–321.

Siva, N., Thavarajah, P., and Thavarajah, D. (2018). The impact of processing and cooking on prebiotic carbohydrates in lentil. *Journal of Food Composition and Analysis* 70: 72–77.

Świeca, M. and Gawlik-Dziki, U. (2015). Effects of sprouting and postharvest storage under cool temperature conditions on starch content and antioxidant capacity of green pea, lentil and young mung bean sprouts. *Food Chemistry* 185: 99–105.

Świeca, M., Baraniak, B., and Gawlik-Dziki, U. (2013). *In vitro* digestibility and starch content, predicted glycemic index and potential *in vitro* antidiabetic effect of lentil sprouts obtained by different germination techniques. *Food Chemistry* 138: 1414–1420.

Thavarajah, D., Ruszkowski, J., and Vandenberg, A. (2008). High potential for selenium biofortification of lentils (*Lens culinaris* L.). *Journal of Agricultural and Food Chemistry* 56: 10747–10753.

Torino, M.I., Limon, R.I., Martinez-Villaluenga, C. et al. (2013). Antioxidant and antihypertensive properties of liquid and solid-state fermented lentils. *Food Chemistry* 136: 1030–1037.

USDA (US Dept of Agriculture) (2022). Food data central. https://fdc.nal.usda.gov (accessed 18 September 2022).

Viadel, B., Barberá, R., and Farré, R. (2006). Uptake and retention of calcium, iron, and zinc from raw legumes and the effect of cooking on lentils in Caco-2 cells. *Nutrition Research* 26: 591–596.

Vidal-Valverde, C. and Frias, J. (1991). Legume processing effects on dietary fiber components. *Journal of Food Science* 56: 1350–1352.

Wang, N., Hatcher, D.W., Toews, R., and Gawalko, E.J. (2009). Influence of cooking and dehulling on nutritional composition of several varieties of lentils (*Lens culinaris*). *LWT Food Science and Technology* 42: 842–848.

WHO (2020). Salt reduction. https://www.who.int/news-room/fact-sheets/detail/salt-reduction (accessed 12 April 2020).

Wiesinger, J.A., Marsolais, F., and Glahn, R.P. (2022). Health implications and nutrient bioavailability of bioactive compounds in dry beans and other pulses. In: *Dry Beans and Pulses: Production, Processing, and Nutrition*, 2e (ed. M. Siddiq and M.A. Uebersax), 505–529. Hoboken, NJ: Wiley.

Xu, B.J. and Chang, S.K. (2007). A comparative study on phenolic profiles and antioxidant activities of legumes as affected by extraction solvents. *Journal of Food Science* 72: S159–S166.

Xu, B. and Chang, S.K. (2009). Phytochemical profiles and health-promoting effects of cool-season food legumes as influenced by thermal processing. *Journal of Agricultural and Food Chemistry* 57: 10718–10731.

Xu, B. and Chang, S.K. (2010). Phenolic substance characterization and chemical and cell-based antioxidant activities of 11 lentils grown in the Northern United States. *Journal of Agricultural and Food Chemistry* 58: 1509–1517.

Xu, M., Jin, Z., Lan, Y. et al. (2019a). HS-SPME-GC-MS/olfactometry combined with chemometrics to assess the impact of germination on flavor attributes of chickpea, lentil, and yellow pea flours. *Food Chemistry* 280: 83–95.

Xu, M., Jin, Z., Simsek, S. et al. (2019b). Effect of germination on the chemical composition, thermal, pasting, and moisture sorption properties of flours from chickpea, lentil, and yellow pea. *Food Chemistry* 295: 579–587.

Yeo, J. and Shahidi, F. (2020). Identification and quantification of soluble and insoluble-bound phenolics in lentil hulls using HPLC-ESI-MS/MS and their antioxidant potential. *Food Chemistry* 315: 126202.

Yin, X., Ma, Z., Hu, X. et al. (2018). Molecular rearrangement of Laird lentil (*Lens culinaris* Medikus) starch during different processing treatments of the seeds. *Food Hydrocolloids* 79: 399–408.

Zare, F., Orsat, V., Champagne, C. et al. (2012). Microbial and physical properties of probiotic fermented milk supplemented with lentil flour. *Journal of Food Research* 1: 94–109.

Zhang, B., Deng, Z., Tang, Y. et al. (2014). Effect of domestic cooking on carotenoids, tocopherols, fatty acids, phenolics, and antioxidant activities of lentils (*Lens culinaris*). *Journal of Agricultural and Food Chemistry* 62: 12585–12594.

Zhang, T., Li, X., Chen, L., and Situ, W. (2016). Digestibility and structural changes of waxy rice starch during the fermentation process for waxy rice vinasse. *Food Hydrocolloids* 57: 38–45.

Zhang, B., Peng, H., Deng, Z., and Tsao, R. (2018). Phytochemicals of lentil (*Lens culinaris*) and their antioxidant and anti-inflammatory effects. *Journal of Food Bioactives* 1: 93–103.

Zia-ul-Haq, M., Ahmad, S., Shad, M.A. et al. (2011). Compositional studies of lentil (*Lens culinaris* Medik.) cultivars commonly grown in Pakistan. *Pakistan Journal of Botany* 43: 1563–1567.

Zou, Y., Chang, S.K., Gu, Y., and Qian, S.Y. (2011). Antioxidant activity and phenolic compositions of lentil (*Lens culinaris* var. Morton) extract and its fractions. *Journal of Agricultural and Food Chemistry* 59: 2268–2276.

14

Antinutritional Factors in Lentils: Their Effect on Bioavailability of Nutrients and Significance in Human Health

Shalini G. Rudra[1], *Akanksha Singh*[2], *Priya Pal*[1], *and Rahul K. Thakur*[1]

[1] *Division of Food Science and Postharvest Technology, ICAR-Indian Agricultural Research Institute, New Delhi, India*
[2] *Amity Institute of Organic Agriculture, Amity University, Noida, India*

14.1 Introduction

Lentil (*Lens culinaris* Medik.) is a cool-season self-pollinated diploid legume crop that was domesticated alongside wheat and barley in medieval Europe, Mediterranean, and Middle East. Lentil cultivation spread westward to the Mediterranean region, the Nile Area, and Central Europe, as well as eastward to Asia, from the fertile crescent. Later, North America and Australia were introduced to the lentil (Cubero 1984; Matny 2015). Lentils are divided into two categories: macrosperma and microsperma types (Cubero 1981). Macrosperma type lentils are primarily grown in Latin and North Americas, North Africa, Southern Europe, have large, flat seeds. Egypt, Afghanistan, South Asia, Ethiopia, and regions primarily grow microsperma type lentils. Lentil is planted in a variety of cropping patterns in South Asia (rice/lentil, fallow/lentil, and pearl millet/maize/cotton/lentil). Lentil is a high-protein, high-carbohydrate, high-fiber, high-vitamin, and high-micronutrient food that ensures food and nutrition security (Kissinger 2016).

Lentil, also known as "poor man's meat," is a complete healthy food source that is rich in proteins, prebiotic carbohydrates, and a variety of essential nutrients. Phytic acid (2.5–4.4 mg/g) and protein inhibitors are very moderate in a 50-g meal, which gives 3.7–4.5 mg of iron, 2.2–2.7 mg of zinc, and 22–34 g of selenium (Thavarajah et al. 2009). Lentils have a short preparation time (10–12 minutes) and need little processing (dehulling for dehulled/split lentil). Lentil breeding efforts around the world have effectively implemented mineral biofortification. Bioavailability of minerals is influenced by a variety of factors, including the composition of the dietary matrix and antinutrient factors (ANFs). Ascorbic acid, prebiotic carbohydrates, phytoferritin, and carotenoids are among the food matrix components used to screen germplasm for iron bioavailability (phytic acid, kaempferol, gallic acid, and chlorogenic acid). Iron bioavailability investigations utilize the *in vitro* Caco-2 cell model or the mice model only, as evidenced by the selected advanced breeding lines that have already been conducted out based entirely on phenotyping data (Thavarajah et al. 2015).

Lentils: Production, Processing Technologies, Products, and Nutritional Profile, First Edition.
Edited by Jasim Ahmed, Muhammad Siddiq, and Mark A. Uebersax.

The presence of antinutrient factors are other important compositional components with nutritional and health significance. While a number of ANFs are inherently present in lentils, appropriate processing techniques can reduce or completely eliminate ANFs, thereby increasing the nutritional benefits. By getting rid of inhibitors like phytic acids in leguminous plants and cereal grains (Raboy 2003), the bioavailability of micronutrients in food grains is improved. To address these bioavailability problems, it is necessary to change the way plants work or create genotypes that can store less antinutrients. Phytic acid makes it harder for minerals to be absorbed by the body. It is also essential for storing phosphorus but phytic acid cannot be completely eliminated from legumes because that would affect the health of the plants. L 4147, Pusa Ageti masoor (iron 65 ppm), L 4717, and IPL 220 (iron 73 ppm and zinc 51 ppm) are notable biofortified varieties in India (Dikshit et al. 2022). Yield of biofortified varieties have shown 35–67% yield advantage over local cultivars. Bangladesh farmers have replaced 99% of lentil seeds with improved varieties. This chapter covers the ANFs found in food legumes, with emphasis on lentils, their effect on nutrients' bioavailability, and methods to reduce or completely eliminate selected ANFs.

14.2 Lentil Protein and Amino Acid Profile

The global need for proteins is expanding rapidly due to population growth, and it is projected to be more than doubled by 2050 (Westhoek et al. 2011). In recent years, there has been a changing trend among consumers to replace animal proteins with new, high-quality plant-based proteins that are produced in an environmentally friendly and sustainable manner. Plant-based sources of protein offer an economical mean to meet the growing need for protein and improve the overall quality of human diets (Aggarwal and Drewnowski 2019). Lentils have an average crude protein content of about 26%, although, as shown in Table 14.1, wide variations exist among lentil verities and cultivars. Proteins from lentils are good for human health since they provide appreciable amounts of most of the essential amino acids that body need for metabolism and nutritional balance.

Table 14.1 Genetic variation for protein concentration in cultivated lentils (dry matter basis).

Protein content (%)	Total number of accessions	References
24.6–30.0	23	Heuzé et al. (2021)
10.5–27.1	45	Kumar et al. (2016)
24.6–30.0	23	Heuzé et al. (2021)
25.3–29.3	35	Alghamdi et al. (2013)
21.8–27.1	14	Zaccardelli et al. (2012)
22.7–31.9	46	Karaköy et al. (2012)
21.8–27.1	14	Zaccardelli et al. (2012)
23.8–29.3	22	Tahir et al. (2011)
24.3–30.2	4	Wang and Daun (2006)
23.0–32.0	—	Hedley (2001)
23.9–26.3	58	Stoddard et al. (1993)
23.4–36.4	1688	Hawtin et al. (1977)
27.5–31.7	—	Barulina (1930)

Lentil seed has been shown to have 12 times higher nutritional quality than the seeds from pea and *Medicago truncatula*, a Mediterranean legume plant (Scippa et al. 2010). Due to the quantitative character of the genes governing protein synthesis, storage protein levels are highly variable (Kumar et al. 2020). Higher genotype environmental interactions (31.31% broad sense heritability) also contribute to the considerable variation in lentil seed storage protein content (Gautam et al. 2018). Non-storage lentil seed proteins have metabolic roles, e.g., governing plant enzymatic activity, structure, and function (Scippa et al. 2010). Barulina (1930) conducted one of the earliest and most important investigations on the protein content of lentil varieties, which showed a range of 27.5–31.7%. Hawtin et al. (1977) analyzed the global germplasm of lentils for protein content and found that the protein content ranged from 23.4% to 36.4% among 1688 accessions. A study by Kumar et al. (2016) reported a wide variability in protein content of wild and cultivated lentils, showing that *L. orientalis* had the greatest average protein content (24.15%), followed by *L. ervoides* (22.99%), *L. odemensis* (19.7%), *L. nigricans* (19.53%), *L. tomentosus* (18.75%), and cultivated lentils (18.7%). Overall, there was a wider range of protein content in *L. orientalis* (18.3–27.75%) and *L. ervoides* (18.9–32.7%).

Endogenous amino acids (AAs), i.e., arginine, aspartic and glutamic acids, and leucine, which account for more than half of total AAs, are abundant in cultivated lentil proteins, but most essential AAs, such as isoleucine, threonine, phenylalanine, methionine, histidine, valine, tryptophan, and leucine, are also present (Table 14.2) (WHO-UNU 2007). Sulfur-containing amino acids, methionine, cysteine, and tryptophan, are found in very

Table 14.2 Amino acid profile of cultivated lentil genotypes.

Amino acids	Concentration (g/100 g of protein)
Alanine	3.31–8.35
Arginine	4.64–13.80
Aspartate/aspartic acid	6.36–13.20
Cysteine	0.60–1.62
Glutamate/glutamic acid	6.12–17.10
Glycine	4.40–10.40
Histidine	1.21–9.15
Isoleucine	2.20–5.00
Leucine	5.21–7.72
Lysine	5.81–9.59
Methionine	0.90–2.23
Phenylalanine	3.85–7.55
Proline	3.50–5.22
Serine	4.90–6.34
Threonine	1.04–4.60
Tryptophan	0.57–1.37
Tyrosine	2.71–7.15
Valine	4.10–5.01

Source: Adapted from Kahraman (2016).

low concentration in lentil seeds. Compared to other legumes, lentils have a relatively better AAs profile. A PCA (principal component analysis) of selected legume species revealed that the AA composition of lentil and pea was highly comparable or superior to the other legumes studied, with significantly higher lysine content (Djemel et al. 2005). The anabolic characteristics of plant-based protein intake can be balanced by consuming higher methionine and lower lysine containing foods, such as wheat and rice. Thus, combining cereals with legume-based diets including lentils are higher in lysine and lower in methionine (van Vliet et al. 2015). It is noteworthy that AA content varies among Lens species, with a range of variability seen for *L. culinaris*, *L. orientalis*, *L. ervoides*, *L. nigricans*, and *L. odemensis*. The concentrations of phenylalanine, methionine, valine, leucine, and isoleucine in wild species are considerably higher than in cultivated lentils (Rozan et al. 2001). Likewise, the nonessential AA concentration of wild species is higher than those of *L. culinaris*.

14.3 Antinutritional Factors (ANFs) in Lentils and Their Properties

A number of ANFs and protein digestibility governs the nutritional value of seed protein. In lentil and other food legumes, these antinutritional compounds (such as phenols, phytic acid, lectins, tannins, peptide inhibitors, and saponin) may reduce the health benefits associated with the consumption of pulses (Nosworthy and House 2017; Urbano et al. 2000). The ANFs, like trypsin inhibitors, phytic acid, phenolics, and tannins, show sensitivity to heat treatments and can be thus decreased in legumes via different processing methods, such as boiling, cooking, and pressure cooking (Patterson et al. 2017). This reduction in the ANFs significantly improves their digestibility, thereby increasing the nutritional profile of legumes.

14.3.1 Protease Inhibitors

Protein inhibitors (trypsin and chymotrypsin inhibitors) dramatically lower the digestibility of proteins if these protease inhibitors are not eliminated or effectively inhibited during the processing (Boye et al. 2010b). In this regard, lentil proteins' nutritional value and digestibility can be greatly improved by different processing procedures that offset the negative effects of ANFs, including protease inhibitors (Nosworthy et al. 2018). Low-molecular-weight proteins called trypsin inhibitors are present in many plant species, including legumes (Savage 1989), and they are known to block the physiological activity trypsin enzyme. They increase the production of the pancreatic enzyme trypsin by inducing hypersecretion, which promotes pancreatic hypertrophy and reduces the digestion and absorption of amino acids, which reduces their bioavailability (Pisulewska and Pisulewski 2000; Savage 1989). The altered or reduced bioavailability leads to decreased nitrogen and sulfur retention, as well as stunted development of plants (Norton 1991). Hefnawy (2011) utilized pressure cooking (121 °C for 35 minutes) and boiling (100 °C for 90 minutes) to reduce tannins by 29–36% and trypsin inhibitors by 93–94% in lentils. Compared to other grain legumes, except pea, lentil has a relatively lower trypsin inhibitor concentration (Shi et al. 2017). Lentil seeds can also be soaked, cooked, and germinated to significantly reduce the trypsin inhibitors.

14.3.2 Tannins and Phenols

Lentil has a relatively high tannin content compared to other grain legumes (Grela et al. 2017). Processing, especially dehulling, can eliminate tannins, which are predominantly present in the seed coat. Tannins decrease the digestibility of proteins by interacting with lysine and methionine, which makes them less readily accessible during digestion (Davis 1981). Lentil with zero-tannin content have been developed (Matus et al. 1993) and the investigation of its inheritance showed that the zero-tannin trait was governed by one recessive gene (tan) (Mirali et al. 2017).

Polyphenols, phenolic acids, and flavonoids are examples of some bioactive dietary phenolics in legumes (Scalbert and Williamson 2000). There has been heightened research interest in phenolic compounds because of their antioxidant activities, which makes them suitable for use in processed foods as natural antioxidants. Twenty different cultivars of lentils were analyzed, and the results showed that flavonoids, such as catechin/epicatechin glucosides, kaempferol glycosides, and procyanidins, were the main phenolic compounds in lentils (Zhang et al. 2015). The phenolic content of lentils is mainly based on the color of their seed coats, with the darker seed coat correlated with the higher tannin concentration (Mirali et al. 2017). Padhi et al. (2017) reported that the total amount of polyphenols in Canadian lentils ranged from 1.22 to 7.45 mg/g, accompanied by a strong correlation between seed coat color and polyphenol content. However, total phenolic content in lentil evaluated by Torino et al. (2013) was found to be significantly higher (32 mg GAE/g) in comparison to other studies as well as 1.9 mg GAE/g reported by Byanju et al. (2021).

Oomah et al. (2011) reported that the total phenolic content of lentil hulls is 3–8 times higher than that of whole lentil seeds, with the hulls having a more diverse phenolic profile than that in the cotyledon. A total of 43 phenolic compounds have been documented in the seed coat of lentils, which included 3 anthocyanidins, 4 pro-anthocyanidins, 6 flavones, 1 stilbene, 1 phenolic acid, 2 flavanones, 7 flavanols, and 15 flavonols (Mirali et al. 2017). Zhang et al. (2015) reported that *p*-hydroxybenzoic acid, syringic acid, *trans*-*p*-coumaric acid, epicatechin gallate, quercetin-3-xyloside, quercetin-3-glucoside, and kaempferol-3-glucoside were identified as predominant phenolics in lentils. Simple phenolic compounds, i.e., phenolic acids, pro-anthocyanidins, anthocyanidins, flavan-3-ols, flavonols, flavones, stilbenes, and flavanones, are the major sub-classes of phenolic compounds found in lentils. The phenolic profile of lentils is dominated by kaempferol glycosides, with the tetraglycoside and triglycoside having the highest concentrations, followed by catechin gallate, catechin glucoside, epicatechin glucoside. Whereas *p*-hydroxybenzoic acid and *trans*-*p*-Coumaric acid are the main free phenolic acids found in lentils. Phenolic compounds in the lentil seed coat are more diverse and abundant than in the cotyledon and mainly comprise of polymeric and oligomeric proanthocyanidins (Mirali et al. 2017).

14.3.3 Phytic Acid

Phytic acid (myoinositol hexakisphosphate or IP6) is an essential compound for human nutrition and health significance. Besides its action as an antinutrient, phytic acid has been shown to play a positive role for humans as an antioxidant, with potential prevention of cancer and some other disorders (Kumar et al. 2010). Phytic acid makes complexes with minerals (iron and zinc), thereby reducing their bioavailability (Tiwari and Singh 2012). It also hampers the digestion of peptide by binding to enzymes like proteases and amylases. Plants metabolism and resistance to pathogens both benefit from phytic acid, despite its very small concentration in the plant cells. Due to its ability to bind heavy metals, some studies have reported benefits of phytic acid, including anticarcinogenic qualities and antioxidant activity (Campos-Vega et al. 2010; Urbano et al. 2000).

The endosperm of grains from leguminous plants stores the phytic acid and its concentration varies from 0.2% to 2.9% in whole grain to >3.7% in the cotyledons (Lestienne et al. 2005). Phytic acid content was found to be higher in the wild species of legumes (0.98–3.14 g/100 g) than in cultivated species. Thavarajah et al. (2009) found that the amount of phytic acid from three distinct market categories of lentils cultivated in two places in Saskatchewan, Canada, varied less (0.3–0.4 g/100 g). The Spanish brown cultivar Pardina, which is grown in Spain, had relatively higher concentration of overall seed phytic acid (1.2 g/100 g). Similarly, Paul and Shang (2004) showed that the mean value of 17 types of lentils grown in Bangladesh was about 0.7 g/100 g. On the other hand, when tested genotypes were exposed to high temperatures, the phytic acid content was shown to increase slightly. Thavarajah et al. (2009) reported that the amount of phytic acid in lentil seeds was significantly higher at elevated growing temperatures. In a recent study, phytic acid content was reported in the range of 8.6–17.1 g/kg in various cultivars of lentils (Shi et al. 2017).

14.3.4 Lectins

Lectins are glycoprotein (carbohydrate binding proteins) predominantly found in the seeds of leguminous plants. These impart resistance to plants against bacteria, fungi, or insects. Lectins in legumes are considered as antinutrient due to their negative impact on the nutritional quality. Lectins can directly bind to intestinal mucosa, interfering with absorption and transportation of 0.01% free gossypol and interacting with enterocytes. The binding ability of lectins with sugars or proteins in cell membrane causes agglutination of red blood corpuscles (RBCs) and thus causes health hazards in humans. The concentration of lectins varies significantly in legumes, e.g., 0.8% in soybean, 2.4–5% and 0.6% of total protein in kidney beans and garden peas, respectively. Grant et al. (1983) suggested using hemagglutinating test to quantify lectins in terms of hemagglutinating units (HU/g). Srivastava and Vasishtha (2013a) reported that the lectins in different cultivars of lentils range from 513 to 617 HU/g.

Lectins present in legumes exhibit sequence homologies and structural similarities despite their differences in sugar moieties and quaternary structures. Lectins when consumed by sensitive individuals can cause *in vivo* effects which can cause physiological reactions such as severe intestinal damage resulting in improper digestion and nutritional

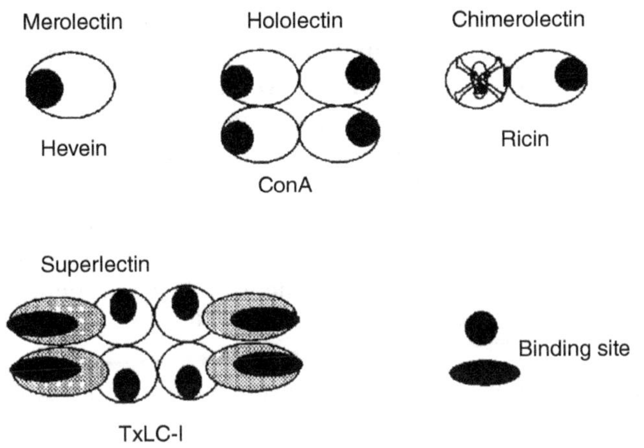

Figure 14.1 Schematic diagram of mercolectin, hololectin, chimerolectin, and superlectins
Source: Adapted from Peumans and Van Damme (1995).

deficiencies, can provoke antibodies such as IgG and IgM causing food allergies and other immune responses (Boehm and Huck 2009; Vasconcelos and Oliveira 2004). Legume lectins are made up of single subunit with one polypeptide chain of 300 amino acids in its primary structure with average molecular weight of 30 kDa (Wales et al. 1991). Lectins are classified in two ways: (i) based on their carbohydrate specificity, or (ii) based on their overall structures (Figure 14.1), which classify lectins into merolectin, chimerolectin, hololectin, and superlectin that contains a carbohydrate binding domain. Merolectins have single carbohydrates binding domain, unable to agglutinate cells. Chimerlectins display catalytic activity by one chain and biological activity by another chain. Hololectins possess two similar domains for carbohydrate binding. Superlectins have at least two unidentical carbohydrate binding domains. As a result, these lectins recognize different sugar moieties structurally (Teixeira et al. 2012).

In recent years, lectins have become focus of intense interest for scientists due to their unique characteristics and wide scope of applications in agriculture and medicine fields. Lectins play a role in the attachment of nitrogen-fixing bacteria to legumes and provide protection against phytopathogens. They are used in various fields of biology as well as therapeutic areas of research. Nowadays, research on lectins is very intensive because of their positive biomedical attributes. Fu et al. (2011) elucidated lectins' potential role against carcinogenic activities. According to Wang et al. (2021) the lectins extracted from lentils were shown to possess antiviral activities against SARS-CoV-2 variants.

14.3.5 Saponins

The main sources of saponins (soy saponins I and VI) in human diet are lentils and chickpeas (Ruiz et al. 1996). Saponins interact with cholesterol and can also cause lysis of erythrocyte membranes and intestinal mucosal cells. The cholesterol binding properties are responsible for controlling lipid composition in blood besides exerting anticarcinogenic effects. In legumes, most saponins are concentrated in the seed coat, plumule, radicle, and hypocotyl. Total saponin concentration in 44 lentil genotypes has been reported to range from 170 to 350 mg/100 g with an average value of 281 mg/100 g (Ahuja et al. 2015). Unprocessed grains of 10 varying lentil genotypes had 198.1–332.8 and 362.9–452.9 mg/100 g of sapogenol A and sapogenol B, respectively (Srivastava and Vasishtha 2013b). Ruiz et al. (1996) estimated total saponin content of 20 Spanish lentil cultivars in the range of 65.4–126.9 mg/100 g and further reported that lentils with green or beige seed coat had higher total saponin content than those with a brown seed coat. Sagratini et al. (2009) reported wide variability in contents of soyasaponin I (2.8–40.7 mg/100 g) and soyasaponin βg (11–124.2 mg/100 g) in 32 lentil cultivars having difference seed size and seed coat color.

Saponins interact with the cholesterol directly but can also cause lysis of erythrocytes and intestinal mucosal cells, thus affecting bioavailability of nutrients. Based on *in vitro* studies in animals, the anticarcinogenic effects exerted by anti-inflammatory, hypocholesterolemic, and immune-stimulatory activities of saponins have been reported by Rao and Gurfinkel (2000). Preliminary studies on hamsters showed that saponins lowered cholesterol levels by inhibiting the intestinal reabsorption of bile acids and enhanced the excretion of bile acids and neutral sterols in the feces. The crude saponin extract inhibited the tumor cell growth by inducing differentiation of cells and curbing the activation of the cell protein kinase C. Soyasapogenol A also prevents liver damage by inhibiting apoptosis of liver cells and elevation of plasma tumor necrosis factor α (TNF-α). Saponins are potent inhibitors of human immunodeficiency virus (HIV) infection and can be potentially used in the treatment of retroviral infection (Singh et al. 2017).

14.3.6 Raffinose Family Oligosaccharides (Total α-Galactosides)

Faris et al. (2013) reported that 53% of the total sugars in lentils are comprised of raffinose, ciceritol, stachyose, and verbascose. Generally, these raffinose family oligosaccharides (RFOs) are known for causing flatulence and gastrointestinal discomfort and are considered somewhat thermostable. However, these RFOs can be reduced significantly by soaking and aqueous processing of lentils. In recent years, RFOs too have been reported to be an important dietary component, due to their unique ability to reduce inflammation, combat chronic diseases, eliminate pathogens, and enhance mineral bioavailability (Johnson et al. 2020; Siva and Thavarajah 2018).

14.4 Nutritional Significance and Implications of Selected ANFs

14.4.1 Tannins

Flavonoids and condensed tannins, widely found in lentils and other food legumes, are reported to positively contribute to weight loss and type 2 diabetes risk reduction, in addition to the antioxidant activities. Total phenolic compounds in lentils range from 4.56 to 8.34 mg GAE/g, dry-weight basis. The presence of tannic acid in lentil proteins modified its cleavage specificity and peptidomics profile. Tannic acid induces aggregation of proteins and reduces proteins' digestibility by obstructing the accessibility of pepsin to the protein network and modifying the released peptides' profile (Boachie et al. 2022).

14.4.2 Protease Inhibitors (Trypsin and Chymotrypsin Inhibitors)

Antinutritional protease inhibitors are categorized into Bowman-Birk inhibitors and the Kunitz inhibitors. The Bowman-Birk inhibitors have molecular weight of around 8 kDa with major portion of disulfide bonds and ability to inhibit trypsin and chymotrypsin at independent binding sites. The Kunitz inhibitors have molecular weight of 21.5 kDa with two disulfide linkages and possess major specificity against trypsin (Kennedy 1998).

Kunitz inhibitors contain 181 amino acid residues with reactive sites located at residues Arg 63 and Ile 64. These inhibitors align with trypsin in stoichiometric fashion, i.e., one molecule of inhibitor inactivates one molecule of trypsin. The complex thus formed is very unstable and analogous to usual enzyme–substrate complexes. The amino acid sequence of Bowman-Birk has two independent binding sites: (i) trypsin reactive (Lys 16 and Ser 17) and (ii) a chymotrypsin reactive site (Leu 43 and Ser 44). The sequence of amino acids that surround the two reactive sites are very similar to each other and a homology has been found between Bowman-Birk inhibitor and a number of other low- molecular-weight inhibitors isolated from other legumes. In contrast to the Kunitz inhibitors, the Bowman-Birk inhibitors are very rich in disulfide bonds, possessing seven disulfide bridges. This particular feature is responsible for a very tight, compact, 3-dimensional structure, as revealed by X-ray crystallography and nuclear magnetic resonance (NMR). In lima beans, cowpeas, some other common beans and lentils, protease inhibitors have been categorized as members of Bowman-Birk family, whereas trypsin/chymotrypsin inhibitors from red kidney beans, Brazilian pink beans, soybeans, and lima beans are closely related with high homology.

The trypsin inhibitor activity-laced purified fraction have been reported to inhibit the growth in rats, mice, and chicks, which is accompanied by a depression in digestibility of dietary proteins. The plausible explanation for the growth inhibition produced by trypsin inhibitors is that it interferes with the digestibility of dietary protein. Furthermore, feeding rats with immobilized trypsin showed improved growth performance when compared to those fed free trypsin (Chan and De Lumen 1982). The simplest explanation for the growth inhibition is that presence of protease inhibitors such as trypsin and chymotrypsin in diet leads to formation of irreversible complex, causing a decrease in trypsin in intestine and reduced digestibility of dietary protein, thus leading to slower animal growth (Gilani et al. 2012).

The presence of trypsin inhibitors, which are one of the secondary metabolites synthesized by plants to protect themselves from pests, reduces nutritive value of foods, such as legumes including lentils. Trypsin inhibitors interfere with the activity of trypsin and chymotrypsin enzymes released by pancreas to digest protein. According to Świeca and Baraniak (2014), trypsin inhibitors bind with enzymes and reduce the required amount of trypsin needed to digest protein. As a result, pancreas tends to produce more enzymes to compensate the protein-inhibitor bindings and end up putting excessive stress on the organ involved. This hyperactivity of the organ often causes hypertrophy and hyperplasia of the

Table 14.3 Trypsin inhibitor activity (TIA) in different legumes.

Scientific name	Common name	TIA (U/mg)
Lens culinaris	Lentil	7.4
Phaseolus vulgaris	Kidney bean	3.10
Phaseolus acutifolius	Tepary bean	11.50–18.00
Phaseolus lunatus	Lima bean	5.89
Vicia faba	Broad bean	4.57
Vigna aunguiculata	Cowpea	7.52
Vigna radiata	Mung bean	1.83
Vigna mungo	Black gram	0.61–1.03
Cicer arietinum	Chickpea	8.10–15.70
Pisum sativum	Pea	2.20
Arachis hypogea	Peanut	5.60
Cajanus cajan	Pigeon pea	4.75
Glycine max	Soybean	94.10
Lupinus spp.	Lupin	1.48–1.74

Source: Adapted from Sara et al. (2017).

pancreas. Thus, it is necessary to completely remove or at least reduce their quantity to below a threshold limit so as to enhance the assimilation of comestible legumes. Trypsin inhibitor activity (TIA) of selected food legumes is presented in Table 14.3.

Trypsin inhibitor, being heat labile in nature, are easily neutralized by means of thermal treatment. However, thermal exposure of food is also associated with loss of nutrients and some functional properties. Thus, there is a need to explore novel processes, e.g., microwave heating, extrusion, micronization, and irradiation, to strike a balance between nutritional quality of legumes and the disadvantages associated with conventional processing methods (Sara et al. 2017).

14.4.3 Lectins

The antinutritional effects of lectins are most likely due to compromising integrity of intestinal epithelium and alter the absorption and utilization of nutrients. The lectins from the field bean are toxic when injected into rats and are known to inhibit growth of rats and cause zonal necrosis of the liver. However, some lectins have been reported to be nontoxic (Grant et al. 1983), such as those from lentils, peas, chickpea, and faba beans. Many researchers have studied a strong correlation between lectin-binding patterns and their biological behavior in various tumors (De Mejia et al. 2003; Gautam et al. 2018). The inclusion of PHA diet from raw kidney bean was reported to reduce the growth of a murine non-Hodgkin's lymphoma tumor in the mouse (Pryme et al. 1999). *Vicia faba* agglutinin from broad beans, stimulated the morphological differentiation and reduced malignant-phenotype colon-cancer cells (Jordinson et al. 1999). Lectins are also being used for discovery of protein markers of cancer using a natural glycoprotein microarray approach, by screening serum samples from pancreatic cancer or pancreatitis patients by selective detection of glycan structures (Dang et al. 2020).

14.5 Effect of Processing Methods on ANFs

14.5.1 Dehulling

The removal or reduction of polyphenols, e.g., tannins, concentrated in indigestible hulls (~80%) eliminates the peculiar, taste of pulses, thereby increasing the acceptability by improving taste, color, and ease of cooking. Most antinutrients in lentils are concentrated in the cotyledon, except phenolic compounds, and thus, dehulling is reported to increase the phytic acid content by 5–20% within the lentil cotyledons (Pedrosa et al. 2021) and also oligosaccharides such as stachyose and verbascose. Thus, dehulling of lentils increases the activity of the common enzyme inhibitors found in pulses such as trypsin inhibitors, chymotrypsin inhibitors, and α-amylase inhibitors. Dehulling of legume grains has no effect on any significant reduction of lectins (Kumar et al. 2022; Srivastava and Vasishtha 2013a).

14.5.2 Soaking and Thermal Treatment

Soaking, the simplest and economic method for inactivation of trypsin inhibitors, hydrates the legume seeds and enhances the removal of water-soluble compounds by leaching into soaking solution. Soaking is affected by temperature, pH, and duration of soaking (El-Adawy et al. 2000). Soaking of lentils with various ratios of lentil: water, different temperatures, and milling of the seeds impart significant effect on the nutritive and antinutritive lentil flour components (Vidal-Valverde et al. 2002a). A maximum reduction of 330%, 83%, 53%, 59%, 57% for trypsin inhibitor activity, raffinose, stachyose, ciceritol, α-galactosidase content was observed. The most effective results were obtained at 42 °C. Pre-soaked seeds showed a 78.7–100% decrease in enzyme inhibitors upon cooking. Shi et al. (2017) showed that long cooking times might be needed for phytic acid inactivation as the density in cotyledons may vary with kind of varieties. Huang et al. (2018) reported that water used for soaking whole green lentils had threefold and fivefold more saponins and total phenolics, respectively. The effect of various processing methods on lentil's saponins is presented in Table 14.4.

Thermal processing is the most widely used and an effective way of improving the nutritional value of legume because it increases protein digestibility mainly by inactivating thermolabile ANFs, e.g., trypsin inhibitors and lectins. Some of the thermal methods include boiling, autoclaving, cooking, roasting, and microwaving. The heat processing results in breakage of intermolecular bonds that are responsible for holding tertiary structure of trypsin inhibitors. Xu and Chang (2009) investigated the effect of different cooking methods on phytic acid content in lentil and reported up to 21.9% decrease in phytic acid content. Whereas a phytic acid reduction of as high as 60.50% after cooking was reported by

Table 14.4 Effect of processing on the saponin content of lentils.

Type of saponin	Saponin content (mg/100 g)	Treatment	Effect on saponins
Saponins A and B	644.9	Soaking	12.6% decrease in sapogenol
Soyasaponins I and βg	31–159.5	Soaking, cooking	61.9% decrease in total sapogenol
Soyasaponins I and VI	65.4–126.9	Cooking	6–14% leaching into cooking solution
Saponins A and B	569–700	Sprouting (6 d)	twofold increase in soyasapogenol B

Source: Adapted from Singh et al. (2017).

Elhardallon and Walker (1994). Hefnawy (2011) used pressure cooking (121 °C for 35 minutes) and boiling (100 °C for 90 minutes) to achieve 36–41% reduction in phytic acid in lentils. Besides research on reducing the phytic acid content via hydrothermal treatments, a significant number of studies have been also done by utilizing other approaches for reducing phytic acid content of food such as roasting of grain, germination, fermentation, by use of different doses or irradiation (Udensi et al. 2008).

Lectins are heat-labile with their stability varying among plant species, some being resistant to inactivation by dry heat and require the presence of moisture for complete destruction (Gemede and Ratta 2014). It is well known fact that cooking at elevated pressure nullify the antinutrients in general. Soaking of lentils in water followed by cooking at >90 °C results in the reduction of hemagglutinating activity by 94% (Coffey et al. 1992). Hence, lentils are safe to consume after cooking as lectins are heat labile and destroyed completely.

14.5.3 Additives Use in Soaking/Cooking

It is common to use food-grade additives in soaking water to facilitate improved water absorption, texture, color preservation, and antinutrient reduction in legumes (Huma et al. 2008; Vadivel and Pugalenthi 2009). Ammonium hydroxide, sodium hydroxide, and sodium bicarbonate additives assist in trypsin inhibitor inactivation along with thermal treatment. This effect is as a result of extremely low or high pH levels, which enhance the loss of enzyme activity due to unfavorable electrostatic interactions between amino acid residues which cause conformational changes in active site. In the case of acidic additive, however, a delayed inactivation effect on trypsin inhibitors is observed, which could be due to the possible unavailability of proteins for heat inactivation under acidic conditions. It is further noted that the application of alkali during food processing causes side reactions that influence treated proteins like racemization of amino acids, which decreases the protein digestibility (Friedman 1999; Gilani and Sepehr 2003). Soaking of hard-to-cook (HTC) lentils in sodium bicarbonate solution has been reported to improve texture and imparts a peculiar taste and also decreases phytic acid and tannin content by 78% and 66%, respectively (Huma et al. 2008).

14.5.4 Extrusion

Extrusion is a continuous thermo-mechanical and short-time process in which usually starchy grains, meals, and flours are passed through die, thereby shaping the extrudate or end product in given configurations (Berrios 2006). Lentils and other food legumes have been evaluated for their feasibility to extrusion with variable process parameters like temperature, screw speeds and number of screws, die longitude whether with or without pretreatments. The starch gets gelatinized and enhanced protein denaturation, e.g., hydrolysis of peptide bonds (Ma et al. 2011). Since this process makes use of high shear forces and temperature, it can deform proteins physically. Therefore, extrusion can also modify the molecules of trypsin inhibitors by changing their secondary and tertiary structures by denaturing them (Rathod and Annapure 2016).

14.5.5 Ultrafiltration

Ultrafiltration is another process that can be used for improving the nutritional qualities of legume products. This can separate trypsin inhibitors into different fractions rather than their inactivation or elimination (Boye et al. 2010a; Mondor et al. 2009). Ultrafiltration uses high selectivity and mild operating conditions to remove the small molecules and particular protein profile of a legume affects the process output. It is important to maintain legume protein in their native forms to avoid their elution as monomers in filtration membranes. This process was used for improving the functional properties of pea, chickpea, and lentil protein concentrates (Boye et al. 2010a; Swanson 1990).

14.5.6 High Hydrostatic Pressure (HPP)

The high hydrostatic pressure (HPP) processing is a nonthermal technology used for food preservation, which has positive impact on nutritional quality and reduction of ANFs (Sharma et al. 2022). The HPP processing also provides an opportunity to develop new products based on legumes with high nutritional value and long shelf life (Porzucek et al. 2002). This HPP process disturbs the intramolecular balance and solvent protein interactions, like electrostatic and hydrophobic with minimal effect on covalent bonds (Estrada-Giron et al. 2005). Apart from inactivation of trypsin inhibitors, when HPP is combined with mild-heat treatment, the retention of isoflavones improved up to 40.28% (Jung et al. 2008).

Dostálová et al. (2009) reported a decrease in raffinose family oligosaccharide (raffinose, verbascose, and stachyose) from 16% to 5% in germinated lentil seeds by HPP treatment at 500 MPa for 10 minutes. Hall and Moraru (2021) compared the effect of HPP processing and thermal treatment on protein quality, *in vitro* protein digestibility, and trypsin inhibitor activity. The HPP processing of lentil protein concentrate at 600 MPa and 5 °C for four minutes resulted in higher gastric proteolysis than thermal treatment at 95 °C for 15 minutes, while the overall *in vitro* protein digestibility was not significantly affected by both the treatment. Also heat treatment reduced trypsin inhibitor activity higher (~86%), as compared to high-pressure processing (~8%). Combining HPP processing with enzymatic hydrolysis of lentil flours using a commercial protease (Corolase 7089) was shown to increase the ACE inhibition activity at 300 MPa (Garcia-Mora et al. 2015).

14.5.7 Instant Controlled Pressure Drop (DIC)

It is a recently developed alternative technology that uses the combination of heat (up to 170 °C) and pressurized steam (up to 8 bar) for a maximum of three minutes. This process eliminates the ANFs in legumes by denaturation, as the seeds develop a porous structure with the application of DIC. The availability of soluble sugars and a reduction in ANFs such as lectins and phytic acid is contributed by seed porosity (Pedrosa et al. 2012). Based on certain studies, application of DIC on lentils greatly reduced the trypsin inhibitors enough to meet the safety limits. DIC has advantages of quick trypsin inhibitors inactivation under a short processing time. It is excellent technology for industrial-level operation (Cuadrado et al. 2011; Pedrosa et al. 2012).

14.5.8 Ultrasound

Ultrasound is an effective technique that uses the phenomenon of cavitation to inactivate different enzymes. This process changes the protein structure by localized increase in temperature, due to rapid formation and subsequent collapsing of gas molecules, which is known as cavitation effect. It is generally used by the inactivation of trypsin inhibitors as synergistic tool to enhance the effectiveness of other treatments. Huang et al. (2008) reported a decrease of 55% in trypsin inhibitors by using 20-minute treatment of 30 kHz ultrasound waves.

14.5.9 Germination/Sprouting

Germination is a natural catabolic process present in all higher order plants in which seed comes out of latency taking available seed nutrients from cotyledons that are used for growth and embryo development (Ampofo et al. 2020; Sangronis and Machado 2007). It is an effective traditional method used to reduce ANFs such as trypsin inhibitors, by breaking down of cellular proteins and enzymes by proteases to get free amino acids for autotrophic growth of seedling. Germination resulted in a substantial increase in thiamine, riboflavin, and niacin content, along with a significant reduction in the trypsin inhibitor activity and elimination of raffinose family oligosaccharides, i.e., raffinose, stachyose, and verbascose (Dhull et al. 2022). Kadlec et al. (2008) also reported a gradual decrease in raffinose family oligosaccharides content of ≥25% after five-days of lentil germination. Bubelová et al. (2018) reported that digestibility of lentil improved by 30% upon germination, which was attributed to the reduction of phytic acid content and proteolytic and amylolytic enzyme inhibitor activities. Germination utilizes the lentils' secondary metabolite reserves using endogenous enzymes for metabolic activities. Vidal-Valverde et al. (2002b) reported that the duration of germination played a crucial role in optimizing the quality of legumes, stating that six days of germination in the presence of light helped in achieving legume flours with enhanced nutritive value.

14.5.10 Fermentation

Fermentation is one of the oldest processes which is used to improve the nutritional quality of food products. Enhanced synthesis of microbial metabolites with biological properties during fermentation is responsible for various functional benefits in human nutrition

(Dhull et al. 2020; Kitum et al. 2022). Solid-state fermentation with *Aspergillus oryzae* for 48 hours could inhibit the α-glucosidase activity of lentils by up to 90% (Magro et al. 2019). The influence of different processes, e.g., fermentation with *Lactobacillus plantarum* and *Pediococcus acidilactici*, precooking, or sonication on the lentil's ANFs, was evaluated by Byanju et al. (2021). Phytic acid content of 0.07 g/kg in raw (control) lentils was significantly reduced to ~0 and 0.03 g/ kg by sonication and fermentation by *L. plantarum* and *P. acidilactici*, respectively, compared to first sonication then fermented samples. It was further observed that two-minute sonication and fermentation were optimal for reducing phytic acid in lentil. Both microorganisms were able to reduce the phytic acid content significantly for precooked and then fermented lentils.

Montemurro et al. (2019) reported that sourdough fermentation of germinated lentil flours reduced raffinose content, trypsin inhibitor activity, condensed tannins, and phytic acid by 94.83%, 87.07%, 57.45%, and 13%, respectively. Further Cuadrado et al. (2002) reported a rapid decline in hemagglutination activity during fermentation while lectins concentration was reduced at 72 and 96 hours, by 98.0% and 97.8%, respectively.

14.5.11 Irradiation Processing

El-Niely (2007) reported that phytic acid content and tannin contents were significantly reduced by 15.1–32.7% and by 7.6–21.7%, respectively, upon radiation treatment of lentils at 5–10 kGy dose levels. Raffinose family oligosaccharides (raffinose and stachyose) were also significantly affected by irradiation treatment but showed no effect after storage for 6–12 months (Aylangan et al. 2017).

14.6 Health Implications of ANFs

The ANFs in foods have deleterious effects in absorption of nutrients and micronutrients, which may interfere with normal body functions (digestion and assimilation). These effects are frequently related to raw, plant based, or vegan diets, including legume-based diets. Nausea, bloating, nutritional deficiencies, headaches, and rashes are some of the most common symptoms from consuming foods rich in ANFs. The sensitivity to antinutrients differ widely among people but an adequate processing of food is necessary to reduce antinutritional factors. Once the antinutrients are introduced in the body, then it is difficult to eliminate (Gemede and Ratta 2014; Graham et al. 2001; Xu and Chang 2009).

Many of the antinutrient compounds are now being reconsidered as possessing important health benefits (Table 14.5), which include antioxidant and anti-inflammatory activities, as well as promoting the bioavailability of micronutrients such as iron (Wiesinger et al. 2022). When used at low levels, phytic acid, lectins, and phenolic compounds as well as enzyme inhibitors and saponins have been shown to reduce blood glucose and/or plasma cholesterol and triacylglycerols. Furthermore, saponins are reported to act effectively in maintaining liver function, preventing osteoporosis as well as platelet agglutination (Campos-Vega et al. 2010; Hart et al. 2020; Wiesinger et al. 2022). Consumers' awareness is crucial especially when abnormal health conditions are established.

Table 14.5 Health benefits of bioactive compounds found in legumes.

Compound	Bioactive – health benefits
Oligosaccharides	Prebiotic
Polyphenols	Antioxidant/prebiotic
Phytic acid/phytate	Antioxidant
Lectins (phytohemagglutinins)	Satiety
Protease/amylase inhibitors	Antidiabetic
Bioactive peptides	Anti-inflammatory
Saponins	Hypocholesterolemic

Source: Adapted from Wiesinger et al. (2022).

14.6.1 Antioxidant Activities and Correlation with Phenolics

The antioxidant activity of lentils varies widely among lentil cultivars. Zhang et al. (2015) determined the antioxidant activity for 20 Canadian cultivars by the DPPH assay (23–35 μmol TE/g, dry-weight basis), FRAP value (18–35 AAE/g DW), and ORAC values (105–168 μmol TE/g, dry-weight basis). Phenolic compounds, especially condensed tannins, are major contributors to the reducing power of lentils.

Lentil consumption has also been reported to be potentially effective in reducing various health conditions, such as hypertension, cardiovascular diseases, diabetes mellitus, and cancer. Flour of some lentil varieties (Blaze and Laird) can bind with bile salts, thereby reducing blood cholesterol levels (Barbana and Boye 2013). Rodrigues Neta and de Castro (2020) reported that among the various legumes, lentil hydrolysates obtained after simulated digestion can serve as potentially good source of bioactive ingredients. Lentils are excellent source of phenolic compounds and condensed tannins. Phenolic compounds identified in the hydrophilic extract of lentils mainly included flavanol and flavonol glycosides, as well as procyanidin oligomers. Flavonols have been found the most important contributors to the inhibitory activities of lentils against α-glucosidase and pancreatic lipase (Zhang et al. 2015).

Various researchers have studied the effect of *in vitro* gastrointestinal simulation and revealed that consumption and digestion of processed lentils can impart a significant inhibitory effect on ACE activity of lentils (Jamdar et al. 2017). Three peptide fractions with potential activity were obtained after the digestion of lentil globulins; among those, fraction 5 (5 F) had the highest ACE inhibitory activity with an IC_{50} value of 0.02 mg/ml (Jakubczyk and Baraniak 2013). Among the different protease enzymes used for simulated digestion studies and comparing their effects, it was found that pepsin (maximum ACE inhibition = 79%) and Savinase (maximum ACE inhibition = 63%) produced the most cost-effective multifunctional hydrolysates from lentil with the highest antioxidant activities (Garcia-Mora et al. 2015). Similarly, Barbana and Boye (2013) conducted a comparative analysis of the ACE inhibitory activity of hydrolysates obtained from sequential digestion of green and red lentil using Alcalase or Flavourzyme that simulated human gastrointestinal digestion, and hydrolysis with papain and bromelain. After gastrointestinal digestion, the ability of phytochemicals to retain their structural integrity and formation of new peptides forms cause inhibition activity of ACE, which forms a major antihypertensive effect in the human body. Peñas et al. (2015) reported that various nonthermal processing techniques have shown to have a remarkable effect on the ACE inhibitory activity of lentils.

Chemopreventive and antitumor effect of lentils is predominantly due to the presence of plenty of polyphenolics. Lectins, bioactive peptides that include protease inhibitors and defensin proteins, saponins glycosides, fermentable fibers, and prebiotic oligosaccharides are other bioactive compounds in lentils with known anticarcinogenic potential. An array of extractable phenolic compounds (flavanones, flavones, proanthocyanidins, anthocyanidins, flavanols, flavonols, phenolic acids, and stilbenes) besides insoluble-bound phenolics (catechin, epicatechin, hydroxycinnamic acids, gallic acid, hydroxybenzoic acids, protocatechuic acid, flavonoids and their derivatives, and 3-hydroxycinnamic acid) have been reported in lentils (Alshikh et al. 2015; Zhang et al. 2018). In lentils, bound phytochemicals were found to contribute higher antioxidant activity than extractable phytochemicals (Han and Baik 2008).

14.6.2 Inhibitory Activities on α-Glucosidase and Lipase

Glucose absorption and post-prandial hyperglycemia can be reduced by α-glucosidase inhibitors, primarily by interfering with the enzyme action in the brush-border of the small intestine. This effect is considered desirable for type 2 diabetes prevention and treatment. The phenolic-rich extracts from lentils were tested for their inhibitory ability on pancreatic lipase and α-glucosidase. The IC_{50} values of the desugared extracts ranged from 23.08 to 42.15 mg/ml (low IC_{50} denotes higher α-glucosidase inhibition). Phenolic compounds, especially flavonols, e.g., quercetin glycosides and kaempferol, are major contributors to the α-glucosidase inhibitory activity (Moreno-Valdespino et al. 2020).

Pancreatic lipase, a key enzyme responsible for absorption of triglycerides in the small intestine, hydrolyses triglyceride into glycerol and fatty acids. Suppression and delay of triglyceride digestion and absorption through lipase inhibition is another approach for control of hyperlipidaemia and obesity. The IC_{50} values of phenolic extracts for pancreatic lipase inhibition were shown to significantly increase from 6 to 9 mg/ml. Flavonols were indicated to be higher contributor to the lipase inhibitory activity. Further, lentil extracts had much stronger inhibitory activity (lower IC_{50} values) toward lipase than α-glucosidase (Zhang et al. 2015). These results suggest that lentils can be an excellent source of pancreatic lipase inhibitors, and thus a useful dietary adjunct for body weight and obesity management.

14.7 Conclusion

Lentils are rich sources of protein, vitamins, and micronutrients like iron, zinc, selenium, folates, and carotenoids. The high content of fiber in lentils helps regularize bowel movements and promotes the growth of healthy gut bacteria. Lentils also contain phytochemicals (some referred as antinutrients), which potentially have functional roles in the human body, e.g., regulating blood sugar levels, anticarcinogenic effect, and for improved cardiovascular health. The phrase *antinutritional factors* was generally used by the food science and nutrition professionals to describe the negative health effects of consuming partially cooked food legumes (common beans and pulses). Some of the antinutritional factors, such as lectins (phytohemagglutinins) and saponins, are mostly degraded during the cooking process. Most other ANFs can be significantly reduced through different processing methods, i.e., dehulling, extrusion, fermentation, germination or sprouting, high hydrostatic pressure treatment, ultrafiltration, and ultrasound treatment. Although antinutrients possess negative health properties, some ANFs may also provide health benefits. For example, many of the ANFs, that interfere with nutrient bioavailability in the upper intestine, also have prebiotic, antioxidant, and anti-inflammatory properties that benefit the health of the lower intestine. This chapter provided an overview of both the positive and negative health effects of consuming bioactive compounds from food legumes, with emphasis on lentils.

References

Aggarwal, A. and Drewnowski, A. (2019). Plant- and animal-protein diets in relation to sociodemographic drivers, quality, and cost: findings from the Seattle Obesity Study. *The American Journal of Clinical Nutrition* 110 (2): 451–460.

Ahuja, H., Kaur, S., Gupta, A.K. et al. (2015). Biochemical mapping of lentil (*Lens culinaris* Medik) genotypes for quality traits. *Acta Physiologiae Plantarum* 37 (9): 1–16.

Alghamdi, S.S., Khan, A.M., Ammar, M.H. et al. (2013). Phenological, nutritional and molecular diversity assessment among 35 introduced lentil (*Lens culinaris* Medik.) genotypes grown in Saudi Arabia. *International Journal of Molecular Sciences* 15 (1): 277–295.

Alshikh, N., de Camargo, A.C., and Shahidi, F. (2015). Phenolics of selected lentil cultivars: antioxidant activities and inhibition of low-density lipoprotein and DNA damage. *Journal of Functional Foods* 18: 1022–1038.

Ampofo, J., Ngadi, M., and Ramaswamy, H.S. (2020). Elicitation kinetics of phenolics in common bean (*Phaseolus vulgaris*) sprouts by thermal treatments. *Legume Science* 2: e56.

Aylangan, A., Ic, E., and Ozyardimci, B. (2017). Investigation of gamma irradiation and storage period effects on the nutritional and sensory quality of chickpeas, kidney beans and green lentils. *Food Control* 80: 428–434.

Barbana, C. and Boye, J.I. (2013). in vitro protein digestibility and physico-chemical properties of flours and protein concentrates from two varieties of lentil (*Lens culinaris*). *Food & Function* 4 (2): 310–321.

Barulina, O.H. (1930). Lentil of USSR and of other countries. *Bulletin of Applied Botany Genetics and Plant Breeding* 40: 1–319.

Berrios, J.D.J. (2006). Extrusion cooking of legumes: dry bean flours. In: *Encyclopedia of Agricultural, Food, and Biological Engineering*, 1e (ed. D.R. Heldman and C.I. Moraru), 1–8. New York, NY: Taylor and Francis Group.

Boachie, R.T., Okagu, O.D., Abioye, R. et al. (2022). Lentil protein and tannic acid interaction limits in vitro peptic hydrolysis and alters peptidomic profiles of the proteins. *Journal of Agricultural and Food Chemistry* 70: 6519–6529.

Boehm, S. and Huck, S. (2009). Presynaptic inhibition by concanavalin: are alphalatrotoxin receptors involved in action potential-dependent transmitter release? *Journal of Neurochemistry* 71: 2421–2430.

Boye, J.I., Aksay, S., Roufik, S. et al. (2010a). Comparison of the functional properties of pea, chickpea and lentil protein concentrates processed using ultrafiltration and isoelectric precipitation techniques. *Food Research International* 43: 537–546.

Boye, J., Zare, F., and Pletch, A. (2010b). Pulse proteins: processing, characterization, functional properties and applications in food and feed. *Food Research International* 43 (2): 414–431.

Bubelová, Z., Sumczynski, D., and Salek, R.N. (2018). Effect of cooking and germination on antioxidant activity, total polyphenols and flavonoids, fiber content, and digestibility of lentils (*Lens culinaris* L.). *Journal of Food Processing and Preservation* 42 (1): e13388.

Byanju, B., Hojilla-Evangelista, M.P., and Lamsal, B.P. (2021). Fermentation performance and nutritional assessment of physically processed lentil and green pea flour. *Journal of the Science of Food and Agriculture* 101 (14): 5792–5806.

Campos-Vega, R., Loarca-Piña, G., and Oomah, B.D. (2010). Minor components of pulses and their potential impact on human health. *Food Research International* 43 (2): 461–482.

Chan, J. and De Lumen, B.O. (1982). Biological effects of isolated trypsin inhibitor from winged bean (*Psophocarpus tetragonolobus*) on rats. *Journal of Agricultural and Food Chemistry* 30: 46–50.

Coffey, D.G., Uebersax, M.A., Hosfield, G.L., and Bennink, M.R. (1992). Stability of red kidney bean lectin 1. *Journal of Food Biochemistry* 16: 43–57.

Cuadrado, C., Hajos, G., Burbano, C. et al. (2002). Effect of natural fermentation on the lectin of lentils measured by immunological methods. *Food and Agricultural Immunology* 14 (1): 41–49.

Cuadrado, C., Cabanillas, B., Pedrosa, M.M. et al. (2011). Effect of instant controlled pressure drop on IgE antibody reactivity to peanut, lentil, chickpea and soybean proteins. *International Archives of Allergy and Immunology* 156: 397–404.

Cubero, J.I. (1981). Origin, taxonomy and domestication. In: *Lentils* (ed. C. Webb and G. Hawtin), 15–38. Norwich, UK: Commonwealth Agricultural Bureaux.

Cubero, J.I. (1984). Taxonomy, distribution and evolution of the lentil and its wild relatives. In: *Genetic Resources and Their Exploitation – Chickpeas, Faba Beans and Lentils* (ed. J.R. Witcombe and W. Erskine), 187–203. Dordrecht: Springer. https://doi.org/10.1007/978-94-009-6131-9_16.

Dang, K., Zhang, W., Jiang, S. et al. (2020). Application of lectin microarrays for biomarker discovery. *ChemistryOpen* 9: 285–300.

Davis, K.R. (1981). Effect of processing on composition and *tetrahymena* relative nutritive. *Cereal Chemistry* 58 (5): 454–460.

De Mejia, E.G., Bradford, T., and Hasler, C. (2003). The anticarcinogenic potential of soybean lectin and lunasin. *Nutrition Reviews* 61: 239–246.

Dhull, S.B., Punia, S., Kidwai, M.K. et al. (2020). Solid-state fermentation of lentil (*Lens culinaris* L.) with *Aspergillus awamori*: effect on phenolic compounds, mineral content, and their bioavailability. *Legume Science* 2: e37.

Dhull, S.B., Kinabo, J., and Uebersax, M.A. (2022). Nutrient profile and effect of processing methods on the composition and functional properties of lentils (*Lens culinaris* Medik): a review. *Legume Science* 2022: e156.

Dikshit, H.K., Mishra, G.P., Aski, M.S. et al. (2022). Lentil breeding. In: *Fundamentals of Field Crop Breeding* (ed. D.V. Yadava, H.K. Dikshit, G.P. Mishra, and S. Tripathi), 1181–1236. Berlin, Germany: Springer.

Djemel, N., Guedon, D., Lechevalier, A. et al. (2005). Development and composition of the seeds of nine genotypes of the *Medica gotruncatula* species complex. *Plant Physiology and Biochemistry* 43 (6): 557–566.

Dostálová, J., Kadlec, P., Bernášková, J. et al. (2009). The changes of α-galactosides during germination and high pressure treatment of legume seeds. *Czech Journal of Food Sciences* 27: 76–79.

El-Adawy, T.A., Rahma, E.H., El-Bedawy, A.A., and Sobihah, T.Y. (2000). Effect of soaking process on nutritional quality and protein solubility of some legume seeds. *Nahrung* 44 (5): 339–343.

Elhardallon, S.B. and Walker, A.F. (1994). Phytic acid content of three legumes in the raw, cooked and fibre forms. *Phytochemical Analysis* 5 (5): 243–246.

El-Niely, H.F. (2007). Effect of radiation processing on antinutrients, in-vitro protein digestibility and protein efficiency ratio bioassay of legume seeds. *Radiation Physics and Chemistry* 76 (6): 1050–1057.

Estrada-Giron, Y., Swanson, B.G., and Barbosa-Canovas, G.V. (2005). Advances in the use of high hydro-static pressure for processing cereal grains and legumes. *Trends in Food Science & Technology* 16 (5): 194–203.

Faris, M.E.A.I.E., Takruri, H.R., and Issa, A.Y. (2013). Role of lentils (*Lens culinaris* L.) in human health and nutrition: a review. *Mediterranean Journal of Nutrition and Metabolism* 6 (1): 3–16.

Friedman, M. (1999). Chemistry, biochemistry, nutrition, and microbiology of lysinoalanine, lanthionine, and histidinoalanine in food and other proteins. *Journal of Agricultural and Food Chemistry* 47 (4): 1295–1319.

Fu, L.L., Zhou, C.C., Yao, S. et al. (2011). Plant lectins: targeting programmed cell death pathways as antitumor agents. *The International Journal of Biochemistry & Cell Biology* 43 (10): 1442–1449.

Garcia-Mora, P., Peñas, E., Frías, J. et al. (2015). High-pressure improves enzymatic proteolysis and the release of peptides with angiotensin I converting enzyme inhibitory and antioxidant activities from lentil proteins. *Food Chemistry* 171: 224–232.

Gautam, A.K., Shrivastava, N., Sharma, B., and Bhagyawant, S. (2018). Current scenario of legume lectins and their practical applications. *Journal of Crop Science and Biotechnology* 21 (3): 217–227.

Gemede, H.F. and Ratta, N. (2014). Antinutritional factors in plant foods: potential health benefits and adverse effects. *International Journal of Nutrition and Food Sciences* 3 (4): 284–289.

Gilani, G.S. and Sepehr, E. (2003). Protein digestibility and quality in products containing antinutritional factors are adversely affected by old age in rats. *The Journal of Nutrition* 133 (1): 220–225.

Gilani, G.S., Xiao, C.W., and Cockell, K.A. (2012). Impact of antinutritional factors in food proteins on the digestibility of protein and the bioavailability of amino acids and on protein quality. *British Journal of Nutrition* 108: S315–S332.

Graham, R.D., Welch, R.M., and Bouis, H.E. (2001). Addressing micronutrient malnutrition through enhancing the nutritional quality of staple foods: principles, perspectives and knowledge gaps. *Advances in Agronomy* 70: 77–142.

Grant, G., More, L.J., McKenzie, N.H. et al. (1983). A survey of the nutritional and haemagglutination properties of legume seeds generally available in the UK. *British Journal of Nutrition* 50 (2): 207–214.

Grela, E.R., Kiczorowska, B., Samolińska, W. et al. (2017). Chemical composition of leguminous seeds: part I – content of basic nutrients, amino acids, phytochemical compounds, and antioxidant activity. *European Food Research and Technology* 243 (8): 1385–1395.

Hall, A.E. and Moraru, C.I. (2021). Effect of high pressure processing and heat treatment on in vitro digestibility and trypsin inhibitor activity in lentil and faba bean protein concentrates. *LWT Food Science and Technology* 152: 112342.

Han, H. and Baik, B.K. (2008). Antioxidant activity and phenolic content of lentils (*Lens culinaris*), chickpeas (*Cicer arietinum* L.), peas (*Pisum sativum* L.) and soybeans (*Glycine max*), and their quantitative changes during processing. *International Journal of Food Science & Technology* 43 (11): 1971–1978.

Hart, J.J., Tako, E., Wiesinger, J., and Glahn, R.P. (2020). Polyphenolic profiles of yellow bean seed coats and their relationship with iron bioavailability. *Journal of Agricultural and Food Chemistry* 68: 769–778.

Hawtin, G.C., Rachie, K.O., and Green, J.M. (1977). Breeding strategy for the nutritional improvement of pulses. In: *Nutritional Standards and Methods of Evaluation for Food Legume Breeders* (ed. J.H. Hulse, K.O. Rachie, and L.W. Billingsley). Ottawa, ON, CA: IDRC.

Hedley, C.L. (2001). *Carbohydrates in Grain Legume Seeds: Improving Nutritional Quality and Agronomic Characteristics*, 1–13. Wallingford: CABI.

Hefnawy, T.H. (2011). Effect of processing methods on nutritional composition and anti-nutritional factors in lentils (*Lens culinaris*). *Annals of Agricultural Science* 56 (2): 57–61.

Heuzé, V., Tran, G., Sauvant, D. et al. (2021). Lentil (*Lens culinaris*). Feedipedia, a programme by INRAE, CIRAD, AFZ and FAO. 18. https://www.feedipedia.org/node/284 (accessed 17 March 2023).

Huang, H., Kwok, K.C., and Liang, H.H. (2008). Inhibitory activity and conformation changes of soybean trypsin inhibitors induced by ultrasound. *Ultrasonics Sonochemistry* 15 (5): 724–730.

Huang, S., Liu, Y., Zhang, W. et al. (2018). Composition of legume soaking water and emulsifying properties in gluten-free bread. *Food Science and Technology International* 24 (3): 232–241.

Huma, N., Anjum, M., Sehar, S. et al. (2008). Effect of soaking and cooking on nutritional quality and safety of legumes. *Nutrition & Food Science* 38 (6): 570–577.

Jakubczyk, A. and Baraniak, B. (2013). Activities and sequences of the angiotensin I-converting enzyme (ACE) inhibitory peptides obtained from the digested lentil (*Lens culinaris*) globulins. *International Journal of Food Science & Technology* 48 (11): 2363–2369.

Jamdar, S.N., Deshpande, R., and Marathe, S.A. (2017). Effect of processing conditions and in vitro protein digestion on bioactive potentials of commonly consumed legumes. *Food Bioscience* 20: 1–11.

Johnson, N., Johnson, C.R., Thavarajah, P. et al. (2020). The roles and potential of lentil prebiotic carbohydrates in human and plant health. *Plants, People, Planet* 2 (4): 310–319.

Jordinson, M., El-Hariry, I., Calnan, D.A. et al. (1999). *Vicia faba* agglutinin, the lectin present in broad beans, stimulates differentiation of undifferentiated colon cancer cells. *Gut* 44: 709–714.

Jung, S., Murphy, P.A., and Sala, I. (2008). Isoflavone profiles of soymilk as affected by high-pressure treatments of soymilk and soybeans. *Food Chemistry* 111 (3): 592–598.

Kadlec, P., Dostalova, J., Bernášková, J., and Skulinova, M. (2008). Degradation of α-galactosides during the germination of grain legume seeds. *Czech Journal of Food Sciences* 26 (2): 99–108.

Kahraman, A. (2016). Nutritional components and amino acids in lentil varieties. *Selcuk Journal of Agriculture and Food Sciences* 30 (1): 34–38.

Karaköy, T., Erdem, H., Baloch, F.S. et al. (2012). Diversity of macro-and micronutrients in the seeds of lentil landraces. *The Scientific World Journal* 2012: 1–15.

Kennedy, A.R. (1998). The Bowman-Birk inhibitor from soybeans as an anticarcinogenic agent. *The American Journal of Clinical Nutrition* 68 (6): 1406S–1412S.

Kissinger, G. (2016). Pulse crops and sustainability: a framework to evaluate multiple benefits. https://iyp2016.org/images/pdf/Pulse-crops_sustainability-framework_final_April-18-2016.pdf.

Kitum, V.C., Kinyanjui, P.K., Mathara, J.M., and Sila, D.N. (2022). Oligosaccharide and antinutrient content of whole red haricot bean fermented in salt–sugar and salt-only solutions. *Legume Science* 4: e110.

Kumar, V., Sinha, A.K., Makkar, H.P., and Becker, K. (2010). Dietary roles of phytate and phytase in human nutrition: a review. *Food Chemistry* 120 (4): 945–959.

Kumar, J., Singh, J., Kanaujia, R., and Gupta, S. (2016). Protein content in wild and cultivated taxa of lentil (*Lens culinaris* ssp. *Culinaris* Medikus). *Indian Journal of Genetics and Plant Breeding* 76: 631–634.

Kumar, S., Gupta, P., Choukri, H., and Siddique, K.H. (2020). Efficient breeding of pulse crops. *Accelerated Plant Breeding* 3: 1–30.

Kumar, Y., Basu, S., Goswami, D. et al. (2022). Anti-nutritional compounds in pulses: implications and alleviation methods. *Legume Science* 4 (2): e111.

Lestienne, I., Caporiccio, B., Besancon, P. et al. (2005). Relative contribution of phytates, fibers and tannins to low iron and zinc in vitro solubility in pearl millet (*Pennisetum glaucum*) flour and grain fractions. *Journal of Agricultural and Food Chemistry* 53: 8342–8348.

Ma, Z., Boye, J.I., Simpson, B.K. et al. (2011). Thermal processing effects on the functional properties and microstructure of lentil, chickpea, and pea flours. *Food Research International* 44: 2534–2544.

Magro, A.E.A., Silva, L.C., Rasera, G.B., and de Castro, R.J.S. (2019). Solid-state fermentation as an efficient strategy for the biotransformation of lentils: enhancing their antioxidant and antidiabetic potentials. *Bioresources and Bioprocessing* 6 (1): 1–9.

Matny, O.N. (2015). Lentil (*Lens culinaris* Medikus) current status and future prospect of production in Ethiopia. *Advances of Plants Agricultural Research* 2: 45–53.

Matus, A., Slinkard, A.E., and Vandenberg, A. (1993). The potential of zero tannin lentil. *New Crops* 279: 282.

Mirali, M., Purves, R.W., and Vandenberg, A. (2017). Profiling the phenolic compounds of the four major seed coat types and their relation to color genes in lentil. *Journal of Natural Products* 80 (5): 1310–1317.

Mondor, M., Aksay, S., Drolet, H. et al. (2009). Influence of processing on composition and antinutritional factors of chickpea protein concentrates produced by isoelectric precipitation and ultrafiltration. *Innovative Food Science and Emerging Technologies* 10 (3): 3427.

Montemurro, M., Pontonio, E., Gobbetti, M., and Rizzello, C.G. (2019). Investigation of the nutritional, functional and technological effects of the sourdough fermentation of sprouted flours. *International Journal of Food Microbiology* 302: 47–58.

Moreno-Valdespino, C.A., Luna-Vital, D., Camacho-Ruiz, R.M., and Mojica, L. (2020). Bioactive proteins and phytochemicals from legumes: mechanisms of action preventing obesity and type-2 diabetes. *Food Research International* 130: 108905.

Norton, G. (1991). Proteinase inhibitors. In: *Toxic Substances in Crop Plants* (ed. J.P.F. DMello, C.M. Duffus, and J.H. Duffus), 68–106. Cambridge, England: Royal Society of Chemistry.

Nosworthy, M.G. and House, J.D. (2017). Factors influencing the quality of dietary proteins: implications for pulses. *Cereal Chemistry* 94 (1): 49–57.

Nosworthy, M.G., Medina, G., Franczyk, A.J. et al. (2018). Effect of processing on the *in vitro* and *in vivo* protein quality of red and green lentils (*Lens culinaris*). *Food Chemistry* 240: 588–593.

Oomah, B.D., Caspar, F., Malcolmson, L.J., and Bellido, A.S. (2011). Phenolics and antioxidant activity of lentil and pea hulls. *Food Research International* 44 (1): 436–441.

Padhi, E.M., Liu, R., Hernandez, M. et al. (2017). Total polyphenol content, carotenoid, tocopherol and fatty acid composition of commonly consumed Canadian pulses and their contribution to antioxidant activity. *Journal of Functional Foods* 38: 602–611.

Patterson, C.A., Curran, J., and Der, T. (2017). Effect of processing on antinutrient compounds in pulses. *Cereal Chemistry* 94 (1): 2–10.

Paul, D.K. and Shang, R.K. (2004). Nutritive values and phytate content of some new varieties of wheat and legumes. *Pakistan Journal of Biological Sciences* 7 (6): 1006–1009.

Pedrosa, M.M., Cuadrado, C., Burbano, C. et al. (2012). Effect of instant controlled pressure drop on the oligosaccharides, inositol phosphates, trypsin inhibitors and lectins contents of different legumes. *Food Chemistry* 131 (3): 862–868.

Pedrosa, M.M., Guillamón, E., and Arribas, C. (2021). Autoclaved and extruded legumes as a source of bioactive phytochemicals: a review. *Food* 10 (2): 379.

Peñas, E., Limón, R.I., Martínez-Villaluenga, C. et al. (2015). Impact of elicitation on antioxidant and potential antihypertensive properties of lentil sprouts. *Plant Foods for Human Nutrition* 70 (4): 401–407.

Peumans, W.J. and Van Damme, E.J. (1995). Lectins as plant defense proteins. *Plant Physiology* 109 (2): 347.

Pisulewska, E. and Pisulewski, P.M. (2000). Trypsin inhibitor activity of legume seeds (peas, chickling vetch, lentils, and soya beans) as affected by the technique of harvest. *Animal Feed Science and Technology* 86 (3, 4): 261–265.

Porzucek, H., Duszkiewicz-Reinhard, W., Piecyk, M. et al. (2002). Changesof flatulence-causing sugars in legume protein samples by high hydrostatic pressure. *Electronic Journal of Polish Agricultural Universities, Food Science and Technology* 5 (2): 7.

Pryme, I.F., Bardocz, S., Pusztai, A., and Ewen, S.W. (1999). The growth of an established murine non-Hodgkin lymphoma tumour is limited by switching to a phytohaemagglutinin-containing diet. *Cancer Letters* 146: 87–91.

Raboy, V. (2003). myo-Inositol-1,2,3,4,5,6-hexakisphosphate. *Phytochemistry* 64: 1033–1043.

Rao, A.V. and Gurfinkel, D.M. (2000). Dietary saponins and human health. In: *Saponins in Food, Feedstuffs and Medicinal Plants* (ed. W. Oleszek and A. Marston), 255–270. Dordrecht: Springer.

Rathod, R.P. and Annapure, U.S. (2016). Effect of extrusion process on antinutritional factors and protein and starch digestibility of lentil splits. *LWT Food Science and Technology* 66: 114–123.

Rodrigues Neta, I.M. and de Castro, R.J.S. (2020). Enzyme-assisted extraction of biocomponents of lentils (*Lens culinaris* L.): effect of process parameters on the recovery of compounds with antioxidant properties. *Biocatalysis and Biotransformation* 38 (1): 15–23.

Rozan, P., Kuo, Y.H., and Lambein, F. (2001). Amino acids in seeds and seedlings of the genus *Lens*. *Phytochemistry* 58 (2): 281–289.

Ruiz, R.G., Price, K.R., Arthur, A.E. et al. (1996). Effect of soaking and cooking on the saponin content and composition of chickpeas (*Cicer arietinum*) and lentils (*Lens culinaris*). *Journal of Agricultural and Food Chemistry* 44 (6): 1526–1530.

Sagratini, G., Zuo, Y., Caprioli, G. et al. (2009). Quantification of soyasaponins I and βg in Italian lentil seeds by solid-phase extraction (SPE) and high-performance liquid chromatography-mass spectrometry (HPLC-MS). *Journal of Agricultural and Food Chemistry* 57 (23): 11226–11233.

Sangronis, E. and Machado, C.J. (2007). Influence of germination on the nutritional quality of *Phaseolus vulgaris* and *Cajan uscajan*. *LWT Food Science and Technology* 40 (1): 116–120.

Sara, A.G., Chuck-Hernandez, C., Sergio, O., and Serna, S. (2017). Inactivation methods of trypsin inhibitor in legumes: a review. *Journal of Food Science* 83 (1): 17–28.

Savage, G.P. (1989). Antinutritive factors in peas. In: *Recent Advances of Research in Antinutritional Factors in Legume Seeds* (ed. J. Huisman, A.F.B. van der Poel, and I.E. Liener), 342–350. Wageningen: Pudoc.

Scalbert, A. and Williamson, G. (2000). Dietary intake and bioavailability of polyphenols. *The Journal of Nutrition* 130 (8): 2073S–2085S.

Scippa, G.S., Rocco, M., Ialicicco, M. et al. (2010). The proteome of lentil (*Lens culinaris* Medik.) seeds: discriminating between landraces. *Electrophoresis* 31 (3): 497–506.

Sharma, N., Sahu, J.K., Joshi, S. et al. (2022). Modulation of lentil antinutritional properties using non-thermal mediated processing techniques – a review. *Journal of Food Composition and Analysis* 109: 104498.

Shi, L., Mu, K., Arntfield, S.D., and Nickerson, M.T. (2017). Changes in levels of enzyme inhibitors during soaking and cooking for pulses available in Canada. *Journal of Food Science and Technology* 54 (4): 1014–1022.

Singh, B., Singh, J.P., Singh, N., and Kaur, A. (2017). Saponins in pulses and their health promoting activities: a review. *Food Chemistry* 233: 540–549.

Siva, N. and Thavarajah, P. (2018). The impact of processing and cooking on prebiotic carbohydrates in lentil. *Journal of Food Composition and Analysis* 70: 72–77.

Srivastava, R.P. and Vasishtha, H. (2013a). Dehusking and cooking effect on dietary fibre, soluble protein and lectin of lentils (*Lens culinaris*). *Indian Journal of Agricultural Biochemistry* 26 (1): 36–40.

Srivastava, R.P. and Vasishtha, H. (2013b). Dietary fiber, protein and lectin contents of lentils (*Lens culinaris*) with soaking and cooking. *Current Advances in Agricultural Sciences (An International Journal)* 5 (2): 238–241.

Stoddard, F.L., Marshall, D.R., and Ali, S.M. (1993). Variability in grain protein concentration of peas and lentils grown in Australia. *Australian Journal of Agricultural Research* 44 (6): 1415–1419.

Swanson, B.G. (1990). Pea and lentil protein extraction and functionality. *Journal of American Oil Chemistry Society* 67: 276–280.

Świeca, M. and Baraniak, B. (2014). Influence of elicitation with H_2O_2 on phenolics content, antioxidant potential and nutritional quality of *Lens culinaris* sprouts. *Journal of the Science of Food and Agriculture* 94: 489–496.

Tahir, M., Lindeboom, N., Båga, M. et al. (2011). Composition and correlation between major seed constituents in selected lentil (*Lens culinaris* Medik) genotypes. *Canadian Journal of Plant Science* 91 (5): 825–835.

Teixeira, E.H., Arruda, F.V.S., do Nascimento, K.S. et al. (2012). Biological applications of plants and algae lectins: an overview. In: *Carbohydrates: Comprehensive Studies on Glycobiology and Glycotechnology* (ed. C.F. Chang), 533–558. Intechopen.

Thavarajah, P., Thavarajah, D., and Vandenberg, A. (2009). Low phytic acid lentils (*Lens culinaris* L.): a potential solution for increased micronutrient bioavailability. *Journal of Agricultural and Food Chemistry* 57 (19): 9044–9049.

Thavarajah, D., Johnson, C.R., McGee, R., and Thavarajah, P. (2015). Phenotyping nutritional and antinutritional traits. In: *Phenomics in Crop Plants: Trends, Options and Limitations* (ed. J. Kumar, A. Pratap, and S. Kumar), 223–233. New Delhi: Springer.

Tiwari, B.K. and Singh, N. (2012). *Pulse Chemistry and Technology*. Royal Society of Chemistry.

Torino, M.I., Limón, R.I., Martínez-Villaluenga, C. et al. (2013). Antioxidant and antihypertensive properties of liquid and solid state fermented lentils. *Food Chemistry* 136 (2): 1030–1037.

Udensi, E.A., Ansa, N.U., and Maduka, M. (2008). Effect of processing methods on the level of some antinutritional factors in *Mucuna flagellipes*. *Nigerian Food Journal* 26 (2): 53–59.

Urbano, G., Lopez-Jurado, M., Aranda, P. et al. (2000). The role of phytic acid in legumes: antinutrient or beneficial function? *Journal of Physiology and Biochemistry* 56 (3): 283–294.

Vadivel, V. and Pugalenthi, M. (2009). Effect of soaking in sodium bicarbonate solution followed by autoclaving on the nutritional and antinutritional properties of velvet bean seeds. *Journal of Food Processing & Preservation* 33: 60–73.

Vasconcelos, I.M. and Oliveira, J.T.A. (2004). Antinutritional properties of plant lectins. *Toxicon* 44 (4): 385–403.

Vidal-Valverde, C., Sierra, I., Frias, J. et al. (2002a). Nutritional evaluation of lentil flours obtained after short-time soaking processes. *European Food Research and Technology* 215 (2): 138–144.

Vidal-Valverde, C., Frias, J., Sierra, I. et al. (2002b). New functional legume foods by germination: effect on the nutritive value of beans, lentils and peas. *European Food Research and Technology* 215 (6): 472–477.

van Vliet, S., Burd, N.A., and van Loon, L.J. (2015). The skeletal muscle anabolic response to plant-versus animal-based protein consumption. *The Journal of Nutrition* 145 (9): 1981–1991.

Wales, R., Richardson, P.T., Roberts, L.M. et al. (1991). Mutational analysis of the galactose binding ability of recombinant ricin B chain. *The Journal of Biological Chemistry* 266: 19172–19179.

Wang, N. and Daun, J.K. (2006). Effects of variety and crude protein content on nutrients and anti-nutrients in lentils (*Lens culinaris*). *Food Chemistry* 95 (3): 493–502.

Wang, W., Li, Q., Wu, J. et al. (2021). Lentil lectin derived from *Lens culinaris* exhibit broad antiviral activities against SARS-CoV-2 variants. *Emerging Microbes & Infections* 10 (1): 1519–1529.

Westhoek, H., Rood, T., van den Berg, M. et al. (2011). The protein puzzle: the consumption and production of meat, dairy and fish in the European Union (No. 500166001). Netherlands Environmental Assessment Agency.

WHO-UNU (World Health Organization and United Nations University) (2007). Protein and amino acid requirements in human nutrition (Vol. 935). World Health Organization.

Wiesinger, J.A., Marsolais, F., and Glahn, R.P. (2022). Health implications and nutrient bioavailability of bioactive compounds in dry beans and other pulses. In: *Dry Beans and Pulses: Production, Processing, and Nutrition* (ed. M. Siddiq and M.A. Uebersax), 505–529. Hoboken, NJ: Wiley.

Xu, B. and Chang, S.K.C. (2009). Phytochemical profiles and health-promoting effects of cool-season food legumes as influenced by thermal processing. *Journal of Agricultural and Food Chemistry* 57: 10718–10731.

Zaccardelli, M., Lupo, F., Piergiovanni, A.R. et al. (2012). Characterization of Italian lentil (*Lens culinaris* Medik.) germplasm by agronomic traits, biochemical and molecular markers. *Genetic Resources and Crop Evolution* 59 (5): 727–738.

Zhang, B., Deng, Z., Ramdath, D.D. et al. (2015). Phenolic profiles of 20 Canadian lentil cultivars and their contribution to antioxidant activity and inhibitory effects on α-glucosidase and pancreatic lipase. *Food Chemistry* 172: 862–872.

Zhang, B., Peng, H., Deng, Z., and Tsao, R. (2018). Phytochemicals of lentil (*Lens culinaris*) and their antioxidant and anti-inflammatory effects. *Journal of Food Bioactives* 1: 93–103.

15

Sensory Properties of Cooked Lentils and Lentil-Based Products

Nazimah Hamid, Natalie Pettitt, Ye Liu, and Kevin Kantono

Department of Food Science and Microbiology, Auckland University of Technology, Auckland, New Zealand

15.1 Introduction

Plant-based proteins are important in future food production because of the negative repercussions of the livestock industry due to energy usage, greenhouse gas emissions, and the depletion of natural resources (Clark et al. 2017). There is now a growing trend toward the use of more environmentally sustainable sources of foods like plant proteins as an alternative to animal proteins (Day 2013). Lentils are considered sustainable due to their ability to fix nitrogen and improve soil conditions (Sandhu and Singh 2007). In addition, lentil cultivation requires a minimum volume of water and can tolerate stress better than other crops (Mishra et al. 2018). In the world, lentils have been ranked fourth after beans, peas, and chickpeas as plant protein sources (Kumar and Pandey 2020).

Lentils have an excellent nutrient profile. They are high in protein and fiber and low in fat content. The protein content of lentils has been reported to be between 20.6 and 31.4 g/100 g (Jarpa-Parra et al. 2014). Lentil protein is judged to be of high quality because of its optimal amino acid profile and digestibility. In addition, they are a good and inexpensive source of protein that contains all essential amino acids, although they are limited in methionine and tryptophan content (Monnet et al. 2019). For these reasons, lentil ingredients are increasingly incorporated into different food products to further improve their nutritional properties. In the food industry, lentils are usually used in the form of flour or protein isolate. Lentil flour is commonly used as a thickener, binder, gelling agent, and stabilizer (Saricaoglu 2020). The functional properties of lentils have been instrumental in improving the quality attributes of different food products. Several studies have been conducted on lentils in food product development as an alternative protein source (Jarpa-Parra 2017; Khazaei et al. 2019; Nosworthy et al. 2018) and to improve digestibility and human health (Drulyte and Orlien 2019; Kumar and Pandey 2020; Romano et al. 2021).

Lentils contain dietary fiber and a range of vitamins and minerals (e.g., iron) (Ganesan and Xu 2017). Following the protein and iron content of lentils, it is anticipated that their inclusion in both omnivorous and plant-based diets can improve diets. Because of the presence of

Lentils: Production, Processing Technologies, Products, and Nutritional Profile, First Edition.
Edited by Jasim Ahmed, Muhammad Siddiq, and Mark A. Uebersax.

antinutrient content in lentils, the legumes cannot be consumed in raw form (Martín-Cabrejas et al. 2009). Lentils require processing to eliminate antinutritive factors, including raffinose and tannins, prior to incorporation into food formulation (Romano et al. 2021). Most traditional lentil-based dishes are either curry or stew-based, with lentils subjected to extensive boiling prior to consumption. The method of preparation and consumption is not the most favorable as boiling can impact the nutritional quality as well as the structural integrity, which can in turn influence flavor. Lentils are now finding applications in foods and has been incorporated into bakery products, extruded food products such as snacks and pasta, as well as meat and dairy products (Chelladurai and Erkinbaev 2020; Jarpa-Parra 2017). Changes in the functional properties of processed lentils can result in changes in texture (Jarpa-Parra 2017) and flavors in the final product (Khazaei et al. 2019). The effects of food processing on the sensory properties of lentils and lentil products are summarized in Tables 15.1–15.7 and will be discussed in Sections 15.2 and 15.3.

15.2 Impact of Lentil Addition on Sensory Properties of Lentil-Based Food Products

The greatest challenge of replacing cereals either partially or wholly with legumes is ensuring that the quality of the legume-based product is uncompromised when the product is developed. The attributes of the new product should resemble the cereal-based food product in terms of sensory properties, which include color, taste, flavor, aroma, and texture to be competitive in the market. Thermally induced Maillard reactions could impart a change in color that further influences the acceptability, and the lysine content (Monnet et al. 2019). The incorporation of legumes into various products can result in a "beany" flavor or off-notes, which are considered "unpleasant."

Many studies have investigated the sensory properties of lentil-derived and lentil-supplemented foods. When formulating food products, texture is a major factor that influences consumer acceptability (Kilcast and Lewis 1990). For a product to be successful, it must maintain a textural profile that is favorable in terms of both physiochemical and sensory properties. Given the possible aversion consumers may have to lentil texture after boiling, incorporation of lentils into texturized foods may create a product that is both nutritionally and texturally favorable. The effects of lentil addition in extruded snacks, snack bars, plant protein substitutes, yogurts, and baked goods on sensory properties are discussed in Sections 15.2.1–15.2.6.

15.2.1 Cooked Lentils

The sensory properties of cooked lentils are influenced by fortification, ingredients added, and plant source. Podder et al. (2018, 2020) conducted a comprehensive study on the sensory properties of uncooked and cooked milled lentils fortified with varying concentrations of iron (Fe) and zinc (Zn). They found significant differences between unfortified and dual-fortified samples for three types of lentils. Liking scores in terms of appearance, odor, and overall acceptability decreased significantly with the increase in Fe and Zn concentration. However, there were non-significant differences between cooked control and

Table 15.1 Summary of sensory testing studies carried out on lentils and lentil-containing products (Cooked lentils).

Food category	Lentil ingredient	Sensory analysis	Summary of findings	References
Cooked lentils	Iron-fortified red lentil	• s146 untrained panelists • 9-point hedonic scale (1 = dislike extremely to 9 = like extremely) to assess liking of appearance and odor; and overall acceptability	• A significant difference in acceptability was observed between the control and NaFeEDTA-fortified lentil samples. • No significant differences were found between the uncooked samples fortified with 800 ppm and 1600 ppm Fe from NaFeEDTA • The 1600 ppm Fe samples in the cooked condition were acceptable, indicating that the 1600 ppm concentration can be used in lentil fortification	Podder et al. (2018)
	Uncooked and cooked milled lentils fortified with varying concentrations of Fe and Zn. Three milled lentil product types (LPTs) - red football, red split, and yellow split	• 196 untrained lentil consumers • Uncooked lentils (odor, appearance, and overall liking)) • Lentil soup unfortified	• Uncooked lentil dal samples revealed significant differences between unfortified and dual-fortified samples for all three LPTs. Liking scores from all three attributes (appearance, odor, and overall acceptability) decreased significantly with the increase in Fe and Zn concentrations • Nonsignificant differences were observed between cooked control and dual-fortified lentil samples for all five attributes (appearance, odor, taste, texture, and overall acceptability) of the three LPTs, except for texture and the overall acceptability of red lentils and the appearance and overall acceptability of yellow split lentils • The fortification process minimally affected cooked dual-fortified lentil samples (fortified with 16 mg of Fe and 8 mg of Zn per 100 g of lentil) compared to control in terms of consumer liking	Podder et al. (2020)
	lentil soup	• 20 trained panels • Nine-point hedonic scales used to assess appearance, flavor, taste, texture, and overall acceptability	• Liking of flavor, taste, and overall acceptability of lentil soups with 1% and 2% turmeric added were significantly higher than control sample	Rozan et al. (2018)
	Lentil seed	• 51 participants • Assess liking of sensory attributes (intensity of aroma, texture, softness, flavor, and overall liking)	• Seeds from Si-supplied drought-stressed plants were the most liked in terms of aroma, flavor, softness, and texture. • For overall liking, the stir-fried seeds from Si-supplied drought-stressed plants were the most liked, followed by boiled seeds from Si-supplied drought-stressed plants • Boiled seeds from control plants were the least liked, followed by stir-fried seeds from control plants	Biju et al. (2021)

Table 15.2 Summary of sensory testing studies carried out on lentils and lentil-containing products (Extruded Snacks).

Food category	Lentil ingredient	Consumer testing	Descriptive analysis	Summary of findings	Reference
Extruded Snacks	Corn and lentil extruded puffs	Consumer testing conducted by 50 untrained panelists for overall acceptance using a nine-point hedonic scale ranging from one (dislike extremely) to nine (like extremely).	10-member trained panel participated in the descriptive analysis of extrudates (appearance, flavor, and texture)	• The sensory porosity (number of open cells) of extrudates decreased with temperature and increased with feed rate • Appearance decreased with the lentil-to-corn ratio, feed moisture content, and temperature but increased with the feed rate • Mealy flavor increased with feed moisture content, while burnt flavor decreased • The increased temperature resulted in decreased mealy flavor and increased burnt flavor • Feed moisture content and raising the extrusion temperature decreased the crunchiness of the samples • The corn extrudates had the highest scores of crispness, followed by lentil extrudates, which contained a 30% lentil-to-corn ratio • The extrusion temperature increased the crispness of snacks • The hardness of the extrudates increased with feed moisture content • The increase in the percentage of lentil flour deteriorates the overall acceptability of the extrudates • The most acceptable snacks were produced at low feed moisture contents • The addition of lentil flour decreased the overall acceptability of the extrudates • Samples containing 30% lentil flour and 13% feed moisture content, processed at low and high temperatures, and medium feed rates, were the most acceptable	Lazou and Krokida (2010)
	Extrudates of red lentils, Faba beans, brown peas, and common bean flour	• Twenty-eight semi-trained panelists evaluated the liking of legume-based extruded snacks according to a ranking test • Panelists ranked the extruded product samples from one (the most-liked sample) to eight (the least-liked sample), according to their degree of liking for four sensory attributes (appearance, texture, taste, and aftertaste)		• The extrudates obtained from red lentil (particularly the star-shaped) were the least liked for all sensory attributes (appearance, texture, taste, and aftertaste). However, red lentil spherical extrudates were liked for their "appearance" and "aftertaste," similar to spherical and star-shaped extrudates from faba bean and brown pea, as well as the star-shaped common bean • Common bean extrudates, particularly the spherical ones, were most liked for all the attributes considered, with puffed and crunchy structure, pleasant taste, and aftertaste	Costantini et al. (2021)

Table 15.3 Summary of sensory testing studies carried out on lentils and lentil-containing products (Snack bars).

Food category	Lentil ingredient	Consumer testing	Descriptive analysis	Summary of findings	Reference
Snack bars	Lentil extrudate	• Ten trained students evaluated the extruded snacks for appearance, flavor, texture, taste, and overall acceptability in triplicates		• Extrudates with good sensory acceptability can be produced from lentils • An extrudate with moisture content of 18% and processed with a die temperature of 160 °C at 200 rpm was found to be the most acceptable	Rathod and Annapure (2016)
	Pulse-based snack bars	• Hedonic evaluation of the chickpea and green lentil snack bars was performed by 14 consumers in terms of liking of the appearance, aroma, texture, taste, and overall acceptability.	• 15 trained panelists • Sensory profiling in terms of appearance, aroma, texture and taste and flavor	• Overall, liking was higher for chickpea snack bar than the green lentil snack bar • Green lentil-based snack bars had a distinctive foreign aroma compared with the chickpea-based snack bars • Panelists described the foreign aroma as unpleasing (beany and cooked pulse flavor from green lentils)	Rajagukguk et al. (2022)
	Snack bar using micronized flaked lentils (MFL) in the formulation Bar 1 – low sweet, high fruit flavor; Bar 2 – high sweet, no fruit flavor; Bar 3 – high oat, low toasted, no fruit flavor; Bar 4 – low oat, high toasted, high fruit flavor; Bar 5 – soft, chewy, moist, moderate fruit flavor; Bar 6 – low softness, low chewiness, low moisture, no fruit flavor	• Six micronized flaked lentil (MFL) snack bars and two commercial bars were evaluated by a consumer panel (n = 62) using the 9-point hedonic scale • JAR scales used to assess samples	• 11 trained panelists in the descriptive panel	• Generally, as sweetness increased, grainy and lentil flavors decreased • Lentil bar samples two, three, five, and six with no fruit or moderate fruit had the highest lentil and grainy flavors • LB1 and LB4 with high fruit flavor had the highest sour/fruity flavor • The commercial samples were sweeter • No significant differences in acceptability were found between LB1, LB3, LB6, and CB2 with the three lentil bars receiving scores corresponding to "like slightly" • "Just right" scales results showed that LB6 had the majority of responses in the "just right" category as CB1 the most liked of the eight samples In both samples, sweetness was the attribute that needed improvement • For the commercial sample, less sweetness was desired • For the lentil samples, more sweetness was desired • LB1 and LB4 had the next highest percentage of attributes in the "just right" category. Improvements for these samples as with the remaining lentil bar samples included decreasing the cereal/grain flavor • An increase in cranberry flavor was noted as an improvement for LB1 with an increase in moistness noted for LB4 • External preference mapping determined that sweetness, grainy and lentil flavors, hardness, cohesiveness, cohesiveness of mass and moistness had the greatest influence on consumer acceptability	Ryland et al. (2010)

Table 15.4 Summary of sensory testing studies carried out on lentils and lentil-containing products (Protein Substitutes).

Food category	Lentil ingredient	Consumer testing	Descriptive analysis	Summary of findings	Reference
Protein Substitutes	Pork meat burgers with the upper pulse flour level (150 g/kg^{-1}) for pea (P5), chickpea (C5), lentil (L5), and bean (B5) flours and an intermediate water/flour ratio (1600 g/kg^{-1})	• 60 panelists • Acceptance testing was conducted using a nine-point hedonic scale (9 = like extremely; 5 = indifferent; 1 = dislike extremely) to assess appearance, color, texture, taste, and overall acceptability		• No significant differences observed in appearance, taste, texture, and overall acceptability	Argel et al. (2020)
	Thermally micronized chickpea and green lentil flours	• Volunteers ($n = 101$) • liking of the aroma, appearance, flavor, texture, and overall acceptability using a 9-point hedonic scale		• For burgers with lentil flour, acceptability was highest for flours micronized at 130 °C for appearance and aroma with very little to no difference in acceptability for the other attributes for both micronized samples • Inclusion of both types of micronized flours at 130 and 150 °C increased acceptability of low-fat beef burgers compared to the burgers over non-micronized flour	Shariati-Ievari et al. (2016)
	Vegetable hamburger patty with a high moisture meat analogue C1 (control) = soy concentrate and soy isolate, T1, T2, and T3 = pea proteins, lentil proteins, and faba bean proteins, respectively	• 80 consumers • Cooked appearance, overall appearance, overall flavor, and overall texture were evaluated using a 9-point hedonic scale	• 10 trained panelists • Taste, flavor, and texture attributes assessed	• The samples containing pulse proteins were significantly higher for flavor attributes that were bean-like, salty, sweet, umami, musty-earthy, and malt-like, and lower for soy and cardboardy compared to control • The samples containing pulse proteins (pea proteins, lentil proteins, and faba bean proteins) were significantly higher for cohesiveness of mass and lower for hardness and springiness compared to the control • Protein sources in meat patties did not significantly affect the liking of the dish. The control sample was significantly higher in overall liking and overall texture compared to faba bean protein	Kim et al. (2021)

Lentil-based milk substitute	• 30 untrained assessors		• Except for the commercial hemp milk, liking of mouthfeel, and flavor, as well as overall liking of the lentil emulsion, containing 3.3% oil, 2.5% (w/w) sucrose, and 0.08% (w/w) salt were not significantly different for all commercial plant-based milk samples (oat, soy, rice, and almond milk) that was less liked	Jeske et al. (2019)
Salad dressings supplemented with raw and thermally treated lentil flours		• 10 trained panelists • A quantitative descriptive analysis method was used for the sensory test, and the following attributes were determined and rated on a categorical scale ranging from zero (attribute not detected) to seven (attribute very strong): overall intensity, legume flavor, vinegar (aroma), acidity (taste), particle size (grittiness), and firmness	• Overall flavor was the highest for salad dressings supplemented with roasted flour compared to roasted seeds and pre-cooked spray-dried lentils, but similar to dressings containing raw lentil flour • Roasted-flour-supplemented dressings had the significantly highest rating for acidity • The grittiness of salad dressing supplemented with roasted seeds was significantly lower • The dressing containing pre-cooked spray-dried lentils had a significantly higher score for firmness • No statistical difference was found for the intensity of vinegar and legume flavors among the samples	Ma et al. (2013)
Lentil-fortified spaghetti	• 15 trained panelists • A nine-point scale (1 = "extremely unpleasant," 9 = extremely pleasant," and 5 = "threshold of acceptability") • Assessed color, homogeneity, break to resistance and odor of uncooked samples and elasticity, firmness, bulkiness, adhesiveness, color, odor, and taste of cooked spaghetti as well as overall sensory quality		• The addition of lentil flour (40% *w/w*) significantly decreased the sensory attributes of in terms of liking of elasticity, firmness, bulkiness, adhesiveness, color, odor, and taste, as well as overall sensory quality of cooked spaghetti compared to control	Di Stefano et al. (2020)

Table 15.5 Summary of sensory testing studies carried out on lentils and lentil-containing products (Incorporation of lentil flour in yogurt).

Food category	Lentil ingredient	Consumer testing	Descriptive analysis	Summary of findings	Reference
Incorporation of lentil flour in yogurt	Lentil (*Lens culinaris*) flour addition to yogurt	• Eight semi-trained panelists • A five-point hedonic scale was used: (1: Do not like at all; 2: Do not like very much; 3: Indifferent; 4: Like a little and 5: Like a lot) • Quality attributes (texture, color, flavor, aroma, viscosity, and overall acceptability)		• The addition of lentil flour did not compromise the sensory characteristics or its acceptance by consumers. Yogurt with unroasted and roasted lentil flour had higher overall acceptability after 21 days of storage compared to control sample	Benmeziane et al. (2021)
	Yogurt supplemented with lentil flour	• 25 untrained panelists • Smoothness, graininess, flavor, overall acceptance, and color evaluated using a 9-point hedonic scale from extremely like (1) to extremely dislike (9)		• No significant differences in terms of overall acceptance as well as liking of color, smoothness, graininess, and flavor between control yogurt and lentil flour supplemented at 1–3% in yogurt	Zare et al. (2011)
	A milk-alternative produced from lentil protein – Lentil-based yogurt		• 13 experienced sensory testers • A scale of 0 to 10 was used to assess the following descriptors: odor intensity, yogurt-like odor, flavor intensity, sourness, bitterness, beany flavor, yogurt-like flavor, sour-cream like flavor, firm mouthfeel, smooth mouthfeel, and aftertaste • Overall acceptability of the samples rated on a scale from 0 to 10	• Both the yogurt-like flavor and overall acceptability of *Lacticaseibacillus paracasei* FST 6.1Lbp6.1 were equal to the control and scored highly in the sensory tests	Boeck et al. (2022)

Table 15.6 Summary of sensory testing studies carried out on lentils and lentil-containing products (bread).

Food category	Lentil ingredient	Consumer testing	Descriptive analysis	Summary of findings	Reference
Bread	Sangak bread with added germinated lentil flour.	• 15 trained panelists • Shape, bread surface, texture, color, chewiness, aroma and flavor were evaluated using the 5 points hedonic method		• Control and 5% germinated lentil flour bread had the highest quality score.	Razavi et al. (2022)
	Bread with lentil sprout flour added	• 20 Mexican bread consumers • 11 attributes (color, porosity, sponginess, hardness, cohesiveness, aroma, stickiness, chewiness, flavor, healthy preferences, and attributes in general) evaluated using a hedonic scale		• Sensory acceptability for porosity, hardness and cohesiveness was higher score for wheat bread without germinated lentil (GL) flour • The bread with the lowest sensory acceptability contained 10% GL • In terms of flavor, the sensory acceptance of was highest for bread with 5% GL • In terms of color, there was higher acceptability of control bread and the least acceptable bread contained 10% GL	Hernandez-Aguilar et al. (2020)
	Sourdough fermentation of whole and sprouted lentil flours		• 10 trained panelists • Sensory attributes: Appearance, texture, taste, and flavor	• The score for crust and crumb color was higher for all sourdough breads, especially sprouted lentil sourdough wheat bread (LSWB) both EPS (exopolysaccharides) positive and negative than control wheat bread (CWB) • The scores for acidic and legume flavor were significantly higher in sourdough breads compared to CWB • A markedly higher perception of legume taste was observed in LSWB compared to sprouted LSWB and CWB	Perri et al. (2021a)

(*Continued*)

Table 15.6 (Continued)

Food category	Lentil ingredient	Consumer testing	Descriptive analysis	Summary of findings	Reference
	Lentil flour and hydrocolloids addition in wholemeal durum wheat bread		• 10 trained tasters • Evaluated for attributes that included color, appearance, crust and crumb firmness, large bubbles and overall quality using a 9-point scale	• Samples with lower content of the lentil flour (LENT 10 and LENT 20) showed the sensory parameters values similar to the control sample (overall quality • Overall quality of the bread samples increased when the amount of structuring agents (carboxymethyl cellulose, guar seed flour, pectin, and tapioca starch) increased from 1% to 2% • Sensory parameters that negatively affected the overall judgment were the appearance, crumb firmness and presence of large bubbles • The bread at the highest and lowest concentration of structuring agents had a very compact crumb with few small bubbles and a low loaf volume • All hydrocolloids showed an organoleptic quality above the threshold acceptability (up to 6.0) at the concentration of 2%, apart from the tapioca starch having the same result at 3% (overall quality of 5.83 and 6.08)	Previtali et al. (2014)
	Pulse flour bread	• Eight trained panelists • Appearance and overall acceptability were also evaluated using 7-point acceptability scales where 1 = not acceptable and 7 = very acceptable	• Eight trained panelists • Aroma, pulse flavor, sweetness, bitterness and aftertaste using 7-point intensity scales where 1 = not intense and 7 = very intense	• Bread made with decorticated red lentil flour stored at 9 and 18 months had decreased bitterness compared to frozen control flour (0-month storage) • Aftertaste increased in breads made from lentil flour stored at 18 and 24 months compared to control bread • Overall acceptability was higher with the decorticated red lentil flour stored for 12 months • Appearance was lower with bread made from flour stored for 24 months, compared to the bread made with the frozen control flour	Sopiwnyk et al. (2020)

Table 15.7 Summary of sensory testing studies carried out on lentils and lentil-containing products (other baked products).

Food category	Lentil ingredient	Consumer testing	Descriptive analysis	Summary of findings	Reference
Other baked products	Extruded-cooked lentil (ECL) flour pizza	• Seventy consumers • The consumers were asked to state if they liked or disliked each pizza sample using a structured scale from 1 to 5 (1 = disliked a little or liked a little; 5 = disliked a lot or liked a lot)	• Eight trained members • Olfactory descriptors (global odor and lentil odor), visual–tactile descriptors (color, consistency, and pliability), taste descriptors (sweetness and saltiness), and flavor descriptors (lentil flavor)	• Compared to control, the ECL pizza had significantly higher intensity of lentil odor and flavor, and in color, with no significant differences in liking • Compared to control, the NL pizza had significantly higher intensity of lentil odor and flavor, consistency, and color, but significantly lower pliability. In addition, it was significantly less liked than control but was not significantly different in liking to ECL	Pasqualone et al. (2022)
	Barley- and lentil protein concentrates addition in doughnuts	• 61 consumer panelists • Appearance, flavor, texture, and overall preference were evaluated by using a 1–9-structured scale (1 = dislike extremely, 9 = like extremely) • Just-about-right (JAR) scales were used to judge the attributes of chewiness, density, and moisture.		• The highest liking rating was for the control doughnut • Incorporation of barley and lentil proteins into the formulations contributed to a decrease in liking for protein-enriched samples in terms of product appearance, flavor, and overall liking • The penalty analysis of doughnut density, chewiness, and moisture showed that these attributes were negatively impacted by the addition of the barley and lentil proteins due to their effect on the crumb structure	Eckert et al. (2018)

dual-fortified lentil samples in terms of liking of all attributes (appearance, odor, taste, texture, and overall acceptability), except for texture and overall acceptability of red football lentils, and the appearance and overall acceptability of yellow split lentils. The authors concluded that the fortification process minimally affected the liking of cooked dual-fortified lentil samples (fortified with 16 mg of Fe and 8 mg of Zn per 100 g of lentil) compared to control cooked samples. In a separate study, Rozan et al. (2018) found that liking of flavor, taste, and overall acceptability of lentil soups with 1% and 2% turmeric added were significantly higher than control samples by a trained panel. Silicon (Si) supplementation has been shown to improve the nutritional and sensory characteristics of lentil seeds harvested from drought-stressed plants (Biju et al. 2021). Seeds from Si-supplied drought-stressed plants were the most liked in terms of aroma, flavor, softness, and texture. For overall liking, the stir-fried seeds from Si-supplied drought-stressed plants were found to be the most liked, followed by boiled seeds from Si-supplied drought-stressed plants. Boiled seeds from control plants were the least liked, followed by stir-fried seeds from control plants.

15.2.2 Extruded Snacks

Extruded snacks are mostly made from rice, soy, wheat, or corn. However, the snacks made from these sources are normally high in calories and low in nutritional value. Despite the negative health effects attributed to overindulgence of extruded snacks, they remain popular among consumers. The incorporation of ingredients obtained from lentils into these products can reduce calories and, at the same time, increase protein and dietary fiber content (Ryland et al. 2010). Commercially available lentil-supplemented extruded snacks are scarce. However, there is a fair amount of research that has been carried out on lentil-supplemented extruded snacks.

The effects of processing conditions on the textural attributes of lentil-enriched extruded snacks have been extensively studied. However, the focus of this chapter is to report changes in the sensory properties of the lentil-based extruded snacks. Lazou and Krokida (2010) studied the effects of extrusion processing conditions [temperature (170–230 °C), feed rate (2.52–6.84 kg/h) and feed moisture content (13–19% weight basis)] on the structural and textural properties of extruded corn and corn–lentil mixtures at a ratio of 10–50% (w/w; legume/corn). Furthermore, they performed descriptive sensory analysis of the extruded product using a trained panel and found that the sensorial diameter of extrudates decreased with temperature and feed moisture content and increased with feed rate. Corn–lentil extrudates showed decreased crunchiness when extrusion temperature and feed moisture content were decreased, while feed rate increased crunchiness. Feed moisture content decreased the crispness of extrudates, while extrusion temperature increased crispiness. The hardness of extruded corn–lentil extrudates decreased with feed moisture content. Feed rate, on the other hand, increased the hardness of extruded products. Extrusion temperature decreased hardness of corn–lentil extrudates. The temperature increase produced more expanded products, crispier and consequently softer.

Several studies have reported the effects of processing on lentil extrudates on consumer acceptability (Costantini et al. 2021) examined the effects of two different die configurations (circular and star-shaped) on the antinutritional compounds as well as physicochemical and sensory properties of extruded breakfast snacks prepared from 100% legume flour (red lentil, faba bean, brown pea, and common bean). The authors reported that extrudates of star-shaped red lentil were the least liked for all sensory attributes (appearance, texture, taste, and aftertaste). However, red lentil spherical extrudates were liked for their "appearance" and "aftertaste" attributes, similar to spherical and star-shaped extrudates from faba bean and brown pea, as well as the star-shaped common bean. It was reported that lentil extrudates had good sensory acceptability (Rathod and Annapure 2016). The authors found that extrudate with a moisture content of 18% and processed with a die temperature of 160 °C at 200 rpm was the most acceptable.

15.2.3 Snack Bars

Because of the high protein content in lentils, manufacturing of snack bars could be one option to improve human health. Rajagukguk et al. (2022) investigated the sensory evaluation of the chickpea and green lentil snack bars using 14 consumers using the Hedonic scale. Overall, liking was higher for chickpea-based snack bar than the green lentil snack bar. Sensory profiling was also carried out using 15 trained panelists, in terms of appearance, aroma, texture and taste, and flavor. Green lentil-based snack bars had a significantly higher undesired aroma compared with the chickpea-based snack bars. Panelists further described the undesired aroma of green lentil snacks as being unpleasantly beany and cooked pulse flavor.

Micronization increases the internal temperature of food samples by exposing them to electromagnetic radiation in the infrared region for a short time. Infrared heating can improve protein efficiency ratios, protein chemical scores, and essential amino acid indexes (Khattab et al. 2009). In addition, it can result in a more open cell structure in lentil seeds, which facilitates the absorption of water during cooking (Arntfield et al. 2001). Ryland et al. (2010) developed a snack bar using micronized flaked lentils (MFLs) and conducted comprehensive sensory testing of the samples. Using descriptive analysis, the authors found that as sweetness increased, grainy and lentil flavors decreased. Lentil bar samples with no fruit or moderate fruit had the highest lentil and grainy flavors, while the bars with high fruit flavor had the highest sour or fruity flavor. The commercial samples were found to be sweeter. In terms of acceptability, there were no significant differences in liking between LB1 (low sweet, high fruit flavor), LB3 (high oat, low toasted, no fruit flavor), LB6 (low softness, low chewiness, low moisture, no fruit flavor), and the commercial bar (CB2). The "just right" scales results further showed that the LB6 had the most responses in the "just right" category as CB1, which was the most liked of the eight samples. However, both these samples needed improvement in terms of sweetness. For the commercial and lentil samples, low and high sweetness were desired. LB1 and LB4 (low oat, high toasted, high fruit flavor) had the next highest percentage of attributes in the "just right" category. Improvements for both these samples and the other lentil bar samples included decreasing the cereal or grain flavor. An increase in cranberry flavor was noted as an improvement for LB1, with an increase in moistness noted for LB4. External preference mapping of samples further showed that sweetness, grainy and lentil flavors, hardness, cohesiveness, cohesiveness of mass, and moistness had the greatest influence on consumer acceptability.

15.2.4 Protein Substitutes

Lentils are good source of plant-based protein. In recent days, there has been an increasing interest in taking advantage of this by either supplementing or developing products using lentil proteins. The two key factors that are driving interest in the use of lentil protein are increasing protein content in an existing food or making a high-protein plant-based alternative to an existing food. Studies have been carried out to investigate the supplementation of meat burgers with pulses. Argel et al. (2020) looked at supplementing pork burger patties with pulse flour to reduce fat content and investigated their changes in physicochemical and sensory properties. They found that all burger formulations with pulse flour have adequate technological properties. Following this, pork meat burgers with a higher pulse flour level (150 g/kg) and an intermediate water/flour ratio (1600 g/kg) for pea, chickpea, lentils, and bean flour were selected for sensory evaluation. The products showed no significant differences of consumer likeliness terms of liking of appearance, taste, and texture, as well as overall acceptability. In another study, the sensory properties of thermally micronized chickpea and green lentil flours as binders in low-fat beef burgers were investigated (Shariati-Ievari et al. 2016). The lentil flour micronized at 130 °C produced acceptable quality burgers. The appearance and aroma showed very little to no difference in acceptability. The inclusion of flour micronized at 130 and 150 °C improved the acceptability of the low-fat beef burgers compared to the burgers with non-micronized flour, which in turn increased the acceptability over the burgers with no binder added.

Meat analogues can also be produced from lentil proteins. Kim et al. (2021) examined the sensory properties of vegetable hamburger patties containing peas, lentils, and faba beans. They found that samples containing pulse proteins were significantly higher for bean-like, salty, sweet, umami, musty-earthy, and malt-like flavors, and lower for soy and cardboardy compared to control. The samples, including pules proteins, were significantly higher for cohesiveness of mass and lower for hardness and springiness compared to the control. In addition, meat analogues did not significantly affect liking in terms of cooked appearance and overall flavor. The control sample was significantly higher in overall liking and overall texture compared to faba bean protein.

Studies have been reported on lentils used in plant milk, salad dressings, and spaghetti. The sensory acceptance between a commercial plant-based milk substitute and a lentil-based formulation was compared (Jeske et al. 2019). The liking of mouthfeel and flavor as well as overall liking of the lentil emulsion, containing 3.3% oil, 2.5% (w/w) sucrose, and 0.08% (w/w) salt, were found not to be significantly different compared to all commercial plant-based milk samples (oat, soy, rice, and almond milk), except for the commercial hemp milk that was less liked. Ma et al. (2013) used descriptive analysis with 10 trained panelists to examine the sensory properties of salad dressings supplemented with raw and thermally treated lentil flour. They found that overall flavor was significantly higher for salad dressings supplemented with roasted flour compared to roasted seeds and pre-cooked spray-dried lentils, but similar to dressings containing raw lentil flour. The roasted flour-supplemented dressings also had the significantly highest rating for acidity. The grittiness of salad dressing supplemented with roasted seeds was significantly lower. The dressing containing pre-cooked spray-dried lentils had a significantly higher score for firmness. The authors concluded that thermally treated pulse flours were suitable value-added ingredients in salad dressings. A recent study by Di Stefano et al. (2020) formulated spaghetti fortified with lentil flour (40% w/w). They found that the addition of lentil flour significantly decreased the sensory attributes in terms of liking of elasticity, firmness, bulkiness, adhesiveness, color, odor, and taste, as the overall sensory quality of cooked spaghetti compared to control.

These examples of lentil protein substitution for animal proteins show potential for wider applications. However, it is important to optimize the formulation in terms of desirable sensory attributes for the product to gain consumer acceptance. The studies emphasize how lentil protein substitution can have useful applications in various food products.

15.2.5 Incorporation of Lentil Flour in Yogurt

Yogurt has many health benefits due to the presence of bioactive compounds. However, its functional attributes can be further improved by the inclusion of lentils. Lentil flour has been reported to improve the functional properties of yogurts, in particular syneresis and sensory attributes. Syneresis is the separation of whey from the curd in milk-fermented products during storage (Jørgensen et al. 2019). The addition of lentil flour can help increase the solid content of yogurt, which will result in a stronger texture, resulting in a decrease in whey separation (Sendra et al. 2010). The high starch content of the lentil flour acts as a stabilizer and assists in decreasing syneresis (Altemimi 2018; Benmeziane et al. 2021).

Lentil flours and lentil protein isolates have been added to yogurt to improve its functional and sensory properties. Benmeziane et al. (2021) assessed the sensory attributes (e.g., texture, color, flavor, aroma, viscosity, and overall acceptability) of yogurt containing 20% roasted and unroasted lentil flour over 21 days of storage. Roasted lentil flour yogurt showed statistically significant differences in the overall acceptability throughout the storage period. The addition of lentil flour did not affect the sensory attributes of consumers. Yogurt with unroasted and roasted lentil flour had higher overall acceptability after 21 days of storage compared to the control sample. Similarly, Zare et al. (2011) reported no significant differences in the overall acceptability and other sensory attributes (e.g., color, smoothness, graininess, and flavor) between the control and lentil flour (1–3%) supplemented yogurt. Boeck et al. (2022) evaluated the sensory attributes of a milk-alternative yogurt produced from lentil protein isolate fermented with three multifunctional bacterial strains of lactic acid, namely *Leuconostoc citreum* TR116, *Leuconostoc pseudomesenteroides* MP070, and *Lacticaseibacillus paracasei* FST 6.1. They have also included one commercial starter culture containing *Streptococcus thermophilus* for the study. Both the yogurt-like flavor and overall acceptability of *L. paracasei* FST 6.1 were found to be similar to the control yogurt and scored high in the sensory evaluation tests. All these studies, however, were conducted with a limited number of panelists, and therefore, it is recommended that at least 50 untrained consumer panelists be included to get more reliable results for consumer testing.

15.2.6 Sensory Properties of Baked Lentil Products

The addition of lentil flour to white bread can improve its nutritional quality by increasing fiber and protein content. However, lentil fortification of bread products may influence flavor, nutrition, and texture. In addition, lentils are gluten free and are increasingly incorporated in gluten-free baked products. However, lentil addition has been found to impact sensory and textural properties of bread.

15.2.6.1 Bread

All breads are high in carbohydrates and low in protein. As a rich source of micro- and macronutrients, lentils have been considered for use in bread fortification (Benayad and Aboussaleh 2021; Gobbetti et al. 2019). Razavi et al. (2022) investigated the effect of adding germinated lentil flour to Sangak bread. A trained panel of 15 panelists evaluated shape, bread surface, texture, color, chewiness, aroma, and flavor characteristics using the 5-point hedonic scale. They reported that control and 5% germinated lentil flour to bread had the highest quality score. Hernandez-Aguilar (2020) also evaluated the addition of germinated lentil flour into bread. Sensory acceptability scores for porosity, hardness, and cohesiveness were higher for wheat bread without germinated lentil (GL) flour. The bread with the lowest sensory acceptability contained 10% GL. In terms of flavor, the sensory acceptance was highest for bread with 5% GL. In terms of color, there was higher acceptability of control bread and the lowest acceptability of bread with 10% GL.

In previous studies, the use of legume flours in bread without exopolysaccharide (EPS) produced by lactic acid bacteria resulted in negative effects on texture and consumer acceptance (Kohajdová et al. 2013). Perri et al. (2021a) investigated the possibility of formulating sourdough bread using fermented lentil flour. In their study, fermented sourdoughs from raw and sprouted lentil flour using EPS-producing lactic acid bacteria were produced, and the chemical, physical, and microbiological characteristics of the sourdough bread were determined. The authors found that *Weissella confusa* SLA4 produced a significantly higher amount of dextran during fermentation, and that the 30% w/w lentil composite sourdoughs had textural, nutritional, and sensory characteristics that were comparable to or even advantageous over the control wheat bread. Sprouted lentil dough also had more total phenols, amino acid availability, and antioxidant activity than raw lentil. Sensory analysis carried out showed that the sprouted lentil improved crust and crumb color, resulted in less beany flavor, and was sweeter than the raw lentil treatment. In another study, Previtali et al. (2014) investigated the effect of lentil flour and hydrocolloid addition on the properties of wholemeal durum wheat bread. Samples with a lower content of the lentil flour had similar sensory parameters to the control sample. The overall quality of the bread increased with an increase in structuring agents (carboxymethyl cellulose, guar seed flour, pectin, and tapioca starch) from 1% to 2%. All samples containing the hydrocolloids were acceptable at a concentration of 2%, apart from the tapioca starch that was only acceptable at 3%.

The storage time of the lentil flour can influence the sensory attributes of the bread. In the studies discussed above, the storage time of lentil flour was not taken into consideration as lentil flour was obtained from supermarkets. Sopiwnyk et al. (2020) studied the effect of pulse flour storage time (1, 2, 3, 6, 9, 12, 18, and 24 months) on the 20% lentil flour bread. Bread made with decorticated red lentil flour stored for 9 and 18 months had decreased bitterness compared to frozen control flour (0-day storage). However, aftertaste increased in lentil flour breads stored for 18 and 24 months compared to control bread. Overall acceptability was higher with the decorticated red lentil flour stored for 12 months. The liking of appearance decreased with bread made from flour stored for 24 months.

15.2.6.2 Other Baked Products

The addition of lentil ingredients has been increasingly researched in other baked products. Pasqualone et al. (2022) added extruded-cooked lentil (ECL) flour to fortify gluten-free pizzas, which was compared to pizza made with rice/corn (control) and native lentil (NL) flour. Compared to the control, the ECL and NL pizza resulted in significantly higher intensities of lentil odor and flavor, as well as color. In terms of liking, ECL was not significantly different from control and NL pizza. However, NL pizza had significantly higher consistency, significantly lower pliability, and was significantly less liked than control pizza. Another study by Jarpa-Parra et al. (2017) that investigated lentil addition to angel cake and muffins used lentil protein concentrate (LPC) as an egg replacer. For angel food cakes, appearance and texture were significantly lower in the lentil protein formulations compared to control. The lentil protein-based muffin (M-100-LPC) was preferred by the consumers and received higher scores for all attributes compared to the control and the high-protein muffin ($P<0.05$). For both 50% (AC-50-LPC) and 75% (AC-75-LPC) replacement of egg white protein with LPC in angel food cake, the just-about-right consumer testing results indicated a reduction in density and chewiness as well as an increase in moistness of samples. In addition, increased chewiness, density, and lack of moisture were found in the high-protein muffin sample, M-100H-LPC.

Eckert et al. (2018) attempted to increase protein content in doughnuts through the substitution of lentil flour. The highest approval (liking) rating was for the control doughnut. The addition of barley and lentil proteins into the formulations resulted in a significant decrease in the liking for protein-enriched doughnuts in terms of product appearance, flavor, and overall liking. The penalty analysis of just-about-right results revealed that doughnut density, chewiness, and moisture were negatively impacted by barley and lentil protein addition due to changes in crumb structure.

15.3 Physical Characteristics of Dough and Sensory Quality of Lentil-Based Baked Products

The sensory studies reported in the literature are summarized in Tables 15.1–15.7, which demonstrate that adding lentil ingredients into baked products can affect texture. A trained sensory panel has been mostly used to investigate the acceptability of textural attributes for lentil-based products. Trained sensory panels, however, should be avoided to assess product liking as their expert knowledge can introduce bias in measurements (Delahunty and Drake 2004). Ideally, a naive consumer panel who are regular consumers of the product type being tested or the representative target market for the product should be used to evaluate consumer acceptance. When carrying out consumer testing of lentil products, it is important for the target market to include participants who already consume lentils or have no aversion to consuming lentils to predict acceptability adequately. Consumer testing to determine acceptability of products involves the use of rating scales that measure relative liking using the hedonic scale. On the other hand, discrimination tests based on consumer preference involve the use of paired preference or ranked preference tests. Consumer testing using just about the right scales is also used to give further information about the sensory characteristics of products that can be improved to make them more acceptable. It is generally recommended that a minimum of 75 targeted consumers should be used for consumer sensory testing (Delahunty and Drake 2004). The use of trained panels and the number of consumers used to assess the acceptability of lentil-containing products are the weaknesses of most of the studies reviewed. Hence, in making decisions regarding the sensory acceptability and sensory characteristics of lentil-containing products, it is important to consider the physical measures of texture as well.

In the following section, the effect of lentil flour (LF) or protein isolate addition in traditionally wheat-based baked products in terms of loaf volume and textural properties are discussed. Baked products are excellent candidates for LF fortification. However, when wheat products are supplemented with LF, the different macronutrient profile has implications for the texture of the doughs and batters (Monnet et al. 2019). Additionally, the lack of gluten in conjunction with differences in starch functionality can impact the texture of the final product (Romano et al. 2021). The effects of LF or protein addition on the textural properties of selected lentil-containing baked goods are summarized in Table 15.8.

15.3.1 Dough Attributes' Effect on Sensory Properties

The rheological behavior of doughs and batters can be influenced by the partial or complete replacement of wheat with LF. Specifically, differences in water absorption (WA) and dough development time (DDT) are often focused on bread formulation studies. WA in bread dough has its effect on mixing time and bread volume (Okuda et al. 2016). Multiple studies have reported conflicting results on WA effects. Bojňanská et al. (2021) noted a decrease in WA at 15% LF replacement, whereas Portman et al. 2018 recorded decreases of 10–40%. Conversely, Razavi et al. (2020) found an increase in WA at 5–20% LF replacement, with Turfani et al. 2017 reporting no changes in WA at all across 6–24% LF replacement. Razavi et al. (2020) employed germinated LF in their study, and germination can result in an elevation of α-amylase activity, leading to a reduction in total starch content. This increased starch damage can enhance water absorption, as demonstrated in previous studies (Perri et al. 2021a) and may account for increased WA. Furthermore, the use of hulled and unhulled LF can affect WA due to the differences in fiber content (Bresciani and Marti 2019). Portman et al. 2018 used hulled lentils in dough and reported use of minimum fiber had the least WA. Additionally, it has been found that flour particle size can also affect WA, with finer particle size typically resulting in lower WA (Zhang et al. 2021). More information on the LF used may help explain the inconsistences in findings between these studies.

In bread baking, DDT is an indication of protein quality within the dough, which is influenced by gluten content (Portman et al. 2018). A wide variation in DDT for LF supplemented doughs has been reported in Table 15.8. Portman et al. (2018) reported a decrease in DDT for the supplementation of 5–40% of LF. Conversely, an increase in DDT has been observed for blended dough supplemented at 6–24% (Turfani et al. 2017; Razavi et al. 2020). The protease content in germinated flour could be the reason for the negatively affected gluten development (Boukid et al. 2019). The decreased fiber content in the hulled lentil flour may have resulted in decreased water uptake by the lentil flour, allowing more water

Table 15.8 Impact of replacing wheat with different percentages of lentil flour or protein on physicochemical and sensory properties of bread, cakes, and doughnuts.

Product	Lentil product	% lentil used	Physico-chemical properties	Color	Sensory attributes	References
Wheat-rye composite bread	Raw lentil flour	5–15%	Water absorption: ↓ at 15% only ↓ Loaf volume Dough viscosity: ↑ at 15% only	Very little color difference found from control at any %	↑ Crust Color ↑ Crumb color Taste: ↑ 5%, ↓ 10+% Elasticity: ↑ 5%, ↓ 10+% Aroma: ↑ 5%, ↓ 10+% Sourness: ↑ 5%, ↓ 10+% Overall appearance.: ↑ 5%, ↓ 10+%	Bojňanská et al. (2021)
Wheat-barley flour doughnuts	Lentil protein conc.	100%	pH ↓ Water activity Storage stability ↓ Hardness ↓ Firmness Cohesiveness Springiness ↓ Gumminess ↓ Chewiness	Internal color: ↓ L^* ↓ a^* ↑ b^* External color:↓ L^*↑ a^* ↑ b^*	↓ Appear. ↓ Flavor Texture ↓ Chewiness ↓ Density ↓ Moisture ↓ Overall liking	Eckert et al. (2018)
Wheat bread	Germ. lentil flour	5–10%	Water activity: ↓ 10% Loaf volume: ↓ 10% ↑ Hardness Elasticity: ↑ 10% Cohesiveness ↑ Chewiness	L^*: ↓ 10% ↑ a^* ↑b^*	Color Porosity: ↓ 10% Sponginess: ↓ 10% ↑ Hardness ↓ Cohesive-ness Aroma Stickiness Chewiness Flavor ↑ Health pref	Hernandez-Aguilar et al. (2019)

(*Continued*)

Table 15.8 (Continued)

Product	Lentil product	% lentil used	Physico-chemical properties	Color	Sensory attributes	References
Wheat-flour angel food cake and muffins	Lentil protein conc.	50–100%	Specific gravity: ↑ for cakes Baked volume ↓ Baking loss % Firmness: ↑ for 100% muffins Cohesiveness Chewiness: ↑ for 75% cakes +100% muffins Springiness	Cakes: L^* ↑ a^* b^* Muffins: L^* ↓ a^* ↓ b^*	Appearance: ↓ for cakes Texture: ↓ for 75% cakes, ↑ for 100% muffins Flavor Overall likeness: ↑ for 100% muffins Chewiness: ↓ for 100% muffins Density: ↓ for 100% muffins Moistness: ↑ for 100% muffins	Jarpa-Parra et al. (2017)
Sourdough composite wheat bread	Raw and germ. Lentil flour	10%	↑ pH Titratable acidity: ↑ for germ. Flour ↑ Baking loss % ↑ Loaf volume ↓ Hardness ↓ Springiness ↓ Cohesiveness ↓ Staling rate	N/A	↑ Crust and crumb color ↑ Elasticity ↑ Consistency ↑ Acidity ↑ Sweetness ↑ Bitterness ↑ Acid odor ↑ Caramel odor ↑ Overall aroma	Perri et al. (2021b)
Sourdough composite wheat-barley bread	Raw and germ. Lentil flour	10%	pH[1] Titratable acidity[1] Baking loss %: ↓ for germ. Dextran fermented ↑ Loaf volume Hardness: ↑ for germ. Non-dextran fermented, ↓ for germ. Dextran fermented Springiness: ↑ for germ. Non-dextran fermented, ↓ for germ. Dextran fermented Staling rate: ↑ for germ. Non-dextran fermented	N/A	↑ Crust and crumb color ↑ Elasticity ↑ Consistency ↑ Acidity Sweetness: ↑ for germ. Dextran fermented ↑ Legume ↑ Bitterness ↑ Acid odor ↑ Caramel odor ↑ Wetness ↑ Overall aroma	Perri et al. (2021b)

Wheat bread	Raw lentil flour	5–40%	Water absorption: ↓ for 10–40% ↓ Dough development time ↓ Loaf volume ↑ Crumb firmness	*L**: ↓ for 40% ↑ *a** *b**: ↓ for 5%, ↑ for 10–40%	N/A	Portman et al. (2018)
Sangak bread	Germ. lentil flour	5–20%	↑ Water absorption Dough development time: ↑ for 10–20% ↑ Moisture content Hardness: ↓ for 5 & 15%, ↑ for 10% ↑ Browning index	N/A	Shape[2] Surface[2] Texture[2] Color[2] Chewiness[2] Aroma[2] Flavor[2]	Razavi et al. (2020)
Wheat bread	Raw lentil flour	6–24%	Water absorption ↑ Dough development time ↓ Viscosity Loaf volume: ↓ for 24%	N/A	↑ Lentil flavor Crunchiness ↑ Chewiness	Turfani et al. (2017)

Accept. = acceptance, Appear. = appearance, Conc. = concentrated, Germ. = germinated, Pref. = preference, ↑ = significant increase, ↓ = significant decrease, no preceding symbol = no significant difference, [1] = control values not given so unable to provide comparison, [2] = data not reported by authors.

to be absorbed by the wheat gluten to assist with development (Portman et al. 2018). The increase in DDT with higher percentages of supplementation by Razavi et al. (2020) suggests that gluten formation was slowed down by the inclusion of lentil proteins.

15.3.2 Loaf Volume Relationship to Sensory Quality

Assessing the final loaf volume of breads with lentil flour is important in understanding the ability of dough to become aerated and maintain structure. Like DDT, loaf volume is associated with the ability of the dough to form gluten networks (Romano et al. 2021). Lentil supplementation often results in the reduction of loaf volume, with some exceptions. Bojňanská et al. (2021) and Portman et al. (2018) reported decreased loaf volume across all percentages of LF supplementation. The addition of LF to wheat breads can weaken the gluten-starch matrix due to the increase in fiber and alternative proteins (Bojňanská et al. 2021). At their highest LF percentage supplementation, Hernandez-Aguilar et al. (2019) and Turfani et al. (2017) observed a decrease in loaf volume. At higher pulse supplementation levels, cell structure becomes compromised, which can lead to lower dough stability and an ability to retain carbon dioxide (Zhang et al. 2021). Therefore, it is possible that variations in formulations in terms of LF particle size, type of lentil, or lentil processing method can compensate for the lack of gluten structure at lower levels of supplementation. Studies by Perri et al. (2021a) and Perri et al. (2021b) reported that loaf volume increased in wheat-lentil composite breads. However, these studies used dextran-fermented sourdoughs developed using germinated lentils. In comparison to control loaves, Perri et al. (2021a) found increased loaf volume using germinated LF compared to raw LF. It is thought that dextran functions similarly to a hydrocolloid, which would negate issues associated with dough stability and cell structure, allowing the dough to retain more carbon dioxide, thus increasing volume (Lynch et al. 2018). Hence problems associated with reduction in loaf volume may overcome by incorporating dextran or other exopolysaccharide fermentation when baking bread using LF.

Lentil protein isolates are another lentil ingredient that has been used in bread. Lentil proteins are extracted from lentil seeds using methods like air classification, alkaline extraction/isoelectric precipitation (IEP), salt extraction, and ultrafiltration (Boye et al. 2010). Jarpa-Parra et al. (2017) found that baking volume for angel cakes increased with the addition of lentil protein isolates. As mentioned earlier, non-wheat proteins can sometimes interfere with dough development due to interference with the gluten network. However, while bread doughs require high gluten development to form networks capable of holding gas, baked products like cakes do not require extensive development (Jarpa-Parra et al. 2017). Thus, differences in volume between breads and other baked products may not be comparable. Lentil supplementation has been shown to result in more viscous batters, which has been hypothesized to retain more gas and increase volume (Paraskevopoulou et al. 2015).

The methods used to determine baked product volume in the different studies employed either the American Association of Cereal Chemists rapeseed displacement method using a volumeter (AACC 2009) or analytical equipment including volume analysers. For Jarpa-Parra et al. (2017), their method for baking volume was not disclosed. Although the rapeseed displacement method is considered accurate for irregular shaped items, such as bakery products, it can crush more delicate items leading to errors in measurement (Whitaker and Barringer 2004). When other analytical methods are compared to rapeseed displacement, they are generally found to be similar, but differences can occur. Thus, volume comparison of breads and baked goods may be more difficult when similar methodology is not used.

15.3.3 Textural Properties That Affect Sensory Acceptance

When developing lentil-based wheat-bread alternatives, measuring the textural properties of the crumb is essential in determining baking quality. Although many textural properties are often investigated, some of the most common properties of baked products include hardness, cohesiveness, and chewiness. Hardness is an important textural quality because it is associated with the freshness of a product (Kadam et al. 2015). It is defined as the amount of force required to cause food deformation (Mousavi et al. 2019). In the studies reviewed, results for the hardness of baked products with the addition of lentil ingredients were somewhat mixed. Hardness increased at 5–10% supplementation, according to Hernandez-Aguilar et al. (2019). However, Razavi et al. (2020) found decreased hardness with 5% and 15% supplementation, and an increased hardness for 10% supplementation. Both studies utilized germinated lentil flour and the flours used had similar proximate compositions. Razavi et al. (2020) noted their WA and hardness results were closely linked. It is conceivable that the interrelationships between gluten and non-gluten proteins as well as the presence and development of lentil starch molecules and matrices have varied effects on WA and in turn, hardness. It is possible that information not given, such as lentil type and flour particle size, may have influenced hardness. Eckert et al. (2018) found that hardness decreased in their lentil-protein-supplemented doughnuts. Increased lentil protein can assist with the formation of more extensive protein networks in the dough that can lead to a lighter texture and softer product (Jarpa-Parra 2018). Studies by Hernandez-Aguilar et al. (2019) and Eckert et al. (2018) did not incorporate a large variety of treatment levels. Therefore, it was more challenging to observe any trends in their data.

Cohesiveness gives an indication of internal bond strength within the food by measuring the extent of deformation before breaking (Chandra and Shamasundar 2014). For the studies that measured cohesiveness (Eckert et al. 2018; Hernandez-Aguilar et al. 2019; Jarpa-Parra 2017; Perri et al. 2021a; Perri et al. 2021b), all results show no significant changes in this attribute across all levels of LF supplementation. This suggests that despite differences in protein and starch networks, LF-supplemented breads can still retain an acceptable structure.

Chewiness is the amount of energy needed to chew the product until it is suitable to be swallowed (Chandra and Shamasundar 2014). It is influenced by the elasticity and cohesiveness of the product (Kasapis and Bannikova 2017). Eckert et al. (2018) noted a decrease in chewiness for their doughnuts, whereas Jarpa-Parra et al. (2017) found an increase for both 75% lentil-protein-supplemented cakes and 100% supplemented muffins. As previously discussed, the addition of protein increased the airiness in dough, which may explain the lack of chewiness. At high lentil protein supplementation levels, increased protein reinforced gluten networks or formed non-gluten protein networks. It is noted by Jarpa-Parra et al. (2017) that treatments that increased chewiness also resulted in higher moisture retention and therefore greater density, likely due to increased overall protein. Hernandez-Aguilar et al. (2019) observed an increase in chewiness for both germinated lentil treatments, which corresponded with increased hardness and elasticity of bread. Germination has the ability to alter proteins from high to low molecular weight, which in turn alters the interactions between the molecular side groups (Ghumman et al. 2016).

15.4 Conclusion

Interest in the development of lentil supplementation in traditionally wheat-based food products has increased substantially in recent years because of the consumers' adoption of sustainable and healthy choices. The addition of LF and protein isolate is increasingly being studied in an extensive array of foods. Studies have shown that supplementing foods with LF or protein can result in products that are technologically and texturally similar to wheat-based products. However, the success is limited to lower supplementation levels. Higher levels of lentil supplementation can influence loaf volume, texture, and sensory properties negatively. Some of these effects could be ameliorated by dextran fermentation, which provides a hydrocolloid-like structural support to make up for the lack of gluten development. Additionally, there are important differences between raw and germinated lentils due to changes in protein and starch molecules. Lentil ratios in food formulations would need to be optimized to increase consumer acceptance and improve sensory quality. Most of the reported results on sensory evaluation suffer from some limitations, such as the number of panelists used as well as the bias of using trained panelists for consumer testing. The incorporation of lentil ingredients may present a challenge since their inclusion may interfere with the sensory and physicochemical properties of the food products. Finally, innovations in the field of flavor science and novel processing could be used to improve flavor attributes in order to promote the production of healthy lentil-containing food products with high consumer acceptability.

References

AACC (2009). *Guidelines for Measurement of Volume by Rapeseed Displacement* (9781891127687). AACC International Approved Methods, Issue.

Altemimi, A.B. (2018). Extraction and optimization of potato starch and its application as a stabilizer in yogurt manufacturing. *Foods* 7 (2): 14.

Argel, N.S., Ranalli, N., Califano, A.N., and Andrés, S.C. (2020). Influence of partial pork meat replacement by pulse flour on physicochemical and sensory characteristics of low-fat burgers. *Journal of the Science of Food and Agriculture* 100 (10): 3932–3941.

Arntfield, S.D., Scanlon, M.G., Malcolmson, L.J. et al. (2001). Reduction in lentil cooking time using micronization: comparison of 2 micronization temperatures. *Journal of Food Science* 66 (3): 500–505.

Benayad, A. and Aboussaleh, Y. (2021). Mineral composition of lentils: Physiological functions, antinutritional effects, and bioavailability enhancement. *Journal of Food Quality* 2021: 1–9.

Benmeziane, F., Raigar, R.K., Ayat, N.E.-H. et al. (2021). Lentil (*Lens culinaris*) flour addition to yogurt: impact on physicochemical, microbiological and sensory attributes during refrigeration storage and microstructure changes. *LWT Food Science and Technology* 140: 110793.

Biju, S., Fuentes, S., Gonzalez Viejo, C. et al. (2021). Silicon supplementation improves the nutritional and sensory characteristics of lentil seeds obtained from drought-stressed plants. *Journal of the Science of Food and Agriculture* 101 (4): 1454–1466.

Boeck, T., Ispiryan, L., Hoehnel, A. et al. (2022). Lentil-based yogurt alternatives fermented with multifunctional strains of lactic acid bacteria: techno-functional, microbiological, and sensory characteristics. *Foods* 11 (14): 2013.

Bojňanská, T., Musilová, J., and Vollmannová, A. (2021). Effects of adding legume flours on the rheological and breadmaking properties of dough. *Foods* 10 (5): 1087.

Boukid, F., Zannini, E., Carini, E., and Vittadini, E. (2019). Pulses for bread fortification: a necessity or a choice? *Trends in Food Science & Technology* 88: 416–428.

Boye, J., Zare, F., and Pletch, A. (2010). Pulse proteins: processing, characterization, functional properties and applications in food and feed. *Food Research International* 43 (2): 414–431.

Bresciani, A. and Marti, A. (2019). Using pulses in baked products: lights, shadows, and potential solutions. *Foods* 8 (10): 451.

Chandra, M.V. and Shamasundar, B.A. (2014). Texture profile analysis and functional properties of gelatin from the skin of three species of fresh water fish. *International Journal of Food Properties* 18 (3): 572–584.

Chelladurai, V. and Erkinbaev, C. (2020). Lentils. In: *Pulses: Processing and Product Development*, 129–143. Springer.

Clark, B., Stewart, G.B., Panzone, L.A. et al. (2017). Citizens, consumers and farm animal welfare: a meta-analysis of willingness-to-pay studies. *Food Policy* 68: 112–127.

Costantini, M., Sabovics, M., Galoburda, R. et al. (2021). Effect of die configuration on the physico-chemical properties, anti-nutritional compounds, and sensory features of legume-based extruded snacks. *Foods* 10 (12): 3015.

Day, L. (2013). Proteins from land plants–potential resources for human nutrition and food security. *Trends in Food Science & Technology* 32 (1): 25–42.

Delahunty, C. and Drake, M. (2004). Sensory character of cheese and its evaluation. *Cheese: Chemistry, Physics and Microbiology* 1 (3): 455–487.

Di Stefano, V., Pagliaro, A., Del Nobile, M.A. et al. (2020). Lentil fortified spaghetti: technological properties and nutritional characterization. *Foods* 10 (1): 4.

Drulyte, D. and Orlien, V. (2019). The effect of processing on digestion of legume proteins. *Foods* 8 (6): 224.

Eckert, E., Wismer, W., Waduthanthri, K. et al. (2018). Application of barley- and lentil-protein concentrates in the production of protein-enriched doughnuts. *Journal of the American Oil Chemists' Society* 95 (8): 1027–1040.

Ganesan, K. and Xu, B. (2017). Polyphenol-rich lentils and their health promoting effects. *International Journal of Molecular Sciences* 18 (11): 2390.

Ghumman, A., Kaur, A., and Singh, N. (2016). Impact of germination on flour, protein and starch characteristics of lentil (*Lens culinari*) and horsegram (*Macrotyloma uniflorum* L.) lines. *LWT Food Science and Technology* 65: 137–144.

Gobbetti, M., De Angelis, M., Di Cagno, R. et al. (2019). Novel insights on the functional/nutritional features of the sourdough fermentation. *International Journal of Food Microbiology* 302: 103–113.

Hernandez-Aguilar, C., Dominguez-Pacheco, A., Palma Tenango, M. et al. (2019). Lentil sprouts: a nutraceutical alternative for the elaboration of bread. *Journal of Food Science and Technology* 57 (5): 1817–1829.

Hernandez-Aguilar, C., Dominguez-Pacheco, A., Palma Tenango, M. et al. (2020). Lentil sprouts: a nutraceutical alternative for the elaboration of bread. *Journal of Food Science and Technology* 57 (5): 1817–1829.

Jarpa-Parra, M. (2017). Lentil protein: a review of functional properties and food application. An overview of lentil protein functionality. *International Journal of Food Science & Technology* 53 (4): 892–903.

Jarpa-Parra, M. (2018). Lentil protein: a review of functional properties and food application. An overview of lentil protein functionality. *International Journal of Food Science & Technology* 53 (4): 892–903.

Jarpa-Parra, M., Bamdad, F., Wang, Y. et al. (2014). Optimization of lentil protein extraction and the influence of process pH on protein structure and functionality. *LWT Food Science and Technology* 57 (2): 461–469.

Jarpa-Parra, M., Wong, L., Wismer, W. et al. (2017). Quality characteristics of angel food cake and muffin using lentil protein as egg/milk replacer. *International Journal of Food Science & Technology* 52 (7): 1604–1613.

Jeske, S., Bez, J., Arendt, E.K., and Zannini, E. (2019). Formation, stability, and sensory characteristics of a lentil-based milk substitute as affected by homogenisation and pasteurisation. *European Food Research and Technology* 245 (7): 1519–1531.

Jørgensen, C.E., Abrahamsen, R.K., Rukke, E.O. et al. (2019). Processing of high-protein yoghurt – a review. *International Dairy Journal* 88: 42–59.

Kadam, S.U., Tiwari, B.K., and O'Donnell, C.P. (2015). Improved thermal processing for food texture modification. *Modifying Food Texture* 115–131.

Kasapis, S. and Bannikova, A. (2017). Rheology and food microstructure. In: *Advances in food rheology and its applications*, 7–46. Woodhead Publishing.

Khattab, R., Arntfield, S., and Nyachoti, C. (2009). Nutritional quality of legume seeds as affected by some physical treatments, part 1: protein quality evaluation. *LWT Food Science and Technology* 42 (6): 1107–1112.

Khazaei, H., Subedi, M., Nickerson, M. et al. (2019). Seed protein of lentils: current status, progress, and food applications. *Foods* 8 (9): 391.

Kilcast, D. and Lewis, D.F. (1990). Structure and texture—their importance in food quality. *Nutrition Bulletin* 15 (2): 103–113.

Kim, T., Riaz, M.N., Awika, J., and Teferra, T.F. (2021). The effect of cooling and rehydration methods in high moisture meat analogs with pulse proteins-peas, lentils, and faba beans. *Journal of Food Science* 86 (4): 1322–1334.

Kohajdová, Z., Karovičová, J., and Magala, M. (2013). Effect of lentil and bean flours on rheological and baking properties of wheat dough. *Chemical Papers* 67 (4): 398–407.

Kumar, S. and Pandey, G. (2020). Biofortification of pulses and legumes to enhance nutrition. *Heliyon* 6 (3): e03682.

Lazou, A. and Krokida, M. (2010). Structural and textural characterization of corn–lentil extruded snacks. *Journal of Food Engineering* 100 (3): 392–408.

Lynch, K.M., Zannini, E., Coffey, A., and Arendt, E.K. (2018). Lactic acid bacteria exopolysaccharides in foods and beverages: isolation, properties, characterization, and health benefits. *Annual Review of Food Science and Technology* 9 (1): 155–176.

Ma, Z., Boye, J.I., Fortin, J. et al. (2013). Rheological, physical stability, microstructural and sensory properties of salad dressings supplemented with raw and thermally treated lentil flours. *Journal of food Engineering* 116 (4): 862–872.

Martín-Cabrejas, M.A., Aguilera, Y., Pedrosa, M.M. et al. (2009). The impact of dehydration process on antinutrients and protein digestibility of some legume flours. *Food Chemistry* 114 (3): 1063–1068.

Mishra, B.K., Srivastava, J.P., and Lal, J.P. (2018). Drought resistance in lentil (Lens culinaris Medik.) in relation to morphological, physiological parameters and phenological developments. International Journal of Current Microbiology and Applied. *Science* 7 (1): 2288–2304.

Monnet, A.-F., Laleg, K., Michon, C., and Micard, V. (2019). Legume enriched cereal products: a generic approach derived from material science to predict their structuring by the process and their final properties. *Trends in Food Science & Technology* 86: 131–143.

Mousavi, M., Heshmati, A., Daraei Garmakhany, A. et al. (2019). Texture and sensory characterization of functional yogurt supplemented with flaxseed during cold storage. *Food Science & Nutrition* 7 (3): 907–917.

Nosworthy, M.G., Medina, G., Franczyk, A.J. et al. (2018). Effect of processing on the in vitro and in vivo protein quality of red and green lentils (Lens culinaris). *Food Chemistry* 240: 588–593.

Okuda, R., Tabara, A., Okusu, H., and Seguchi, M. (2016). Measurement of water absorption in wheat flour by mixograph test. *Food Science and Technology Research* 22 (6): 841–846.

Paraskevopoulou, A., Donsouzi, S., Nikiforidis, C.V., and Kiosseoglou, V. (2015). Quality characteristics of egg-reduced pound cakes following WPI and emulsifier incorporation. *Food Research International* 69: 72–79.

Pasqualone, A., Costantini, M., Faccia, M. et al. (2022). The effectiveness of extruded-cooked lentil flour in preparing a gluten-free pizza with improved nutritional features and a good sensory quality. *Foods* 11 (3): 482.

Perri, G., Coda, R., Rizzello, C.G. et al. (2021a). Sourdough fermentation of whole and sprouted lentil flours: *in situ* formation of dextran and effects on the nutritional, texture and sensory characteristics of white bread. *Food Chemistry* 355: 129638.

Perri, G., Rizzello, C.G., Ampollini, M. et al. (2021b). Bioprocessing of barley and lentil grains to obtain in situ synthesis of exopolysaccharides and composite wheat bread with improved texture and health properties. *Foods* 10 (7): 1489.

Podder, R., Khan, S.M., Tar'an, B. et al. (2018). Sensory acceptability of iron-fortified red lentil (*Lens culinaris* Medik.) dal. *Journal of Food Science* 83 (3): 804–813.

Podder, R., Hassan Al Imam, M., Jahan, I. et al. (2020). Sensory acceptability of dual-fortified milled red and yellow lentil (*Lens culinaris* Medik.) dal in Bangladesh. *Foods* 9 (8): 992.

Portman, D., Blanchard, C., Maharjan, P. et al. (2018). Blending studies using wheat and lentil cotyledon flour-effects on rheology and bread quality. *Cereal Chemistry* 95 (6): 849–860.

Previtali, M.A., Mastromatteo, M., De Vita, P. et al. (2014). Effect of the lentil flour and hydrocolloids on baking characteristics of wholemeal durum wheat bread. *International Journal of Food Science & Technology* 49 (11): 2382–2390.

Rajagukguk, Y.V., Arnold, M., Sidor, A. et al. (2022). Antioxidant activity, probiotic survivability, and sensory properties of a phenolic-rich pulse snack bar enriched with *Lactiplantibacillus plantarum*. *Foods* 11 (3): 309.

Rathod, R.P. and Annapure, U.S. (2016). Effect of extrusion process on antinutritional factors and protein and starch digestibility of lentil splits. *LWT Food Science and Technology* 66: 114–123.

Razavi, S.N., Hojjatoleslamy, M., Molavi, H., and Boroujeni, L.S. (2020). The effect of germinated lentil flour on the physicochemical and organoleptic characteristics of Sangak bread. *Journal of Culinary Science & Technology* 20 (3): 253–265. https://doi.org/10.1080/15428052.2020.1840473.

Romano, A., Gallo, V., Ferranti, P., and Masi, P. (2021). Lentil flour: nutritional and technological properties, in vitro digestibility and perspectives for use in the food industry. *Current Opinion in Food Science* 40: 157–167.

Rozan, M.A., Bayomy, H.M., and Boriy, E.G. (2018). Effects of turmeric addition on chemical composition, antioxidant activity and sensory evaluation of lentil soup. *Alexandria Science Exchange Journal* 39 (January–March): 1–6.

Ryland, D., Vaisey-Genser, M., Arntfield, S.D., and Malcolmson, L.J. (2010). Development of a nutritious acceptable snack bar using micronized flaked lentils. *Food Research International* 43 (2): 642–649.

Sandhu, J.S. and Singh, S. (2007). History and origin. In: *Lentil: An Ancient Crop for Modern Times*, 1–9. Springer Dordrecht.

Saricaoglu, F.T. (2020). Application of high-pressure homogenization (HPH) to modify functional, structural and rheological properties of lentil (Lens culinaris) proteins. *International Journal of Biological Macromolecules* 144: 760–769.

Sendra, E., Kuri, V., Fernández-López, J. et al. (2010). Viscoelastic properties of orange fiber enriched yogurt as a function of fiber dose, size and thermal treatment. *LWT-Food Science and Technology* 43 (4): 708–714.

Shariati-Ievari, S., Ryland, D., Edel, A. et al. (2016). Sensory and physicochemical studies of thermally micronized chickpea (*Cicer arietinum*) and green lentil (*Lens culinaris*) flours as binders in low-fat beef burgers. *Journal of Food Science* 81 (5): S1230–S1242.

Sopiwnyk, E., Young, G., Frohlich, P. et al. (2020). Effect of pulse flour storage on flour and bread baking properties. *LWT* 121: 108971.

Turfani, V., Narducci, V., Durazzo, A. et al. (2017). Technological, nutritional and functional properties of wheat bread enriched with lentil or carob flours. *LWT Food Science and Technology* 78: 361–366.

Whitaker, A. and Barringer, S. (2004). Measurement of contour and volume changes during cake baking. *Cereal Chemistry* 81 (2): 177–181.

Zare, F., Boye, J.I., Orsat, V. et al. (2011). Microbial, physical and sensory properties of yogurt supplemented with lentil flour. *Food Research International* 44 (8): 2482–2488.

Zhang, Y., Hu, R., Tilley, M. et al. (2021). Effect of pulse type and substitution level on dough rheology and bread quality of whole wheat-based composite flours. *Processes* 9 (9): 1687.

16

Global Consumption and Culinary Trends in Lentil Utilization

Tahira M. Ali[1], *Natasha A. Butt*[2], *and Abid Hasnain*[1]

[1] *Department of Food Science and Technology, University of Karachi, Karachi, Pakistan*
[2] *Department of Biomedical Engineering, Faculty of Engineering, Science, Technology and Management, Ziauddin University, Karachi, Pakistan*

16.1 Introduction

The history of lentil cultivation is approximately 8000–10,000 years old. The archeological studies indicate its cultivation in different regions of the world as lentil remains were reported in Greece's Franchthi cave, Syria, villages of Neolithic (Helbaek 1959), Iran, Ceramic Hacilar (Turkey), Macedonia, Egypt, Jordan, and Iraq (Sandhu and Singh 2007). According to Harlen and Jati (2018), the history of lentil consumption and cultivation is as old as 8500 BCE. Though lentil existed in different regions, historians suggest that it actually originated from Central Asia and Middle East from where it was brought to other regions of the world including Europe, Africa, and Latin America (Sonnante et al. 2009). The lentil consumption by humankind has been reported since the beginning of the agricultural revolution and was important part of diet along with cereal grains specifically wheat and barley (Zohary 1976).

The cultivated varieties of lentils are different in terms of their seed color and nutrition. Lentils are considered a high protein source food available in different colors ranging from yellow/red to tan, green-, and yellow-colored varieties. The different color varieties suggest presence of varying anthocyanins with different antioxidant capacities (Xu and Chang 2010). Historically, the consumption pattern of lentil has been very diverse compared to other legumes and was eaten since ancient times to meet both dietary and medicinal needs (Duke 2012). Though it is widely cultivated in Canada, India, Australia, Turkey, and Nepal (FAO 2020), in terms of consumption, it is generally considered a staple food of South Asia and Middle East (Joshi et al. 2017).

This chapter provides an overview of lentil's nutritional significance and consumption trends, preparatory/culinary steps for lentil-based cuisines, regional lentil cuisines worldwide, and lentils in food service and restaurants.

Lentils: Production, Processing Technologies, Products, and Nutritional Profile, First Edition.
Edited by Jasim Ahmed, Muhammad Siddiq, and Mark A. Uebersax.

16.2 Global Production and Consumption Trends

Canada is the leading producer and exporter of lentils in the world. Canada's total lentil production in 2020 reached 2.87 million tons, followed by India with total production of 1.18 t (FAO 2020). India was the traditional leader in global lentil production, but some other countries have emerged as major producers of lentils in the last two decades, e.g., since 2000, Australia, Canada, and United States have shown 223%, 214%, and 145% increase in lentils production, respectively. Besides Canada, Australia, Turkey, and United States have emerged as major exporters of lentils. The top five lentil importing countries are India, Turkey, Bangladesh, United Arab Emirates (UAE), and Pakistan (FAO 2020). These statistics are suggestive of the extensive consumption of lentil-based cuisines in Asia and the Middle East. Interestingly, India being the second highest producer of lentils is also the second biggest importer of lentils so as to meet its extensive consumption demand for lentils.

The lentil consumption is mostly concentrated in Asia and Middle East region. The total yearly consumption of lentils is the highest in India, followed by Bangladesh, Turkey, Canada, and Nepal (FAO 2019). However, as shown in Figure 16.1, UAE leads in per capita daily intake of calories from lentils with 77 kcal, followed by Sri Lanka (66 kcal), Nepal (52 kcal), Syria (52 kcal), and Canada (51 kcal). In Caribbean region and Africa, the countries with the highest lentil consumption are Saint Lucia and Seychelles, respectively.

16.3 Utilization of Lentils in Diverse Cuisines

Lentils have played an important role in increasing plant-based proteins in daily diets. The most common lentils available for household uses and for food services are (i) whole green and brown lentils and (ii) red-colored dehulled, split lentils, also referred

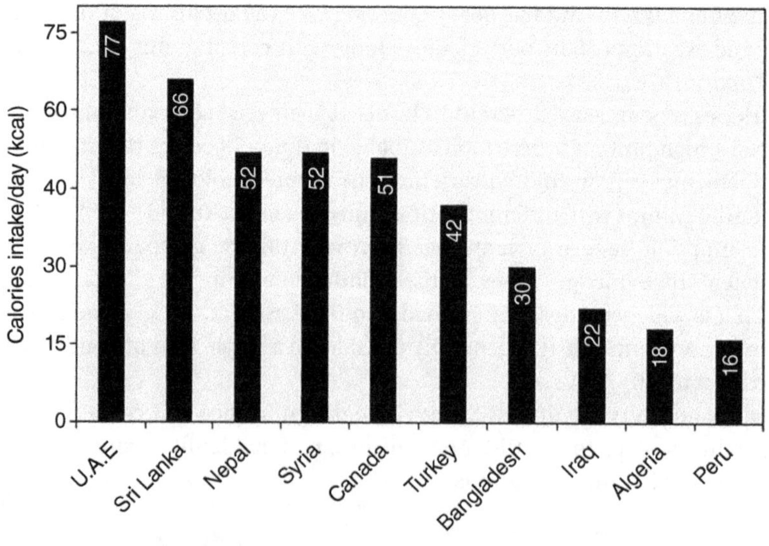

Figure 16.1 Top 10 countries with the highest daily per capita calorie consumption from lentils in 2019. *Source:* Data from FAO (2020).

to as *dal* or *dhal* in Asia. The whole lentils are usually used in salads and take slightly longer time to cook than the split lentils. The split lentils or dal have lower antioxidant capacity compared to whole lentils due to decortication/dehulling (Parikh and Patel 2018). However, split lentil or dhal is more popular in Asia and the Middle East, which are frequently used in preparing curries, soups, gravies, patties, and some other products. Besides the green and brown lentils, black-colored beluga, whole lentils are also consumed in the Mediterranean region and few other Western countries. In uncooked form, both whole and split lentil seeds have a very long storage life up to 12 months similar to other pulses.

16.3.1 Preparatory/Culinary Steps for Lentil-Based Cuisines

The common preparatory steps for lentils are presented in Figure 16.2. Compared to other legumes, like chickpea and most other beans (kidney, pinto, navy, lima, great northern, and faba beans), the cooking time of lentils is significantly less, i.e., around 18–20 minutes. Given the shorter cook time, lentils (and green gram or mung bean) are categorized as "quick-cooking" legume (Polak et al. 2015; Sashikala and Narasimha 2010). For harder-to-cook legumes, usually baking soda (sodium bicarbonate, $NaHCO_3$) or other additives are used in soaking water for softening beans, which significantly reduces the cooking time. Though no additives are required for lentils, Prodanov et al. (2004) reported an increase in riboflavin content when $NaHCO_3$ is used in soaking water. Lentils are used for the preparation of different cuisines like soups, salads, dhal, or rice-based dishes, and lentils are first washed in water to remove any adhered soil.

Soaking of lentils is preferred but not mandatory; however, Queiroz et al. (2002) reported that soaking of legumes before cooking reduces flatulence-causing oligosaccharides (e.g., raffinose and stachyose) without any negative impact on the nutritional value. Soaking is followed by boiling in a pot containing almost 1 : 3 ratio of lentils to water. Salt may be added at the start of boiling but, traditionally, in South Asia, it is added in the last stage of boiling or even after boiling. The lentils after boiling are simmered for few minutes (Erskine 2009). The boiled lentils can then be added to any cuisine. The preparation/boiling time of whole lentil seeds is four times greater compared to that of split lentils (Polak et al. 2015).

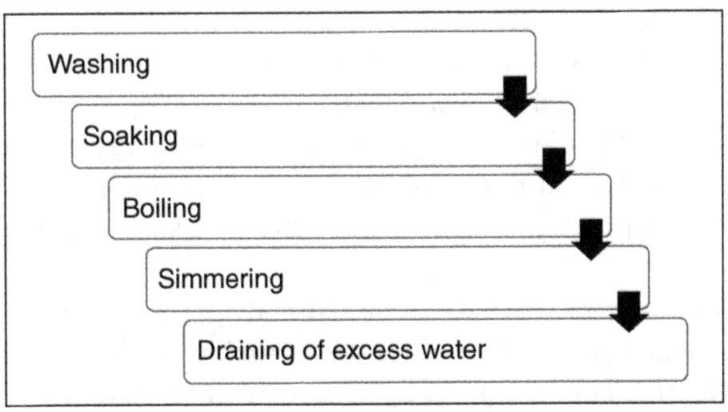

Figure 16.2 Typical preparatory steps used in culinary applications of lentils.

Table 16.1 Lentil-based cuisines around the world.

Curries/soups	Rice-based	Snacks/others
• Fakes Soupa • Dal/soup • Misir (Egypt) Wat • Ethiopian Wat • Lentil Raghout • Cajun lentil stew • Iranian Adasi • Shorbat Adas • Mercimek Köftesi • Trinidad stewed lentils	• Mujaddara • Koshari • Khichdi • Lentil/Masoor Pilaf • Pulav • Adas Polow (Polo) • Fakorizo	• Dosa • Papad • Fritters • Hlalem • Matani bread • Lentil burger patties

16.4 Regional Lentil Cuisines Worldwide

Lentil is considered a nutrition "powerhouse" in both developing and under developed regions of the world as it is a source of both macro and micronutrients (Thavarajah et al. 2011). Among vegetables, it is the second richest source of protein after soybean (Bhattacharya et al. 2005). Some lentil cuisines are specific to different regions of the world, while there are some other cuisines, which is commonly consumed in every region of the world, e.g., soups and salads. Table 16.1 lists popular lentil cuisines or products worldwide. A detailed discussion of common lentil-based regional cuisines is provided in the following sections.

16.4.1 South Asian Cuisines

16.4.1.1 Masoor/Lentil Dal

Lentils are called *Masoor* in most South Asian countries. *Masoor dal* is a curry-based cuisine that is commonly consumed as lunch as well as dinner meal. It is consumed both with and without wheat flat bread (*roti*, *chapati*, and *naan*, a leavened flat bread) and is also frequently consumed by pouring over boiled rice. In Nepal, dal is served with boiled/steamed rice along with pickles and chutney (sauces) as sides. This cuisine is locally called *dal bhat* in Nepal (rice is called *bhat*) where it is considered a staple food. The same cuisine in Pakistan is called *dal chawal* since rice is called *chawal*. The method of dal preparation varies from country to country, including the choice of spices used. Occasionally, when prepared in butter and it is called *dal makhani* in Pakistan, but it is more commonly prepared in vegetable oils. Dal makhani is a premium form of dal, which is usually sold at higher prices in Pakistani restaurants.

Masoor dal/curry can be prepared both from whole or dehulled and split lentils; however, the former takes longer time to cook, typically 45 minutes versus under 30 minutes for the latter. Moreover, split lentil dal gives a thicker, soup-like appearance compared to whole lentil where seeds stay separate even after cooking to complete tenderness. Both forms of dals are usually seasoned with curry leaves, garlic, ground green chili paste, coriander powder, roasted cumin powder, and salt. Sour dals are also prepared in some area after adding freshly squeezed lemon or tamarind paste and garnished with coriander leaves, mint leaves, and freshly thin-cut green chilies. The masoor dal is also prepared with the addition of vegetables like bottle gourd and bamboo shoots. *Masoorchi amti* is another form of *dal* popular in Maharashtra, India, which is prepared from sprouted whole (red) lentils along with a mix of spices, tamarind, *Jaggery* (raw sugar nuggets), and grated coconut (Datar 2014).

Another variant of *masoor dal* popular in Sri Lanka and India is called masoor dal sambar. Sambar is a lentil and vegetable stew in which small amount of oil is first heated with green chilies, curry leaves, mustard seeds followed by addition of onion and shallots until it turns light brown in oil (Arthi 2010). This step is followed by addition of squeezed tomatoes, salt, and *sambar* powder, which is a mixture of spices mainly coriander, cumin, fenugreek, black pepper, cloves, dried whole red chili, cinnamon, and coconut powder (Doshi 2002). The split lentils previously cooked with turmeric and asafoetida is then added along with water to get the desired consistency. Lastly, tamarind paste is added to give sourness. In South India, *sambar* prepared by adding equal amounts of red lentils and split pigeon pea (Doshi 2002). Lentils are also added in a mixed *dal* cuisine in Pakistan, which is prepared by adding split lentil, mung bean, and split chickpea in equal amounts. *Peeti* is another dal-based cuisine, which is prepared only with mixture of mung bean and masoor (Yasmeen et al. 2019). Occasionally, chicken and beef chunks with or without bones are also added to give more rich taste to *dals* occasionally.

Lentil *dal*, being frequently consumed, can also be used as a source of iron after fortification to meet the dietary needs of iron among general population. Podder et al. (2018a) observed significant increase in Fe content and its bioavailability due to use of iron-fortified lentils in dal prepared by traditional Bengali recipe containing onion, canola oil, salt, turmeric powder, and water. The lentils fortified with 1600 ppm of ferric sodium ethylene diamine tetraacetic acid (NaFeEDTA) were found to have insignificant difference in terms of sensory traits when compared to cooked lentil *dal* prepared from unfortified lentils (Podder et al. 2018b). This study proposed the use of lentils in mass nutrition programs related to iron fortification along with wheat so as to avert Fe malnutrition particularly in women in developing countries.

16.4.1.2 Khichdi

Khichdi, a rice/pulse-based cuisine, is a popular dish in India, Bangladesh, and Pakistan (Singh and Singh 2014). It is prepared using a combination of split lentils and rice and is often eaten with vegetable curries. Because of its soft texture and efficient digestibility, it is frequently consumed by elderly people and is also an important part of weaning menus. Rice chosen for *khichdi* preparation are the small kernel size and not the long-kernel premium quality basmati rice. Moreover, it is common to prepare *khichdi* using partially broken or damaged rice.

The four mandatory ingredients of khichdi are split lentil or dal, oil, rice, salt, and water (Singh and Singh 2014). The *khichdi* recipe varies slightly in different countries of South Asia. In Pakistan, it is made by using split lentils and rice in the ratio of 4 : 1 to 2 : 1 while in Bangladesh and India, it is prepared by using rice and lentil at 1 : 1 ratio. The rice and lentils are soaked together for about 15 minutes before cooking. Besides salt, a variety of spices are added (turmeric powder, whole cumin, and sometimes all spice powder or *garam masala*). Variations can be done by adding diced vegetables and/or chicken but, most often, it is eaten plain with only curries. The *khichdi* is given a finished look by simmering it for five minutes after removing most of the water.

Prasad et al. (1995) reported the proximate composition of precooked dehydrated *khichdi* (easy to reconstitute form) with 4.6% moisture, 12.9% protein, 19.7% fat, 56.4% carbohydrates, 1.8% fiber, and 4.5% ash. *Khichdi* is a very light food in terms of nutritional balance and is also chosen during dietary management of diarrhea (Shah et al. 2012).

16.4.1.3 Lentil Papad

Lentil *Papad* is a popular South Asian snack food, which is also used as an appetizer. The papad is a fried lentil cracker or wafer type product, often served as a side dish with dal and vegetables curries (Saraswat and Jain 2014). Besides frying, it may be roasted on an iron griddle or could also be roasted directly over open fire stove (Chowdhury et al. 2009). *Papad* is famous in Nepal, Bangladesh, India, and Pakistan. In unfried form, it has a longer shelf life due to its extremely low moisture and no fat content. *Papad* is usually packed and marketed in sealed polyethylene bags to maintain its crisp texture. Lentil *papad* is prepared from lentils, salt, spices, vegetable oil, sodium carbonate, and sodium bicarbonate (Renu and Waghray 2016) but un-spiced lentil *papads* are also available.

Lentils are also used in preparing mixed legumes *papad* by adding cowpea and mung bean. The flour of pulses is used for the preparation of *papad,* which are turned into dough by addition of water, oil, and seasonings after which the dough is rolled into a circular disc of 0.6–0.8 mm thickness using a wooden roller pin. These discs are then dried under sun heat to a moisture content of 8–10% (Garg and Dahiya 2003). Alkalinity of *papad* dough is necessary to give it rollability, expansion, and crispiness for which *papad-khadis* (composed of sodium carbonate and sodium bicarbonate) is traditionally used in India (Renu and Waghray 2016; Shaikh et al. 2009). *Papad* whether deep fried or roasted has a very short cooking time of less than 60 seconds. For deep frying, papad must be added when the oil is fully hot. The consumption of *papad* is popular with entrees or solely as snack in both rural and urban areas of South Asia. The lentil crisps or *papads* are rich in protein and gluten free (Xu et al. 2020) and thus are considered a healthy snack particularly if it is roasted, as frying increases caloric value due to absorbed fat.

16.4.1.4 Dosa

Dosa, originally a South Asian wafer type, thin crepe product made with split lentils, has also become popular in other parts of the sub-continent. Similar to pancakes, it is made from a thin batter and not dough; therefore, it is also referred as a spicy pancake or crepe. *Dosa* is commonly sold in restaurants as well as by cart vendors and is a popular as a street food, especially in South India. Originally, *dosa* was prepared from black gram and rice in a 1 : 3 ratio (Sozer et al. 2016). However, culinary trends continue to evolve regionally, as new ingredients both from health and taste point of view are incorporated from time to time. Lentils being high in protein and fiber but low in fat are now added commercially in combination with other pulses to make *dosa* (Roopa et al. 2017). *Dosa* prepared from only lentil is more popular as a homemade cuisine than in restaurants or by street vendors. However, traditional *dosa* is made by overnight soaking (six to eight hours) of split red lentils either alone or with some other split pulses (black gram, chickpea, black-eye pea, white peas, mung beans) followed by blending with water and salt to make a batter of thin to medium consistency. The batter is then allowed to ferment for eight to nine hours, which expands batter's volume as it becomes less dense and appears "fluffier." Appropriate quantity of this batter is then spread over heated flat pan with one to two table spoons of oil and is cooked only from one side.

Dosa is typically consumed during breakfast or as lunch and is served with spicy diced potatoes along with variety of coconut-based chutneys (sauces). The *dosa* is not a side cuisine rather it is a complete main course meal and is served in large plates. In recent years, due to increasing consumption trends and less time availability among working women, instant *dosa* mixes are also available, which can be readily mixed with water to make a batter, thereby offering convenient cooking. The instant *Dosa* mixes are pre-fermented flour blends of lentils and other pulses offering great convenience for busy households.

16.4.1.5 Masoor (Lentil) Pulav

Masoor (lentil) *pulav* is a rice-based dish in which boiled lentils are added. It is a mild spicy dish and is usually consumed in lunch or dinner as a main course meal along with salads. Besides rice and whole boiled lentils, potatoes, spices, and cumin seeds are other ingredients used. Only green chili spice in moderate quantity is used during its preparation in Pakistan, while in India red chili and other spices are also added. Whole boiled lentils are added after diced potatoes become tender, followed by addition of soaked rice. The rice along with lentils are then further cooked so that rice become soft. It is then simmered for 10 minutes and often garnished with fried onions and thinly cut green chilies. Masoor *pulav* is usually served with a yogurt and cumin sauce (*Raita*). This cuisine has a good balance of legume and rice and is therefore very popular among vegetarians. Though the traditional method of preparing masoor *pulav* is very simple, new cooking trends continue to provide some variations in the recipe. In Pakistan, women are now also adding boiled minced beef during its preparation to add more protein to this cuisine. The common serving size of *pulav* is 200 g, which contains with 280 cal (Lentils.org 2015).

16.4.1.6 Lentil Pakoras (Fritters)

Pakora, originating from the Indian subcontinent, is a spiced and fried fritter prepared with different variations of pulses and vegetables. The original type of fritter, prepared with chickpea flour and spices, is usually served as an appetizer in restaurants. However, it is more popular as a street snack food in India and Pakistan and is served with traditional spicy sauces. Traditionally, it is fried just before serving otherwise the *pakoras* lose their crispy texture and become soggy. Lentil fritters are mainly famous in Middle-East's gulf region and have found a place in restaurant menus. *Pakoras* are gluten free and completely vegetable based. Compared to chickpea flour fritters, lentils flours are much crispier, with on serving four fritters providing about ~220 kcal. Lentil and onion-based fritters (called Piaju) are very popular in Bangladesh (Mamun et al. 2013). Typically, these fritters are prepared by soaking split red lentils for two to four hours and, after draining water, ground into a thick consistency batter. Salt, crushed coriander, curry leaves, chopped onions, cumin, coriander, and mustard seeds are added. Adding finely chopped fresh green chilies and red pepper powder, which enhances spicy flavor, is optional. Thus, prepared batter is then spooned into hot oil for deep frying, which takes about five minutes. Lentil *pakoras* are commonly served with tamarind and green chili sauces for added flavor.

16.4.2 Middle-Eastern Cuisines

16.4.2.1 Mujaddara

Mujaddara is popular dish in most Arab countries (Jordan, Lebanon, Iraq, Palestine, UAE, Syria, and Saudi Arabia) and equally popular in some Mediterranean countries. It is typically prepared from lentils, rice, olive oil, and onions (Haddad et al. 2021). In Bahrain, it is also known as Muaddas (Al-Khusaibi et al. 2019). This cuisine offers a good balance between protein and carbohydrates as it is prepared from both pulses and cereals. Being completely a non-meat dish, it is also becoming popular among vegans beyond the Middle-Eastern region. A standard serving of *Mujaddara* (120–130 g) 250–312 kcal and is a rich source of fiber (7.2 g). *Mujaddara* provides about 20–25% of the daily requirement of fiber, which is 28 and 36 g for adult women and men, respectively (DeSalvo et al. 2016). The average proximate composition (g/100 g) for protein, fat, fiber, and carbohydrates is 6.8, 6.0, 7.2, and 35.6, respectively (Haddad et al. 2021).

Typically, *Mujaddara* is prepared by first cooking lentils in water till tender, followed by addition of salt and olive oil. Rice is separately cooked in sufficient water and once boiling starts, it is to simmer at medium heat for 10 minutes, followed by the addition of pre-cooked lentils. The contents are then simmered for additional 20 minutes on low to medium heat until all water is absorbed by rice. A gentle tossing is done to prevent mashing of lentils and rice. Before serving, the *Mujaddara* is garnished with fried, crispy thinly sliced onion (Al-Jebrin et al. 1985). Depending upon different regional customs, the *mujaddara* is served with creamy hummus, mint sauce, falafel, Baba Ganoush (cooked and mashed eggplant), and tahini sauce (smooth paste of sesame seeds). Traditionally, brown rice is used for preparing *mujaddara*, but other rice varieties may also be used.

16.4.2.2 Lentil Soups (Shorbat Adas)

Lentil soups/stews, though common to different parts of the world, varies in formulations and preparations In the Middle East, lentil soup is known as *Shorbat Adas* where *Shorbat* and *Adas* meaning soup and lentils, respectively (Ismael et al. 2014). Lentil soup may be served in breakfast or as side dish in lunch/dinner main course meal. It is considered a healthy food and is also popular as a main course meal since it gives the feeling of fullness upon eating. It may be prepared from different types of whole or dehulled lentils (green, brown, black, red, or yellow). However, *Shorbat Adas* prepared from dehulled lentils is thicker and smooth in consistency compared to that prepared with whole lentils. In Iraq, this soup is consumed with pita bread after adding freshly squeezed lime juice (Ismael et al. 2014). The Arab lentil soups are somewhat similar to South Asian lentil *dal*/curry. The simplest lentil soup in levant region is prepared from lentils, broth, cumin, and spices. The cumin seeds add flavor and also aids in digestion of soups as lentils frequently cause flatulence due to presence of oligosaccharides (Abumaizar 2020).

16.4.2.3 Matany Bread

Matany bread is made from a blend of barley and lentil flour, with a higher ratio of lentil flour. The dough is prepared with water, lentil and barley flours, and salt (Amri et al. 2005; Kang and Priyadarshan 2008). This bread is eaten in rural households of Yemen and is similar to *Maloonga* bread, which is prepared from barley and wheat flours. *Maloonga* is a fermented bread prepared from yeast and are baked in a taboon oven, a traditional method of baking on a bed of small hot stones in the oven (Potdar 2015).

16.4.3 African Cuisines

16.4.3.1 Koshari

Koshari, a traditionally popular Egyptian dish, is a lentil- and rice-based dish (similar to *Mujaddara*), prepared with lentils, rice, pasta, chickpea, fried onions, and tomatoes (Iriti and Varoni 2017). However, its method of preparation is slightly different from *mujaddara*. It is served as ready to eat street food (Mohamedin et al. 2015) and is also available in restaurants as a main course meal. Like *mujaddara*, it is also completely a vegan dish (Al-Khusaibi et al. 2019). *Koshari* preparation involves brown lentils, medium long rice kernels, and elbow pasta. The preparation of *koshari* is a bit time consuming as it involves separate cooking of pasta, rice, chickpea, and lentils. Besides these, tomato sauce and fried onions must also be prepared separately before all the ingredients are combined together. Typically, *koshari* is assembled in the serving platter by first making a base of rice and lentils, which is then topped with half of prepared tomato sauce (containing onion and vinegar), elbow pasta and chickpeas, and finally garnished with fried onion. The remaining tomato sauce and fried crisp onions are served separately with *Koshari* (Karadsheh 2019). The tomato sauce gives a rich tart flavor to this Egyptian cuisine. The single serving size of *koshari* is 160–250 g and provides 210–270 cal (Myfitnesspal 2021).

16.4.3.2 Hlalem

Hlalem is a traditional Tunisian pasta dish, which means "pinion-like." *Hlalem*, which is sun-dried pasta, is prepared by using durum semolina, water, and salt which is kneaded into a thick paste after which it covered with a damp cloth (Zaroual et al. 2019). After few hours, it is again kneaded and shaped into pinion-shaped pasta and sun-dried. The sun-dried pasta is shelf-stable and can be stored for several weeks in an air-tight containers or sealed polyethylene bags. This pasta is then cooked with equal quantity of lentils to make *Hlalem*. This cuisine is considered to be high in protein and fiber content, thus offering a nutrient-rich food option for health-conscious people (Zaroual et al. 2019). This cuisine is prepared in the form of soup by first sautéing onions, carrots, and garlic in oil, followed by addition of crushed tomatoes. These ingredients are cooked until vegetables become mushy in appearance and then pasta and canned lentils are added. Boiled lentils are also used in the place of canned lentils if the latter are not easily available. Since raw pasta is added to this soup-like contents; therefore, it is given enough time to cook so that pasta becomes tender. One serving of *Hlalem* can easily meet approximately 36% of daily fiber needs (Kauffman 2018).

16.4.3.3 Bsissa

Bsissa, a North-African cuisine particularly popular in Tunisia and Libya, is prepared from lentils, wheat, barley, and chickpeas (Amri et al. 2005). The other names of *Bsissa* are Zammita in Libya, Rowina and Tamina in Algeria and Zrira and Sellou in Morocco (KAWA 2021). This dish is served in Tunisian restaurants at breakfast and lunch. A variation of *Bsissa* in Mediterranean region is traditionally prepared from only roasted barley and wheat (Zaroual et al. 2019). However, in the Djerba Island in south of Tunisia, lentil is the primary ingredient in *Bsissa*. *Bsissa* is a nutrient-rich dish and is usually eaten in the holy month of Ramadan both at *Suhor* (pre-dawn meal) and *Iftar* (meal after sunset).

The preparation of *Bsissa* mix involves three steps, i.e., roasting ingredients, followed by grinding and sifting. All seeds including barley, lentil, wheat, chickpea, coriander, fennel, sesame, and cumin are roasted and then ground to a fine powder, and the resulting powder is sifted to ensure uniform particle size by excluding any coarse particles (Ghazouani et al. 2021; Smati 2006). The roasting time is usually 30 minutes at 230 °C for both lentils and cereals but less than three minutes for spices. *Bsissa* may be eaten as a savory or as a dessert in North of Africa while it may be added to milk or water to make a drink, called Rowina. Dates or figs are also added along with water to make *Bsissa* an energy-rich and refreshing drink. This beverage is served warm in winters or chilled with ice in summers. Another way of eating *Bsissa* is to make it into a thick paste with olive oil, honey or maple syrup and sugar, and topping it off with nuts in order to make it a semi-solid dessert. It is a custom in Tunisia that mothers, after giving birth to a child, are given *Bsissa* to regain their vitality and also improves their milk production during lactation period. Gluten-free *Bsissa* is also marketed for coeliacs. *Bsissa*, on a per 100-g basis, provides 283 kcal and is a rich source of fiber (14.7 g) and potassium (693 mg) (Al-Khusaibi et al. 2019)

16.4.3.4 Ethiopian Wat (Lentil Stew)

This is a popular dish in Ethiopia and is basically a stew prepared from lentils and is served with *Injera* (a fermented flat bread) and salads. In addition, these stews also include vegetables and chickpea flour. Ethiopian spice blend called *Berbere* (roasted spice blend containing cloves, fenugreek, coriander, cumin, green cardamom, black pepper corns, dried red chili peppers, and whole all spice berries) is also used in the preparation of this wat (Killebrew 2013). This lentil stew is also known as *Misir Wat* locally (Alyssa 2021; Mackean 2019). The preparation time of this cuisine is relatively shorter, and it is easy to prepare. One serving provides around 284 kcal when traditionally prepared with butter (Alyssa 2021).

16.4.4 European/Mediterranean Cuisines

16.4.4.1 Lentil Raghout

This is a classical lentil-based French cuisine that is mostly served as a side dish. Lentil *Raghout* is prepared from French green lentils and is also known as *Le Puy Lentil* since the origin of green lentils is a southern town in France, Le-Puy-en-Velay. These lentils are known to have a nutty flavor and is therefore commonly used in making side dishes and salads (Jamieson 2011). Aromatic herbs and vegetables are first sautéed in oil, and then lentils are cooked in vegetable stock instead of water until tender and saucy. Lentils absorb flavors from vegetable stock during cooking, giving it a very rich taste. For making this special side dish, canned lentils are not preferred as the aroma and flavor of *Raghout* primarily depends on the absorption of volatiles by lentils from vegetable broth. Besides lentils, the major ingredients of this cuisine include garlic, onion, carrots, thyme, parsley, and bay leaves. In order to thicken the sauce or gravy, tomatoes are used, which also add sour or tart flavor to *Raghout*. A single serving of this dish provides about 415 kcal, out of which 52% are derived from proteins. At meals, *Raghout* could be paired with roasted chicken, duck confit, bistros, or braisseries (Nagi 2021).

16.4.4.2 Lentil Balls (Mercimek Köftesi)

Mercimek Köftesi is a Turkish cuisine, which is equally popular in whole of Mediterranean region. Adana province of Turkey is specifically famous for this dish's popularity. Lentil balls are prepared from split red lentils along with tomato paste, fine bulgur, parsley, onion, and other spices. The first preparation step involves cooking of lentils in water to a mushy texture, followed by addition of bulgur and additional cooking to make bulgur soft and palatable. The onions, tomatoes, spices, and herbs are for flavor. This thick consistency batter is then given a cylindrical shape and served with lettuce and traditional herbs (Güzeler et al. 2016). *Mercimek Köftesi* may also be served as a wrap with lettuce and tahini sauce, but it is then called Lentil *Falafel*.

16.4.4.3 Fakes Soupa

Soups occupy an important place in Mediterranean cooking; however, preparation methods may vary in from region to regions. In Greece, lentil (fakes) soup known as *Fakes Soupa* is very popular in winters. Brown lentils are boiled for making Fakes Soupa, which are then simmered on a pan along with water, bay leaves, chopped onions, and garlic for almost half an hour followed by addition of vinegar and olive oil. Typically, a small amount of tomato paste is added at the end for flavor and to give a reddish color. This *Fakes Soupa* is also topped with shredded feta cheese. Overall, it is a nutritious soup, which has been liked by generations of the Greeks. One bowl of *Fakes Soupa* provides around 348 kcal (Giannopoulos 2022).

16.4.4.4 Cotechino con Lenticchie

The Greek and Roman populations have considered lentils as a sign of fortune since ancient times. Since lentils are coin shaped and increase in its volume after cooking, Italians consider it as a sign of "abundance," and on the midnight of new year, their dinner includes lentil-based cuisines called *Cotechino con Lenticchie*. As per Roman tradition, the more you eat lentils at the beginning of new year, the luckier you will be. Lentils were even gifted on new year as a part of their old custom and tradition (Speak 2020). *Lenticchie* means "lentils," whereas *cotechino* is a pork sausage and as the name suggests, the pork sausage is served along with lentils. The lentils are however cooked along with diced vegetables, e.g., celery, carrots, and onions in olive oil along with rosemary and bay leaves. These two herbs are separated after lentils become tender. This lentil curry is then served with cooked and diced *Cotechino* sausages (Zafferano 2020).

16.4.4.5 Fakorizo

Fakorizo, meaning "Lentil rice" in Greek, is a protein-rich vegan dish, which is somewhat similar to *mujaddara*. *Fakorizo* can be prepared with both brown or green or French lentils as these retain their shape even after cooking and do not become mushy as is the case with the red lentils. That is why red lentils are avoided in *Fakorizo*. This cuisine is prepared by taking equal parts of uncooked rice and lentils prior to soaking/cooking. The preparation of *Fakorizo* involves frying of onions to golden brown followed by the addition of garlic, tomatoes, salt, and pepper. Drained lentils, previously boiled with bay leaves and soaked rice along with water are then added and the contents are simmered for 15 minutes. The *Fakorizo* is topped with freshly cut parsley and is served with feta cheese and freshly cut lemon. In case of vegan *Fakorizo*, cheese is excluded from the recipe (Khan 2020).

16.4.4.6 Salads

Salads are an integral part of Mediterranean diet and contributes to their good health and longevity compared to other parts of the world. Lentils are also added in different Mediterranean salads; however, there is no strict recipe to follow (Trichopoulou and Vasilopoulou 2000). Lentils are added in salads to give a longer feeling of satiety and thus are also considered as a filler salad. The famous Greek lentil salad combines black beluga or French green lentils with spinach, olives, sundried tomatoes, and parsley and is tossed with Greek salad dressing. Similarly, other Mediterranean lentil salads may contain lettuce with other vegetables, e.g., carrots, onions, cucumbers, tomatoes along with spices (black pepper or paprika) and herbs (bay leaves, oregano, and basil). The dressings used are Dijon mustard sauce, French dressing, hummus, tahini sauce, and mayonnaise (Hill 2021; Izzy 2018).

16.4.5 Caribbean Cuisines

16.4.5.1 Trinidad Stewed Lentils

Stewed lentils, which may be offered as a main course meal in lunch or dinner, is prepared from boiled and drained lentils. The unique feature is that the cooked lentils are added to sugar caramelized in heated oil along with ketchup, garlic, tomatoes, scallions, carrots, pumpkin, thyme, and black pepper. The ingredients are brought to a boil and then simmered for about 30 minutes over low heat. One serving of Trinidad stewed lentils provides around 187 kcal (Ria 2021).

16.4.5.2 Lentil Curries

Lentil curries are also very popular in Caribbean region and are prepared in different styles. Typically, whole brown lentils are used to make these curries and are served with rice or salad as a main course meal. Occasionally, these lentil curries are also prepared in coconut milk to give a creamy texture. The lentil curries are prepared from brown lentils in Caribbean along with different vegetables like potatoes, tomatoes, onions, celery, and carrots and are seasoned with salt, curry powder, cumin, and garlic (Blackwood 2019; Lee 2013).

16.4.6 American and Canadian Cuisines

16.4.6.1 Cajun Lentil Stew

Cajun, an ethnic group of primarily Irish and Spanish ancestry, represents a significant portion of population in the U.S. state of Louisiana (Mulcahy et al. 2019). *Cajun* lentil stew, prepared mainly by *Cajuns*, is also popular among other Americans. The Cajun lentil stew is served with salad or bread and is usually prepared to welcome winters. This stew contains celery, onion, and green bell peppers in equal parts. *Cajun* seasoning is used to add flavor to this cuisine which is easily available in the US markets. *Cajun* lentil stew is prepared with either green or brown lentils. Sweet potatoes are added to this stew, which makes it different from stews popular in other regions of the world. The fire-roasted tomatoes are also added, providing a unique character to *Cajun* lentil stew. Similar to other stews preparation, first vegetables are sautéed in heated oil followed by addition of garlic. Lentils, sweet potatoes, tomatoes, and vegetable broth are then added along with *Cajun* seasoning, salt and pepper and the contents are cooked until sweet potatoes and lentils are fully cooked. *Cajun* lentil stew, packed with good balance of macro and micronutrients, can also cater to vegans and consumers opting out for gluten-free diets (Esposti 2021).

16.4.6.2 Lentil Burger Patties/Nuggets

Though less common, lentil-based burger patties are marketed in frozen form by some Canadian companies. Lentil burger patties are specifically prepared as a meat substitute of beef patties for vegetarians and vegans. The Canadian based *Yves Veggie Cuisine*® company introduced lentil burger patties formulated with quinoa, bulgur, brown rice, wheat gluten, seasonings, and hydrocolloids. This burger patty is claimed to be trans-fat and cholesterol free, thus promoting healthier lifestyle. Compared to calorie-loaded beef patties, lentil-based burger patty provides only 110 kcal (Yves Veggie Cuisine 2022). Canada's Food Guide in collaboration with Health Canada also suggested lentil-based nuggets prepared with rolled oats addition. The lentil nuggets are prepared by blending lentils, rolled oats, cumin, lemon juice, dried basil, and garlic powder in a blender followed by shaping in the form of disc-shaped nuggets. These lentil nuggets are then shallow fried in olive oil on both sides. The brown-colored fried nuggets can then be served with any dressing of choice (Health Canada 2021).

16.4.6.3 Lentil and Rice Pilaf

The American version of Mujadara or Fakoriza is called Lentil and rice pilaf. However, this US cuisine contains higher amount of whole brown lentils than rice. Similar to *mujaddara* it is also garnished with golden-brown fried onions. The separately cooked and drained lentils are added to uncooked but soaked rice with sufficient quantity of water. This mixture of rice and lentils is brought to boil and then simmered till the rice completely absorbs water and becomes tender. Salt and pepper are added as seasonings and the pilaf is topped with previously caramelized onions (Stewart 2004). Compared to *mujaddara*, one serving of this pilaf provides lesser calories, around 160 kcal (USDA 2021).

16.4.7 Persian Cuisines

16.4.7.1 Adas Polo/Adas Polow

Adas Polo or *Polow* is a traditional rice and lentil-based cuisine popular in Persia (Iran). It is somewhat similar to *mujaddara* but is usually served with lamb or chicken meat (Shirazi 2005). *Adas Polow* may also be cooked together with ground lamb meat. Brown whole lentils are used to make this dish since whole lentils remain intact by retaining their shape and, like split lentils, do not turn mushy upon cooking. Lentils are boiled separately while rice is half-cooked in boiling water such that the texture is hard in center but soft on outside. The main feature of this cuisine is its assembling of rice and lentil in layers, which is similar to *Biryani*, a popular South Asian cuisine.

The process of preparing *Adas Polow* begins with chopped onions, which are sautéed in oil followed by addition of turmeric powder, raisin, dates, and a piece of cinnamon to give a strong aroma. Then, the half-cooked rice is first layered on the bottom of pan followed by two to three alternate layers of boiled lentils and rice, and saffron water is poured over the top rice layer. After covering the pan with a lid, it is cooked on medium–high heat until the rice produces steam. At that stage, heat is lowered to medium–low level, which eventually leads to complete cooking of rice. *Adas Polow* is garnished with raisins, dates, and fried onions (Mehran 2009). This dish is served with cooked ground meat, a cucumber-yogurt dip (*Masto-Khiar*), and *Shirazi* salad, a popular Iranian salad prepared from diced/chopped tomatoes, cucumbers, onions, and mint tossed with olive oil and lemon juice (Mehran 2008a,b).

16.4.7.2 Iranian Adasi

Adasi is a traditional lentil-based Iranian stew/soup prepared from green lentils but can be prepared with other lentils too and is generally served at breakfast (Soltani 2021). Unlike Arab Shorbat-Adas, it does not have a mushy appearance rather than lentils retain their shape after cooking and no blending is done. It is cooked on a very low heat for up to eight hours. Iranian *Adasi*, except long cooking step, is otherwise easy to prepare and requires few ingredients (onions, cumin seeds, pepper, turmeric, salt, and oil). These ingredients are sautéed in oil for about five minutes and then washed/rinsed lentils are added, followed by slow cooking for eight hours. Water is added few times during cooking to prevent drying of lentils. This dish is served with lemon, fried onions, herbs, and eggs as a complete breakfast (Walker 2020).

16.4.8 Chinese and Japanese Cuisines

Lentils are not generally very popular in China and Japan and are usually available at ethnic grocery stores in these countries. However, with growing awareness about lentils and their health benefits, some traditional cuisines have been modified to add lentils. For example, Japanese *Miso* soup is modified to make lentil miso soup through addition of red lentils. Japanese *Mochi* prepared from adzuki beans can also be prepared with lentils. Similarly, some Chinese salads and noodles are now topped with lentils. Lentils are also used in “Pork lentil stuffed buns” – a famous cuisine in Beijing. This cuisine separately makes dough by adding sugar and baking powder and is left for wild fermentation until its volume becomes double. The stuffing of these buns is prepared using boiled lentils, ground pork meat (pre-marinated with oil, monosodium glutamate, wine, white pepper, sugar, salt, and soy sauce) and diced shallots. The fermented dough is then rolled into a thin circular discs and stuffing is added between two dough discs, which are then is molded into the shape of a bun. The buns are then steamed for about 15 minutes (Leo 2020). Other Chinese cuisines prepared with lentils include lentil braised noodles, lentils with tahini sauce, *Kung Pao* lentils (prepared from *Kung Pao* sauce), and braised pork with dried lentils.

16.5 Potential for Enhancing Market Opportunities for Lentils

Consumers are increasingly seeking foods that meet a broad agenda of personal interests that transcend many of the traditional criteria for selection. These choices include aspects of environmental impact, social, or lifestyle needs. Lentils provide excellent opportunities in today's changing marketplace. Lentils are particularly well suited to meet many of these current trends for improved food crop sustainability and plant-based dietary patterns.

16.5.1 Social/Environmental/Ethical Considerations

Lentils are gaining significant visibility in lifestyle diets well beyond the traditional lentil consuming countries. The major traditional reasons for lentil selection have been based on high protein, low cholesterol, and high fiber contents. More recent trends for selection have included the relatively shorter cooking time and improved taste profile. According to a study by Warne et al. (2019), the common driving factors among North American people regarding lentil consumption were related to its plant origin, nutritional profile, and its ability to give satiety. From agronomic point of view, the lentil crop also offers sustainability in terms of its ability to grow in arid environments and requires low input costs. Further, versatile adaptability to integrate into crop rotation programs where it reduces the need for nitrogen fertilizer for the subsequent crop growth through symbiotic nitrogen fixation (Peoples et al. 2015; Warne et al. 2019). These agronomic factors associated with food crop sustainability will be of increased importance in the marketplace. Lentils specifically align with consumer diet trends. The health claims of lentils are attracting consumers, while less hectic preparation and shorter cooking times are allowing chefs to easily offer lentils in additional cuisines.

Meat consumption continues to be a significant part of human diets worldwide particularly among high- and middle-income consumers. However, recently an increase in ethical food choices have challenged reliance on animal-derived products. The high level of meat consumption, besides having negative impact on health, is also bringing adverse impact on the environment in terms of energy emissions and higher water consumption. In contrast to animal-derived foods, plant-based foods are acceptable to all segments of people beyond ethical and religion boundaries and also have low energy emissions, thus making plant-based foods a more sustainable option for the future (Godfray et al. 2018).

16.5.2 Food Service and Restaurants

Chefs around the world are exploring uses of lentils in salads, main course meals, sauces, side dishes, and desserts. Thus, a new era of opportunities has opened up, and many restaurants, particularly in the West, have added innovative items to their menus, whereas traditional lentil cuisines like koshari, *mujaddara*, pulav, dal/soup, and Shorbat Adas continue to be served in the Middle Eastern and South Asian restaurants. Lentils can also cater to the plant food/vegan market because of their versatility and are featured in different restaurants offering vegan food.

Sweetgreen, a casual fast-food restaurant in the United States, with over 150 locations in 13 states, introduced a *Super Green Goddess Salad* bowl with black lentils as one of the ingredients. This salad is served with ranch dressing and is designated as a low carbon impact salad by sweetgreen (Sweetgreen 2021). CAVA, a Mediterranean casual restaurant chain offers customized meal option for its customers. Customers first choose a base (e.g., bowl, salad, pita, RightRice®) followed by selection of dip, spread, dressing, protein option, and topping. Lentils can be found in their base option under the category of RightRice, which is claimed to be high fiber and protein and is prepared from lentils, chickpea, and peas. Similarly, CAVA also offers black lentils in another base option, i.e., grain bowl. Besides the lentil base, one of the toppings includes lentil tabbouleh, which is basically a salad prepared from green lentils, tomatoes, mint, fresh herbs, and cucumbers (CAVA 2021). The B.GOOD restaurant chain in the United States also introduced a Mediterranean bowl in which lentils are being introduced (B.GOOD 2021). Starbucks, a company with global presence, also launched a certified vegan diet containing black lentils by the name of "Lentils and Vegetable Protein Bowl with Brown Rice" served with lemon tahini sauce. The Starbucks claims that one serving of this bowl provides 23g of protein (Starbucks 2021). These culinary developments demonstrate that, besides traditional cuisines, healthy cuisines containing lentils are being developed and served in different food chains worldwide.

16.5.3 Lentil-Based Flexitarian Diets

Many consumers are reluctant to completely switch from an animal-based diet to a complete vegetable-based diet; therefore, the concept of flexitarian or semi-vegetable diet has gained some traction over complete vegetable-based diets. Those who occasionally add meat and fish in their diet are called "flexitarians" and basically, they try to add more plant foods and reduce but not eliminate foods of animal origin in their diet (Derbyshire 2017). Lentils provide an opportunity to be added to any food plan and as presented in Table 16.2 provide a comparable or better nutritional profile than meat or the widely consumed staple, wheat.

Lentils can play an important role in the development of highly favored flexitarian diets. Chaudhary and Tremorin (2020) experimented with beef patties in which 26% of beef was

Table 16.2 Comparison of macronutrients, water content, and calories derived from raw lentils and cooked lentils, wheat, and beef (per 100 g).

	Raw lentils	Cooked lentils	Boiled/braised beef	Cooked whole wheat
Carbohydrates (g)	63.4	20.1	Nr[a]	10.4
Proteins (g)	24.6	9.02	30.8	1.55
Fats (g)	1.06	0.38	13.2	0.28
Dietary fiber (g)	10.7	7.9	Nr	1.3
Water (g)	8.26	69.6	54.4	87.3
Calories (kcal)	352	116	251	47

[a] Not reported.
Source: USDA (2019a,b, 2020).

replaced with cooked lentils, which led to 20% increase in nutrient density while increasing the fiber content by 60%. A burger patty prepared only with lentils may not be that flavorful; therefore, it is more feasible and sustainable to substitute a portion of meat in burger patties with lentils. Thus, vast opportunities exist where lentils and animal meat can be blended to prepare innovative dishes with improved consumer acceptability and a reduced environmental footprint. Different kebabs, burrito bowls, taco fillings, and salads can be prepared with acceptable sensory quality, by partially replacing meat with lentils.

16.6 Conclusion

Lentil consumption is increasing worldwide, and globally new culinary trends are quickly penetrating "fast moving consumer good" (FMCG) markets and food service sectors. Lentils not only have a healthier profile but also have a reduced "beany flavor" among legumes and thus exhibit higher consumer preferences. Culinary professionals are continuously innovating through different cuisines, particularly bowls and blends (mix of lentils with meat). Lentils store well and are an economical, nutrient-dense food choice. Being versatile, lentils easily fit into breakfast, lunch, and dinner menus. Further, lentils, being an environmentally sustainable legume, can play a crucial role in orienting plant-based dietary options. The potential for lentils to meet many aspects of changing criteria for food selection by the consumer is most apparent. As consumer awareness of the versatility lentils to meet specific social needs, worldwide consumption will increase. Thus, lentils are well positioned for potential consumption increases due to shifts in consumer preferences within the emerging marketplace and health consciousness.

References

Abumaizar, F. (2020). Middle eastern lentil soup – 30 min, healthy and yum- every little crumb. https://everylittlecrumb.com/middle-eastern-lentil-soup (accessed 23 March 2022).

Al-Jebrin, A., Sawaya, W.N., Salji, J.P. et al. (1985). Chemical and nutritional quality of some Saudi Arabian dishes based on cereals and legumes 1. Proximate composition, amino acid contents and nutritive value. *Ecology of Food and Nutrition* 17: 157–164.

Al-Khusaibi, M., Al-Habsi, N., and Rahman, M.S. (2019). *Traditional Foods: History, Preparation, Processing and Safety*. Cham, Switzerland: Springer.

Alyssa (2021). Ethiopian lentils (misir wat). https://www.simplyquinoa.com/ethiopian-lentils (accessed 12 May 2022).

Amri, A., Ouammou, L., and Nassif, F. (2005). Barley-based food in Southern Morocco. In: *Food Barley: Importance, Uses and Local Knowledge* (ed. S. Grando and H.G. Macpherson), 53–82. Aleppo, Syria: ICARDA.

Arthi (2010). Masoor dal sambar – lentil soup. http://www.spicytasty.com/veggie-entrees-sides/masoor-dal-sambar/ (accessed 23 March 2022).

B.GOOD (2021). Lentils.org partners with B.GOOD on new mediterranean bowl. https://www.lentils.org/lentils-org-partners-with-b-good-on-new-mediterranean-bowl/#_ftn2 (accessed 23 March 2022).

Bhattacharya, S., Narasimha, H.V., and Bhattacharya, S. (2005). The moisture dependent physical and mechanical properties of whole lentil pulse and split cotyledon. *International Journal of Food Science & Technology* 40: 213–221.

Blackwood, M. (2019). Coconut lentil curry (Jamaican style) – healthier steps. https://healthiersteps.com/recipe/coconut-lentil-curry (accessed 15 April 2022).

CAVA (2021). Build it. https://cava.com/menu (accessed 23 March 2022).

Chaudhary, A. and Tremorin, D. (2020). Nutritional and environmental sustainability of lentil reformulated beef burger. *Sustainability (Switzerland)* 12 (17): 1–18.

Chowdhury, M.G.F., Miaruddin, M., Rahman, M.M. et al. (2009). Study on the effect of preservative on the storage quality of spiced papads. *Journal of Innovative Development Strategy* 3: 30–33.

Datar, S. (2014). Masoorachi amti or red lentil curry – CKP recipe. https://snehasrecipe.blogspot.com/2014/09/masoorachi-amti-or-red-lentil-curry-ckp.html (accessed 13 May 2022).

Derbyshire, E.J. (2017). Flexitarian diets and health: a review of the evidence-based literature. *Frontiers in Nutrition* 3 (January): 1–8.

DeSalvo, K.B., Olson, R., and Casavale, K.O. (2016). Dietary guidelines for Americans. *JAMA* 315: 457–458.

Doshi, M. (2002). *Cooking Along the Ganges: The Vegetarian Heritage of India*. iUniverse.

Duke, J. (2012). *Handbook of Legumes of World Economic Importance*. Springer Science & Business Media.

Erskine, W. (2009). *The Lentil: Botany, Production and Uses*. Wallingford, UK: CABI.

Esposti, D. (2021). Cajun lentil stew (Vegan) – our salty kitchen. https://oursaltykitchen.com/cajun-lentil-stew (accessed 23 March 2022).

FAO (2019). Calories consumption data. https://www.fao.org/faostat/en/#data/SCL (accessed 15 May 2022).

FAO (2020). Crop production data. https://www.fao.org/faostat/en/#data/QCL (accessed 15 May 2022).

Garg, R. and Dahiya, S. (2003). Nutritional evaluation and shelf life studies of ppads prepared from wheat-legume composite flours. *Plant Foods for Human Nutrition* 58: 299–307.

Ghazouani, T., Atzei, A., Talbi, W. et al. (2021). Occurrence of acrylamide, hydroxymethylfurfural and furaldehyde as process contaminants in traditional breakfast cereals: Bsissa. *Food Control* 124: 107931.

Giannopoulos, E.K. (2022). Greek lentil soup recipe (fakes soupa). https://www.mygreekdish.com/recipe/lentil-soup-fakes-soupa (accessed 23 March 2022).

Godfray, H.C.J., Aveyard, P., Garnett, T. et al. (2018). Meat consumption, health, and the environment. *Science* 361 (6399): eaam 5324.

Güzeler, N., Yıldırım, Ç., and Arıdıcı, A. (2016). The culinary culture and traditional foods of Adana province. *Journal of Agricultural Faculty of Uludag University* 30: 538–545.

Haddad, M.A., Yamani, M.I., Abu-Romman, S.M., and Obeidat, M. (2021). *Chemical Profiles of Selected Jordanian Foods*. Cham, Switzerland: Springer.

Harlen, W.J. and Jati, I.R.A.P. (2018). Antioxidant activity of anthocyanins in common legume grains. In: *Polyphenols: Mechanisms of Action in Human Health and Disease* (ed. R.R. Watson, V.R. Preedy, and S. Zibadi), 81–92. London, UK: Elsevier.

Health Canada (2021). Lentil nuggets – Canada's food guide. https://food-guide.canada.ca/en/recipes/lentil-nuggets (accessed 15 April 2022).

Helbaek, H. (1959). Domestication of food plants in the old world. *Science* 130 (3372): 365–372.

Hill, M. (2021). Mediterranean lentil salad. https://www.culinaryhill.com/mediterranean-lentil-salad (accessed 23 March 2022).

Iriti, M. and Varoni, E.M. (2017). Pulses, healthy, and sustainable food sources for feeding the planet. *International Journal of Molecular Sciences* 18 (2): 255.

Ismael, D.S., Vollmannová, A., Timoracká, M., and Harangozo, Ľ. (2014). Legumes utilised in traditional foods in Iraq. *Journal of Microbiology, Biotechnology and Food Sciences* 3: 218–220.

Izzy (2018). Mediterranean lentil salad. https://www.shelikesfood.com/mediterranean-lentil-salad (accessed 11 May 2022).

Jamieson, P. (2011). Learning to love lentils. *Tufts University Health & Nutrition Letter* 29 (5): 6.

Joshi, M., Timilsena, Y., and Adhikari, B. (2017). Global production, processing and utilization of lentil: a review. *Journal of Integrative Agriculture* 16: 2898–2913.

Kang, M. and Priyadarshan, P.M. (2008). *Breeding Major Food Staples*. Oxford, UK: WileySons.

Karadsheh, S. (2019). Egyptian Koshari recipe. https://www.themediterraneandish.com/egyptian-koshari-recipe (accessed 9 June 2022).

Kauffman, T. (2018). Tunisian pasta with lentils (Hlalem) recipe. https://www.allrecipes.com/recipe/276242/tunisian-pasta-with-lentils-hlalem (accessed 11 May 2022).

KAWA (2021). Discover bsissa, this typical Tunisian dish. https://kawa-news.com/en/discover-bsissa-this-typical-tunisian-dish (accessed 23 March 2022).

Khan, D. (2020). Lentils & Rice Pilaf: Fakorizo. https://www.dimitrasdishes.com/lentils-rice-pilaf-fakorizo (accessed 13 May 2022).

Killebrew, K. (2013). Best authentic Berbere (Ethiopian spice blend). https://www.daringgourmet.com/berbere-ethiopian-spice-blend (accessed 23 March 2022).

Lee, K. (2013). Jamaican curry lentil recipe. https://jamaicans.com/currylentil (accessed 28 April 2022).

Lentils.org (2015). Big book of little lentils. https://www.lentils.org/resource_library/big-book-of-little-lentils (accessed 23 March 2022).

Leo (2020). Pork lentil stuffed buns. https://misschinesefood.com/pork-lentil-stuffed-buns (accessed 12 May 2022).

Mackean, C. (2019). 5 must eat dishes in Harar, Ethiopia. https://travelforfoodhub.com/5-must-eat-dishes-in-harar-ethiopia (accessed 23 March 2022).

Mamun, M.A., Rahman, S.M.M., and Turin, T.C. (2013). Knowledge and awareness of children's food safety among school-based street food vendors in Dhaka, Bangladesh. *Foodborne Pathogens and Disease* 10: 323–330.

Mehran, A. (2008a). Turmeric and saffron: Salad Shirazi. https://turmericsaffron.blogspot.com/2008/12/salad-shirazi.html (accessed 17 April 2022).

Mehran, A. (2008b). Turmeric and saffron: Mast-o-Khiar - persian yogurt and cucumber dip. https://turmericsaffron.blogspot.com/2008/12/mast-o-khiar-persian-yogurt-and.html (accessed 23 March 2022).

Mehran, A. (2009). Turmeric and saffron: Adas Polow – rice with lentils. https://turmericsaffron.blogspot.com/2009/08/adas-polow-rice-with-lentils.html (accessed 10 May 2022).

Mohamedin, A., Michel, M., and Tolba, M. (2015). Assessment of the microbiological quality of public restaurants and street vended ready-to-eat Koshari meals. *International Journal of Applied Sciences and Biotechnology* 3: 452–458.

Mulcahy, K.V., Naquin, T.C., and Harper, J.N. (2019). Cajun Louisiana: culture as a way of life. *The Journal of Arts Management, Law, and Society* 49: 14–29.

Myfitnesspal (2021). Egyptian Koshari recipe. https://www.themediterraneandish.com/egyptian-koshari-recipe (accessed 23 March 2022).

Nagi (2021). Lentil Ragout – French Puy lentil side dish. https://www.recipetineats.com/lentil-ragout-french-puy-lentils-side-dish (accessed 10 April 2022).

Parikh, B. and Patel, V.H. (2018). Total phenolic content and total antioxidant capacity of common Indian pulses and split pulses. *Journal of Food Science and Technology* 55: 1499–1507.

Peoples, M., Swan, T., Goward, L. et al. (2015). *Legume Effects on Soil N Dynamics-Comparisons of Crop Response to Legume and Fertiliser N.* Canberra, Australia: Grains Research and Development Corporation, Government of Australia.

Podder, R.M., DellaValle, D.T., Tyler, R.T. et al. (2018a). Relative bioavailability of iron in Bangladeshi traditional meals prepared with iron-fortified lentil dal. *Nutrients* 10 (3): 354.

Podder, R.M., Khan, S.M., Tar'an, B. et al. (2018b). Sensory acceptability of iron-fortified red lentil (*Lens culinaris* Medik.) dal. *Journal of Food Science* 83: 804–813.

Polak, R., Phillips, E.M., and Campbell, A. (2015). Legumes: health benefits and culinary approaches to increase intake. *Clinical Diabetes* 33: 198–205.

Potdar, A. (2015). Maloonga ~ a flat bread from Yemen. https://themadscientistskitchen.com/maloonga-flat-bread-from-yemen (accessed 23 March 2022).

Prasad, N.N., Khanum, F., Siddalingaswamy, M., and Santhanam, K.N. (1995). Proximate composition and dietary fibre content of various foods/rations processed to suit the Indian palate. *Food Chemistry* 52: 373–378.

Prodanov, M., Sierra, I., and Vidal-Valverde, C. (2004). Influence of soaking and cooking on the thiamin, riboflavin and niacin contents of legumes. *Food Chemistry* 84: 271–277.

Queiroz, K.D.S., de Oliveira, A.C., Helbig, E. et al. (2002). Soaking the common bean in a domestic preparation reduced the contents of raffinose-type oligosaccharides but did not interfere with nutritive value. *Journal of Nutritional Science and Vitaminology* 48: 283–289.

Renu, R. and Waghray, K. (2016). Development of papads: a traditional savoury with purslane, *Portulaca oleracea*, leaves. *Health Scope* 5 (1): 1–5.

Ria (2021). Trinidad stewed lentils. https://cookingwithria.com/2021/02/trinidad-stewed-lentils.html (accessed 15 March 2022).

Roopa, S.S., Dwivedi, H., and Rana, G.K. (2017). Development and physical, nutritional and sensory evaluation of instant mix (dosa). *Technofame-A Journal of Multidisciplinary Advance Research* 6: 109–113.

Sandhu, J.S. and Singh, S. (2007). History and origin. In: *Lentil – An Ancient Crop for Modern Times* (ed. S.S. Yadav, D.L. McNeil, and P.C. Stevenson), 1–9. Cham, Switzerland: Springer.

Saraswat, S.S. and Jain, P. (2014). Income effect on consumers' buying preferences for food supplements (Papad). *SVIM E-Journal of Applied Management* 2 (2): 54–58.

Sashikala, V.B. and Narasimha, H.V. (2010). An objective method and a device for determining the end point of cooking of dhal and rice. *International Journal of Food Properties* 13: 877–889.

Shah, M.S., Ahmad, A., Khalique, N. et al. (2012). Home-based management of acute diarrhoeal disease in an urban slum of Aligarh, India. *The Journal of Infection in Developing Countries* 6: 137–142.

Shaikh, M.B., Tarade, K.M., Bharadwaj, V.R. et al. (2009). Effect of an alkaline salt (papad khar) and its substitute (2:1 sodium carbonate: sodium bicarbonate) on acrylamide formation in papads. *Food Chemistry* 113: 1165–1168.

Shirazi, F. (2005). The sofreh: comfort and community among women in Iran. *Iranian Studies* 38: 293–309.

Singh, K.M. and Singh, A.K. (2014). Lentil in India: an overview. *SSRN Electronic Journal* 2014: 2510906.

Smati, N. (2006). Patrimoine culinaire en Tunisie. Le festival de la Bsissa dans le Sahel tunisien, entre valorisation du patrimoine et risque de folklorisation. *Horizons Maghrébins-Le droit à la mémoire* 55: 114–120.

Soltani, N. (2021). Dietary practices of Iranian migrant women in the United Kingdom. *Appetite* 162: 105144.

Sonnante, G., Hammer, K., and Pignone, D. (2009). From the cradle of agriculture a handful of lentils: history of domestication. *Rendiconti Lincei* 20: 21–37.

Sozer, N., Holopaien-Mantila, U., and Poutanen, K. (ed.) (2016). *Traditional and New Food Uses of Pulses*, 1–33. Espoo, Finland: VTT Technical Research Centre of Finland Ltd.

Speak, C. (2020). Here's why Italians eat lentils on New Year's Eve. https://www.thelocal.it/20201230/why-do-italians-eat-lentils-on-new-years-eve (accessed 12 May 2022).

Starbucks (2021). Lentils & vegetable protein bowl with brown rice. https://www.starbucks.com/menu/product/2122324/single?parent=%2Ffood%2Flunch%2Fprotein-bowls (accessed 18 April 2022).

Stewart, M. (2004). Lentil-and-rice pilaf. https://www.marthastewart.com/317896/lentil-and-rice-pilaf (accessed 23 March 2022).

Sweetgreen (2021). Sweetgreen – menu. https://www.sweetgreen.com/menu (accessed 27 May 2022).

Thavarajah, D., Thavarajah, P., Wejesuriya, A. et al. (2011). The potential of lentil (*Lens culinaris* L.) as a whole food for increased selenium, iron, and zinc intake: preliminary results from a 3 year study. *Euphytica* 180: 123–128.

Trichopoulou, A. and Vasilopoulou, E. (2000). Mediterranean diet and longevity. *British Journal of Nutrition* 84: S205–S209.

USDA (2019a). FoodData central: lentils, raw. https://fdc.nal.usda.gov/fdc-app.html#/food-details/172420/nutrients (accessed 23 March 2022).

USDA (2019b). FoodData central: lentils, mature seeds, cooked, boiled, without salt. https://fdc.nal.usda.gov/fdc-app.html#/food-details/172421/nutrients (accessed 23 March 2022).

USDA (2020). FoodData central: beef, pot roast, braised or boiled, NS as to fat eaten. https://fdc.nal.usda.gov/fdc-app.html#/food-details/1098198/nutrients (accessed 23 March 2022).

USDA (2021). FoodData central: rice and lentil pilaf. https://fdc.nal.usda.gov/fdc-app.html#/food-details/1893882/nutrients (accessed 23 March 2022).

Walker, C. (2020). Iranian Adasi – lentil stew. https://www.proportionalplate.com/adasi (accessed 9 June 2022).

Warne, T., Ahmed, S., Byker Shanks, C., and Miller, P. (2019). Sustainability dimensions of a North American lentil system in a changing world. *Frontiers in Sustainable Food Systems* 3: 88.

Xu, B. and Chang, S.K.C. (2010). Phenolic substance characterization and chemical and cell-based antioxidant activities of 11 lentils grown in the Northern United States. *Journal of Agricultural and Food Chemistry* 58: 1509–1517.

Xu, J., Zhang, Y., Wang, W., and Li, Y. (2020). Advanced properties of gluten-free cookies, cakes, and crackers: a review. *Trends in Food Science & Technology* 103: 200–213.

Yasmeen, I.K.F., Ahmad, J., Abdullah, A. et al. (2019). Developing a meal-planning exchange list for commonly consumed Pakistani dishes. *Progress in Nutrition* 21: 421–429.

Yves Veggie Cuisine (2022). Lentil and quinoa patties. http://yvesveggie.ca//en/products/burgers/lentil-and-quinoa-patties (accessed 9 June 2022).

Zafferano, G. (2020). Cotechino (Pork sausage) with lentils. https://www.giallozafferano.com/recipes/cotechino-pork-sausage-with-lentils.html (accessed 26 January 2022).

Zaroual, H., Boughattas, F., and Karoui, R. (2019). Traditional foods in Maghreb: production and research progress. In: *Traditional Foods* (ed. M. Al-Khusaibi, N. Al-Habsi, and M.S. Rahman), 51–113. Cham, Switzerland: Springer.

Zohary, M. (1976). *A New Analytical Flora of Israel*. Tel Aviv, Israel: Am Oved Publications.

Index

Lentils: Production, Processing Technologies, Products, and Nutritional Profile, First Edition.
Edited by Jasim Ahmed, Muhammad Siddiq, and Mark A. Uebersax.

t

www.ingramcontent.com/pod-product-compliance
Lightning Source LLC
LaVergne TN
LVHW062204200926
848484LV00006B/1315